High-Speed Flight Propulsion Systems

Edited by
S. N. B. Murthy
Purdue University
West Lafayette, Indiana

E. T. Curran
Wright Laboratory
Wright-Patterson Air Force Base, Ohio

Volume 137
PROGRESS IN ASTRONAUTICS AND AERONAUTICS

A. Richard Seebass, Editor-in-Chief
University of Colorado at Boulder
Boulder, Colorado

Published by the American Institute of Aeronautics and Astronautics, Inc.
370 L'Enfant Promenade, SW, Washington, DC 20024-2518.

1-56347-011-X/91 $2.00+.50

ISSN 0079-6050

Table of Contents

Preface

Hypersonic flight vehicles, whether intended wholly for cruise or essentially for acceleration, are complex, and demand total integration from various points of view. This integration is critically important for vehicles operating at high Mach numbers, where the energy available from airbreathing chemical propulsion is marginally adequate for the overall mission. For this flight regime it is essential to conserve available energy and, correspondingly, minimize losses. Although hypersonic flight vehicles had been designed from time to time, and continue to be designed in single and multiple stage configurations with various forms of propulsion systems, it seemed worthwhile to re-examine the propulsion issues and the energy aspects of such vehicles.

To this end, a two-session colloquium was organized in January 1989 by the editors of this volume, during the 27th AIAA Aerospace Sciences Meeting, under the broad title of "Energy Analysis for High Speed Propulsion." The Colloquium consisted of invited papers by P. Czysz and S.N.B. Murthy, P.E. Dimotakis, M.G. Dunn, R.J. Stalker, and J.S. Swithenbank. These papers addressed the then-current state of developments in energy analysis, energy addition in propulsive ducts, turbulent-mixing of compressible, reactive flows, and high-speed test facilities.

Later in 1989, it was determined that the papers provided the basis for a volume that would deal with the subject of propulsion for high speed flight vehicles in a comprehensive manner, if additional contributions were included. In particular, two subject areas needed to be included: first, propulsion systems, including combined cycle engines for use in various parts of the mission; and second, numerical prediction schemes for turbulent mixing. Regarding propulsion systems, papers were invited from F.S. Billig, E.T. Curran, and F.A. Hewitt. Concerning numerical analysis of mixing, a paper was invited from J.P. Drummond. In addition, a paper was invited from V.K. Baev on testing in the short duration facilities at Novosibirsk, to provide a glimpse at such developments in the Soviet Union. The outcome is this volume with an introduction by E.T. Curran and a total of nine other contributions.

Interest in combined cycle engines and hypersonic flight has greatly increased in the time that has elapsed since the initial conception of this volume. All of the papers presented in the 1989 colloquium have been rewritten, elaborated, and updated for this volume. The invited papers have been written, and organized along with the initial papers to provide a cohesive treatment of the complex issues in high speed propulsion. The papers in the volume, both individually and collectively, give reasonably comprehensive introductions to the current capabilities for addressing several fundamental aspects of high speed vehicle propulsion development.

We have been helped a great deal by the reviewers who read and commented on the various contributions. We are especially grateful to Dr. Martin Summerfield for his kind advice and encouragement.

It has been a great pleasure to work with the staff of the Editorial Department of AIAA, in particular Jeanne Godette and Amy Hoeksema.

S.N.B. Murthy
E.T. Curran
September 1991

Introduction

E. T. Curran*
Wright Laboratory, Wright-Patterson Air Force Base, Ohio

Prolegomena

The early decades of aviation were characterized by the pursuit of speed and altitude, range, and extended range at high flight speeds. A major push for higher speed was initiated by the X-series of aircraft that explored the supersonic flight regime in the 1950s and 1960s. The X-series culminated in the rocket-powered X-15-A2 which propelled man to his highest speed in an airplane, namely Mach 6.7. In the same time period some exploratory flight experience was gained with both airbreathing and rocket engine elements and several other interesting projects were pursued, for example, the rocket-airbreathing powered fighters such as the Republic XF-91 (1949) and the Saunders Roe SR53 (1957). The turboramjet engine also appeared as the power plant of the outstanding Nord Aviation Griffon 02 aircraft in France, and as the power plant of choice for the Republic XF-103 fighter; this latter aircraft was later canceled. The work of René Leduc in pioneering ramjet and turboramjet propulsion must also be noted: the flight programs involving the ramjet powered Leduc 0.10, Leduc 0.21 (with Turbomeca turbojet), and Leduc 0.22 (with Atar turbojet) were significant steps in aviation progress. However, at the time of writing, maximum speeds of commercial and military aircraft appear to have reached a plateau with Concorde flying in the region of Mach 2 and the SR71 in the Mach 3 region.

In regard to cruising flight the potential of high-speed commercial vehicles is being evaluated in the U.S. under NASA's High-Speed Civil Transport (HSCT) program. A general discussion of the probable evolution of high-speed commercial transports is difficult, because the foundational technologies such as aerodynamic configuration, power plant, structure, and fuel type change markedly with increasing flight speed. In the last major U.S. effort to build a supersonic transport, the need to accommodate both low-speed and high-speed flight requirements led to the evolution of a

 (This paper is an expanded version of Ref. 1.)

*Associate Chief Scientist.

significant new engine technology, namely that of variable-cycle engines. For even higher maximum flight speeds it is to be expected that advanced variable-cycle engines, combined-cycle or mixed engine types, and also novel engine cycles will emerge to bridge the wide speed range. Thus, up to the present time some limited experience has been gained with both the rocket and airbreathing elements that might be assembled in a combined-cycle engine.

Turning next to the potential performance of conventional engine cycles, using, for example, hydrogen fuel, the approximate performance levels are shown in Fig. 1. It is apparent that as flight speed increases the turboaccelerator class of engine is supplanted first by the subsonic combustion ramjet and second by the supersonic combustion ramjet; rocket propulsion may also be needed, primarily for the higher flight speeds. Consequently, for a hypersonic flight vehicle operating at a maximum speed above, say, about Mach 5, a multimode propulsion system will be required. Bearing in mind the limitations of materials, such an engine system might operate as a turboaccelerator to speeds approaching Mach 4, then transition to subsonic ramjet operation up to speeds of about Mach 6, and then operate totally as a supersonic combustion ramjet engine for speeds above about Mach 7. It must be pointed out that the right choice of terminology and classification of such engine systems is not a trivial problem. Thus, in the literature one finds terms such as; mixed mode, multimode, hybrid, compound, composite, and combined-cycle used to describe various engine configurations. It is not intended in this paper to address the correct choice of terminology. The interested reader is directed to Ref. 2-5. For the purposes

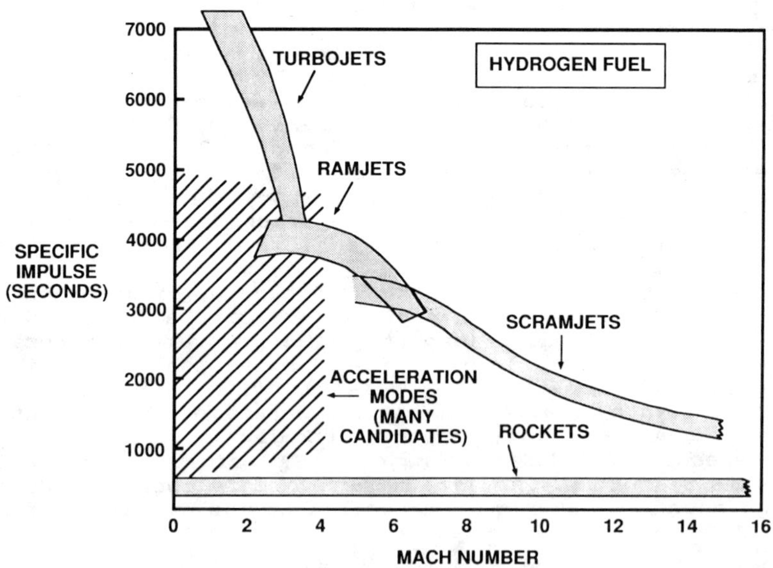

Fig. 1 Approximate Performance of Hydrogen Fueled Engine.

of this paper the term combined-cycle is used, without further justification, to indicate an airbreathing engine system whose main element is the ramjet engine (with subsonic and/or supersonic combustion) that is boosted to ramjet takeover speed by means of a turboengine (turboaccelerator) or rocket- based system, and that uses ramjet propulsion at the higher speeds.

In recent years there has been a proliferation of international studies of airbreathing vehicles such as hypersonic transports, first-stage space launchers, and single-stage-to-orbit (SSTO) vehicles. Such studies, at various levels of investment, include efforts by the United States, the United Kingdom, Germany, France, Japan, and the USSR. Although these various studies can arbitrarily be divided into cruising vehicles and accelerator vehicles, this is not a hard and fast distinction. Thus the acceleration phase of a predominately cruising hypersonic vehicle is a significant segment of the total flightpath, and similarly a dedicated accelerator vehicle may require substantial cruise capability, for example, to obtain access to desired orbit planes, or for subsonic-ferry capability.

However, from these studies, one interesting aeronautical challenge emerges, which is the achievement of orbital conditions with a single-stage vehicle using primarily airbreathing propulsion. This elegant but elusive approach to orbit is a classical aerospace concept. For this vehicle the total flight spectrum from horizontal takeoff to orbital speed is encompassed, and a combined-cycle engine system is required for efficient operation over this wide speed range. Such a propulsion installation will probably require significant variable geometry. However, the vehicle aerodynamic configuration will presumably not incorporate variable geometry in the interest of low structural weight.

It is also important to note that with increasing maximum flight speed, the fuel of choice typically changes from the conventional kerosene fuels for the lower speeds, to cryogenic fuels such as methane or liquid hydrogen for hypersonic speeds, with endothermic fuels currently emerging for possible use at intermediate speeds. The introduction of liquid hydrogen with its high cooling capacity and its high work capability has introduced major new opportunities for the synthesis of efficient aeropropulsion systems. Hydrogen may be used to cool both internal (engine) and external flow surfaces, or to cool the engine working fluid (air) and thus impact the engine thermodynamic cycle. Also, hydrogen may be used as an additional working fluid in its own right, providing opportunities for additional energy interactions as an intrinsic part of the aeropropulsive system. Of course, it may be desirable to utilize more than one fuel in a given vehicle to offset the large fuel volume required for a vehicle using solely hydrogen fuel.

From this point forward the discussion will primarily apply to the accelerator class of vehicle. There is, of course, much that is also applicable to hypersonic cruising vehicles.

Overall Propulsion Considerations

As noted earlier, Fig. 1 affords a simple comparison of various engine cycles based on fuel specific impulse (Isp) and it is therefore generally appropriate to steady cruise flight performance, where the Breguet equation may be used to compare ranges. For cruising flight, the engine fuel economy is of primary importance. However, for acceleration missions, the engine weight becomes a strong concern, and the performance of the engine may be evaluated initially on the basis of both its specific impulse and its corresponding installed thrust-to-weight ratio.

It is clear, from Fig. 1, that for flight speeds in excess of Mach 4 the use of a conventional ramjet cycle is imperative, and for flight speeds in excess of about Mach 6 the scramjet is currently the appropriate high speed engine. To form a truly combined-cycle engine, a lower-speed engine cycle must be integrated with the appropriate ramjet element(s). In the following discussion the term low-speed engine applies typically to turbo-accelerator or rocket-based systems used to accelerate the vehicle up to speeds of roughly Mach 4.

For purposes of discussion, let us consider a hypersonic vehicle with maximum speed in excess of, say, Mach 7, and consequently using a supersonic combustion ramjet as its dominant propulsion system. As noted earlier, a plausible overall propulsion installation for such a vehicle is one which uses a low-speed engine for take-off (and landing) and for acceleration to speeds of about Mach 4, followed by an initial transition to ramjet operation. A subsequent transition to scramjet operation takes place at about Mach 6, and the vehicle then accelerates to a higher Mach number on scramjet power. The installation of the combined-cycle engine elements into the airframe is a complex problem. If it is assumed that such an engine is fed by a single inlet and exhausts through a common nozzle, then the ramjet/scramjet engine flowpath may be incorporated, in general, in either a series (tandem) or parallel (bypass) arrangement with the low-speed engine system. In the parallel installation a complex mechanical valving arrangement may be required to effect transition from low-speed operations to ramjet operation. However, ramjet to scramjet transition may be handled aerodynamically by fuel-injection sequencing or staging.

It is appropriate to point out that, because of the complexities of transitioning between the various engine modes, there is a substantial payoff to eliminating such transitions. For example, a nominal turboaccelerator maximum speed capability might be stretched to permit direct transition to a scramjet mode. Alternatively, for, say, a Mach 7 system, the conventional ramjet engine might be stretched to avoid transition to a true scramjet mode: alternatively the scramjet takeover Mach number may be reduced or "stretched" down to eliminate the ramjet mode. Similarly, in relation to fuels it is possible, where desirable, to stretch the performance of hydrocarbon-class fuels to higher Mach numbers, say, by improving the fuel thermal stability, and thus defer the logistical problems of operating with cryogenic fuels. Consequently, the propulsion engineer must seek novel solutions such as stretching capability wherever appropriate, and where mission requirements will not be compromised.

Low-Speed Propulsion Trade-Offs

For maximum payload capability it is desirable for a candidate low-speed engine to possess high specific impulse, a high installed thrust-to-weight ratio, and the capability to function over a broad Mach number range. For a nominal acceleration mission the tradeoff curve between specific impulse and thrust-to-weight ratio, for a given payload, is usually of the form shown in Fig. 2. Typically, the class of high specific impulse engines, such as turbojets, are mechanically complex and therefore heavy, whereas lightweight engines such as liquid fuel rockets generate relatively low specific impulses.

Note that for each class of engine the overall thrust loading of the vehicle must be chosen to optimize the appropriate payload fraction. Different aircraft thrust loadings are required for each class of engine, and consequently the corresponding initial vehicle accelerations are different. Thus, the rocket class of engine requires a higher initial acceleration for optimum payload fraction than the turboaccelerator class. (Of course a pinch point in excess thrust may exist along the flight trajectory, requiring additional thrust loading.) In regard to Fig. 2 it is unfortunately true that for typical candidate propulsion systems there is no single class of engine which simultaneously possesses high specific impulse and high

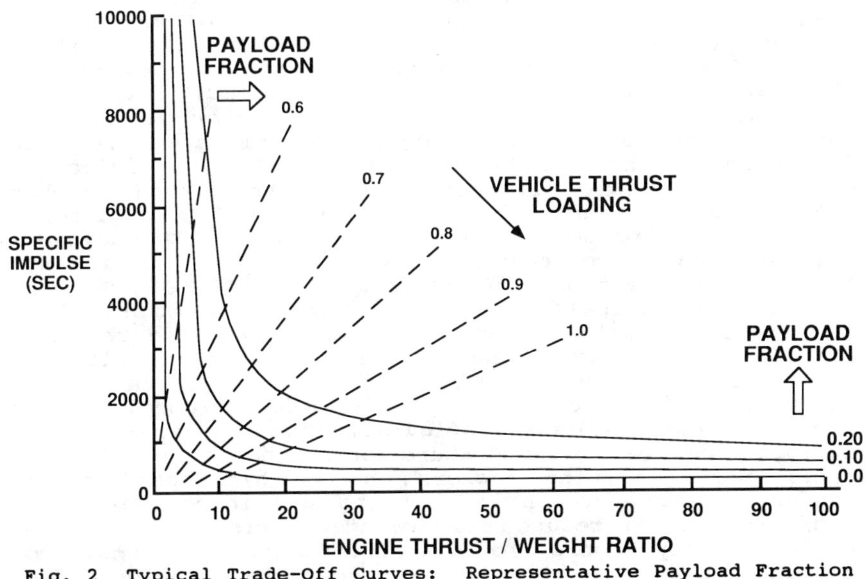

Fig. 2 Typical Trade-Off Curves: Representative Payload Fraction as a Function of Specific Impulse and Engine Thrust-to-Weight Ratio.

installed thrust-to-weight ratio. Barrere[4] has recently drawn attention to the generic relationship between engine thrust-to-weight ratio and basic specific impulse; a similar functional relationship was also utilized by Builder and Cuadra[6] in evaluating candidate engines for acceleration through the transonic regime.

In order to develop engine systems with improved payload potential, three simple observations are appropriate: first, for the turboaccelerator class of engine, the appropriate approach is to strive to increase the engine thrust-to-weight ratio; second, for the rocket class of engine, attention should be focused on improving the specific impulse; and third, it would be expected that an engine type intermediate between the turboaccelerator and the rocket engine, for example, the air turborocket, might offer a performance superior to its progenitors. This latter expectation, of course, accounts for the current renewal of interest in combined-cycle engines. Extended discussions of various types of combined-cycle engines and their performance potentials are given in the following papers in this volume authored by Billig and by Hewitt.

Candidate Low-Speed Engines

One stream of development is, as noted above, to improve the thrust-to-weight ratio of turboaccelerators. The specific thrust of the engine can be increased by a modest but useful amount by increasing the turbine inlet temperature, and also by increasing the efficiency of the internal flow processes. However, there are potentially large benefits available by significantly reducing engine weight through the use of new materials allied with innovative structural concepts. In this regard a major initiative is underway in the U.S. to double the thrust-to-weight ratio of advanced turbine engines by about the year 2000; this program is known as the Integrated High Performance Turbine Engine Technology (IHPTET)[7] initiative. The rotating machinery of a turboaccelerator is a major contributor to the engine weight; it also dictates the basic air swallowing characteristics of the propulsion device and imposes mechanical limitations on the Mach number capabilities of such engines. The IHPTET program is anticipated to produce significant gains in thermodynamic performance, with reduced weight and size of such machinery. Other engine development efforts similar to IHPTET are being pursued internationally. Also, with increasing Mach number, the size and complexity of the ducting in which the core engine is housed become increasingly heavy and result in a significantly reduced installed thrust-to-weight ratio. In this regard, there is a driving need to devise novel compact inlet and exhaust nozzle configurations. Additionally, the internal pressure and temperature loads imposed on such inlet ducts become limiting factors on the high-speed flight boundary. One approach to reducing the inlet weight and complexity is to reduce the amount of flow diffusion by maintaining supersonic flow either throughout the engine or through a fan stage; this latter concept has been worked on intermittently over the years. Recent studies[8] of the performance of podded air-turborocket (ATR) installations using a variable geometry supersonic through-flow fan (STFF) have shown significant gains in potential performance compared to a conventional "over/under" ATR-ramjet arrangement. Furthermore, the installation of the STFF is much simpler, leading to a lighter-weight, reduced-drag power plant. Much more work needs to be done to develop high-efficiency, rugged supersonic fans and this work is currently in progress.

It is also well known that for the conventional turbojet the overall engine total pressure ratio (turbine exit to compressor face) decreases rapidly at higher Mach numbers, for a given turbine temperature. Typically, the performance capability of the basic turbojet can be increased either by use of afterburning or by use of an integrated ramjet system. (Such turboramjets may be of tandem or of parallel design, but, of course, the mechanical integration of the turbojet core engine and the ramjet may result in a relatively heavy and complex installation.)

An alternative approach to avoid the turbomachine-based compression process is the use of nonmechanical pumps and/or crypto-steady compressors: such devices frequently produce very low overall pressure ratios and are therefore very sensitive to losses in the flow process. However, the combination of a ramjet (or scramjet) with a rocket (liquid or solid) driven ejector-pump has received major attention in the literature and many combined-cycle engine concepts have emerged. Examples of such engines are the ejector ramjet (ERJ), the supercharged ejector ramjet (SERJ), the ejector scramjet (ESCRJ), and the rocket-ramjet-scramjet (ROCSCRAM), which utilizes two rocket engines. The evolution of the first two engine systems is described in Ref. 5 and the ROCSCRAM in Ref. 9. It is also interesting to note that an early multimode engine, named the "Hyperjet," was actually flight tested by the Marquardt Corporation, under USAF sponsorship, in 1959/1960. This engine was basically a ramjet engine with a valved intake, and was capable of operation as a rocket (inlet valve closed, and with propellent injection), yielding thrust at either static or high-altitude conditions, or as a ramjet (valve open, and ramjet fuel only).

Returning to our initial discussion, we have noted that for turboaccelerator engines the payload potential was best improved by seeking increased thrust-to-weight ratio; on the other hand, the capability of the rocket class of engine was best improved by increasing specific impulse. In this latter case it is interesting to note that, historically, improvements in rocket efficiency have been sought by extracting work from the thrust chamber to drive external propellers or fans. As early as 1932, R. H. Goddard was reported as working on a "turbine rocket plane" in which a rocket-driven turbine was used to drive two propellers. A related concept is the engine described by Sloop[10] for the REX-1 aircraft system (1954). This concept initially used a series of three gas-generator-powered turbine units to drive a propeller system; later the propeller was replaced by a compressor system. This concept evolved to the REX II engine, a form of turborocket, and subsequently to the REX

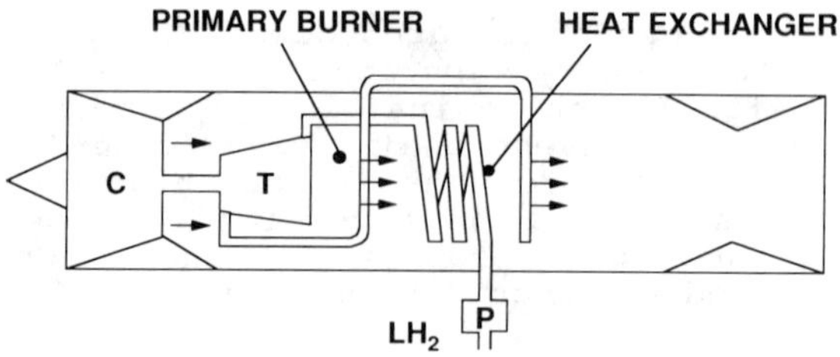

Fig. 3 Schematic of a Hydrogen Expander Engine.

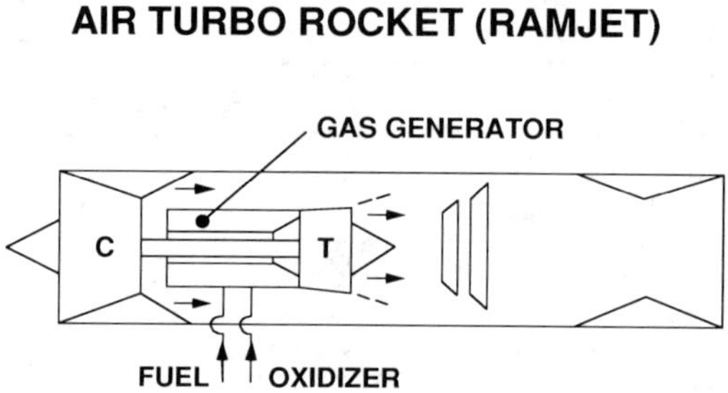

Fig. 4 Schematic of an Air-Turborocket Engine.

III engine, a form of hydrogen-expander engine. The hydrogen-expander class of engine, illustrated in Fig. 3, was extensively studied in the U.S. in the mid-1950s, as also discussed by Sloop: these initial designs involved complex, and therefore heavy, heat exchangers. The overall performance was also constrained by the inability to reach a high hydrogen temperature due to materials limitations. Also, in these engines the fan-turbine work balance may impose limitations on the cycle performance achievable without added mechanical complexity in the form of multistage turbines or gearboxes.

A more familiar engine, which largely decouples the turbine mass flow from that of the compressor, is the so-called air-turborocket illustrated in Fig. 4. In this case, the gas generator can be fed by bipropellant combinations, with a significant increase in gas-generator temperature compared to the expander cycle, or by a monopropellant: the ATR cycle is, however, also limited by fan-turbine matching constraints. Obviously, the basic expander engine cycle and air- turborocket cycle can be combined to yield new engine cycles. Another variation on the air-turborocket theme is the rocket fan,[11] in which the fan is driven directly by small rocket nozzles at the fan blade tips. The hydrogen that is exhausted through these nozzles is then burned with the fan discharge air in an afterburner.

An interesting approach to the problem of improving the specific impulse of rocket type engines during atmospheric

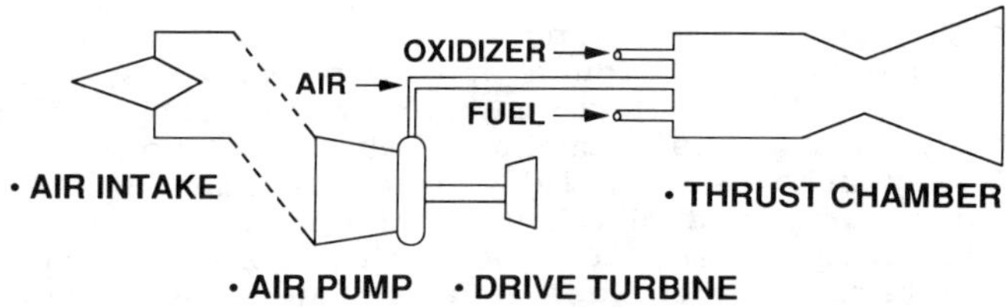

Fig. 5 Schematic of an Air Fed Rocket Engine.

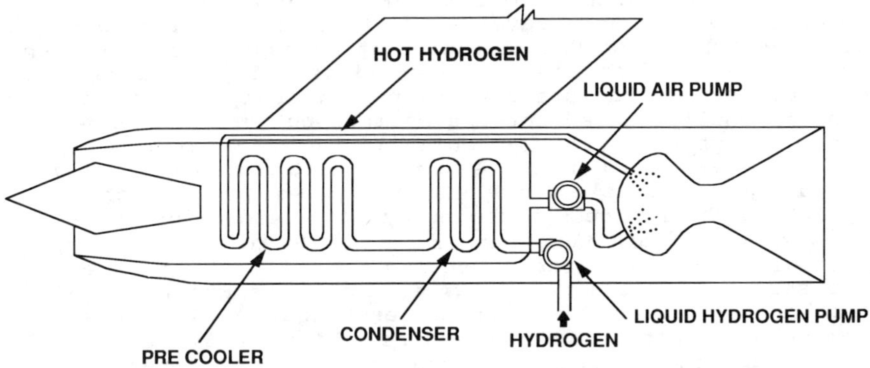

Fig. 6 Basic Liquid Air Cycle Engine.

flight is to replace, either partially or completely, the conventional liquid oxidizer by atmospheric air[12]: the immediate problem is the compression of such air from freestream to combustion chamber pressure levels (see Fig. 5). One approach to this, which avoids heavy turbomachinery, is to pump the air in the liquid phase[13]; obviously a direct approach to condensing the engine airflow is to use the onboard cooling capacity of cryogenic fuels and oxidizers,[14] or cryogenic diluents, such as liquid nitrogen (see, for example, Fig. 6). An intermediate approach is strong precooling of the inlet flow without actual liquification: this effect permits substantial reduction in turbomachinery mass but of course introduces the mass of a heat exchanger.

Precooling has also received attention in regard to conventional turboaccelerator designs, initially from the point of view of extending the flight Mach number capability of such engines. In cases where the maximum Mach number of the engine is limited by the compressor exit temperature, then precooling of the air can permit significant increases in flight speed.[15] However, it was soon realized that additional potential performance advantages exist and that the precooled turboaccelerator can be a significant contender for high-speed flight. The precooling concept has also been extended to include both air-hydrogen and air-air heat exchangers in the air cooling system[16]. In an overall sense the introduction of precooling can potentially improve engine thrust performance and reduce engine size; also, because of the increased flight Mach number capability it may be possible to defer transition to other engine modes to higher speeds. Nevertheless, inlet flow matching with minimal variable geometry is a challenge. However, the ground test problem is simplified in that the core engine, which experiences temperatures less severe than those corresponding to the flight Mach number, may be developed in available test facilities. The potential performance advantages are tempered, however, by the practical difficulties of fabricating lightweight, reliable heat exchangers and avoiding the fouling problems of flight operations. The potential performance advantages of air precooling have led to a re-emphasis on the role of heat exchanger processes and components in advancing the capabilities of jet engines. Of course, heat exchangers have been previously considered for conventionally fueled turbine engines but are now receiving renewed emphasis with cryogenically fueled systems.

The above discussion does not exhaust the innovative uses of cryogenic fuels in new engine cycles. One such novel cycle[17] is the inverted turbojet cycle intended for operation in the range of Mach numbers from 0 to 6; this cycle initially received wide attention in 1961. In this engine the sequence of the core-engine components was arranged in the order: turbine - heat exchanger - compressor - combustor. Although this engine did not give acceptable cycle performance at low speeds, it did lead directly to a variant cycle, namely the precooled turbojet cycle discussed earlier. It should also be noted that the inverted Brayton cycle has also been considered for application to stationary power generation systems in the role of a "bottoming" cycle (see discussion in Ref. 18).

An improved version of the inverted cycle engine (ICE) has recently been proposed[19] in France that, as shown in Fig. 7, incorporates an additional burner installed between the

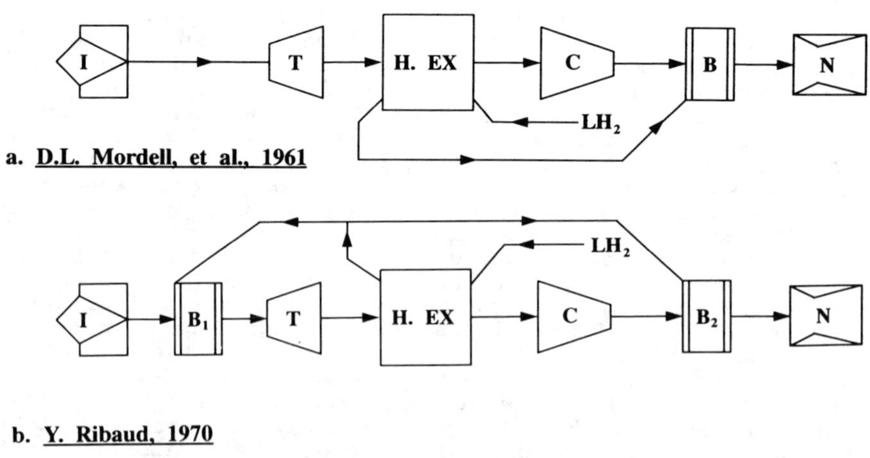

Fig. 7 Inverse Cycle Engine Arrangements.

inlet and the turbine. At low speeds this additional burner allows a heat input upstream of the initial turbine to augment thrust. The fuel-rich mode of operation is particularly beneficial; as speed increases, the equivalence ratio of this burner can be lowered and eventually reduced to zero. A bypass duct can also be incorporated in this engine.

In summary, the search for superior engine concepts which satisfy the many requirements for the low-speed engine continues, and it is to be expected that innovative solutions will emerge as more attention is directed to high-speed vehicle design.

High-Speed Propulsion

For speeds in excess of Mach 4 it is generally necessary to transition from the turboaccelerator to the conventional ramjet mode. However, as outlined in the previous section, it may be possible in the future to extend the capability of the turboaccelerator class[20] to speeds in excess of Mach 4. In this case direct transition to a scramjet mode may become possible. The characteristics of the conventional subsonic-combustion hydrogen-burning ramjet are well understood and combustor performance has been explored in ground tests corresponding to speeds in excess of Mach 6. As is well known, the performance of the subsonic-combustion ramjet deteriorates at hypersonic speeds due to real gas effects and internal losses, and the structural design of the engine becomes very difficult due to the high internal temperatures and pressures.

Thus, for really high speeds it is necessary to utilize the supersonic combustion engine (scramjet). The scramjet engine potentially offers outstanding specific impulse performance to high hypersonic Mach numbers. Such performance estimates are, of course, theoretical, and the upper practical speed limits of such engines have not been established on an engineering basis.

The speed range between the conventional ramjet and the scramjet can be bridged by the dual mode engine, as illustrated in Fig. 8. The dual mode engine is one in which, initially, subsonic combustion is employed at lower speeds (typically Mach 3-6) with transition to the supersonic combustion mode at speeds of about Mach 6. This class of engine has undergone extensive analysis and ground test in the U.S. and abroad. It is probable that this class of engine will be the next workhorse engine for future hypersonic vehicles.

As previously shown in Fig. 1, the supersonic combustion ramjet engine offers progressively higher performance than the conventional engine at speeds in excess of about Mach 6. It is difficult to be precise about the point of transition to supersonic combustion because mixed flow conditions can exist in the combustor until substantially higher Mach numbers are attained, and, in any event, one-dimensional flow concepts and criteria are difficult to apply. The phenomenon of supersonic flow through the engine introduces totally new technological challenges compared to the conventional ramjet. These challenges are principally associated with the wave

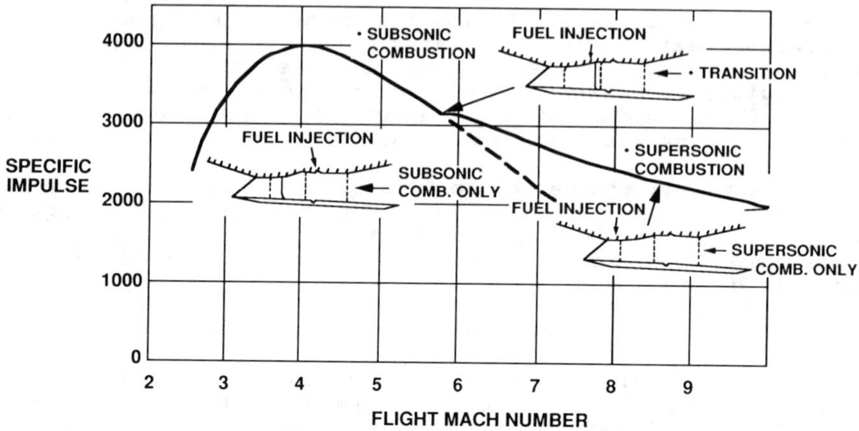

Fig. 8 Dual Mode Ramjet Performance.

interaction phenomena generated by the basic engine processes of diffusion, fuel injection, mixing, combustion, and expansion.

In one-dimensional analysis an optimum amount of diffusion is required to maximize engine performance at a given flight condition. Simplistically, this optimum may be said to determine the engine capture area ratio and associated compression ratio, and it also determines the level of temperature and pressure at entry to the combustor. However, this is not the end of compression effects in the engine because, for a given combustor configuration, the processes of fuel injection, mixing, and heat release will all exert an influence on the overall evolution of the flow. Thus, the presence of fuel injector struts or other protrusions into the combustor will generate shock/expansion fields. In general, complex wave patterns will be transmitted through multiple interactions downstream through the engine from their point of origin. These phenomena are discussed in this volume in a paper by Stalker. The art of scramjet design thus lies in the effective utilization of these wave processes to enhance combustion and thrust generation, while avoiding critical phenomena such as major flow separations, excessive thermal loading, and unstart of the combustor or inlet.

The degree of diffusion also falls between certain bounds. A minimum diffusion is required primarily to ensure appropriate pressure and temperature levels for efficient combustion. However, too much diffusion results in excessive thermal and mechanical loads, and can produce dissociation and nonequilibrium effects in the combustor, leading to severe thrust losses in the nozzle. It is also appropriate to note that, because of the accumulation of losses associated with shock wave propagation, skin friction, and heat transfer, engine length is always at a premium. Opposing this requirement is the need for adequate length for the processes of fuel injection, mixing and combustion.

These important processes are treated in the papers of Dimotakis, Swithenbank, and Drummond presented in this volume. It should also be noted that to avoid combustor-inlet interaction a significant length may be required for an isolator section between the combustor and inlet.

As noted earlier the degree of diffusion is a significant performance parameter. The overall diffusion process is carried out partially by the vehicle forebody (precompression), or in some installations by a wing surface, and partially by the engine module intake.

These precompression and postcompression processes may be nonadiabatic and possess large wall boundary layers. Therefore the usual one-dimensional parameters for assessing intake losses and starting criteria are not rigorously applicable at the higher speeds, and new performance parameters have to be derived to embrace this situation. Once again the engine compression process and the vehicle aerodynamics are inextricably coupled, not only through the flow turning processes, but also through the forebody flow field, which potentially involves shock/boundary layer interactions with laminar, transitional, or turbulent boundary-layer flows. In the case of an actively cooled forebody there is a further coupling through the aerothermodynamics of the cooling process. Once again the challenge is to use all such flow couplings to overall vehicle advantage.

The supersonic combustor represents a major technological challenge in engine development. Once again the existence of supersonic throughflow presents the problem of adding significant energy, usually in a fixed geometry configuration, while avoiding the pitfalls of undesired thermal choking at low speeds and excessive loss mechanisms, or thermal loadings, at any speed. Early efforts at demonstrating supersonic combustion have addressed geometries such as; constant-area cylinder, cylinder-cone and step-cylinder configurations and also included two-dimensional versions of these geometries. Injection techniques have included both wall and strut elements with injection at normal, tangential, and angled directions to the airflow. In addition to the problem of adequate fuel penetration to give the required transverse fuel distribution, the fuel injection must of course be staged to give appropriate longitudinal coverage. Following ignition and stabilization, spatial control of the heat release, as a function of flight speed, has to be maintained by means of the fuel injection system and its control mechanism. A considerable data base on the mixing and heat release characteristics of various injector arrangements has been developed, and the modeling strategy for such complex flows has been well documented in the pioneering efforts of Anderson[21] and Billig.[22] As a result of such efforts, together with many international contributions, a good understanding of the generic behavior of supersonic combustion systems exists. However the establishment of a mature technology base is dependent on the detailed understanding and enhancement of the component processes of fuel injection, flow interaction, mixing, heat release, and final production of translational gas energy. Fortunately, the emergence of powerful computational fluid dynamics (CFD) techniques has permitted a more structured approach to modeling the elemental combustor processes: the CFD approach, combined with complementary experiments, should permit relatively confident synthesis of initial combustor designs. It should be noted that the terminology "hypersonic combustion" has recently been introduced[23] to describe combustors operating with hypersonic through-flow Mach numbers, i.e., with operation exceeding Mach 5, corresponding to flight Mach numbers of about 20.

The exhaust nozzle is a critical component of the engine in that the final stream thrust increment is generated by the nozzle flow, and the net thrust of the engine is very sensitive to the level of nozzle performance. This performance is closely coupled to conditions at the combustor exit flow surface (not usually planar in the one-dimensional sense) and the initial expansion segment of the nozzle may be developed as part of the combustor-nozzle combination. The development problem of the nozzle is well understood: from an internal flow viewpoint the effects of entry flow nonuniformity, chemical non-equilibrium, wall friction and heat losses, and nonaxial exit flow vectors must be addressed. Similarly, the overall aerodynamic effects of two-dimensional asymmetric nozzle flow on force generation and moments, mismatched expansion, and external-internal flow interactions must, when appropriate, be elucidated.

Once again it is evident that the exhaust nozzle is an integral part of the overall engine design and is also tightly coupled to the overall performance of the vehicle. The performance of high-area-ratio, two-dimensional, asymmetric nozzles has not yet received the same degree of attention in the literature as other engine components.

The scramjet engine is currently the most promising engine for hypersonic flight. However, the application of both steady and nonsteady detonation waves to high-speed propulsion schemes has had a long history and recently there has been a renewal of interest in the oblique detonation wave engine. The problems traditionally associated with this engine include the uniqueness and stability of such detonation waves; the mechanism of premixing the fuel with the inlet airstream; and the prevention of "flashback" through this mixing zone. It is hoped that the application of numerical simulation and modeling will resolve some of the issues.

Up to this point, only heat addition to the internal flow has been considered. Heat addition to the external flow is of relevance primarily to relieve base drag in the transonic speed range. It is also possible to modify the basic lifting and propulsive forces associated with a high-speed vehicle by selective external heat addition. Thus the approach used in internal flows, namely to synthesize a multidimensional flow where the basic wave structures constructively interact with regions of heat addition may also be applied to external flowfields. See, for example, the works of Broadbent[24].

Integration

For optimum flight performance the aerodynamic, structural, and propulsion elements of the vehicle have to be integrated into a cohesive whole, taking every opportunity to synergistically enhance performance. For a single-stage vehicle, where vehicle integration is critical, some technological relief could be obtained by refueling or by air collection. In the acceleration case it should be noted that the overall propulsive parameter is the "effective" specific impulse, which couples the basic impulse of the flight engine to the overall vehicle thrust/drag characteristics. For an airbreathing engine such characteristics are of course dependent on the flight trajectory. There are conflicting flight path requirements: to maintain high engine thrust a high dynamic pressure trajectory is desirable; however, the converse is true in regard to heating and pressure loads. Thus, trajectory optimization is also an integration issue. In regimes of minimum thrust margin a sharp loss in effective impulse can occur. This can be relieved by thrust augmentation typically by simple mass and/or heat addition. Thus, the use of water injection to augment turbojet and ramjet thrust is well documented. Similarly, augmentation by simple fuel and/or oxidant injection can also be an effective way to increase thrust and enhance overall performance. The overall spectrum of such airbreathing elements utilizing thrust augmentation, or direct rocket-augmentation, is well covered in an early foundational paper by Lindley.[25] Such techniques may be an attractive alternative to oversizing an acceleration engine simply to overcome regimes of minimum thrust margin.

As flight speed increases the engine size also grows and soon dominates the vehicle configuration. The only efficient way to accommodate this reality is by total integration of the vehicle and propulsion systems, which is well described by Townend.[26] Essentially, the entire undersurfaces of the forebody and afterbody provide propulsion functions, and their contributions to the thrust, drag, and control of the vehicle are of paramount importance. Considerable ingenuity is required to tailor the vehicle configuration and its associated flowfields, and the engine installation and its mass flow characteristics. Typically, matching of the engine flow characteristics and the available airflow over a wide speed range requires variable geometry. Although it is not always possible to separate rigorously the airflow associated with a hypersonic vehicle into external and internal airflows, it is still conventional to treat such flows as separate aerothermodynamic systems. Such an approach cannot continue, since the efficient production of the overall aeropropulsion forces on an actively cooled vehicle is inextricably linked through the cryogenic fuel circuits and many other coupling mechanisms (see, for example, Fig. 9). The overall energy bookkeeping of the total aerothermodynamic system that constitutes the vehicle can be rationally addressed by the explicit use of available energy methods which have been successfully applied to many complex thermodynamic systems in recent years.

Also, in addition to the basic engine installation, the thermodynamics of both the overall active-cooling circuits and the engine fuel system must be such as to avoid unnecessary losses of available energy. The energy of the

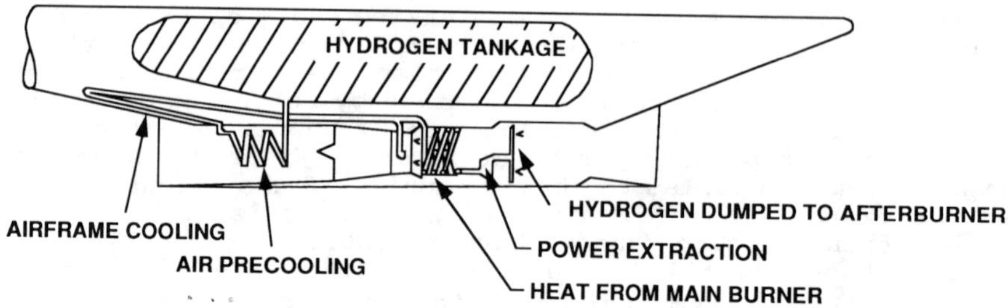

Fig. 9 Engine Installation: Potential Uses of Hydrogen.

basic hydrogen-air system is marginal for the single-stage-to-orbit (SSTO) mission; in this case, energy must be conserved, and utilized in the most effective manner. Once again, an available energy (exergy) based approach can be used to address total vehicle thermodynamics. In this volume the paper by Czysz and Murthy addresses the energy aspects of high speed vehicle synthesis.

Regarding minimization of the structural weight fraction, the design problems are severe. The vehicle will be subject externally to both high dynamic pressures and high temperature levels during ascent, while, internally, low temperature cryogenic storage must be maintained. The engine structure is also subject to even higher internal pressures and temperatures, possibly in boxlike actively cooled structures: integration of the basic engine and airframe structures is desirable to avoid duplicative weight elements. Fortunately, new high-temperature materials are emerging to alleviate the design problems; nevertheless, the achievement of a reusable high-temperature structure of minimum structural fraction is a major challenge. As in the case of the engine, ground- test validation of candidate structural approaches and assemblies is a demanding requirement.

A similar challenge is the evolution of an intrinsic overall vehicle control system philosophy and system. This is a significant architectural and engineering problem.

Test Facilities

In previous years the testing of high-speed flight engines has evolved to a disciplined, sequential, ground-test process leading to flight qualification and subsequent flight demonstration. Up to now the full simulation of supersonic flight conditions has involved large-scale plant, handling substantial mass flows at high-enthalpy and high-pressure conditions. The process of producing a large, high-quality, supersonic stream tube at the correct simulated flight Mach number is not trivial. Correctly designed axisymmetric nozzles can be used for discrete Mach number simulation. Such conventional facilities can currently simulate conditions that approximately correspond to typical Mach 7/8 hypersonic flight speeds. Traditional criteria for both tunnel starting, with various blockages, and the avoidance of excessive transient loads can be applied. One new requirement for multimode engines will be the convincing demonstration of stable mode transitions. Such transitions will generally occur at Mach numbers below about 8 and, thus, such tests can be undertaken in suitably modified conventional facilities. Also the testing of installed modular systems will be required to validate integrated airframe/engine performance.

For significantly higher speeds, ultrahigh enthalpies (and of course correspondingly high pressures) are required, and although many schemes have been postulated to achieve such enthalpies, the currently available test facilities are limited to relatively short duration systems. The use of such short-duration facilities to test scramjet engine components has been well established over the last 25 years in the U.S., the USSR, and other locations. In this volume there is a U.S. contribution on facility requirements by Dunn and a report on related work in the USSR by Baev.

Such test facilities, supported by increasingly sophisticated instrumentation and diagnostics, have been invaluable in estimating the potential aerothermodynamic performance of inlets, combustors, and nozzles through tests lasting only a few milliseconds. The application of the supercomputer to the analysis of both aerothermodynamic flows and structural design, can make maximum use of such short-duration test data. However, the process of validating the performance and durability of a full-scale flight engine is somewhat removed from the elemental tests performed in short-duration facilities, although these latter tests are essential. Currently an incremental flight test procedure appears to be the only logical course to proceed safely with engine certification.

In the future, extended duration ground tests at high-speed conditions will require novel facilities. Such facilities must avoid generating the simulated stagnation (or reservoir) conditions in the facility both because of engineering limitations and because the test gas may often be in a dissociated state. Consequently one must look to generating the required test enthalpy by progressive acceleration of the test gas by magnetohydrodynamic (MHD) or other energy transfer processes.

In any event, a fundamental need exists for the incremental flight testing of hypersonic engine systems.

Conclusions

Currently we stand at a crucial point in aeronautical progress. Aeronautics has advanced along the traditional flight corridor to achieve relatively conventional flight to speeds exceeding Mach 3. Such cruising flight may be regarded as an established stable boundary condition at the lower speed end of the flight corridor. At higher speeds another established steady condition exists, namely orbital "flight." Between these established stable states, a wide range of transatmospheric flight paths exist. Currently, passage through the transatmospheric region is of a transient nature, involving insertion to, and re-entry from, orbital conditions, or transient, "zoom" operations. However, in the longer term, ability to operate in the transatmospheric region may be of considerable interest as the fields of aeronautics and space continue to merge: once again, combined-cycle propulsion will be key.

It is clear that entry into the hypersonic flight regime means that long-term commitments must be made to embark on new areas of technology in engines, fuels, aerothermodynamics, materials, and structures. Similarly, the whole infrastructure supporting operation of hypersonic cryogenically-fueled vehicles must be addressed. Such changes will tend to be more revolutionary than evolutionary and should be recognized as a major shift in traditional aeronautical development. As previously noted, we stand at an exciting point in aeronautical, or rather aerospace, history: new vehicle configurations and engine concepts are emerging, and these opportunities need to be addressed in a cohesive, imaginative fashion, particularly in a time of diminishing resources. Fortunately, increasingly powerful supercomputers are now available to help in addressing both engineering and management problems. However, the overall strategy for evolving efficient high-speed vehicles is a complex and major multidisciplinary challenge - a point first stressed by Kuchemann.[27] Although such major changes in aeronautical development have been latent for some decades, the current emphasis on hypersonics now requires dedicated attention so that the required engineering disciplines and methodologies can be evolved through coordinated programs and appropriate management structures, probably at an international level.

Acknowledgment

Much of the material in this article was first published by the Advisory Group for Aerospace Research and Development, North Atlantic Treaty Organization (AGARD/NATO) in Ref. 1. Permission to reproduce the original material is gratefully acknowledged.

References

[1]Curran, E. T., "The Potential and Practicality of High Speed Combined Cycle Engines," Hypersonic Combined Cycle Propulsion, AGARD Conference Proceedings No. 479, 1990, AGARD, Neuilly-Sur-Seine, France, pp. K, 1-9.

[2]Gordon, H. S., and Johnston, R. B., "Composite-Engined Aircraft as a Basic Conception," Aircraft Engineering, Sept. 1946, pp. 299-301.

[3]Fishbein, B. D., "On the Development of a Flight Vehicle Engine Classification," Izvestiya VUZ Aviatsionnaya Teknika (Soviet Aeronautics), Vol. 32, No. 1, 1989, pp. 70-74.

[4]Barrere, M., "Quelques Remarques Sur Les Systems De Propulsion Multifonctions Ou 'Combines'," Acta Astronautica, Vol. 15, No. 11, 1987, pp. 931-935.

[5]Bendot, J. G., "Composite Propulsion Systems for an Advanced Reusable Launch Vehicle Application," Proceedings of the Second International Symposium on Air Breathing Engines, Paper 9, Royal Aeronautical Society, London, 1974.

[6]Builder, C. H., and Caudra, E., "An Elemental Approach to Propulsion Selection for the Supersonic Transport," Institute of the Aerospace Sciences Paper No. 62-72, Jan. 1962.

[7]Dix, D. M., and Petty, J. S., "Aircraft Engine Technology Gets a Second Wind," Aerospace America, July 1990, pp. 36-39.

[8]Kepler, C., and Champagne, G., "Performance Potential of Air Turboramjet Employing Supersonic Through-Flow Fan," AIAA Paper 89-0010, Jan., 1989.

[9]Von Entreb-Fursteneck, L. H., and Konig, O., 'ROCSCRAM' An Exceptional Airbreathing STS-Engine with Well Known Technology," International Astronautical Federation, Paper IAF-90-259, Oct. 1990.

[10]Sloop, J. L., Liquid Hydrogen as a Propulsion Fuel, 1945-1959, NASA SP 4404, 1978.

[11]Kerr, W. B., and Marra, J., "Rocket Fan - A Hybrid Air-Breathing Hydrogen-Fueled Engine," AIAA Paper 87-2000, June 1987.

[12]Benham, Charles B., and Nottage, H. B., "Increasing Rocket Burnout Velocity by the Introduction of Compressed Ambient Air," American Society of Mechanical Engineers, ASME Paper 64-WA/AV-6, New York, 1964.

[13]Jeffs, R. A., and Beeton, A. B. P., "Liquid Air Cycle Engines for High Speed Aircraft," Journal of the British Interplanetary Society, Vol. 19, 1963-64, pp. 484-490.

[14]Kajita, M., Ito T., Hasegawe, K., and Togowa, M., "Air Condensation Type Airbreathing Propulsion System," International Astronautical Federation Paper IAF-86-180, Oct 1986.

[15]Kunkler, H., "The Influence of Air Precooling before Compression in Air Breathing Engines of a Space Launcher," Proceedings of the Second International Symposium on Air Breathing Engines, Paper 8, Royal Astronautical Society, London, 1974.

[16]Hewitt, F. A., "Air Breathing Propulsion for Advanced Orbital Launch Vehicles," International Aeronautical Federation IAF Paper No. IAF-87-266, Oct. 1987.

[17]Mordell, D. L., Eyre, F. W. and Sreenath, A. V., "The Inverted Turbojet," Proceedings of the Eighth Anglo-American Aeronautical Conference, The Royal Aeronautical Society, London, Sept. 1961, pp. 159-180.

[18]Wilson, D. G., The Design of High-Efficiency Turbomachinery and Gas Turbines, MIT Press, Cambridge, MA, 1988.

[19]Y. Ribaud, "Inverse Cycle Engine for Hypersonic Air-Breathing Propulsion," Proceedings of the Ninth International Symposium on Air Breathing Engines, Vol. 2, AIAA, Washington, DC, 1989, pp. 1044-1050.

[20]Stricker, J. M., and Essman, D. J., "Turbojet Potential for Hypersonic Flight," Hypersonic Combined Cycle Propulsion, AGARD Conference Proceedings No. 479, 1990, AGARD, Neuilly-sur-Seine, France, Paper 9, pp. 1-10.

[21]Anderson, G. Y., "An Outlook on Hypersonic Flight," AIAA Paper 87-2074, June 1987.

[22]Billig, F. S., "Combustion Processes in Supersonic Flow," Seventh International Symposium on Air Breathing Engines, AIAA, Washington, DC, 1985, pp. 245-256.

[23]Anderson, G., Kumar, A., and Erdos J. "Progress in Hypersonic Combustion Technology with Computation and Experiment," AIAA Paper 90-5254, Oct. 1990.

[24]Broadbent, E. G., "Flows with Heat Addition," Progress in Aerospace Sciences, Vol. 17, No. 2, 1976, pp. 93-108.

[25]Lindley, C. A., "Performance of Air-Breathing and Rocket Engines for Hypervelocity Airacraft," ICAS Fourth Congress Proceedings, Spartan Books, Washington, DC, 1965, pp. 941-976.

[26]Townend, L. H., "Research and Design for Hypersonic Aircraft," Philosophical Transactions of the Royal Society, Vol. 335, No. 1637, April 15, 1991, pp. 87-224.

[27]Kuchemann, D., The Aerodynamic Design of Aircraft, Pergamon, Oxford, UK, 1978.

Chapter 1

Propulsion Systems from Takeoff to High-Speed Flight

F. S. Billig*

Johns Hopkins University, Applied Physics Laboratory, Laurel, Maryland

Abstract

Potential applications for missiles and aircraft requiring highly efficient engines serve as the basis for discussing new propulsion concepts and novel combinations of existing cycles. Comparisons are made between rocket and airbreathing powered missiles for anti-ballistic and surface-to-air missions. The properties of cryogenic hydrogen are presented to explain the mechanics and limitations of liquid air cycles. Conceptual vehicle designs of a transatmospheric accelerator are introduced to permit examination of the factors that guide the choice of the optimal propulsion system.

Nomenclature

A — area
$C_{D\ ADD}$ — additive drag coefficient
C_F — net thrust coefficient
C_T — thrust coefficient
C_{TAB} — thrust coefficient of airbreather
C_{TR} — thrust coefficient of rocket
CR — condensation ratio
D — drag, diameter
ER — equivalence ratio
f — fuel/air weight ratio
f_s — stoichiometric fuel/air weight ratio
$\mathfrak{F}$ — stream thrust = $PA+\rho u^2A$
g — acceleration due to gravity
h — specific enthalpy
H — enthalpy
I_{AB} — specific impulse of airbreathing engine

*Associate Department Supervisor and Chief Scientist.

I_{eff}	effective specific impulse = I_{SP} (1-T/D)
I_R	specific impulse of rocket
I_{SP}, I_{SPN}	specific impulse
ℓ	length
L	lift
$\dot{m}$	propellant flow rate
M	Mach number
OV	overall
p	pressure
q	dynamic pressure
Q	heat
Q_{FUS}	heat of fusion
r	radius
R	range
t	time
T	thrust, temperature
u	velocity
$\dot{w}_a$	weight flow of air
$\dot{w}_f$	weight flow of fuel
$\dot{w}_p$	weight flow of propellants
W	weight, work
x	axial coordinate
X_A	downrange
Y	initial mass fraction of slush
Z	altitude, sonic flow parameter
α	angle of attack
β	local flow angle with respect to vehicle axis
γ	ratio of specific heats
δ	angle between thrust vector and vehicle axis
Σ	structural coefficient
η	efficiency
θ	launch angle, flight path angle with respect to horizontal
Π	product of function
ρ	density
ψ_a	elevation angle with respect to horizontal

Subscripts

a,b,c,d, d',e,f,j	flow stations, points on trajectories, points in cycle diagram
AIR	air
AB	airbreather
b	burnout
c	cruise
D	design
ex	exit
i	individual stage, inlet
ij	injector
L	payload, LACE

max	maximum
N	nozzle
P	propellants
0	initial, overall, freestream conditions
R, Ref	reference
s	structural, satellite
t	total, stagnation
x	remaining at end of cruise, vehicle axis

Superscripts

*	sonic conditions
-	average

Introduction

The choice of the propulsion system for a high-speed vehicle is dependent on numerous factors in addition to its efficiency as measured by the specific fuel consumption or its reciprocal, the fuel specific impulse. Among these factors are weight, complexity, variability, longevity and cost of the components, and the density, rheology, stowability, handling, combustion characteristics, cost of the fuel. A comprehensive coverage of all of these issues is beyond the scope of this paper; the items covered will emphasize engine and vehicle performance. Moreover, subjects that already enjoy adequate coverage in the literature, e.g., the design and operating characteristics of turbojets, turbofans and rockets, will be referred to only in the context of comparative performance or as they may be appropriate in describing a tandem or combined-cycle engine.

Missions

The following list gives some insight into the types of missions and vehicles that are considered "high-speed" in the context of this discussion. In the broadest sense, these vehicles could be categorized as aircraft and expendable missiles. The missions for high-speed vehicles include:

1) Surface-to-air fleet or continental defense weapon.

2) Air-launched hypersonic antiship or antisurface missile.

3) Surface (or subsurface) launched hypersonic tactical or stragetic weapon system.

4) Hypersonic aircraft for reconnaissance, missile launch/support, research test bed, or transport.

5) Transatmospheric accelerator.

6) Antiballistic missile.

7) Maneuvering re-entry vehicle.

8) Hypersonic target drone.

To limit the scope of the discussion and to focus on the applications that are in the forefront of current efforts, the propulsion system and flight characteristics for particular examples of the missions in the preceding list, namely, a Mach 6-8 tactical missile, a transatmospheric accelerator, and a hypersonic cruise aircraft will be examined. Table 1 lists some of the characteristics of these

Table 1 Characteristics of hypersonic vehicles

Mission	Tactical missile	Transatmospheric accelerator	Hypersonic cruiser
Speed (Mach)	0 to 6-8 Surface-launched 0.7-3.5 to 6-8 Air-launched	0-25	0-6 0-15
Propulsion Cycle	Dual-combustor ramjet and/or rocket	Dual-mode ramjet/scramjet + numerous low-speed options	M6 turboramjet (others) M15, add a scramjet
Description	Fixed-geometry passively cooled	Variable-geometry actively cooled	
Fuel	Storable liquid, slurry, or solid	LH_2 (LO_2-augmented) CH_4	M6 hydrocarbon M15 LH_2; CH_4
Flight duration	< 10 min expendable	20-30 min > 100 cycles	M6 1-3 h M15 1 h 1000 cycles
Lengths			
Overall	5 - 15 ft	100 - 200 ft	
Combustor	2 - 5 ft	2 - 6 ft	
Nozzle	2 - 5 ft	50 - 80 ft	

vehicles. The tactical missile application is an example wherein the requirement of Mach 6-8 does not exceed the practical limit for the operation of a passively cooled air-breathing engine. Moreover, this speed is above that which can be reached by most other air-breathing cycles, e.g., turbojets and subsonic combustion ramjets. This tactical missile and the transatmospheric accelerator could be single or two-stage vehicles. Most of the remainder of the discussion will be devoted to the development of the arguments that lead to the preference of a particular propulsion cycle for each of these three applications. Two approaches will be used; the first will be by example, where much of the rationale for the selection of the engine design is omitted and the comparison is based on flight performance. The second will emphasize the methodology used to determine the performance of candidate cycles.

Antiballistic Missile or Surface-to-Air Missile

The first comparison of the propulsion systems is for application wherein a target is to be intercepted on its incoming trajectory by a missile on its ascending trajectory. Figure 1 depicts a two-stage, land-based continental defense system, which typifies this application. The more general problem includes one- or two-stage defensive systems and the targets and interceptors could be land or sea based. The interdependence of the design and operation of the propulsion system and the optimization of trajectory is the

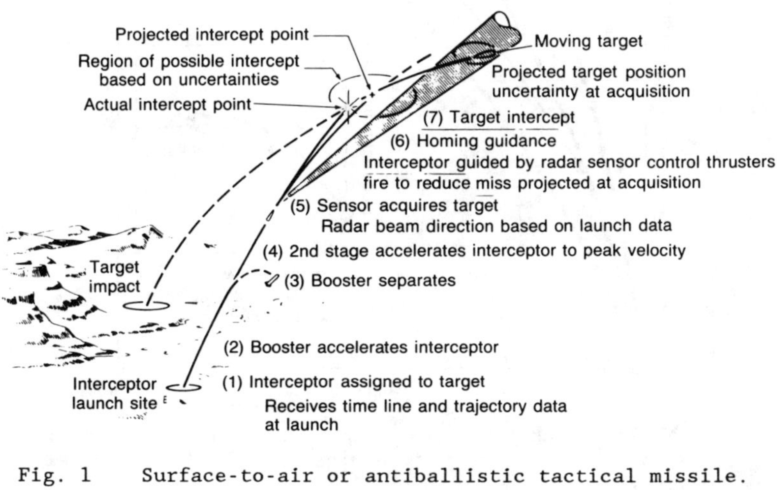

Fig. 1 Surface-to-air or antiballistic tactical missile.

issue. The intercept missile is launched vertically (θ = 90 deg.) and will climb and accelerate to a point on the intercept trajectory having an altitude Z_A and downrange X_A. The criterion for selecting the best propulsion system is maximum velocity at point A which, in general, corresponds to minimum flight time and maximum potential capability for subsequent intercept. Let us hypothesize that for a solid rocket it will be possible to select propellant grains and design grain shapes and nozzle sizes to produce the desired flow rates. Figure 2 shows trajectories for three different propellant flow rates. Curve 1 in the figure corresponds to a high flow rate, a short-duration turn from the vertical, and a zero-lift ascent trajectory, i.e., no normal aerodynamic force is applied. Thus, the vehicle flies ballistically. Maximum velocity is reached when the propellant burns out and the missile coasts with some slowdown from burnout to altitude Z_A.

Curve II is for a lower flow rate. Burnout occurs later at a higher altitude, and the trajectory "falls over"

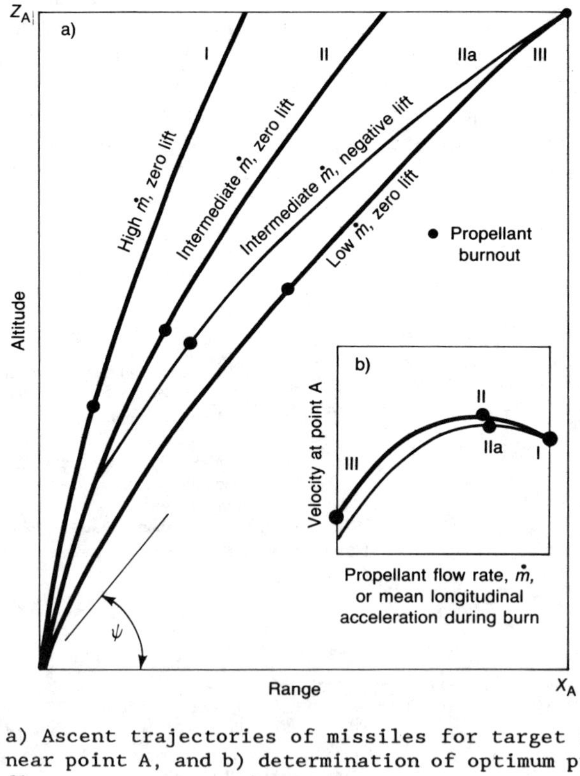

Fig. 2 a) Ascent trajectories of missiles for target intercept near point A, and b) determination of optimum propellant flow rate.

farther from the vertical because the velocity during the early phase of the climbout is lower than for curve 1. Curve III is for the lowest flow rate covered in this example. Burnout occurs nearer X_A, Z_A, and the trajectory falls over somewhat farther.

By introducing negative aerodynamic lift it is possible to force trajectories I and II to also pass through X_A and Z_A, also, as typified by curve IIa for the intermediate value of flow rate. Similarly, trajectories II and III could be forced to pass through the terminal point of trajectory I by introducing positive lift.

Regardless of whether a comparison is made for ballistic ascent trajectories or for lifting trajectories forced to pass through the same point in space, there exists a propellant flow rate that yields a maximum mean velocity, as shown in Fig. 2. The optimum trajectory corresponds to the case where the integrand in

$$U_b - U_0 = gI_{sp} \ell n \frac{W_0}{W_b} - \int_0^b \left(g \sin\theta + \frac{D}{W} \right) dt \ , \qquad (1)$$

is minimized. Here, W_0 is the initial weight ($W_S + W_P + W_L$), where W_S is the structural weight, W_P the weight of the propellant, W_L the payload weight, $W_b = W_S + W_L$ the burnout weight, g the acceleration due to gravity, and θ the local flight path angle with respect to the horizontal. This can best be understood by examining the limiting case of $\dot{m} \rightarrow \infty$ which, from the equation, would produce the maximum velocity immediately after launch, However, the vehicle would experience rapid slowdown as a result of high drag at low altitude. With a lower flow rate, the missile would climb through the dense lower atmosphere at a lower velocity with lower drag impulse, taking the same time to arrive at an intermediate altitude but with higher velocity.

As an example, consider the case of a single-stage rocket weighing 4000 lb at liftoff, of which two-thirds is propellant that produces an average I_{sp} of s lbm/lbm. The rocket is 18 in. in diameter, is 230 in. long, and has a low-drag nose. Taking the unrealistic limiting case of instantaneous burning of the propellant, the burnout velocity would be 9190 ft/s. For a vertical climb, the rocket would reach an altitude of 95,300 ft in 11.99 s, with a velocity of 7500 ft/s and an altitude of 204,100 ft at a velocity of 7000 ft/s. For ascent trajectories to targets having elevation angles less than the zenith, the corresponding altitudes of these velocities would be lower because the drag impulse would be higher.

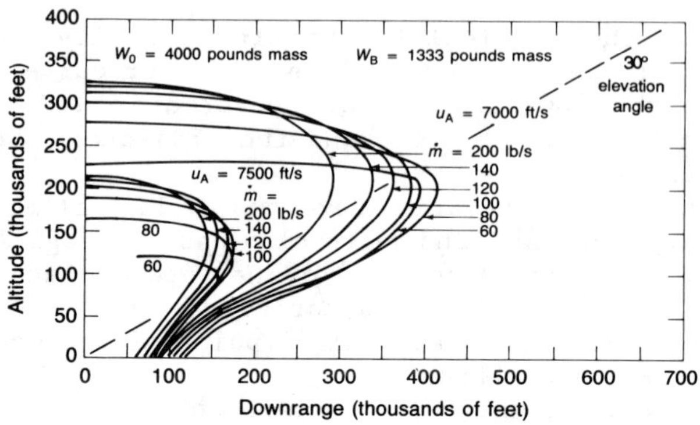

Fig. 3 Effect of rocket propellant flow rate on ascent trajectories.

Figure 3 shows that there is a range of finite values of propellant flow rates that yields a higher altitude Z_A than $\dot{m} = \infty$ for a corresponding velocity on a vertical climb. For clarity, the $\dot{m} = \infty$ curve $Z_A = 204{,}100$ ft is not included. Curves drawn through the loci of points on ascent trajectories corresponding to velocities of 7500 and 7000 ft/s are shown for flow rates from 60 to 200 lb/s. It is clear that only when targets are near the zenith are high flow rate values desirable. For targets below an elevation angle of about 30 deg., the optimal flow rate is about 80 lb/s. For the case with a constant flow rate of 80 lb/s the propellant burn time is 33.3 s, and the burnout velocities vary from 7260 ft/s for low elevation angles to 7858 ft/s for elevation angles of about 35 deg. For an elevation angle of 90 deg., the burnout velocity drops to 7770 ft/s which again shows the interaction of the terms within the integrand in the previous equation. This range of burnout velocities corresponds to a mean longitudinal acceleration during burning (U_b/gt_b), of 6.76 to 7.32 g, where t_b is the time at burnout. Although each missile design and mission would have a unique flow rate value, the value that yields a mean acceleration of 5-10 g is typical of a wide variety of cases.

Granting that a constant flow rate of 80 lb/s is a reasonable design point from which departures can be made to examine other effects, the case of a nonconstant flow rate can be studied. Two cases are considered: a constant flow rate with an intermediate period without burning and a linear variation of flow rate with time. Figure 4 shows the effect of interrupting the burning of the propellant for periods of 5-40 s once the vehicle reaches a velocity of 5000 ft/s. For the range of elevation angles of primary interest, nonburning pulses up to 30 s lead to increases in performance, i.e., the

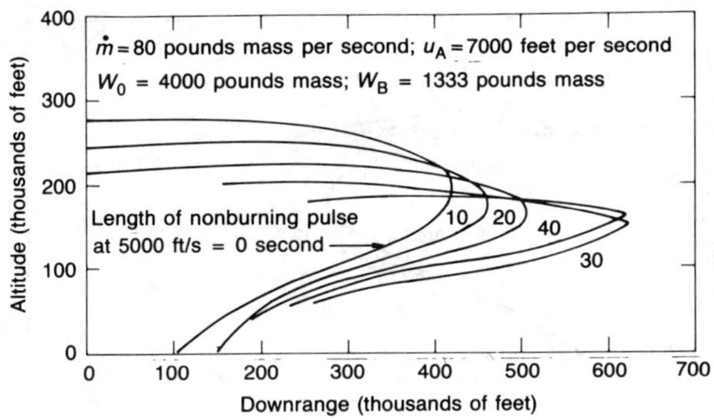

Fig. 4 Performance of pulsed rocket motors.

U_A = 7000-ft/s contour is moved farther from the launch point. The shift of the curve for altitudes below 120,000 ft for the 30-s pause corresponds to a slowdown of about 400 ft/s for the steadily burning case. Lengthening the pulse beyond 30 s causes too great a slowdown to move the contour further and, indeed, if the pulse is long enough, the vehicle cannot reach 7000 ft/s. The choice of 5000 ft/s for the point at which the zero flow rate pulse is initiated is about optimum for extending the U_A = 7000-ft/s contour for pulse lengths of 30 s. Had the pause been delayed until U_A = 6000 ft/s, only about half of the gain would have been realized. Whether the added complexity to the motor design to accommodate pulsing is warranted would depend on how crucial the additional performance is in meeting mission requirements. The amount of payload that would have to be replaced by additional propellant to produce an equivalent extension of the 7000-ft/s contour would be about 85 lb.

Figure 5 compares the performance of linearly varying flow rates with constant flow rates. The 80-lb/s constant flow rate case is again used as reference. Four cases were examined: linearly increasing flow rates from 60 to 100 lb/s and from 70 to 90 lb/s and linearly decreasing flow rates over the same two ranges. As shown in Fig. 5, increasing flow rates degrade performance, whereas decreasing flow rates extend the 7000-ft/s contour. Unfortunately, from the point of view of grain design, a so-called regressive burning rate requires an intricate shape. Reference 1 gives a rather extensive discussion on the selection of propellants and the factors that influence the design of grains that could yield the desired characteristics used in these examples.

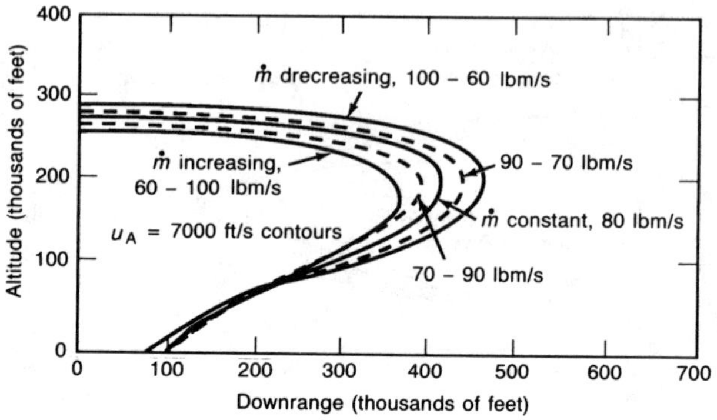

Fig. 5 Effect on rocket motor performande of linearly varying the weight flow of propellant.

Dual-Combustor Ramjet

Granting that the preceding arguments provide a basis for selecting an optimal rocket, it is now appropriate to discuss the design and operating characteristics of the candidate air-breathing propulsion system for the tactical missile application listed in Table 1. One of many possible configurations is the dual-combustion ramjet (DCR). It is schematically depicted in Fig. 6 (see Refs. 2,3). In this sketch an axisymmetric forebody serves as the initial compression surface of the supersonic inlet. In the plane of the cowl lip, the flow is subdivided into four small sectors and four large sectors. The smaller sectors direct flow to a dump-type subsonic combustor through a duct whose cross-sectional area increases with streamwise direction. This provides stable operation over a range of Mach numbers by controlling the position of the normal shock structure. These inlets are operated supercritically, i.e., the normal shock is swallowed because expulsion of the normal shock would produce detrimental interactions with the flow entering the larger flow passages.

The major portion of the air is turned supersonically toward the engine axis by the outer cowl compression surface. In this design the flow in the four ducts is spread circumferentially to form an annulus of flow in the dump plane of the subsonic combustor. The aft portion of these supply ducts is shaped so as to provide a constant or slightly increasing cross-sectional area in the streamwise direction, which serves as a combustor-inlet isolator. The shock train structure in these ducts supports a pressure rise equivalent to a normal shock when the vehicle is operating at a high equivalence ratio (ER) and low flight Mach number. For lower

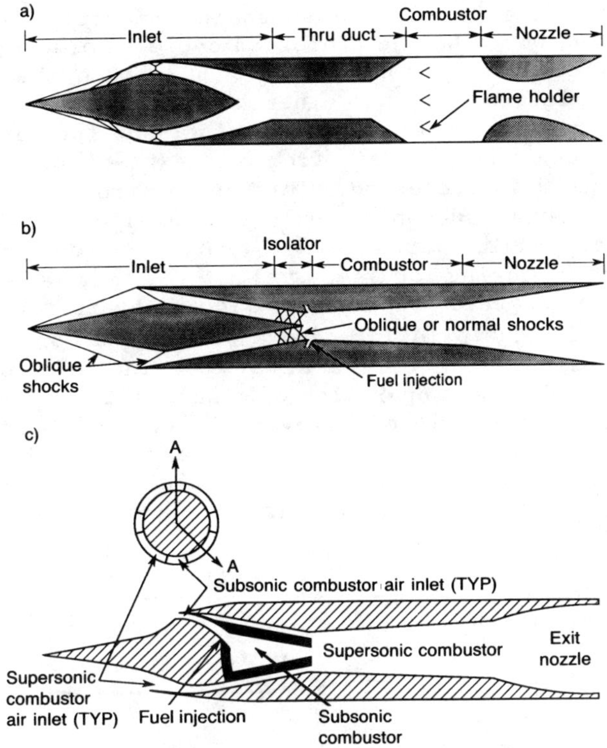

Fig. 6 Tactical high-speed ramjet engines: a) subsonic combustion ramjet; b) dual-mode scramjet; c) dual-combustion ramjet.

ER and/or higher M_0, the shock train pressure rise corresponds to an oblique wave structure. In the "normal shock" operating mode, the mean Mach number at the combustor entrance is subsonic, and the mean Mach number in the combustor exit is either sonic or supersonic. During operation in the "oblique shock" mode, the mean flow at all stations throughout the combustor is supersonic. This gives rise to the term "dual-mode" operation, which has been discussed in considerable detail in the literature (e.g., Refs. 4-6).

Superficially, it would appear that this engine concept is just an arbitrary combination of two ramjet cycles, a subsonic combustion engine (Fig. 6a) and a scramjet (Fig. 6b), and it would be expected to yield performance that would lie within the range associated with single-cycle ramjets of the two types and to have no particular advantages over one or the other. In fact, if the range of operating Mach number

M_0 does not exceed 5-6, the subsonic combustion integral-rocket ramjet would be the preferred cycle. Moreover, if the ramjet had to function only at $M_0 > 6$-7, then the scramjet would be preferred. On the other hand, if the air-breather must use a conventional hydrocarbon fuel and function over a wide Mach number range, say, from below $M_0 = 4$ to above $M_0 = 6$, then the DCR is preferred. This arises from the fact that a fixed-geometry subsonic combustion ramjet designed to produce acceptable performance at $M_0 = 4$ operates very inefficiently during cruise at $M_0 > 6$, and a scramjet, designed to operate solely on liquid hydrocarbon fuels without auxiliary piloting, has not been demonstrated at conditions simulating flight at $M_0 \leq 4$ in the tropopause. Heretofore, efficient operation of a scramjet engine at M_0 4-5 has been restricted to hydrogen[7], alkylated boranes[8]. or with a supplemental oxidizer (chlorine trifluoride by the Marquardt Co.[9], or with a separately fueled near-stoichiometric pilot (by the United Aircraft Research Laboratories[10].

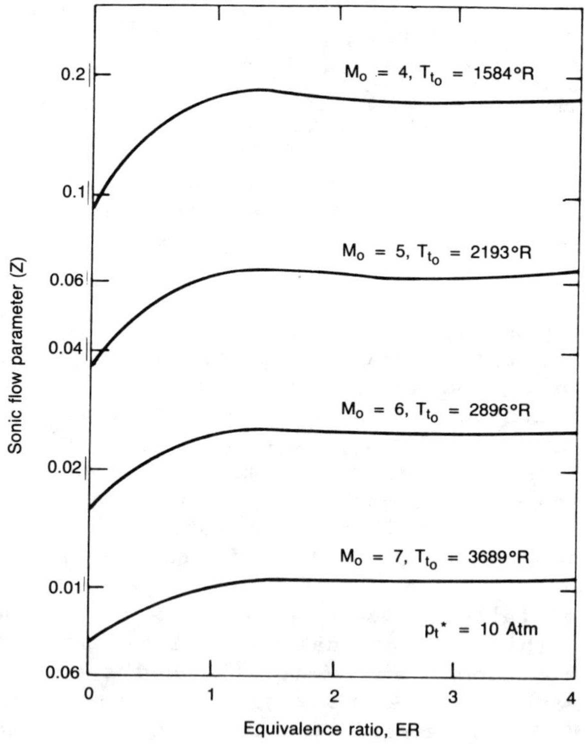

Fig. 7 Sonic flow parameter Z.

Therein lies the efficacy of DCR: the dump combustor can act as either a pilot or a gas generator to assure that heat can be efficiently released in the supersonic combustor, even when M_0 is low. If it is operated as a pilot, fuel is added to both streams; if it operated as a gas generator, all, or nearly all, of the fuel is added within it, and the main combustor becomes a supersonic "afterburner." The latter approach requires the dump combustor to operate very fuel-rich.

For a missile application, cost complexity and weight considerations generally dictate the use of fixed-geometry components. Thus, a nozzle throat must be specified for a subsonic combustion ramjet, as well as for the gas generator of the DCR. Figure 7 will help to exlain why this constraint leads to an upper bound on the practical operating M_0 of a subsonic combustion ramjet and how the effects of the constraint are mitigated when operating the gas generator of the DCR with ER > 1.

Values of the sonic flow parameter defined as $Z = A^*P_t^*/A_0P_{t0}$ are shown in Fig. 7 for the range of $0 \leq ER \leq 4$ for freestream temperatures of 1584, 2193, 2896 and 3689°R, which correspond to $M_0 = 4$, 5, 6, and 7. Here, A^*/A_0 is the ratio of the cross-sectional area of the combustion products passing through the sonic point in the exhaust to the area of the airflow in the freestream and P_t^*/P_{t0} is the ratio of the respective total pressures at the same points in the flow. These calculations are for the products of combustion of Shelldyne H and air (see Ref. 11). The curves are for $P_t^* =$ 10 atm but curves for the range of pressures (i.e. $1 \leq P_t^* \leq$ 20 atm) typical of this engine would be indistinguishable. Shelldyne H, with a gravimetric heating value of 17,890 Btu/lbm and a specific gravity of 1.08, is one of a class of special fuels that have been tailored for use in volume-limited systems.

The near-constant value of Z for ER > 1 compared to about a factor of 2 between ER = 0.25 and ER = 1 is the basis for the large difference in the efficiency of the DCR and the subsonic combustion ramjet when throttling is required. A DCR engine can be designed to split the inlet airflow such that 25% of the flow goes to the gas generator. Thus, throttling the overall engine from ER = 1 to ER = 0.25 corresponds to throttling the gas generator from ER = 4 to ER = 1. If Z is constant, then P_t^*/P_{t0} is constant; thus, for a given M_0 there is no decrement in total pressure required to throttle the DCR. Conversely, in a subsonic combustion engine with a fixed exhaust area, a 50% loss in total pressure must be accepted if the engine is throttled from ER = 1 to ER = 0.25. The large variation in Z for different M_0

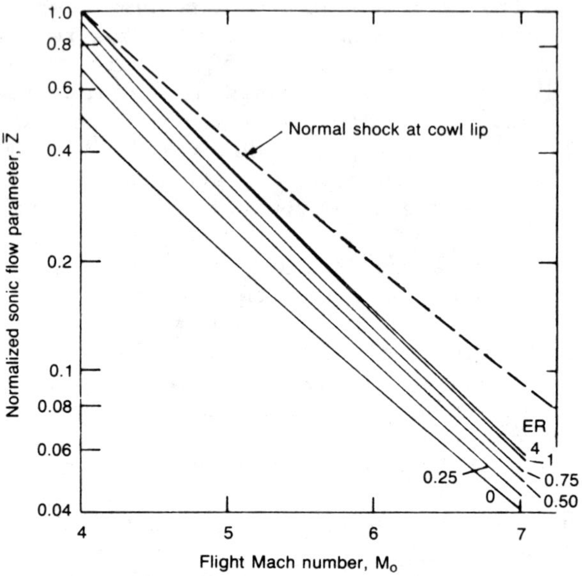

Fig. 8 Normalized sonic flow parameter.

is the source of the penalty in performance that must be accepted in a fixed-geometry engine.

In a variable-exit engine the nozzle area could be adjusted to maximize P_t^* at any operating condition. For most inlets this corresponds to the "critical" operating point. The ratio of the total pressure entering the exhaust nozzle to the total pressure that could be obtained if the nozzle area could be changed is given as

$$\frac{P_t}{P_{t_{MAX}}} = \frac{Z}{Z_{REF}} = \bar{Z} = \left[\frac{A_0/A_i}{A_0/A_{i\ REF}}\right] \frac{P_t^*/P_{t_{0\ REF}}}{P_t^*/P_{t_{0\ MAX}}} \tag{2}$$

where the subscript REF refers to conditions that correspond to the point chosen to size the exhaust nozzle. The ratios of air capture and critical total pressure are dependent on a particular inlet design.

For the results shown in Fig. 8, the reference condition was taken as Mach 4 and ER = 1, and the inlet operating conditions were extracted from Ref. 2. These inlets were a family of axisymmetric forebodies having a 12.5-deg. half-angle cone followed by varying degrees of isentropic turning. All of the surfaces were designed to have full capture (A_0/A_i = 1) at $M_0 \geq 6$. Interestingly, the quotient $(A_0/A_i)/(P_t^*/P_{t0})_{MAX}$ is nearly invariant for the entire family. The

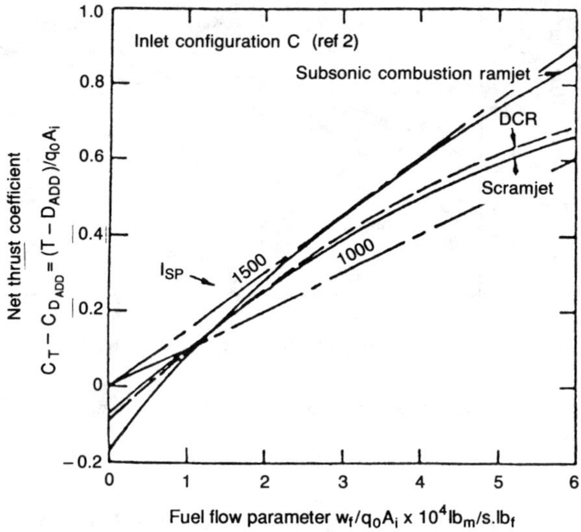

Fig. 9 Performance comparison at $\alpha = 0$ deg., $q_0 = 5000$ lbf/ft^2, $M_0 = 4$.

figure shows that a fixed-geometry subsonic combustion ramjet would have to accept a 51% loss in total pressure when throttled to an ER = 0.25 at $M_0 = 7$. The loss in the DCR would be 31% and, moreover, only 25% of the airflow would be involved.

The basis for the selection of the DCR over either of its genera, the supersonic combustion ramjet and the ramjet, for the tactical missile are substantiated by the data shown in Figs. 9 and 10. Results from Ref. 2 compare the three engine cycles wherein all three have the same degree of external inlet compression, namely, a 12.5-deg. conical shock followed by 8.5 deg. of isentropic turning. In the supersonic flow passages of the DCR and the scramjet, the flow was further contracted by 1/6, with an additional loss of 10% in total pressure. Net thrust coefficient C_F is plotted vs the fuel flow parameter $\dot{w}_f/q_0A_i$, where $C_F = C_T - C_{DADD}$ and

$$C_T = \left(1/q_0A_i\right)\left[\left(\rho U^2 A\right)_{ex} - \left(\rho U^2 A\right)_0 + A_{ex}\left(P_{ex} - P_0\right)\right] \quad (3)$$

$$C_{D_{ADD}} = \frac{1}{q_0 A_i} \int_{A_0}^{A_i} \left(P - P_0\right) dA \quad (4)$$

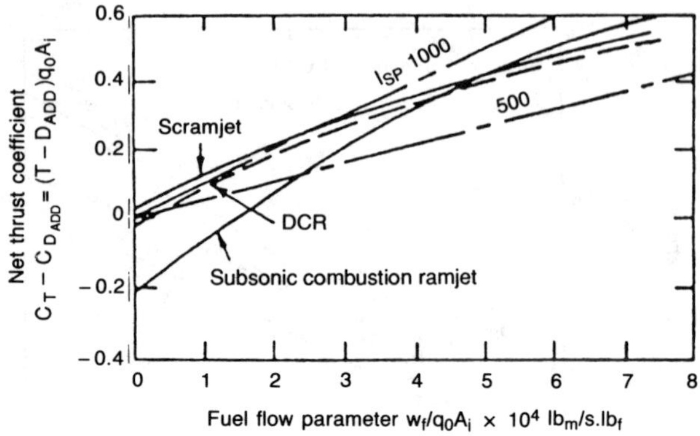

Fig. 10 Performance comparison at $\alpha = 0$ deg., $q_0 = 5000$ lbf/ft^2, $M_0 = 7$.

Note that the engine inlet area A_i is used to normalize the thrust and drag coefficients and that the engine exit area A_{ex} is assumed to be 25% larger than A_i, i.e., $A_{ex}/A_i = 1.25$. Since the engine efficiency is determined by the fuel flow rate for a given thrust (rather than ER), $\dot{w}_f$ is the appropriate parameter to hold constant when comparing the net thrusts of various configurations. Accordingly, C_F has been plotted against the normalized fuel flow parameter $\dot{w}_f/q_0A_i$ in the figures, which show the result for the three engine cycles at $M_0 = 4$ and $M_0 = 7$. Grids of net specific impulse defined as $I_{spn} = C_Fq_0A_i/\dot{w}_f$ have been added to facilitate quantitative comparison of engine efficiencies.

At $M_0 = 4$, the curves for DCR and scramjet cross the ramjet curve at $\dot{w}_f/q_0A_i = 1.3 \times 10^{-4}$ s lbf/lbm, which corresponds to ER = 0.25, $C_F = 0.14$. At $M_0 = 7$, the crossover point with the DCR is at $\dot{w}_f/q_0A_i = 4.37 \times 10^{-4}$ s lbf/lbm or ER = 0.59 and $C_F = 0.37$ and, with the scramjet, at $\dot{w}_f/q_0A_i = 4.92 \times 10^{-4}$ s lbf/lbm or ER = 0.67 and $C_F = 0.43$. This trend toward an increasing ER with increasing M_0 at the crossover point occurs at different ER values for other fuels and/or engine configurations. Generally, for $M_0 > 7.5$, engines with supersonic combustion have higher efficiency than subsonic combustion ramjets at all ER values. It should be noted that the selected mass fraction of 0.25 to the gas generator and the setting of the geometries by conditions at $M_0 = 4$ with overall ER = 1 lead to the better performance of DCR at lower M_0 and poorer performance at high M_0 relative to the scramjet. Had these parameters been modified, the trend with M_0 could have been adjusted to suit the application.

Most applications, however, that require the high efficiency of air-breathing engines at $M_0 > 5$ include long cruise ranges corresponding to low C_F; hence, the DCR or the scramjet turn out to give higher overall performance. For example, at $M_0 = 7$, a typical C_F at cruise is 0.2. From Fig. 10 the values of $\dot{w}_f/q_0A_i$ are 1.7×10^{-4}, 2.1×10^{-4}, and 2.9×10^{-4} s lbf/lbm for the scramjet, the DCR, and the ramjet, respectively; thus, DCR would get 38% and scramjet 71% more cruise range than the ramjet for the same amount of fuel.

Staging

A complete examination of the differences in performance of two-stage vehicles in which flow rate is varied for the two stages is an arduous task; however, some exemplary results can be discussed that show the general trends that would result from a more thorough investigation. Classical theory, which will be discussed, shows that for vehicles having multiple stages, the single-stage equation previously given must be revised to determine the velocity change for an accelerating vehicle. In this case, the equation becomes

$$\left(U_b - U_0\right) = g\, I_{SP}\, \ell n \left[\prod_1^n \left(W_{0i}/W_{bi}\right)\right] - \int_0^b \left(g \sin\theta + \frac{D}{W}\right) dt \quad (5)$$

where the bracketed term is the product of the n weight ratios of the i individual stages. In general, the burnout velocity of the nth stage U_{bn} is greater than the burnout velocity for a single-stage vehicle having the same ratio of payload to initial weights, W_L/W_0. The higher velocity is a result of discarding the structural weight of the burnout stage at staging, thereby saving the propellant that would have been used to accelerate this mass. For the large velocities required for long-range ballistic missiles and space vehicles, staging is mandatory for high W_L/W_0. Three- and four-stage vehicles are typical for these missions. For tactical missiles the velocity increment is typically one-third to one-half of that required for ballistic missiles, and no more than two stages are required.

Analytical solutions to maximize the burnout velocity for particular propulsive and structural characteristics will be discussed that show, for the idealized case of drag-free/gravity-free climb-out of a vehicle having equal I_{SP} and structural coefficients $\epsilon = W_{si}/(W_{si} + W_{pi})$ for all stages, that each stage should have the same weight fraction and therefore produce the same velocity increment as every other stage. Here, W_{pi} is the weight of propellant in the ith stage. When drag, gravity, varying I_{spi}, and ϵ, are considered, the optimum U_{bi} for rocket-powered vehicles occurs

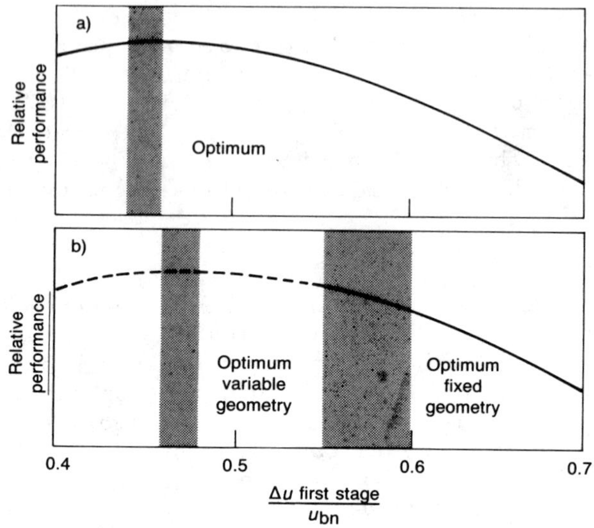

Fig. 11 Optimal staging of tactical missiles: a) two-stage rocket; b) rocket booster, air-breathing second stage.

when the velocity increment of the stage increases with each succeeding stage. Figure 11 shows that optimum performance for two-stage rockets is obtained when the velocity increment of the first stage is 44-46% of the total ΔU.

The same trend would also be observed when the second stage is an air-breathing missile. However, this cannot be fully exploited with the fixed geometry engines generally used in tactical missiles. Usually it is not possible to design a fixed geometry engine (i.e., the inlet area and the throat of the exhaust nozzle cannot be varied) that will operate over a velocity range below 55 to 60% of the maximum velocity. If variable geometry is practical, the velocity increment of the first stage can be reduced to about 46 to 48% for optimal performance.

Rocket/Dual-Combustor Ramjet Comparison

Figure 12 compares the single-stage rocket with a family of optimally designed air-breathers for the missile launched vertically to a point in space. The downrange scale has been compressed. The rocket is the preferred engine for all elevation angles above 10 deg. The region for the rocket is terminated since, at high altitude, control of the missile would have to be provided by a means other than aerodynamic

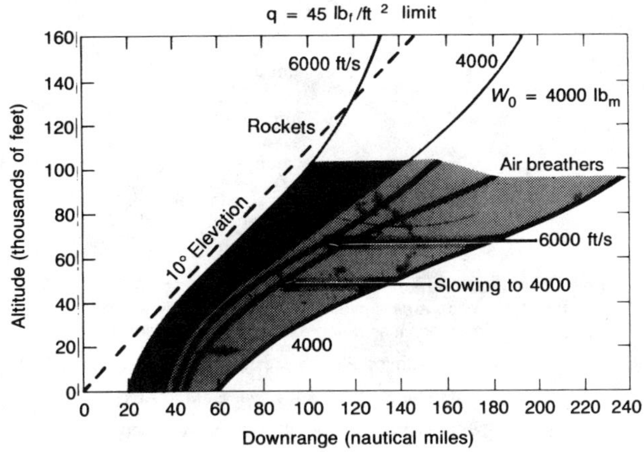

Fig. 12 Comparison of rockets and ramjets for area defense mission.

surfaces. The 160,000-ft limit shown corresponds to a dynamic pressure q of 45 psf for a velocity of 6000 ft/s. Control at lower dynamic pressure would have to be provided by auxiliary thrusters; this not only greatly complicates the missile design, but thrusters generally are not available in tactical missiles. The criterion for selection of the better engine is maximum velocity, which corresponds to minimum time to target. At these higher target elevations, the rocket is capable of reaching at least 7000 ft/s, and the air-breather does not have sufficient accelerative capability to exceed about 6500 ft/s. At elevations below about 10 deg., the terminal speeds are less than 7000 ft/s. High total drag impulse limits the rocket for these cases. Structural constraints imposed by passive cooling limits the airbreathing systems. In the region of "block speeds," (line-of-sight range/time of flight) below about 6500 ft/s, the air-breather is preferred. There are two assumed air-breathing systems, each having a different maximum terminal velocity. One is designed to accelerate to about Mach 7 and thereby has the capability of block speeds somewhat in excess of 6000 ft/s. It would require a booster weighing about 1960 lb, which means that the DCR stage would weigh 2040 lb to meet the total weight limitation of 4000 lb. The other is designed to have a maximum speed of 4000 ft/s and would weigh 2741 lbm because the booster weight would have to be only 1259 lbm.

Another defensive application is the protection of high-value surface combatants from coordinated airborne attacks. In this scenario, shown in Fig. 13, a large number

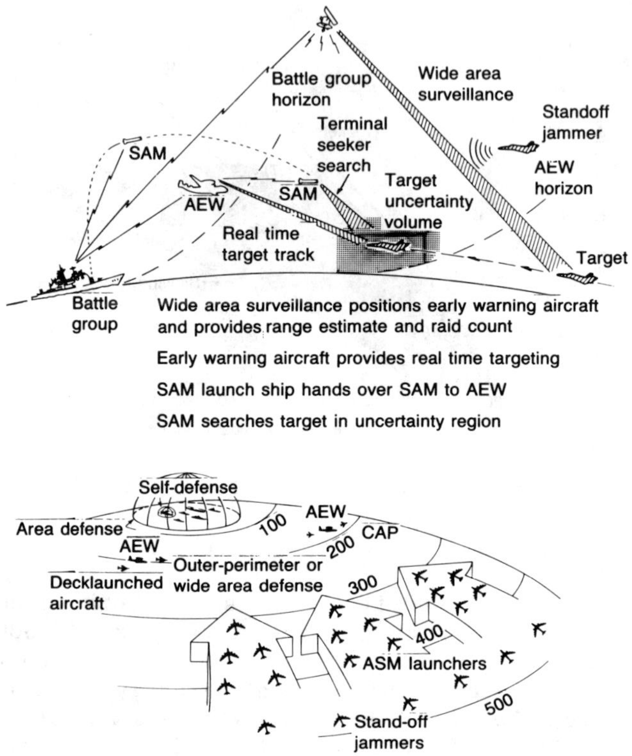

Fig. 13 Wide-area defense missiles.

of hostile aircraft are dispersed over a broad sector. The aircraft penetrate to a point at which antiship missiles are launched under the shield of high-density electronic jamming. The defensive system is composed of high-acceleration missiles that must intercept either the antiship missiles at relatively short ranges or the launch and jamming aircraft at longer ranges.

The missiles are but part of the assets that must be brought to bear in defending the ships. Targeting can be provided by space- and/or airborne platforms. Generally, the information provided by the targeting systems is sufficient to guide the defensive missile only to the region of the target. When the missile reaches the general area, terminal seekers onboard the missile provide the terminal acquisition. Moreover, requirements generally stress the capabilities of the defensive missile, often imposing propulsive requirements that may dictate the selection of the optimal engine cycle. The trajectory that must be flown to intercept these targets is quite different from those previously examined.

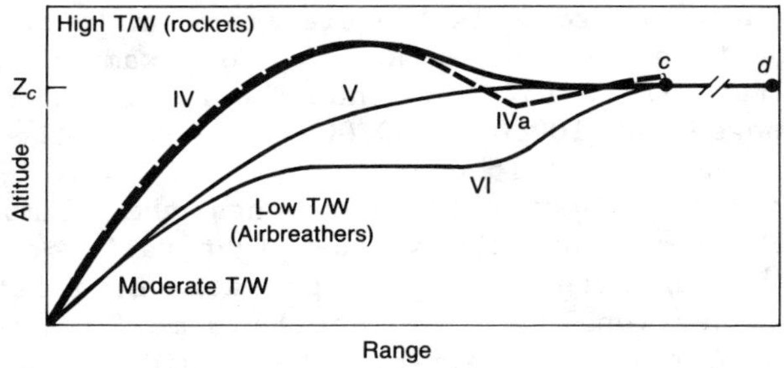

Fig. 14 Effect of thrust-to-weight ratio on ascent trajectories; missiles designed for near-level flight at an altitude Z_c.

Trajectories representative of this generic class of missions are shown in Fig. 14. Here the objective is to obtain maximum range by accelerating and climbing to an altitude Z_c and cruising in near-horizontal flight from points c to d. As in the previous example, the rocket has a maximum speed greater than the air-breather. The propellant is consumed relatively early in the climb-out, i.e., 33.3 s into a climb-out of about 100 s, and the vehicle decelerates following burnout. For the vehicle to obtain maximum range, the flight path during climb-out and approach to the cruise condition must be optimally shaped. In general, the optimal path corresponds to launch near the vertical (θ = 90 deg.) and continual turndown (negative lift) up to the apogee. If near-vertical launch is not achievable, positive lift (along with thrust vector control) is used during the early phase of the climb-out to bring the vehicle close to the $\theta = f(Z)$ characteristic of the optimal vertical launch case.

The types of trajectories that must be examined to obtain optimal range of a high thrust-to-weight (T/W) ratio vehicle (e.g., the rocket) are typified by curves IV, IVa, and V. A high T/W vehicle has the acceleration capability to "overshoot" Z_c and then glide from the apogee back to Z_c by a number of possible paths with the introduction of positive lift. Curve IV corresponds to a case with a lift-to-weight ratio (L/W) > 1 as Z_c is approached. With L/W = 1 at Z_c, the vehicle will oscillate about Z_c with a decreasing excursion as time progresses, as typified by curve IVa. With L/W < 1, the vehicle can approach Z_c at $\theta = 0$ asympotically from below (not shown). The permissible altitude at apogee is limited to the aforementioned requirement to maintain control but, in general, optimal range is obtained on trajectories having lower apogees. These constraints significantly restrict the region of permissible climb-out trajectories to a narrow corridor in ψ_a, where ψ_a is the elevation angle of the apogee with respect to the launch point. For example, given that the flow rate is 80 lb/s and that the rocket is constrained to an apogee of 100,000-160,000 ft, then the range in elevation angle is 11-15 deg.

Curve V schematially represents the climb-out of a rocket with a moderate thrust-to-weight ratio or of an air-breather having a relatively large ratio of inlet areas to body cross-sectional area. For the same Z_c, however, the optimal climb-out for an air-breather would lie below that of a rocket. That is a consequence of the need to optimize thrust and fuel flowrate simultaneously with the minimization of the integrand in the acceleration equation for the climb-out of an air-breather. Typically, those trajectories require modulation of the normal force that produces turn-down following launch, followed by positive lift at low and intermediate altitudes, and negative lift as Z_c is approached.

Curve VI is representative of the climb-out of a low thrust-to-weight ratio engine. The accelerative capability is inadequate to climb to Z_c and accelerate to the desired U_c simultaneously. Instead, the vehicle must climb to a lower altitude, use lift to maintain this altitude, accelerate to U_c, then turn up with $L > W$, and finally approach Z_c with $L < W$. Usually this type of climb-out results in poorer performance than a case V climb-out, which is one of the factors that influence the choice of the proper thrust-to-weight ratio propulsion system.

Optimal cruise range for the air-breathers or glide range for the rockets usually results when the vehicle flies from points c to d at an angle of attack corresponding to $(L/D)_{max}$. The maximum lift-to-drag ratio varies with the aerodynamic configuration, velocity, and altitude but, for tactical missiles in the range of 4000-7000 ft/s, $(L/D)_{max} \sim 23$, and the corresponding angle of attack α is between 8 and 12°. During cruise the rocket is lighter than the air-breather, and the optimum altitude at the same velocity is somewhat higher. As the rocket slows down during the glide, the optimum altitude decreases. On the other hand, as the air-breather burns fuel to maintain constant velocity, the optimum altitude increases with the decreasing weight unless that altitude is too high to permit efficient operation of the combustor.

Figure 15 shows the altitude-range envelope for points d for the same tactical missiles used for the flight envelope of the last example. For the rocket, boundaries are shown for slowdown to U = 4000 and 6000 ft/s. For air-breathers, three boundaries are shown. The area above the dashed line

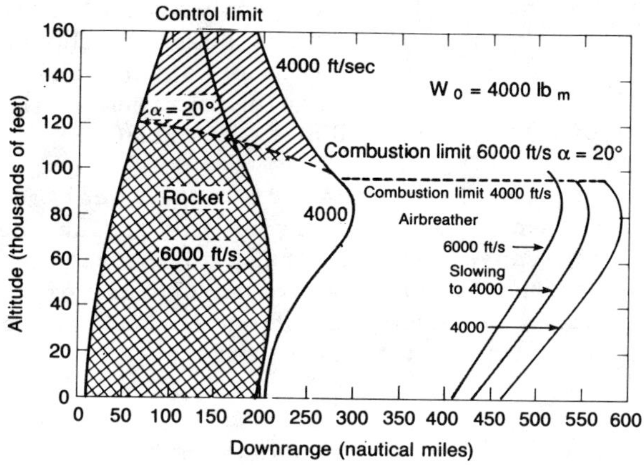

Fig. 15 Comparison of rockets and ramjets for long-range cruise missions.

requires angles of attack greater than 20 deg. to maintain level flight of a coasting rocket. Indeed, at altitudes above 130,000 ft, level flight is not possible for this vehicle. The U = 6000 ft/s air-breather design is for a missile designed to cruise at 6000 ft/s. Two 4000 ft/s air-breather boundaries are also shown; the one having the greatest extent is for a missile designed for cruise at 4000 ft/s and the other is for the 6000 ft/s design, coasting to 4000 ft/s following fuel exhaustion.

As in the previous example, the range that can be reached by the air-breathing missile is considerably greater than that reached by the rocket. On the other hand, for ranges less than that corresponding to equal velocities of the rocket or air-breather (i.e., 6000 or 4000 feet per second) for this example, the rocket could be the preferred propulsion system for a tactical missile, the argument being that the mean velocity would be greater, hence the time to target would be less for the rocket. This, in turn, could perhaps permit more time for assessment prior to firing or time for more salvos in a shoot-look-shoot scenario.

The optimum cruise altitude for the rocket, assuming $C_L = L/q_0A_R \approx 1$ at $(L/D)_{max}$, where $A_R = \pi D^2/4$, would be about 111,000 ft at U = 6000 ft/s and 93,000 ft at U = 4000 ft/s. Maximum range, however, occurs at lower altitudes because cruise range for the coasting rocket is small compared to the ranges of the optimal paths to point c. For the air-breathers, the corresponding altitudes for flying at $(L/D)_{max}$ would be 2000-4000 ft lower at the same velocity, but better performance would be achieved as a result of the aforementioned problems of low inlet and combustor efficiency at low pressure levels. Thus, the cruise phase of the air-breather occurs at a somewhat lower angle of attack than that which yields $(L/D)_{max}$. For this example, optimum cruise altitude is about 95,000 ft for U = 6000 ft/s and 80,000 feet for U = 4000 ft/s.

This mission is typical of a tactical missile that provides for its own guidance during the cruise phase and is, in fact, a long-range weapon. Thus, in the parlance currently in vogue, the performance shown is typical of a "wide-area defense" missile, whereas that shown in the previous example is typical of an "area defense" missile.

Transatmospheric Accelerator-Hypersonic Cruiser/Aircraft

Engine Cycles

A different approach will be taken to explore the possible choices of the propulsion systems for transatmospheric accelerators and hypersonic cruisers. A more fundamental understanding of the thermodynamics of cycles and the physical characteristics of the candidate propellants will be developed. The cycles of interest include rockets, ram/scramjets, turbomachines, and a multitude of combinations and variants of the basic cycles. The discussion on turbomachinery will be brief since many of excellent textbooks are available (see, e.g., Refs. 12, 13).

Air-breathing propulsion systems function as a modified Joule or Brayton cycle, shown in Fig. 16; i.e., the incoming fluid is compressed, heat is added at near-constant pressure, and the gas is expanded to high velocity. Force on the vehicle is directly proportional to the momentum increase in the stream. The increase in momentum arises from the addition of the mass of the propellants and an overall

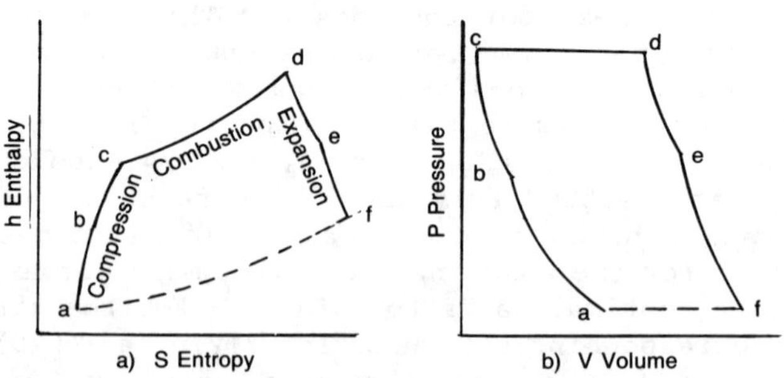

Fig. 16 Modified Joule or Brayton cycle.

increase in the velocity of the propulsive stream. The cycle diagram is helpful in explaining the role that pressure rise plays in increasing the stream velocity. The pressure rise is provided by aerodynamic and/or mechanical mechanisms. In the turbojet operating at subsonic speeds, the pressure rise mechanism is primarily mechanical, e.g., in an axial flow compressor. Thus, points a and b are nearly superposed. In a ramjet the compression is aerodynamic; thus, points b and c are superposed. To a first approximation, the pressure ratio in the compression and expansion processes are equal. The Mach number following expansion is somewhat lower than the Mach number preceding compression because of the increase in the ratio of specific heats and losses in total pressure, i.e.,

$$M_f^2 = \frac{2}{\gamma_{ef}-1}\left(\frac{P_{t_f}}{P_{t_a}}\right)^{\frac{\gamma_{ef}-1}{\gamma_{ef}}}\left(1+\frac{\gamma_a-1}{2}M_a^2\right)^{\frac{\gamma_{ef}}{\gamma_a}\left(\frac{\gamma_a-1}{\gamma_{ef}-1}\right)} \tag{6}$$

However, the sound speed at f, a_f is much larger than at a_a; thus, $U_f \geq U_a$ and the momentum is increased by $\dot{w}_a$ $[(1+f)U_f-U_a]$. When compression is obtained by mechanical work, as in an axial flow compressor in the turbojet cycle, there is an additional contribution to the exit momentum resulting from the increase in total pressure in the cycle. The work to compress the incoming gas is proportional to the total temperature rise $T_{tc}-T_{tb}$, whereas the total pressure increase is proportional to the temperature ratio. The work of the turbine in the expansion process matches that in the compression ratio; thus, $(1+f)\ (T_{te}-T_{td}) = (T_{tc}-T_{tb})$, but the ratio $(T_{td}/T_{te}) < (T_{tc}/T_{tb})$; thus, the total pressure drop in the turbine is less than the total pressure rise in the compressor.

The principal candidate propulsion cycles can be placed in two major categories: 1) Mach 0 to about 6 devices that are turbomachinery, and 2) Mach 0-16 to 26 engines. Those in category 1 would be the engines for the lower-speed hypersonic cruiser and the first stage of a two-stage transatmospheric accelerator (TSTO). In the second category are candidates for a high-hypersonic-speed cruiser and the single-stage-to-orbit (SSTO) vehicle.

Each of the combined cycles in Fig. 17 avoids a fundamental limitation of the simple basic cycles. The turboramjet is simply a hybrid that contains some of the best features of the two genera cycles. The adjustable bifurcated inlet provides the means for optimally splitting the airflow to the respective engines. At low flight speeds most of the inlet flow is ducted to the turbojet compressor, and the remainder is "pumped" through the ramjet duct by ejector action. As speed increases, the proportion of air to the ramjet duct is increased. At high Mach numbers the entire inlet air is diverted around the turbojet duct, thereby circumventing the problem of excessive turbine inlet temperatures, generally the factor that limits the maximum operating M_0 of the turbojet.

The air-turborocket eliminates the constraint of high turbine inlet temperatures at high M_0. The turbine is driven by a gas generator which presupposes the inclusion of an oxidizer in the propellant. The substitution of a stored oxidizer for air leads to a reduction in specific impulse. Nonetheless, the cycle can be competitive with the turboramjet because the rotating engine components are smaller and lighter. Since the allowable compressor discharge conditions are not governed by consideration of the maximum permissible turbine inlet temperature, it can be operated at

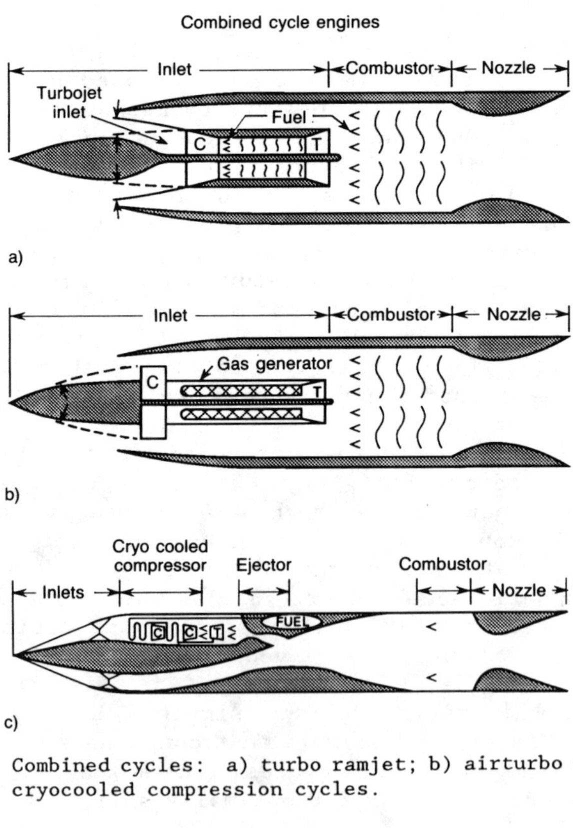

Fig. 17 Combined cycles: a) turbo ramjet; b) airturbo rocket; c) cryocooled compression cycles.

higher M_0. Flows to the compressor can be diverted, as shown, to permit the highest operating ER, presuming that sufficient fuel can be provided to the ramjet burner with the gas generator shut down.

Figure 17c is representative of a number of possible engine cycles that use the cooling capacity of cryogenic hydrogen to precool captured air prior to and/or during the compression process. Typically, precoolers are inserted among the compression stages. Several benefits can accrue: The density of the air is increased; thus, the rotating machinery is smaller. The compressor outlet temperature is lower; therefore, the turbine inlet temperature is correspondingly reduced. The work required to produce a given pressure ratio is significantly reduced, and the engine ER is not constrained by the condensation ratio as it is in the air liquefaction cycles. In this cryocooled compressor cycle, the compressed air is burned with part of the hydrogen in the driver motors that provide ejector pumping to the remainder of the captured air.

Another viable means of mechanical compression is a liquid turbopump, wherein the cooling or refrigeration capacity of cryogenic propellants is used to liquefy the air. Aerodynamic compression can also be provided by ejector pumping, but the pressure rises are so low that this path through the Brayton cycle is practical only as an

Table 2. Methods of compression in air-breathing propulsion systems

Method	Range of Compression Ratios	Remarks
Liquid turbopump	5-400	Optimal for stored cryogens (or normal liquids) Non-optimal match for liquefying air
Gas-phase multistage	4-60	Efficient, relatively heavy and large Weight/size reduction with precooling limited to $M = 3$ turbojet, $M = 6$ airturbo rocket
Ejector	1.1-2	Modest thrust augmentor
Shock-supersonic turning	2 @ $M = 2$ 20 @ $M = 3$ 50 @ $M = 7$ 500 @ $M = 24$	No subsonic capability Ineffective at subsonic speeds

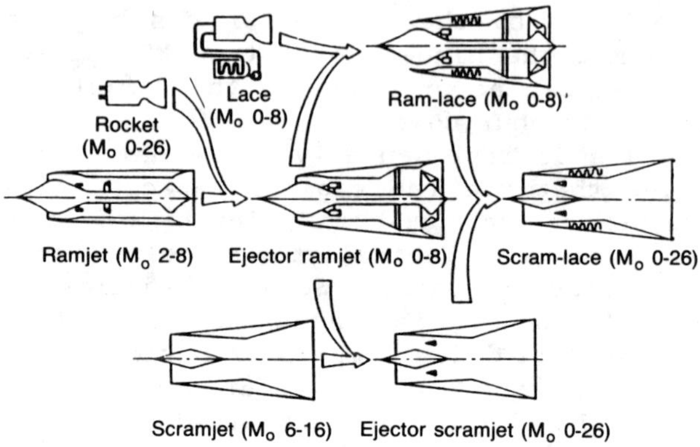

Fig. 18 Evolution of advanced propulsion systems.

augmentor of thrust. Table 2 summarizes the characteristics of these four methods of compression.

The genera for a number of rocket-ramjet combined cycle engines are shown in Fig. 18 taken from Ref. 14. The conventional subsonic combustion ramjet with a typical range of application of M_o = 2-8, and the supersonic combustion ramjet (M_o = 6-16) are shown as distinct genera. The subsequent discussion will focus on a combination of the two into a dual-mode ram/scramjet that already has been demonstrated by several organizations and should be able to operate over the entire M_o = 2-26 range. These cycles all depend on supersonic wave compression to raise the pressure prior to combustion to permit a viable propulsion cycle. Since this mechanism is not available at subsonic speeds and is of insufficient strength at low supersonic speeds, either 1) mechanical compression in the liquid phase, e.g., with a positive displacement pump, or 2) aerothermal compression, e.g., in an ejector, is used. A simple example of method 2 using a rocket to compress air is schematically illustrated as the ejector ramjet in the center of the figure. The low-speed pumping capability extends the range of engine operation to M_o = 0-8. When combined with the scramjet (or the dual-mode ram/scramjet), the rocket can provide not only the low-speed compression but also augmentation at high speed to permit high levels of thrust over the entire speed range for a transatmospheric accelerator or hypersonic cruiser. Other paths to a candidate system based on compression method 1 are also shown. This system, having the acronym LACE (liquid air cycle engine), can operate from M_o = 0 to about M_o = 8. The upper bound on Mach number operation arises from two constraints: 1) the refrigerative capability of the cryogenic propellants and 2) the inherent drag of the inlet system when compressing the air from a high velocity to a stagnation state. At $M_o = 8$, the air inlet could be closed, and the rocket could then convert to operation on stored LOX and LH_2 to operate to $M_o = 26$ (not shown). The LACE rocket could also be combined with either the ramjet or scramjet to give a $M_o = 0$-8 capability or the full required transatmospheric acceleration, respectively. In these cycles a portion of the air captured by the inlet (or in a separate inlet) passes into the heat exchanger, is subsequently compressed, and then flows to the rocket motor ejector augmentors. These simplified sketches omit the details of the bifurcated flow path.

Cryogenic Propellants

For the orbital vehicles and the high-Mach-number hypersonic cruiser, hydrogen is the candidate fuel of choice but, as the subsequent discussion will show, H_2 may have to be complemented with oxygen and a tripropellant in an optimal system. The gravimetric heating value h_v of 51,570 Btu/lbm for hydrogen is about two to three times that of all other fuels. Engine specific impulse is to the first approximation directly proportional to h_v. Consequently, hydrogen offers the greatest potential performance in most engine cycles. Stowability is one of hydrogen's greatest limitations because the density requirements for all viable propulsion systems demand that the propellants be stored in either the liquid or solid phase and, for hydrogen, this requires very low temperatures. Even when hydrogen is stored as a cryogenic liquid, its density is less than one-tenth of most other propellants, which greatly complicates vehicle design. However, the additional refrigerative capacity that is made available can be of great benefit both as an airframe and engine coolant and as a means for

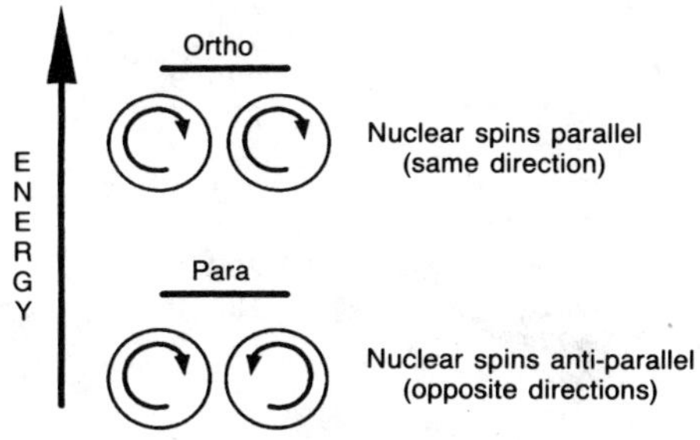

Fig. 19 Ortho-para-hydrogen.

providing liquefaction of the oxidizer in a LACE cycle. In the subsequent discussion, LACE will refer to either the genera or the derived cycles. To develop the arguments necessary to provide insight into the operational characteristics of these propulsion cycles, it is necessary to examine in detail the physical properties of hydrogen and other propellants.

At room temperature, gaseous hydrogen (in equilibrium) is a mixture composed of about 75% ortho and 25% para. This 3:1 mixture is generally referred to as "normal" hydrogen. The direction of the spin of the protons in the hydrogen molecule characterizes the two forms as shown in Fig. 19. In ortho, the spins are parallel; in para, the spins are in the opposite directions. An equilbrium mixture of liquid hydrogen at its boiling point (36.5°R, @ 1 atm) is composed of about 99.8% para. Thus, the equilibrium composition varies significantly between the boiling point and room temperature. Moreover, the heat to convert from one form to the other is strongly dependent on temperature and independent of pressure. Table 3 lists the equilibrium composition, heat of conversion from para to normal and from para to equilibrium as a function of temperature. Figure 20 is a plot of the same data. An important feature of the

Table 3 Properties of cryogenic hydrogen

T°R	Ortho, %	Heat of conversion (Btu/lbm) Para to Normal, Btu/lb	Para to Equilibrium, Btu/lb
18	0.0001	227.0	0.00
36	0.179	227.0	0.54
36.5	0.211	227.0	0.64
54	2.979	227.0	9.02
59.4	4.966	227.0	15.03
72	11.273	227.0	34.12
90	22.946	226.9	69.42
108	34.431	226.3	103.89
126	44.009	224.7	131.85
144	51.463	221.3	151.85
162	57.118	215.6	164.20
180	61.380	207.0	169.41
216	67.041	184.0	164.48
270	71.397	138.9	132.22
360	74.026	70.5	69.58
450	74.736	30.3	30.19
540	74.928	11.9	11.89

data is the region in which a significant quantity of heat can be tied up in the para-to-equilibrium transition. In a propulsion system in which the "refrigerative capacity" of the propellants, i.e., hydrogen in this case, is used to chill the air, the rate of conversion of para to equilibrium can be an important consideration in the system design and performance. This is of particular significance when a "pinch point" in the heat-exchanger system occurs in the region of the peak in the para-to-equilibrium conversion curve in Fig. 20. Typically, the rate of conversion from para to equilibrium is significantly slower than the residence time in a noncatalytic heat exchanger. If there is no pinch point in the heat-exchanger system and the final temperature of the hydrogen is in excess of about 500-600°R, then this issue does not arise. An example could be a gas-phase cryocooled compressor operating at supersonic flight speeds. In these cycles the "enthalpy" path of the hydrogen from tank conditions to room temperature could avoid the "dome" of the para-equilibrium curve in Fig. 20 without any impact.

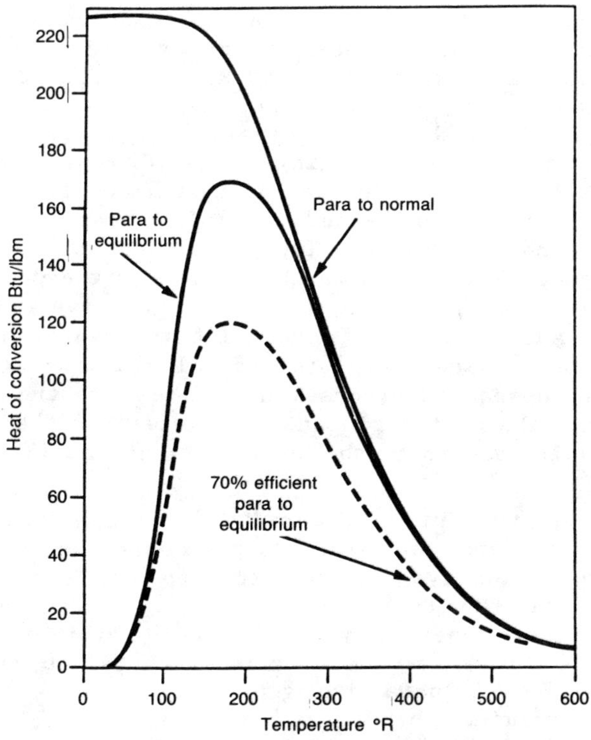

Fig. 20 Heat of conversion of para-hydrogen.

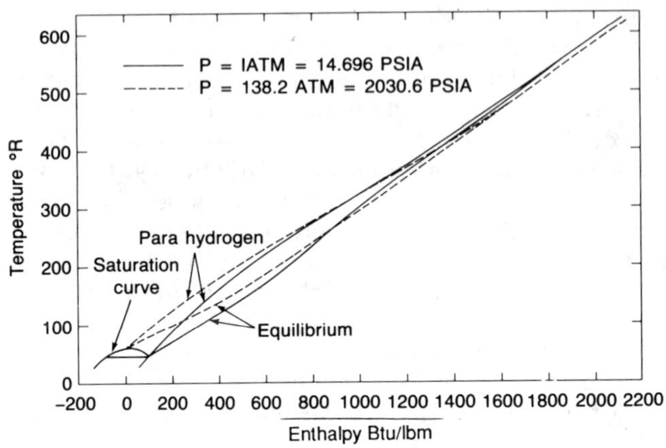

Fig. 21 Enthalpy of hydrogen.

Conversely, when the propulsion cycle depends on liquefaction of the air, the impact is large. In a liquid air cycle that does not include a gas-phase compressor, the air condensation occurs near or below ambient pressures, thus at temperatures of about 140°R. From Fig. 20 the heat of conversion from para to equilibrium at 140°R amounts to about 147 Btu/lbm, which is equivalent to 30% of the cooling capacity available in bringing para-hydrogen from storage conditions to 140°R. To avoid losing this cooling capacity, the para-hydrogen is passed through a catalytic bed to accelerate the conversion. Typical efficiencies of conversion are about 70%, which is the basis for the dashed curve in Fig. 20. Thus, some loss in refrigerative capacity is inherent. Additionally, a nonequilibrium hydrogen composition must be considered in precise calculations of the flow path. The subsequent discussion will provide the arguments to show that the most stringent requirement on the refrigerative capacity occurs in the condenser in a liquid air cycle engine.

Whereas the curve of para-to-ortho conversion serves only as a reference in regard to the design and performance of the propulsion system, the ortho-to-para conversion has importance in the logistics of vehicle operation. This arises in the preparation of the fuel and in filling and maintaining the tanks, where it is essential to assure that all of the fuel is para-hydrogen.

To determine the balance between the refrigerative capacity of the fuel and that needed to condense and compress the air, the state conditions leaving the propellant tank, at the liquid air passage discharge and at the air inlet, hydrogen outlet of the condenser, need to be defined.

Figure 21 shows the enthalpy of para- and equilibrium hydrogen, for the temperature range $24.8 \leq T \leq 630°R$ for two pressures, 1 and 138.2 atm. The saturation curve is also shown as a solid line. The para and equilibrium curves for 138.2 atm lie just above the saturation curve for $T < 60°R$ but cannot be distinguished at this scale. Properties on the saturation curve are listed in Table 4. Details of the saturation region are relatively unimportant in propulsion system analysis, however, because the hydrogen has already been pumped to high pressure at lower temperatures. Note again the large divergence of the para and equilibrium H_2 in the important region of 120-150°R. Note also the crossing of the pressure lines at higher temperatures. Thus, for cycles not constrained by an air condenser pinch point, slightly higher refrigerative capacity is available at higher pressures. Moreover, the para-equilibrium enthalpy difference at $T > 400°R$ is so small that the degree of conversion is not an issue. Thus, a catalyst in these systems would be superfluous.

To avoid the specifics of heat-exchanger design at this point in the discussion, the temperature of the inlet air is assumed to be 10°R above the temperature of the hydrogen at the condenser face. In an actual design the minimum difference in temperatures is a compromise between weight and refrigerative capacity. If the hydrogen and air components, including the condenser, pumps and turboexpander, could be considered as a closed system, satisfaction of the energy equation would require both a heat and power balance. In concept, the work required to compress both the liquid air and the hydrogen could be provided by a hydrogen turboexpander, with the power extracted from the hydrogen somewhere prior to its exit from the condenser. The subsequent analysis will show that this may not be advantageous in practical system designs. This is generally the case when the cycle demands high-pressure liquid air. However, for systems with modest requirements for pressure in the hydrogen discharge, i.e., between 500 and 700 psia, the work provided by the turboexpander can be matched to the work required to drive only the hydrogen pump. This is an especially simple case to analyze the pinch point issue.

From the energy equation for a thermal and power balanced system

$$\Delta Q = \Delta H_{AIR} - W_{AIR} = \Delta H_{H2} + W_{AIR} + W_{H2} - W_{H2} \tag{7}$$

Table 4 Thermodynamic properties of saturated liquid and vapor of para-hydrogen

T, °R	Pressure psia	Enthalpy Liquid Btu/lbm	Enthalpy Vapor Btu/lbm	Latent Heat of Vaporization Btu/lbm	Entropy Liquid Btu/lbm	Entropy Vapor Btu/lbm	Density Liquid lbm/ft³	Density Vapor lbm/ft³
24.84	1.02	-132.81	60.32	193.12	1.186	8.961	4.8084	0.00783
25.2	1.15	-132.20	61.09	193.29	1.208	8.870	4.7979	0.00867
27.0	1.95	-129.28	64.96	194.24	1.320	8.515	4.7443	0.01391
28.8	3.13	-126.14	68.62	194.76	1.431	8.193	4.6885	0.02113
30.6	4.77	-122.79	72.06	194.85	1.541	7.909	4.6303	0.03072
32.4	6.99	-119.21	75.24	194.45	1.652	7.653	4.5694	0.04309
34.2	9.89	-115.39	78.12	193.50	1.763	7.422	4.5054	0.05865
36.0	13.56	-111.31	80.65	191.96	1.875	7.210	4.4378	0.07787
36.48[a]	14.69	-110.15	81.39	191.54	1.905	7.158	4.4191	0.08352
37.8	18.13	-106.97	82.98	189.95	1.988	7.015	4.3662	0.10106
39.6	23.70	-102.28	84.78	187.06	2.102	6.829	4.2901	0.12929
41.4	30.40	-97.29	86.20	183.49	2.218	6.653	4.2086	0.16306
43.2	38.36	-91.92	87.10	179.02	2.337	6.484	4.1209	0.20319
45.0	47.69	-86.16	87.53	173.69	2.458	6.320	4.0261	0.25078
46.8	58.52	-79.92	87.32	167.24	2.583	6.158	3.9226	0.30724
48.6	70.95	-73.13	86.33	159.46	2.712	5.996	3.8085	0.37456
50.4	85.15	-65.69	84.48	150.17	2.848	5.830	3.6812	0.45559
52.2	100.48	-57.44	81.47	138.91	2.994	5.656	3.5362	0.55477
54.0	119.29	-48.02	76.87	124.89	3.153	5.467	3.3668	0.67967
55.8	139.63	-36.89	69.91	106.80	3.334	5.249	3.1582	0.84534
57.6	162.41	-22.52	58.51	81.03	3.563	4.971	2.8712	1.09238
58.193[a]	187.51	16.46		0	4.195		1.9620	

[a]Two Phase Boundary

For this thermally balanced but power deficient portion of the cycle,

$$\Delta Q = \Delta H_{AIR} - W_{AIR} = \Delta H_{H2} \tag{8}$$

which can be further simplified if the total enthalpy difference for the air is taken as H_1-H_3, where H_1 and H_3 are the enthalpies of the air entering the condenser and the pump, respectively, and H_8 and H_{14} are the total enthalpies of the hydrogen entering the condenser and in the storage tank, respectively. Thus,

$$\Delta Q = H_1 - H_3 = H_8 - H_{14} = W_{AIR}(h_1 - h_3) = W_{H2}\left(h_8 - h_{14}\right) \tag{9}$$

and the condensation ratio CR

$$CR = \frac{\text{Airflow}}{\text{Hydrogen flow}} = \frac{W_{AIR}}{W_{H2}} = \frac{h_8 - h_{14}}{h_1 - h_3} \tag{10}$$

Two items should be noted for cycles of this design, namely:

1) The CR is independent of the pressure level of the discharge air.

2) The CR is somewhat lower than it would be if some or all of the work to compress the air is extracted from the hydrogen prior to the condenser exit.

In the condenser, the air is brought from the saturated vapor state to the saturated liquid state. It is then subcooled to assure that it remains in the liquid phase while being pumped to a suitable pressure for injection. Table 5 lists the properties of saturated liquid and vapor for air. For saturated vapor entering a condenser at 14.7 psia and 147.2°R and under the assumption of no pressure drop, the outlet temperature would be 141.8°R. (Note that isobaric condensation is not isothermal in multicomponent fluids.) With pressure drop, the outlet temperature would be lower. For this sample calculation, the outlet pressure P_2 is assumed to be 7.35 psia, and the outlet temperature is 131.7°R (see Table 2). Assuming 2°R of subcooling, conditions entering the air pump would be T_3 = 129.7°R and h_3 = -59.97 Btu/lbm. Assuming that the hydrogen is supplied from storage at the triple point, T_{14} = 24.84°R, and h_{14} = -132.81 Btu/lbm (see Fig. 22 and Table 6). Considerable effort and explanation are needed to justify the remaining term h_8 since it is a function of pressure and composition as well as temperature.

Table 5 Thermodynamic properties of saturated liquid and vapor of air

Atm	Psia	Temperature °R (Bubble)	Temperature °R (Dew)	Entropy Btu/lbm °R Liquid	Entropy Btu/lbm °R Vapor	Enthalpy Btu/lbm Liquid	Enthalpy Btu/lbm Vapor	Latent Heat of Vaporization Btu/lbm	Density lbm/ft^3 Liquid	Density lbm/ft^3 Vapor
0.5	7.35	131.7	137.4	0.676	1.358	-59.07	32.17	91.24	56.458	0.148
1	14.7	141.9	147.3	0.711	1.324	-54.27	33.86	88.13	54.889	0.281
2	29.4	153.8	158.2	0.750	1.291	-48.57	35.78	84.35	52.900	0.535
3	44.1	161.7	166.0	0.773	1.272	-44.80	36.84	81.64	51.527	0.783
5	73.5	173.0	177.0	0.806	1.247	-39.22	37.99	77.21	49.418	1.274
7	102.9	181.7	185.8	0.830	1.231	-34.80	38.74	73.54	47.670	1.759
10	147.0	191.6	195.3	0.857	1.212	-29.61	38.94	68.55	45.546	.520
15	220.4	204.0	207.2	0.891	1.188	-22.59	38.33	60.92	42.520	3.883
20	293.9	213.8	216.5	0.918	1.167	-16.57	36.88	53.45	39.755	5.426
25	367.4	222.0	224.2	0.943	1.147	-10.90	34.64	45.54	36.968	7.255
30	440.9	229.2	230.8	0.967	1.125	-5.02	31.33	36.35	33.799	9.603
35.25	518.2	236.8		1.094		25.37		0	13.510	

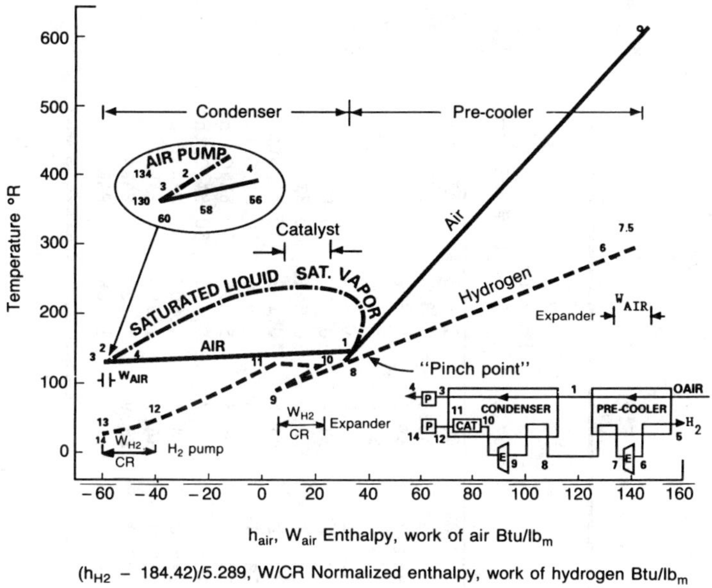

Fig. 22 Energy-temperature dagram for LACE cycle.

The composition can be obtained if the temperature and efficiency of the catalyst bed are defined. Ideally, the catalyst is operated at the highest temperature to obtain the maximum conversion. Whether it is located ahead of, or subsequent to, the turboexpander has an insignificant effect on h_8 if the operating temperature is the same. For this example, the catalyst is placed ahead of the turboexpander, operates isothermally at T = 126°R, and has an enthalpy conversion efficiency of 70%. With reference to Table 3 and Fig. 20, an equilibrium conversion would yield an increase in cooling capacity of 131.85 Btu/lbm and a composition of 44.009% ortho and 55.991% para. With an efficiency of 70%, the loss of cooling would be 0.3 x 131.85 = 39.56 Btu/lbm, which would correspond to a composition of 30.8% ortho. This composition would then remain constant during the remainder of the cycle and therefore would be in nonequilibrium at all temperatures other than 106.7°R (Table 3).

For this example, the pressure at the hydrogen pump discharge is set at P_{12} = 2050.4 psia. Assuming a pump efficiency η_p = 0.8, the outlet enthalpy and temperature would be h_{12} = 34.10 Btu/lbm and T_{12} = 43.9°R. Conditions entering the catalyst, assuming 15 psia drop in the first stage of the condenser, would be P_{11} = 2035.4 psia, h_{11} =

Table 6 Conditions at key points in LACE cycle (see Fig. 22)

Code	Location	Pressure Psia	Pressure P	T, °R	h Btu/lbm	S Btu/lbm °R	Composition
0	AIR - PRECOOLER ENTRANCE	22.0	0.152	610.0	146.0	1.643	AIR
1	AIR - CONDENSER ENTRANCE	14.7	0.101	147.3	33.9	1.324	AIR
2	AIR - SATURATED LIQUID	7.35	0.051	131.7	-59.1	0.676	AIR
3	AIR - PUMP ENTRANCE	7.35	0.051	129.7	-60.0	0.669	AIR
4	AIR - PUMP DISCHARGE	1000.0	6.895	133.5	-56.0	0.675	AIR
5	H_2 - PRECOOLER DISCHARGE	580.0	3.999	309.2	993.9	10.417	30.8%/Ortho
6	H_2 - EXPANDER II DISCHARGE	581.0	4.006	303.3	972.0	10.341	30.8%/Ortho
7	H_2 - EXPANDER II ENTRANCE	632.2	4.359	309.2	993.8	10.152	30.8%/Ortho
8	H_2 - CONDENSER DISCHARGE	652.2	4.497	137.3	390.5	7.496	30.8%/Ortho
9	H_2 - EXPANDER I DISCHARGE	657.2	4.531	91.0	213.5	5.895	30.8%/Ortho
10	H_2 - EXPANDER I ENTRANCE	2034.4	14.000	126.0	308.7	5.776	30.8%/Ortho
11	H_2 - CATALYST ENTRANCE	2035.4	14.034	126.0	218.6	4.909	PARA
12	H_2 - PUMP EXIT	2050.4	14.138	43.9	-34.1	1.790	PARA
13	H_2 - PUMP ENTRANCE	1.95	0.013	27.0	-128.9	1.320	PARA
14	H_2 - STORAGE	1.02	0.007	24.8	-132.8	1.186	PARA

218.6 Btu/lbm, and T_{11} = 126°R. Assuming 5 psia pressure drop in the catalyst, conditions at its exit would be P_{10} = 2030.4 psia, h_{10} = 308.7 Btu/lbm, and T_{10} = 126°R. The work required to pump the hydrogen is $W_{H2} = h_{12} - h_{11}$ = [-34.10 - (-129.28)] = 95.18 BTU/lbm. To provide this amount of work in a turboexpander at an efficiency η_T = 0.9, the expander outlet conditions would be P_9 = 657.2 psia, h_9 = 213.5 Btu/lbm, and T_9 = 91.0°R. The hydrogen would then be heated in the second stage of the condenser to T_8 = 137.3°R, consistent with the minimum heat-exchanger ΔT = 10°R assumption. Taking a small pressure drop of 5 psia in this part of the condenser yields outlet pressure and enthalpy of P_8 = 652.2 psia and h_8 = 390.5 Btu/lbm.

The resulting condensation ratio is CR = $(h_8 - h_{14})/(h_1 - h_3)$ = [390.5 - (-132.8)]/[33.86 - (-59.97)] = 5.577. The corresponding equivalence ratio is ER = $1/f_s$CR = 34.28/5.577 = 6.147, where f_s is the stoichiometric hydrogen/air weight ratio = 0.029171.

To complete the cycle diagram shown in Fig. 22, it is necessary to define the condition at the outlet of the liquid air pump, the initial condition of the air entering the system, and the conditions at the entrance of the second hydrogen turboexpander. The liquid air is assumed to be pumped to 1000 psia at an isentropic efficiency, η_P = 0.8 giving an outlet temperature and enthalpy of T_4 = 133.5°R and h_4 = -59.97 Btu/lbm, respectively. Assuming flight at M = 0.8, near sea level, the conditions of the air entering the first heat exchanger, generally called a precooler, would be T_0 = 610°R, P_0 = 22 psia, and h_0 = 146.0 Btu/lbm. In principle, the second turboexpander could be located either at an intermediate point or at the outlet of the precooler. Here it is located at a point where the entrance temperature is equal to the discharge temperature of the precooler, i.e., $T_7 = T_5$ = 309.2°R. Condenser pressure losses of 20 and 1 psia were assumed upstream and downstream of the expander. The work of the expander is simply the W_{AIR} x CR = 3.938 x 5.577 = 21.962 Btu/lbm of hydrogen. With an expander efficiency η_T = 0.9, the expander discharge conditions are P_6 = 581.0 psia, T_6 = 303.3°R and h_6 = 972.0 Btu/lbm. The hydrogen outlet temperature from the precooler is obtained by a heat balance with the air, by an iterative method, using the expander pressure drop and enthalpy reduction. Flow properties at all of the major stations in the cycle are listed in Table 6.

The pinch point depicted in Fig. 22 corresponds to the location where the temperature difference between the two fluids is a minimum. To depict this in a simple chart, it is necessary to reference the hydrogen "enthalpy" such that the two fluids are thermally balanced, i.e., there is no net heat flux from or to the condenser or precooler. This is accomplished by subtracting 184.42 Btu/lbm from the hydrogen enthalpy and dividing by CR. Similarly, the work in turbo-expander II is depicted as W_{AIR} x CR. It is apparent that considerable cooling capacity of the hydrogen is not used in this cycle.

It is important to emphasize that the preceding example was intentionally constrained to avoid addressing several practical issues that lead to lower values of CR in actual systems. Figure 23 is the first of several figures that will be interdigitated into the discussion to introduce, by example, some of these issues that need to be considered in more refined studies.

Figure 23 is a schematic illustration of a so-called baseline system of an air liquefaction cycle. Several differences between the "ideal" example cited previously and this real system need to be noted with respect to the condenser pinch point. They are:

1) The catalyst is operating at a lower temperature; hence, the percentage of para-hydrogen at the outlet of the condenser is greater.

2) The hydrogen outlet temperature of one of the condensers is 94°R and 118°R on the other as compared to 137.2° in the example; hence, the cooling capacity is lower.

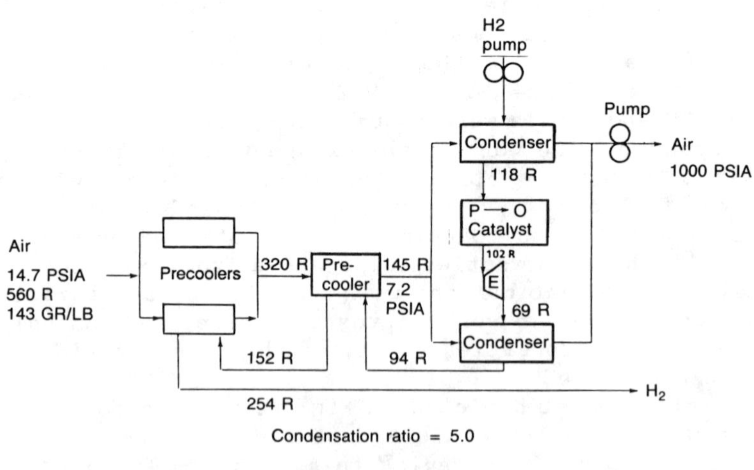

Fig. 23 Baseline system.

3) The air entering the condensers is at lower pressure (7.2 psia vs 14.7 psia); thus, the work required to compress would be slightly larger.

Whereas these changes lead to a lower CR in the actual system, the weights of the condensers would also be lower. The compromise between CR and weight obviously entered into the particular design shown. Note that, in the actual system, the precooling is done in two stages. In stage 1 the 143 grain/lb of water in the incoming air are removed to avoid icing in the second stage and in the condensers. Parallel units are used alternatively, thereby permitting an intermittent ice-removal capability. Note also that a single turboexpander is shown, which indicates that a power balance is achieved prior to the pinch point in the hydrogen flow path. For the temperature drop shown, the work for pumping would be about 106 Btu/lbm of hydrogen which closely approximates the total work in the example case, i.e., $(h_{10} - h_9) + (h_7 - h_6) = (308.7 - 213.5) + (993.8 - 972.0) = 117.0$ Btu/lbm.

Pressure levels in the hydrogen system were not shown in Fig. 23. However, as shown in the following calculations, the hydrogen system pressures downstream of the first hydrogen expander are limited to a very narrow range. Figure 24 shows expander discharge pressures P_9 for a range

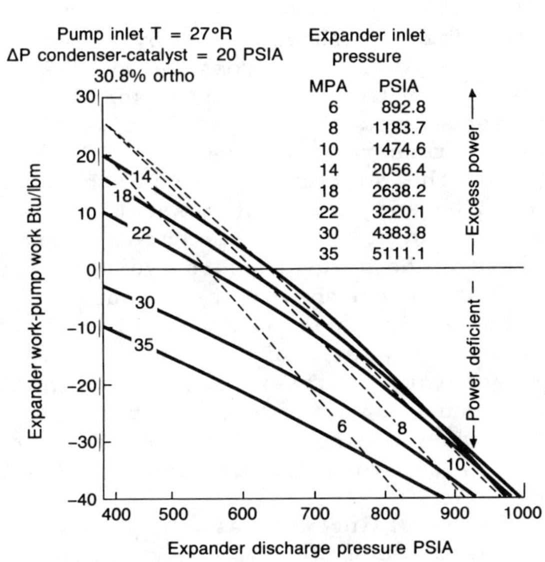

Fig. 24 Power balance between hydrogen expander and pump; T expander inlet = 126°R.

of expander inlet pressures $892.8 \leq P_{10}$ 3220.1 psia that yield a power balance between the turboexpander and the pump. The values of P_9 lie between 540 and 620 psia. The intercepts for the 30 and 35 P (4383.8 and 5111.1 psia) curves are at P_9 values below 400 psia. Another consideration in the system design is the number of turbine stages needed to produce the work that is directly related to the expander pressure ratio. Figure 25 shows the expander pressure ratios corresponding to the cases examined in Fig. 24.

Since the expander discharge pressure is nearly invariant, the enthalpy at point 8 is also nearly invariant for a specific T_1-T_8. Consequently, the condensation ratio is nearly independent of the hydrogen pump pressure. However, in many systems it would be desirable to have higher hydrogen pressures. Two methods for obtaining higher hydrogen pressures need to be considered. The first is to split the hydrogen into two flow paths wherein a portion of the total flow bypasses the turboexpanders. The second method uses an alternative source, e.g., a gas generator to provide the energy to drive an auxiliary turbine. Consideration of the first method influenced the choice of pressure levels selected in the example. In the example, the ER was equal to 6.147. If enough hydrogen to burn at stoichiometric conditions bypassed the turboexpanders, then the work/lbm of the remainder would have to increase by the ratio of 6.147/5.147 = 1.194. This would increase the work/lbm in the first turboexpander by 18.47 Btu/lbm and would decrease the discharge pressure by 212.2 psia from 632.2 to 420 psia. In the second turboexpander the pressure drop would increase from 51.2 to 60.8 psia. Thus, the sum of the increased pressure drops would be 273 psia, and the exit pressure in the smaller and larger hydrogen streams would be about 2003 and 307 psia, respectively.

This would be an acceptable method of operating a ram-LACE cycle, wherein the high-pressure hydrogen would be used in the ejector motors and the lower-pressure hydrogen would be used in the afterburner (ramburner). For the LACE rocket, it would be far more advantageous to run an auxiliary gas-generator turboexpander. A fraction of the compressed air would enter a second-stage pump and then would be burned with a portion of the compressed hydrogen. Another method is to store the oxidizer needed to operate the gas generator.

Whereas the preceding methods can provide relief from the problems due to a power imbalance, the performance losses due to the very high ER needed to provide a thermal balance are excessive. Consequently, one or more of the

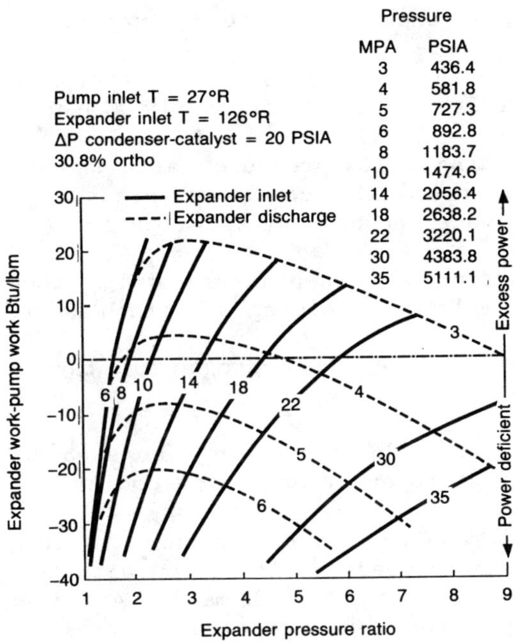

Fig. 25 Expander pressure ratio.

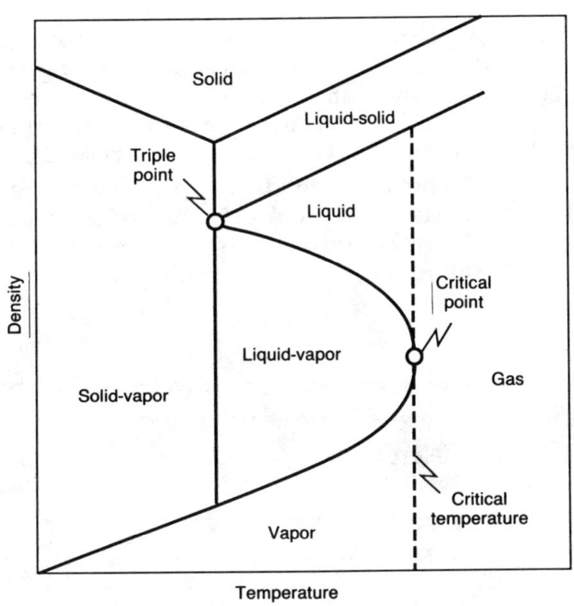

Fig. 26 Density-temperature single-component substance.

following systems changes need to be examined:

1) Storing the hydrogen as a two-phase slush and recirculating during early stages of flight.

2) Adding a low-pressure ratio fan or compressor to raise the condenser pressure and thereby reduce the heat of vaporization (see Table 5).

3) Adding a high-pressure ratio compressor and completely avoiding the pinch point.

4) Using the excess hydrogen in a tandem propulsion system.

5) Passing the air through an expander (turbine) prior to entering the precooler.

Methods 1 and 3 require further explanation; method 2 can be assessed from the preceding discussion with the aid of the forthcoming material on method 3; method 4 can be handled by superposition of the genera cycles, and method 5 is the subject of Ref. (15).

Two-Phase Storage

Two potential benefits can accrue by storing the fuel as a liquid-solid mixture: 1) increased propellant density (see Figs. 26 and 27 and Table 7, and 2) increased condensation ratio by recirculation of a portion of the heat-exchanger flow back into the fuel tank. Systems in which only a small portion of the total hydrogen stored is used to fuel the liquid air cycle are ideally suited to exploit the latter. The previous example can be used to examine the magnitude of the benefit.

Table 7 Density of slush hydrogen at triple-point temperature

Solid by weight, %	Density lbm/ft^3	Solid by weight, %	Density lbm/ft^3
0	4.81	60	5.16
10	4.87	70	5.22
20	4.93	80	5.28
30	4.99	90	5.34
40	5.05	100	5.40

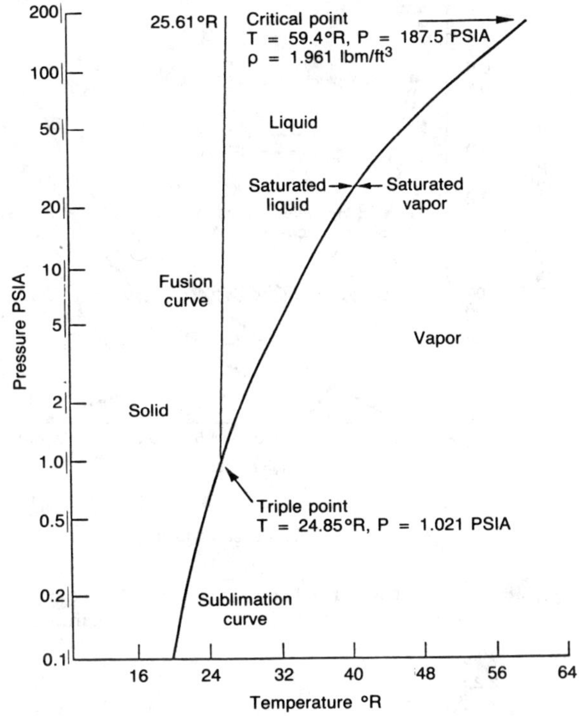

Fig. 27 Equilibrium phase diagram for para-hydrogen.

With the same assumptions at the pinch point, namely, Δh_{H2} = 523.3 Btu/lbm, Δh_{AIR} = 93.83 Btu/lbm, and the work to pump the air provided in the second turboexpander, the condensation ratio can be obtained from

$$CR = \dot{W}_{AIR}/\dot{W}_{H2} = \left\{ \Delta h_{H2} + \left[\left(\frac{1 - W_L/W_T}{W_L/W_T} \right) + 1 \right] YQ_{FUS} \right\} + \Delta h_{AIR}$$

$$= \left[523.3 + \frac{W_T}{W_L} \; Y \; 24.7 \right] / \; 98.83 \qquad (11)$$

where Y is the initial mass fraction of slush in the hydrogen tank and W_L/W_T is the ratio of hydrogen used during LACE operation to the total hydrogen stored. Figure 28 shows the effects of recirculation on the pinch point values of ER and CR for three values of the solid fraction in the

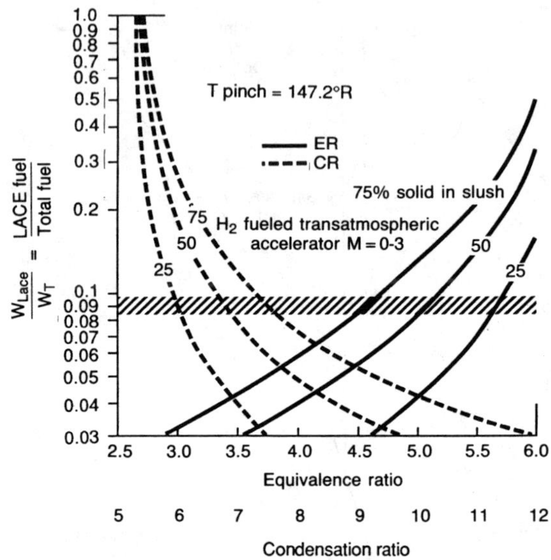

Fig. 28 Effect of recirculation on limiting fuel air ratio in LACE heat exchanger.

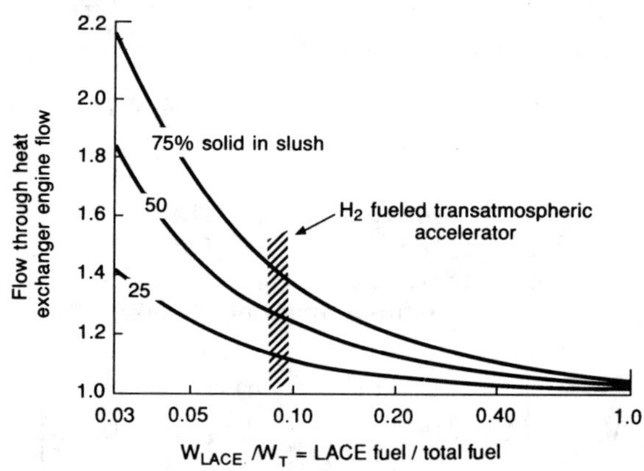

Fig. 29 Increased fuel flow in heat exchangers with recirculation.

slush mixture, 0.75, 0.50, and 0.25. Provided that the "duty cycle" of the LACE is short, useful reductions in ER can be obtained. Duty cycle, in this case, is measured by the ratio of the hydrogen used during LACE operation to the total hydrogen stored. Unless this ratio is small (i.e., < 0.1), the benefits would probably not be worth the added system complexity and weight penalty. A possible candidate application would be a combined cycle engine that operates with air liquefaction at low speeds prior to becoming a hydrogen- fueled ramjet/scramjet at high speeds. An example is the single-stage, transatmospheric accelerator, with LACE operation from takeoff to Mach 3 and ramjet-scramjet propulsion to orbital speeds. In this system, 8.5-9.5% of the fuel would be used during LACE operation. Thus, the CR could be increased from about 5.58 to 6.3-7.9, with a corresponding decrease in ER in an ideal system. Figure 29 shows that the fuel flow through the heat exchangers would correspondingly increase by 12.5-42%.

The trend toward improved performance that accrues with decreasing overall ER can be more explicitly stated with the aid of actual system designs. Figure 30 depicts three air liquefaction propulsion cycles. In the basic LACE cycle, the outflows of the condenser are fed to a liquid air-H_2 rocket motor. The ER_o = 8 corresponds to CR = 4.29, which is somewhat lower than that of the system shown in Fig. 23. Presumably, the difference can be attributed to a less efficient but perhaps lighter-weight heat-exchanger system. The corresponding specific impulse, I_{SP} = 1000 lbf/lbm for sea level static conditions, is about one-half that which could be realized if the pinch point in the heat exchanger could be avoided.

If, instead, the output of the condenser is used in a combined cycle RAMLACE, the overall equivalence ratio ER_o can be significantly reduced and the I_{SP} increased. In the cycle depicted, part of the hydrogen from the condenser is combined with the compressed liquefied air in small driver motors that serve as the primaries in an ejector system, and

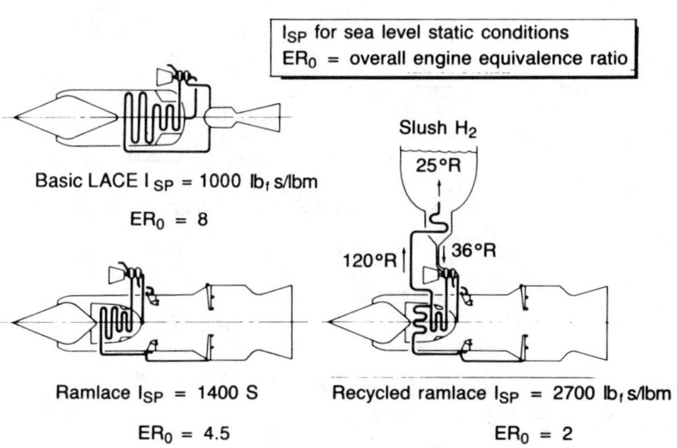

Fig. 30 Air liquefaction cycles (Ref. 14).

the remainder is used to fuel an afterburner. At low flight speed, the driver motors provide all of the pumping. At high flight speeds, the drivers augment the ram compression. The reduction in ER_o to 4.5 due to the increased airflow results in an increase in I_{SP} to 1400 lbf/lbm. The increase in I_{SP} is not directly proportional to the decrease in ER_o, primarily because the exhaust nozzles of the two systems operate at much different pressure levels.

In the recycled RAMLACE, the recirculation feature with slush H_2 storage has been added. The reduction in ER_o that is greater than a factor of 2 from the RAMLACE cycles is somewhat higher than would be anticipated from the results shown in Fig. 30. It suggests a high slush fraction, a low duty cycle, or perhaps a different assumption for the H_2 storage temperature in the two cycles. The I_{SP} at sea level static in the RECYCLED RAMLACE increases to 2700 lbf/lbm.

CryoCooled Compressor Cycles

Figures 31 and 32 show cycles wherein gas-phase compressors are used to provide part or all of the pressure rise of the air. In the "supercritical compression cycle" (Fig. 32), the air is initially cooled by recirculated high-pressure air and then by the hydrogen. The air enters the compressor at the saturated gas condition and exits at supercritical conditions of 598°R, 576 psia, where it is subcooled to 200K and then pumped to high pressure. It then recirculates through the precooler exiting at 350°R and 1000 psia. The hydrogen is routed through the precooler, the para-ortho catalyst, and a subcooler, raising the temperature to 510°R. Waste heat, which implies a pass through a heat exchanger to extract heat from the ejector ramjet

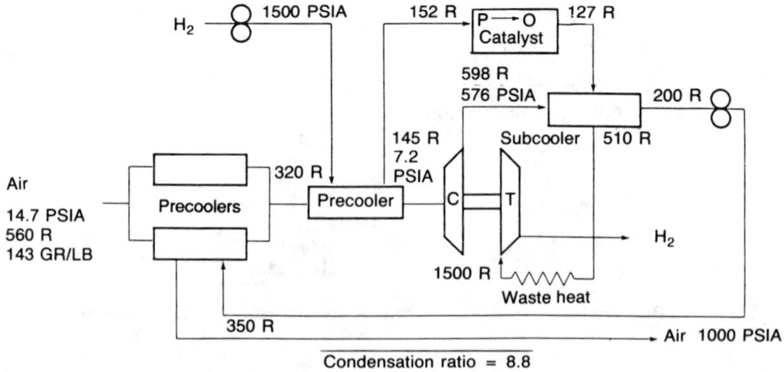

Fig. 31 Supercritical compression.

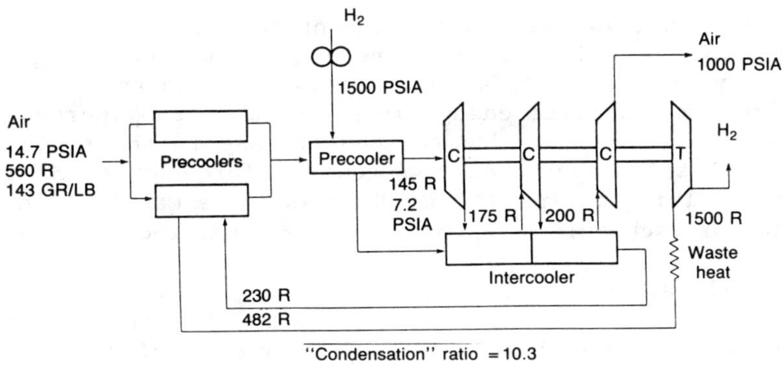

Fig. 32 All air pumping by compression.

burner, is used to raise the temperature to 1500°R prior to its entry into the turbine. By this means the condensation ratio is increased to 8.8 and ER_{ij} reduced to 3.894.

In Fig. 32, all of the air pumping is in the gas phase through a three-stage compression process. Hydrogen passes through the air precooler and then into an intercooler, where heat is extracted from the air between the compressor stages. Waste heat is again needed to raise the hydrogen temperature prior to its entering the turbine. This method raises the C_R to 10.2 and drops the ER_{ij} to 3.328. The equivalence ratio of the overall cycle, presuming that the hydrogen and air are then delivered to the injectors of an ejector ramjet, depends on the bypass ratio β (airflow into the ramjet duct/airflow plus hydrogen flow from these cycles).

Ram/Scramjet Cycle

Figure 33 shows schematic sketches of the operating modes of the ram/scramjet engine cycle at low, intermediate, and high flight speeds. The charts on the right show the desired engine geometries that correspond to the various operating modes based on cycle analysis results. Air from the freestream is compressed on the vehicle forebody through a series of oblique compression waves and is then turned and further compressed by waves emanating from the cowl. At low speeds the amount of area contraction in the compression process is not large, which means that the engine duct must be as large as practical to permit the capture of an adequate air supply. As the flight speed increases, the amount of contraction increases, and the through-duct wants to be smaller. The top curve on the right shows a typical variation in this desired inlet contraction ratio A_0/A_4 as a function of flight Mach number M_0. Whereas the graph shows a variation in A_0/A_4 of about 14 over the range of M_0, the desired geometric change in A_4 is smaller by perhaps a factor of 3. This is due to the decrease in the angle of the compression waves on the vehicle forebody as M_0 increases. This folding of the shock waves is manifest in an air capture schedule A_0/A_i for the inlet that increases with M_0.

Full capture of the inlet, i.e., $A_0/A_i = 1$, means that all of the air contained in a streamtube with a cross-sectional area equal to the projected area of the inlet enters the engine. This occurs when the forebody compression waves fall inside the cowl lip. Within limits, the designer can choose M_D, the value of M_0 at which $A_0/A_i = 1$, by placement of the engine cowl lip. A more favorable air capture characteristic can be obtained by selecting a lower M_D, but high inlet efficiency at $M_0 > M_D$ may be difficult to obtain. Movement of the cowl is one of the attractive candidate choices to provide the desired A_0/A_4 and perhaps to obtain optimal wave cancellation in the internal portion of the inlet but, when meeting these objectives, the movement has only a relatively small impact on A_0/A_i.

Whereas the flow exiting these external/internal compression inlets can be supersonic at M_0 as low as 3, it

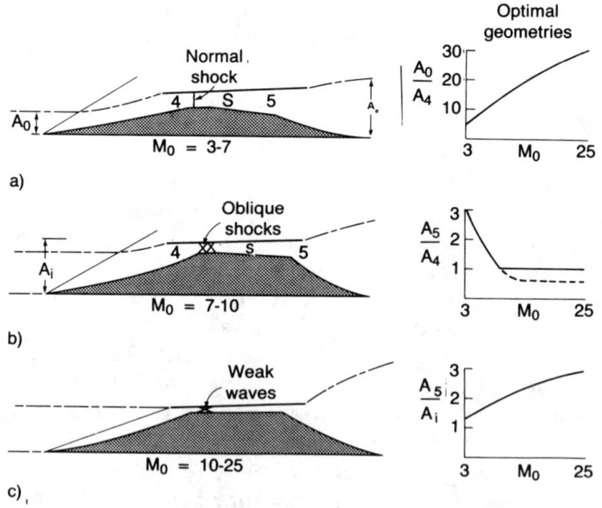

Fig. 33 Ramjet/scramjet engine cycle: a) $M_S < 1$; $M_5 = 1$; $A_5/A_4 >> 1$; $P_e < P_0$; b) $M_S > 1$; $M_5 = 1$; $A_5 > 1$; $P_e \approx P_0$; c) $M_S >> 1$, $M_5 > 1$; $A_5 = A_4$; $P_e > P_0$.

is advantageous for the combustion process to initiate in subsonic flow at M_0 = 3 to about 6. Thus a normal shock structure, i.e., a shock train, is located at the downstream end of the inlet. A length of duct is required to anchor the shock train and thereby prevent combustor/inlet interaction. This shock-stabilizing section is called an isolator. Two types of internal duct area variations downstream of the isolator can be used in the subsequent heat-addition process. In the conventional subsonic ramjet, the duct area is expanded prior to combustion, and a converging-diverging nozzle is added following the combustor. In the dual-mode engines shown in the Fig. 33, the combustion process begins just downstream of the isolator and continues in a diverging area duct. The heat release and duct area variation are tailored such that the flow accelerates from a subsonic to a sonic or supersonic condition at the combustor exit (see Ref. 4). Thus, there is no need for, nor possibility of, adding a converging-diverging nozzle. The right section of Fig. 33b shows that the desired combustor area ratio decreases rapidly from $A_5/A_4 > 3$ at $M_0 = 3$ to A_5/A_4, which can be as small as 1 at $M_0 = 7$. Consideration of kinetic losses in the combustor and exhaust nozzle and the maximum permissible pressure rise in the precombustion shock train determine the optimum value of A_5/A_4 at $M_0 > 6$. Figure 33b shows how the shock structure in the isolator accommodates the conditions at M_0 above the point where $A_5/A_4 = 1$ is first reached. The strength of the shock train decreases, and the mean flow conditions entering, and throughout, the combustor are supersonic. At very high speeds the wave structure in the isolator is quite weak, and the pressure rises are correspondingly small.

The desired nozzle area ratios are shown in Fig. 33c. At M_0 below about 2.5, the exit area would have to be smaller than the inlet to avoid overexpansion of the flow from the engine. Conversely, at high M_0 a very large area ratio A_6/A_i would be needed to obtain all of the thrust capability of the engine. Therein lies a principal challenge to the vehicle designer because there is no practical means to provide the desired variation in A_6/A_i. Several approaches are feasible to obtain the variation in A_0/A_4 (in reality A_4/A_i) including movable cowls and adjustable body-side compression ramps. The combustor area ratio variation could be provided either by mechanical means or by adjusting the locations of the fuel injectors.

Ducted Rocket Cycle

The ram/scramjet could be the propulsion system for the second stage of a TSTO vehicle or would be the principal propulsion path of a combined-cycle engine for either an SSTO vehicle or a hypersonic cruiser. In the simplest combined- cycle concept, the propellant is composed of both the fuel and an oxidizer. Variants of this engine cycle have been labeled "ejector scramjet" (Fig. 18), ducted rocket, air- augmented rocket, etc. When sufficient air cannot be ingested into the engine and/or when the compression ratio is too low to produce useful thrust in a Brayton cycle, the oxidizer is added. Moreover, the subsequent discussion will show that oxidizer addition can be beneficial when high acceleration is needed in spite of a decrease in engine specific impulse. Figure 34 schematically illustrates this point. In this candidate propulsion scheme for an SSTO application, the fuel (H_2) to oxidizer (O_2) equivalence ratio of the injector (rocket) is varied to yield maximum net force specific impulse $I_{eff} = I_{SP}$ (1-T/D). For $5 < M_0 < 13$, only hydrogen is injected, and so the reciprocal of the injector equivalence ratio $1/ER_{ij} = 0$. For $0 \leq M_0 < 5$ and $M_0 > 13$, oxygen is added. At low Mach numbers the rocket motor serves as a low-pressure ratio pump or ejector to raise the efficiency of the air-breathing Brayton cycle. Moreover, the injector rocket produces most of the total engine thrust. At high Mach number the thrust contribution of the rocket injector again becomes important, which explains the need for oxidizer addition. The desirability of operating the combined cycle at greater than stoichiometric, i.e., overall $ER_0 > 1$, is due to several factors. At low Mach numbers it is simply a matter of higher I_{eff}; at high Mach numbers the cooling requirement of the vehicle demands higher H_2 flow rates, and nonequilibrium

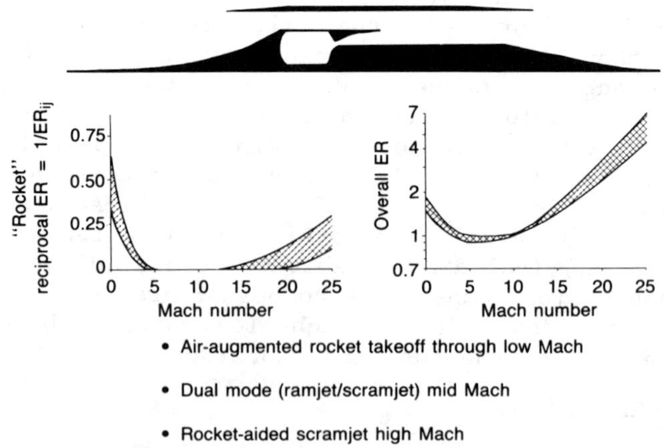

Fig. 34 Variable-equivalence ratio ducted rocket/scramjet cryofueled (H_2-O_2).

losses in the combustor and nozzle are mitigated by the addition of excess hydrogen.

Reference Trajectory for SSTO, TSTO and Hypersonic Cruiser

A prerequisite for the assessment of the various engine cycles for the orbital and hypersonic cruiser application is the adoption of a formal force accounting system, the definition of a reference trajectory, and the methodology for cycle analysis. Figure 35 shows the reference trajectory selected for this study, albeit for an orbital mission or one in which an initial segment would be the acceleration phase of the hypersonic cruiser. The orbital trajectory is composed of a powered phase that terminates at an altitude Z_2 = 175,000 ft and velocity U_2 = 26,600 ft/s and of an unpowered climb on an elliptic transfer orbit to an altitude $Z_S = 6.08 \times 10^5$ ft (100 nm) and U_S = 25,578 ft/s. The value of U_S corresponds to a circular orbit at Z_S. The conditions at Z_2 are determined by adding a velocity increment of 497 ft/s to the velocity that would be required at Z_2 to be at the perigee of the drag-free elliptic transfer orbit. According to Keppler's laws, this velocity $U = U_S (Z_S + r_0) | (Z + r_0)$ = 26,103 ft/s. For an easterly orbit with a near-equatorial launch, the terminal velocity requirement would be lessened by the rotational speed of the Earth, i.e., by 1522 ft/s, U_2 would be 25,078 ft/s and Z_2 = 161,240 ft.

For comparison, the ascent trajectory for the Shuttle is shown (i.e., mission 3A) in Fig. 35. Note that a very rapid climb is used to minimize drag and aerodynamic loads.

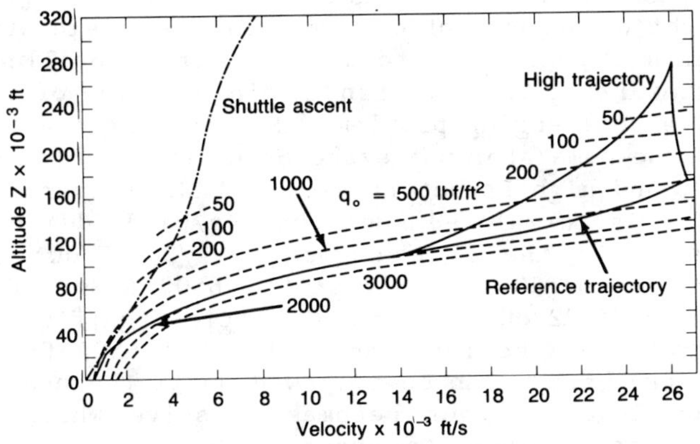

Fig. 35 Typical trajectories for SSTO vehicles and Shuttle.

The powered phase of the trajectory comprises segments, each of which can be expressed as an analytic function. They are:

$0 \leq U \leq 500$ ft/s	$Z = 0$
$500 \leq U \leq 1200$ ft/s	$Z = 2.035 \times 10^{2}\left(U^{2} - 500^{2}\right)$ ft
$1200 \leq U \leq 6500$ ft/s	$U = 500 + 7.23\text{x}10^{-3}Z + 8.95 \times 10^{-7}Z^{2}$ ft/s
$6500 \leq U \leq 14{,}000$ ft/s	$q = \rho\ (U^{2}/2) = 2000$ lbf/ft^{2}
$14000 \leq U \leq U_{2}$	$U = -\ 39{,}256 + 0.66472Z - 1.64 \times 10^{-6}Z^{2}$ ft/s

The first segment on the runway is for takeoff at 500 ft/s, which corresponds to constant accelerations of 0.518 and 0.389 g for runway lengths of 7500 and 10,000 ft, respectively. Takeoff speeds have to be high because the wing area must be kept to a minimum to minimize drag at hypersonic speeds. The basis for the choice of the velocity-altitude relationship for the next two segments requires examination of the fundamental performance parameters for an accelerator. This parameter is a restatement of Eq. (1), where the substitution $dZ = U \sin\theta dt$ is made; thus,

$$\frac{W_{0}}{W_{2}} = \exp \int_{0}^{2} \frac{[1 + (g/U)(dZ/dU)]}{g\dfrac{T}{\dot{w}_{P}}\,[1 - D/T]} = \frac{W_{0}}{W_{0} - W_{P}} \tag{12}$$

where W_0 is the initial weight, W_2 the final weight, and W_P the weight of propellant consumed in the acceleration. The designer's objective is to minimize the propellant, which means minimizing the integrand. Minimization of the numerator is a challenging problem. Suppressing the trajectory reduces the numerator but exacerbates the problem of noise, flutter, and high structural loads. The segment $500 \leq U \leq 1200$ ft/s is formulated such that $(g/U)(dZ/dU) = 1.309$, a constant. In the next segment, $(g/U)(dZ/dU)$ decreases rapidly from 0.530 to 0.337 at $U = 6500$ ft/s. From there on, the $(g/U)(dZ/dU)$ term is of little significance in the minimization of the integrand. For $U > 6500$ ft/s, a constant q trajectory can be followed until the heat-transfer rates on leading edges become excessive which, in this modeling, is presumed to occur at $U = 14{,}000$ ft/s. The trajectory at $U > 14{,}000$ ft/s is intended to approximate a constant heat load.

For the cruise portion of the flight of a hypersonic airplane, the measure of vehicle performance is cruise range as formulated by the well-known Breguet equation, namely, level flight, constant velocity, lift equals weight, and thrust equals drag, i.e.,

$$\text{RANGE} \equiv R = \frac{gU[T\cos(\alpha+\delta)]\left(\dfrac{L+T\sin(\alpha+\delta)}{D}\right)\ell n\, W_i/\left(W_i/W_P\right)}{\dot{w}_p\left[1-\left(U/U_c\right)^2\right]} \text{ ft} \tag{13}$$

where

U = velocity in ft/s

T = thrust in lbf

L = aerodynamic lift in lbf

D = vehicle drag in lb

$\dot{w}_p$ = propellant flow rate in lbm /s

α = angle of attack

δ = angle between the thrust vector and the vehicle axis

W_P = total weight of propellants consumed in cruise lbm

$W_1 = W_e + W_P + W_S + W_X$ = total weight of vehicle at initiation of cruise, which is composed of engine weight W_e; propellant weight W_P; structural weight W_S, and the remaining weight at the end of cruise, W_X.

Here, $U_c = r_0 g_0/(r_0 + Z_c)$ is the velocity required for a circular orbit of the Earth at an altitude Z_c. For example, taking Z_c = 150,000 ft, g_0 = 32,174 ft/s, and r_0 = 3963 x 5280 = 20.919 x 10^6 ft, U_c = 25,857 ft/s. The velocity ratio in the denominator, which is often neglected, becomes quite important at hypersonic speeds.

Force Accounting

Whereas the solution to Eqs. (12) and (13) must be independent of which of the forces on the vehicle are labeled thrust and which drag, many simplifications accrue with a judicious choice for the force accounting system. Figure 36 will be used to develop this argument. The sketch shows a vehicle powered by an air-breathing engine operating at below the inlet design speed, i.e., $M_0 < M_D$. Consequently, the shock from the leading edge of the vehicle lies ahead of the engine cowl lip. The air that is captured in the inlet is in streamtube A_0. The cross-sectional area labeled A_4 corresponds to conditions following the inlet

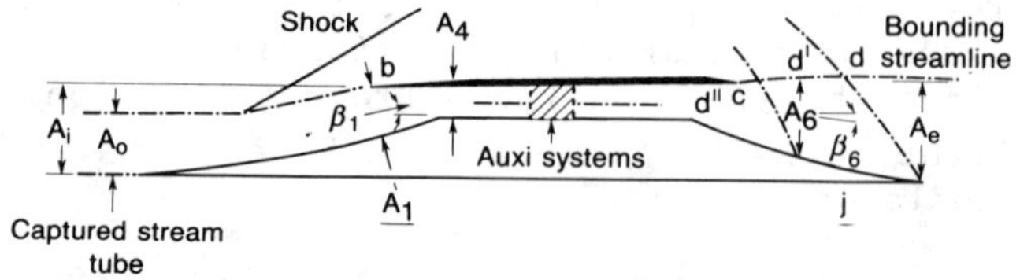

Fig. 36 Model for force accounting.

compression process. The subsequent discussion will argue that the maximum value of A_4 governs the amount of engine air captured at low M_0. At higher M_0, air capture is governed by the position of the forebody wave pattern relative to the cowl lip. In this article the demarcation is, for convenience, set at $M_0 = 3$ to coincide with the flight speed at which the engine transitions from whatever low-speed mode is used to a ram/scramjet. The elements that make up the low-speed system all lie within a control volume that is prescribed such that forces can be deduced from change in momentum. These elements are schematically shown as a crosshatched box. The dashed streamline entering and leaving the box is simply a schematic depiction of a splitting of the captured airflow into substreams that could have quite different flow paths but that ultimately mix before passing from the engine.

The expansion of the flow in the exhaust nozzle depends on both the ratio of the total pressure of the expanding gas to the static pressure in the freestream, P_{t6}/P_0, and the nozzle geometry. For high P_{t6}/P_0, typical of flight at $M_0 > 7$, the nozzle is "underexpanded," i.e., the nozzle flow has expanded to the end of the vehicle, $P_6 > P_0$, and the adjustment in pressure in the expanding jet to P_0 occurs beyond the confines of the vehicle. For this situation the end of the control volume is the isobaric surface $P_6 = P_e$. In the sectional view of the flow model in Fig. 36, the point of intersection of this isobaric surface with the outboard streamline (stream surface) is labeled d. For low P_{t6}/P_0, a more complex flow structure exists. In the expansion the nozzle flow reaches $P_6 = P_0$, well upstream of station *e*, it then overexpands to $P < P_0$. For most situations the overexpansion ends with a compression shock that separates the surface boundary layer. The shock compression raises the pressure back to approximately $P = P_0$ at station j, upstream of the end of the body, and the pressure remains nearly constant from j to *e*. The end of the control volume in this case is the upstream isobaric surface $P_6 = P_0$, which may occur within the confines of the cowl. Thus, the control volume for engine thrust accounting ends at point d" (inside cowl), d' (downstream of cowl exit), or d. The corresponding projected cross-sectional area in the x direction is A_6. The character of the portion of flow lying between the upstream and downstream isobaric surfaces $P = P_0$ determines the base drag of the vehicle.

Whereas the upstream surface of the control volume can be positioned in the freestream, it is more convenient to define it as the isobaric surface passing through the cowl lip (point b in Fig. 36). If the axial thrust T_x is defined as the "gauge-corrected" change in momentum between stations 1 and 6, then

$$T_x = \left(\rho U^2 A\right)_6 \cos\left(\beta_6 + \alpha\right) - \left(\rho U^2 A\right)_1 \cos\left(\beta_1 + \alpha\right) - \left[P_0\left(A_6 - A_1\right)\right] \quad (14)$$

When station d (or d') lies downstream of c, a small portion of T_x or pseudoforce, i.e., ${}_c\int^d p dA_x$ called "plume drag," does not act on the body; thus, a minor correction in T_x must be made to obtain a precise determination of the forces on the body. The axial forces on the remainder of the body are considered as drag. For underexpanded nozzle flows, $A_6 = A_e$, and there is no base drag on the vehicle. For overexpanded flows, there is a base drag denoted as

$$D_b = \left(A_j - A_6\right)\left(\bar{P}_{6,j} - P_0\right) \quad (15)$$

where $\bar{P}_{6,j}$ is the mean pressure on the surface between A_6 and A_j. This force can be large at low flight speeds and will be an important discriminate in the selection of the propulsion scheme at low flight speeds.

A preliminary conceptual design is needed to quantify T and the terms in Eq. (14). Cycle analysis is the first step in this process.

Cycle Analysis

Cycle analysis plays an important role in the development of vehicle designs. As shown in Fig. 37, integral methods are used; i.e., the integral values of the flow variables are defined on convenient elemental surfaces of the control volume shown in Fig. 36. The elemental surfaces are those wall surfaces that are acted on by the pressure and shear of the captured airflow and in lateral planes representing the freestream, the combustor inlet, the nozzle entrance, and the vehicle base. The equations of motion are solved simultaneously with a suitably defined equation of state, with restrictive assumptions on the thermochemistry.

Details of the structuring and operational characteristic of the computer code (RJPA) used to solve the equations are covered in Ref. 16. Figure 37 shows that the input parameters to RJPA are treated as independent variables. Modeling, experimental data bases, and higher-order analysis are used to reduce the number of variables and their respective ranges of interest. In addition to generating the engine performance data, cycle analysis provides the input for multistreamtube analyses, wherein the flowfield is subdivided into streamlets and each streamlet is handled by a separate RJPA run. The methodology that uses RJPA to provide boundary conditions for finite-difference methods (CFD) is discussed in detail in Ref. 17. The streamline following technique is an economical approach used to gain insight into the effects of finite-rate chemistry in the combustor and nozzle. Solutions from RJPA are used to define the nozzle entrance conditions, and the results serve as revisions to the engine performance from the simpler equilibrium expansion results (see Ref. 18). The subsequent discussion will show how RJPA and its complementary analyti-

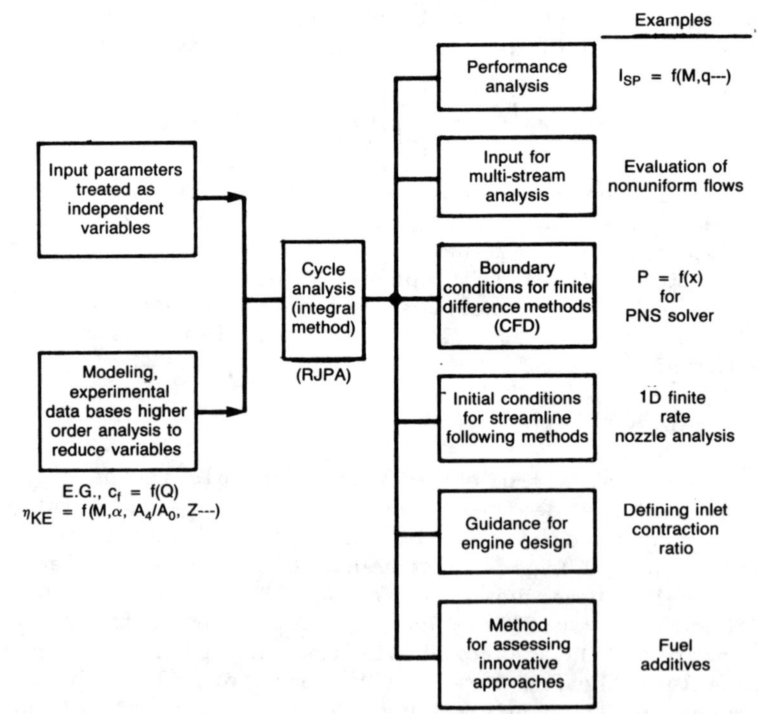

Fig. 37 Role of cycle analysis in the development of advanced air-breathing engines.

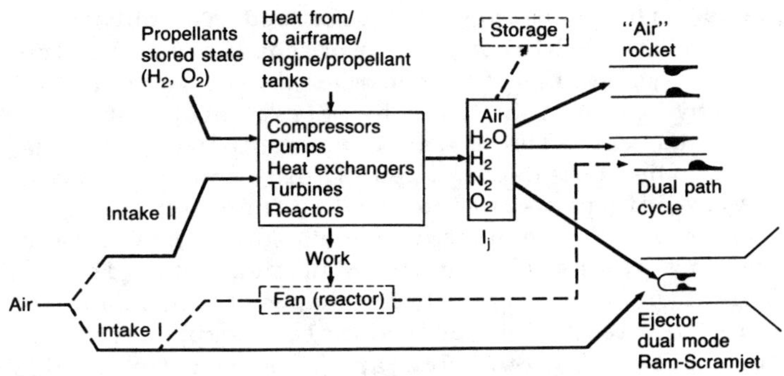

Fig. 38 *2Schematic illustration of composite air-breathing engine cycles (H_2-O_2 air systems).

cal procedures provide guidance for engine designs and disciplined methods for assessing innovative design approaches.

Figure 38 shows the method used to analyze composite engine cycles with RJPA. Intakes I and II represent the bifurcated flow path shown in Fig. 36. In most of the lower-speed cycles, the flow into intake II is processed by one or more of the items located in the box containing compressors, etc. Propellants and air enter this "processor," and products exit. For nearly all cycles of interest, the species exiting the processor are in their ground state and therefore can be described by a simple form of the state equation. The analysis of the "processor" is handled as it was in the development of the results summarized in Fig. 22 and Table 6. The processor can accept or reject heat and produce net work if needed to provide auxiliary power or drive a fan, as in a bypass turbojet. Storage of one or more of the exiting species is permitted. An example would be the an air collection and storage system (see, e.g., Ref. 19). The box containing the exit species is then treated as the "fuel" package in a simple, single-flow engine cycle using RJPA. Performance maps are generated, where the grids on a "carpet plot" are the ER_{ij} and ER_{OV}. The detailed analysis of the processor provides the ER_{ij} which, in turn, provides the criteria for selecting the optimal value of ER_{OV} for a particular application.

Conceptual Design of Vehicle

A conceptual vehicle design is needed to proceed in the assessment of engine cycles. The engine flow path dictates the vehicle geometry in hypersonic airplanes. Since the ram/scramjet cycle must operate over a broad range of speeds, the geometry is selected to obtain maximum performance from this cycle. Consequently, the low-speed propulsion system has to accommodate the constraints that are thereby established. The first step in conceptual design is to vary the inlet contraction (or compression) ratio and the combustor area ratio and to examine the sensitivity of performance as a function of component efficiencies. The results serve to guide the selection of the geometry of the flow path. The next step is an investigation of possible geometric shapes to produce the wave structure to yield the desired inlet compression and contraction ratios. Flowfields are computed for families of compression surfaces, with simplifying assumptions on the gas chemistry. These simplifications are removed when CFD analyses of specified inlet designs are made. Figure 39 shows typical results for a family of inlets wherein the compression is provided by either a finite number of oblique shocks or on a surface that is shaped to compress the air without shock losses, a so-called isentropic ramp. The wave angles of each set of multiple oblique shock inlets are adjusted to yield maximum efficiency for a given compression ratio. The compression ratio P_4/P_0 is shown as a function of M_0. Results from the cycle analysis provided the rationale for the values of P_4/P_0 of interest. The prominent features of the results are the large variation in the desired amount of total flow turning from low to high Mach numbers and the relative insensitivity to the details (i.e., number of shocks) in the compression process.

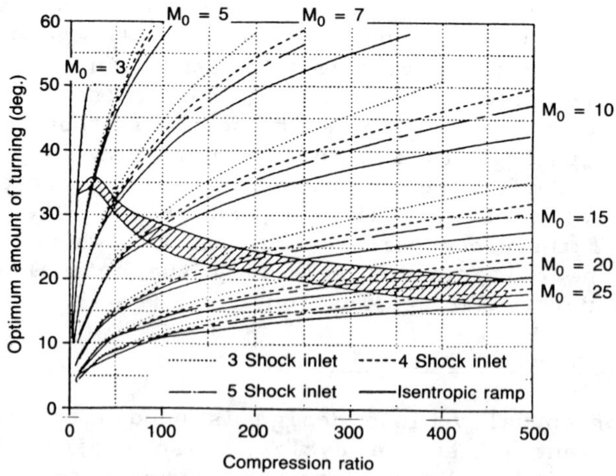

Fig. 39 Optimum scramjet inlets.

The required flow turning can be obtained by waves that turn the flow either away from or toward the vehicle axis. To obtain near coaxial flow at the discharge of the inlet, the outward turning must equal the inward turning. A reasonable strategy is to shape the vehicle forebody to provide one-half the total required turning at the highest operating M_0 and an equal amount of turning toward the vehicle axis from the engine cowl surfaces. From the figure this would suggest a forebody design with 8-11 deg. of outward turning. To obtain the larger outward turning at lower M_0, several options exist, such as: 1) adjustable compression ramps on the body side, 2) rotating cowls, 3) adjustable sidewall compression surfaces, 4) translating cowl, and 5) insertable multipurpose struts to produce additional lateral compression and serve as instream fuel injectors.

All of the techniques introduce mechanical complexities and sealing problems and add weight. Moreover, none of the methods are exempt from generating unwanted distortion in the flowfield, and the desired schedule of compression ratio with M_0 can only be approximated.

Given that a particular variable-geometry concept is viable, either the direct results from CFD analyses of flowfields or a suitable analytical model can be used to limit further the range of parameters needed to be examined in the cycle analysis. The latter method has considerable merit in the early stages of conceptual design, especially if the model contains features that permit realistic examination of design variables, e.g., the design Mach number of the inlet M_D. An example of this modeling for $3 \leq M_0 \leq 25$ is:

Air Capture Ratio

$$\frac{A_0}{A_i} = 1 - \left(M_D - M_0\right)\left(0.155 - 0.94 \times 10^{-2} M_D + 1.8 \times 10^{-4} M_D^2\right) \tag{16}$$

Inlet Contraction Ratio

$$\frac{A_0}{A_4} = -3.5 + 2.17\, M_0 - 0.017\, M_0^2 \tag{17}$$

Inlet Compression Ratio

$$\frac{P_4}{P_0} = -8.4 + 3.5\, M_0 + 0.63\, M_0^2 \tag{18}$$

Table 8 Conditions in the freestream and the combustor inlet in a typical ram/scramjet engine

	FREESTREAM CONDITIONS						ISOLATOR ENTRANCE CONDITIONS				
M_0	Z_0 k ft	P_0 psia	T_0 °R	U_0 ft/s	q_0 lbf/ft^2	h_{t_0}[a] Btu/lbm	$\frac{A_0}{A_4}$	$\frac{P_4}{P_0}$	P_4 psia	T_4 °R	U_4 ft/s
3	47950	1.868	390.0	2904	1694	133.3	2.86	7.8	14.51	744	2034
4	57480	1.183	390.0	3872	1910	264.3	4.91	15.7	18.57	930	2885
5	65720	7.978^{-1}	390.0	4840	2011	432.8	6.92	24.9	19.86	1102	3799
6	73300	5.569^{-1}	394.0	5839	2020	646.7	8.91	35.3	19.65	1279	4770
7	80077	4.049^{-1}	397.7	6844	2000	902.3	10.85	47.0	19.03	1451	5757
10	95500	1.984^{-1}	406.1	9879	2000	1918.2	16.49	89.6	17.78	1958	8744
15	114250	8.577^{-2}	424.8	15155	1945	4561.0	25.23	185.9	15.94	2880	13908
20	137760	3.194^{-2}	460.8	21040	1287	8824.6	33.11	313.6	10.02	4074	19648
24	178210	6.759^{-3}	480.5	28865	425.8	16629.8	40.12	472.9	3.20	5187	27205

[a]h = 0 @ 536.7°R

[b]Reference Airbreathing Trajectory (Fig. 5.1)

Note that the independent variable in Eq. (16) is M rather than U as in the preceding discussion. This is in recognition of the fact that the inlet parameters are, in part, governed by the wave structure in the flowfield, which is Mach-number-dependent. On the other hand, U is the more fundamental parameter for the flight mechanics, engine thrust, and drag is U.

With the modeling of Eq. 16-18 and an assumed 1% loss in total enthalpy from the inlet surfaces by radiation to space, the mean flow conditions at the combustor inlet can be calculated. Table 8 lists values of these conditions for the reference trajectory. Although P_4/P_0 increases by a factor of about 60 over the range of M_0, the static pressures at the isolator entrance vary by only a factor of 6. Additional pressure rises in the shock train depend on engine fuel/air equivalence ratio ER and combustor area ratio A_5/A_4. Maximum rises occur at Mach 6-7 and can be as high as a factor of 5 -10; i.e., local pressure can be in the 100-200 psia range. The structural design of the engine is governed by these conditions. Maximum heat transfer and, hence, the maximum cooling requirements are dependent primarily on the driving enthalpy and local dynamic pressure. The combined effects are most severe in the Mach 15-20 range. The mean static temperature T_4 increases by about 200°R for a unity increase in M_0 and reaches a level ($T_4 >$ 4000°R) at which dissociation of O_2 becomes significant at about $M_0 = 17$-20. Within the boundary layer the temperature can be significantly higher, especially at high M_0. Note that, for $M_0 < 9$, the static temperature prior to the shock train is below 1800°R, which is the minimum temperature required to autoignite hydrogen. This suggests that an ignition source will be needed at low M_0 and, with ignition, considerable heat will have to be released to insure the establishment of a precombustion shock structure to raise the temperature throughout the flow. The mean values of M_4 increase monotonically with M_0, and the ratio of M_4/M_0 decreases from about 0.5 at $M_0 = 3$ to 0.3 at $M_0 = 25$. The velocity decrements $U_0 - U_4$ increase from about 870 ft/s at $M_0 = 3$ to about 2660 ft/s at $M_0 = 25$. However, the velocities in the combustor remain very high at high M_0, which means that the mixing and combustion must be extremely rapid. For example, at $M_0 = 20$, the residence time in a 5-ft-long duct would be only 10 μs.

With conditions at station 4 specified, engine cycle results are used to determine the mean flow conditions and geometry requirements at the combustor entrance and exit and at the end of the nozzle expansion. With these results, the problem of nonideal nozzle expansion can be addressed.

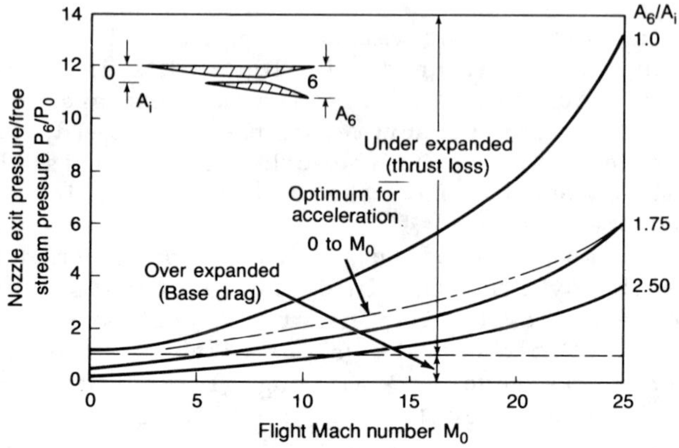

Fig. 40 Effect of nozzle exit-to-inlet area ratio on vehicle performance.

Figure 40 will explain the procedure used to specify the nozzle area ratio. The ratio of nozzle exit to freestream pressure P_6/P_0 is shown as a function of flight Mach number for three values of A_6/A_i, namely, 1.0, 1.75, and 2.5. Pressure ratios increase with M_0. When $P_6/P_0 < 1$, the nozzle flow overexpands, which produces base drag; when $P_6/P_0 > 1$, the full potential of the engine exhaust to produce thrust cannot be realized. A compromise in the choice of A_6/A_i must be made in the design of a vehicle that operates over a wide range of M_0. With these results a range of A_6/A_i values could be used to specify several conceptual vehicle designs, and then trajectories could be flown. This is a rather tedious task and so, for this study, a value of A_6/A_i = 1.75 is used. This value is in accord with the historical data that formed the basis for the curve labeled "optimum."

The engine airflow characteristics for low-speed cycle condition are governed by the geometric limits imposed on A_4. With consideration of the practical structural and actuator constraints and limitations on fuel distribution, let us suppose that it is possible to vary the geometry such that A_4 for low-speed engine operation can be 4.92 times larger than at high speed. Moreover, at high speed, $A_0 = A_i$ and the values of A_0/A_4 from Table 8 are to be matched. Assuming that an increase in A_4 also results in an increase in A_i, e.g., from outward translation of the cowl, A_4/A_i = 0.0249 at M_0 = 25, and $A_4/A_i = A_{4max}/A_i$ = 4.92/(1+40.12) = 0.120 at $M_0 \leq 3$. From Table 1, A_0/A_4 at M_0 = 3 = 2.86; thus, A_0/A_i = 2.86 x 0.120 = 0.342 at M_0 = 3.

For low-speed cycles the air capture can then be calculated for a desired Mach number at station 4 of the inlet total pressure recovery as specified. The model adopted for total pressure recovery for $0 \leq M_0 \leq 3$ is

$$\frac{P_{t_{4'}}}{P_{t_0}} = 1 - 0.0291\, M_0 - 0.0206\, M_0^2 \tag{19}$$

where the prime on 4 means that the conditions at the end of compression are subsonic. The modeling was incumbent on matching of $P_{t4'}/P_{t0}$ at $M_0 = 3$ with the value that would be obtained through a normal shock at M_4 set by Eqs.(17) and (18) given in Table 8. With $P_{t4'}/P_{t0}$ given by Eq. (19), the limits on A_0/A_i for $M_0 \leq 3$ can be established for low-speed cycles by specifying $M_{4'}$. Two classes of low-speed engines establish the limit curves shown in Fig. 41. The first class comprises engines that use rotating machinery to compress the air (e.g., a turbojet). In these engines the Mach number entering the compressor should be low, so that $M_4 = 0.3$ is taken as the lower bound. The other class uses ejector pumping to compress the air (e.g., a ducted rocket). These cycles optimize with high entering Mach numbers so that $M_4 = 0.8$. The results shown in Fig. 40 are quite disconcerting to the designer of an accelerator because, in the transonic region, the air capture ratios are so that low either the transonic "barrier" cannot be crossed or an enormous amount of fuel would be used to accelerate through

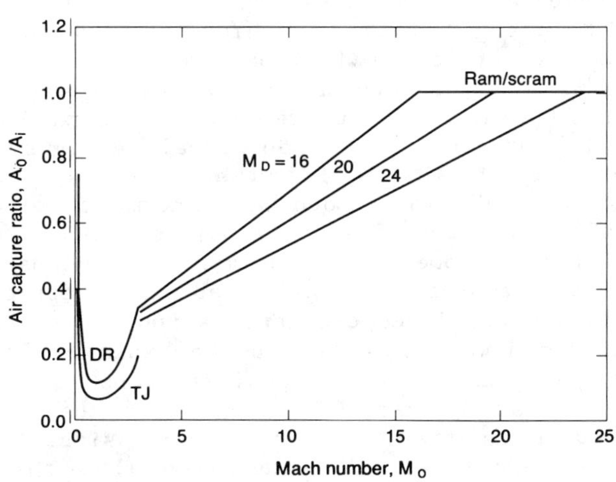

Fig. 41 Engine air capture ratio.

the barrier. The only solution is to augment the thrust with rocket propulsion, albeit a tandem system or embedded within the airflow path. Curves of A_0/A_i for $M_0 > 3$ calculated from Eq. (16) are shown for $M_D = 16$, 20, and 24. The subsequent analysis will show that the increased air capture for the $M_D = 16$ case results in a lower fuel requirement to reach the cruise condition or orbit.

With the air capture characteristics defined, the performance for air-breathing cycles can be computed and expressed in terms of the engine specific impulse I_{AB} and thrust coefficient C_{TAB}, which are given as

$$C_{TAB} = T/q_0A_i = \Big\{\eta_N \mathcal{F}_6 \cos\beta_6 - \mathcal{F}_1 \cos\beta_1 - P_0\left(A_6 \cos\beta_6 - A_1 \cos\beta_1\right)\Big\}\Big/ q_0A_i \tag{20}$$

$$I_{AB} = \frac{T}{\dot{w}_P} = \frac{\eta_c C_T U_0 A_i}{64.35\, A_0} \frac{\dot{w}_a}{\dot{w}_p} \tag{21}$$

where

η_N = nozzle efficiency

η_c = combustion efficiency

$\dot{w}_P/\dot{w}_a$ = propellant flow/airflow

For these calculations, $\eta_N = 0.98$, $\eta_c = 0.95$, and $\dot{w}_P/\dot{w}_a = 0.02917$ ER for engine cycles in which the total propellant is hydrogen. For cycles in which oxygen is part of the propellant, $\dot{w}_P/\dot{w}_a$ includes both the hydrogen and oxygen.

At this point, the rudiments of a data base are available to proceed with conceptual designs. The information was the basis for the two generic engine designs[20] shown in Fig. 42. Both have maximum inlet capture areas of 100 ft^3 at Mach 25 and maximum body cross-sectional areas of 175 ft^2; one is a two-dimensional planar geometry (PGE), and the other is a so-called axisymmetric sector engine (ASE). Overall vehicle lengths and cross-sectional area distribution were selected on the basis of fuel tank volume requirements and drag considerations. The inlet cowls were set at $x/\ell \approx 0.67$, where x is the axial coordinate and ℓ is the fuselage length, since previous studies (see, e.g., Ref. 21) suggested that the engine modules should be at about this location to provide stability and control with low induced drag.

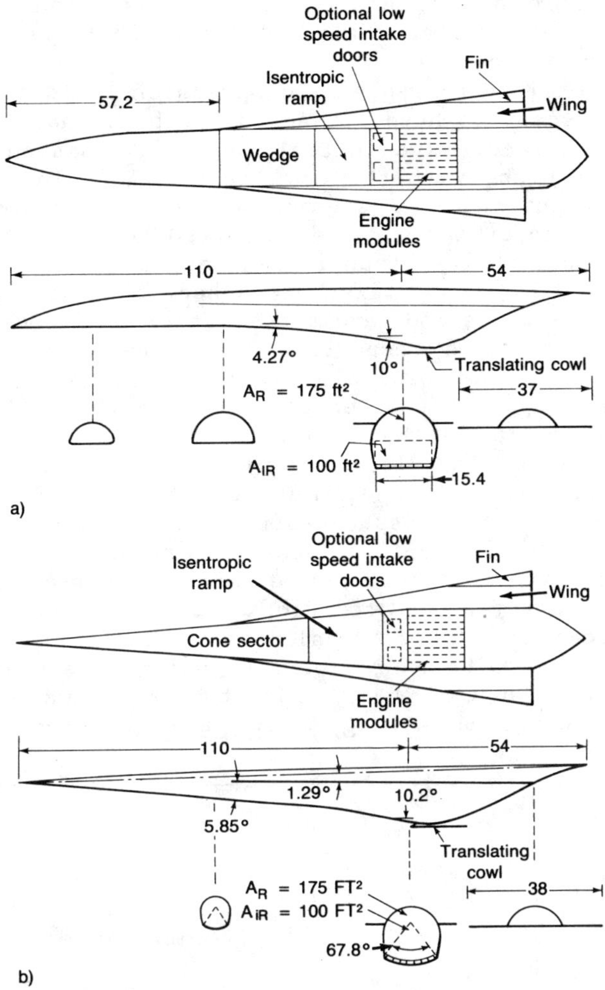

Fig. 42 Schematic illustrations of vehicle configurations: a) planar geometry engine PGE; b) axisymmetric sector engine ASE.

Both inlets are of the external-internal compression type, wherein the flow is turned outboard by the cone sector (ASE) or wedge (PGE) and then returned coaxially by the straight cowl. In the ASE, the forebody is a 1.5-in. radius, 5.85 deg. half-angle blunted cone, followed by a curved surface that turns the flow outboard reaching an angle of 10.2 deg. at x = 1221 in. The sector angle of 67.8 deg., the cowl lip radius of 13 ft, and the location of the axis of symmetry were selected to yield the desired air capture ratio and to provide forebody section geometries that include circular fuel tank cross sections with very little wasted space.

To obtain a near-planar compression field in the PGE, it is necessary to have a tapered forebody that is flat on the undersurface. The initial wedge compression angle is 4.27 deg. which, with the cowl at x = 1320 in., yields a capture height of 78 in. and the desired capture area of 100 ft^2. The compression surface is curved from x = 1063 in. to x = 1199 in., where the angle reaches 10 deg.

To provide the desired cross-sectional area change in the internal ducts and wave cancellation at the shoulder of the innerbody, the respective cowls and cowl flaps are translatable. For orbital flight and re-entry, the cowl flaps close against the vehicle body. Both configurations show interior walls that subdivide the internal flow into individual engine modules. The planform areas of the underbodies are 2425 ft^2 and 3160 ft^2 for the ASE and PGE, respectively. The gross takeoff weights are 160,000 lb.

With the conceptual designs specified, reasonable estimates of vehicle drag can be made. Figure 43 shows drag coefficients referenced to frontal area based on relatively simple modeling and historical data bases. The nominal drag curve includes a base drag penalty due to overexpansion of the engine exhaust at $M_0 = 7$. If, by some means such as external burning, this drag could be eliminated, the sig-

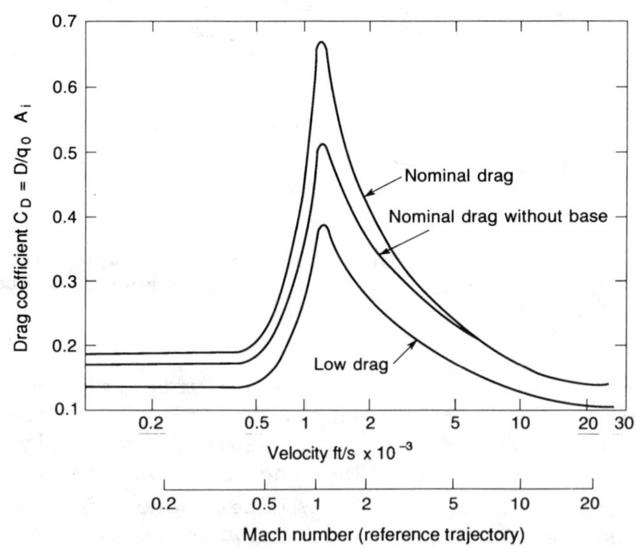

Fig. 43 Engine specific impulse: vehicle drag coefficient based on inlet area; $A_{i\ REF} = 100\ ft^2$.

nificant reductions in C_D would be realized as shown in the intermediate drag curve. The lowest curve is included to assess the benefits that could accrue in the event that an extremely efficient aerodynamic configuration could be designed.

Cycle Selection for Vehicle Assessment

The preceding discussion has provided the basis for a disciplined assessment of a broad spectrum of lower-speed propulsion cycles. However, the following arguments will show that the counterbalance between thrust and specific impulse leads to the conclusion that factors other than engine performance per se will dictate the selection of the engine for SSTO and hypersonic cruiser applications. This is a consequence of the severe constraints placed on the operation of the lower-speed propulsion system when it must be integrated with the ram/scramjet. The principal constraint is the very limited air capture that leads to a small engine relative to the frontal area and, in turn, the drag of the vehicle. The engine cycles that have the higher specific impulses tend to have lower specific thrust, and vice versa.

To quantify these arguments, two limiting cases will be compared: the turbojet and the ducted rocket. This choice avoids the complexity that is inherent when all of the nuances of using the cooling capacity of the propellants to enhance the compression in the Brayton cycle have to be included. It is argued that a more thorough examination of the other cycles will have little impact on the major conclusions. To accentuate the argument, the performance of the turbojet cycles was based on an optimal compressor pressure ratio cycle. That is, at each flight speed the pressure rise in the compressor was selected such that the specific impulse of the engine was vacuum. Had "off design" losses been considered, I_{SP} would have been lower. The optimum compressor pressure ratio is

$$\frac{P_4}{P_3} = \left(\frac{1}{G+1}\right)\left[\frac{C_{P_{10}} T_{t_{10}}}{C_{P_3} T_{t_3}}(1+f)\ \eta_T + \eta_c + 1\right]^{\frac{\gamma_c}{\gamma_c - 1}} \tag{22}$$

where

$$G = \left(\frac{\gamma_c - 1}{\gamma_c}\right)\left(\frac{\gamma_T}{\gamma_T - 1}\right) \tag{23}$$

and

γ_c, γ_T	=	ratio of specific heats in compressor and turbine, respectively
η_c, η_T	=	compressor and turbine efficiency, respectively
C_{P3}, C_{P10}	=	specific heats at constant pressure at the compressor and turbine inlets, respectively
T_{t3}, T_{t10}	=	total temperature at the compressor and turbine inlets respectively

Values of $\eta_c = 0.85$, $\eta_T = 0.90$, and $T_{t10} = 3000°R$ were used in this analysis. Calculations were made for the ER corresponding to $T_{t10} = 3000°R$ with no afterburning and with afterburning with ER = 1.

The subsequent analysis will show that either of the turbojet cycles would need the assistance of an adjunct rocket to meet the acceleration requirements for takeoff and to avoid the use of excessive amounts of fuel to accelerate through the transonic region. Selection of the size of the adjunct rocket is a trade between the benefits that accrue with a large thrust increment one the one hand vs the difficulty in packaging the system, the volume and weight on the other. Trade studies suggest a design producing 600

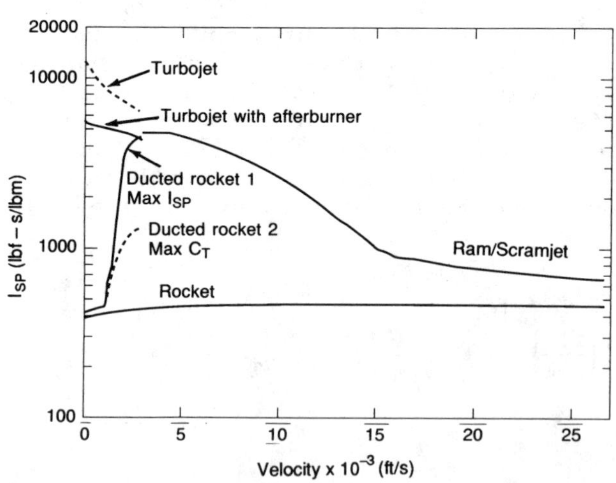

Fig. 44 Computed values of specific impulse.

lbf/ft^2 of engine frontal area is about optimum. For these vehicles with A_i = 100 ft^2 the thrust requirement is therefore 60,000 lb at sea level. For this study a chamber pressure of 1500 psia and an O/F (O_2/H_2 weight ratio) = 5.29 were assumed with a "c star" efficiency of 0.95. The resulting throat area is 24.04 in^2.

Figure 44 shows the computed values of specific impulse for these cycles. The turbojet operating without an afterburner has, by far, the highest I_{AB}, with values decreasing from 12,400 lbf/lbm at U = 0 to 6218 lbf/lbm at U = 2904 ft/s (M=3). With the afterburner, the corresponding values are 5360 lbf/lbm and 4770 lbf/lbm, respectively. The specific fuel consumptions (sfc) of 0.290 and 0.579 lbm/lbf hr are low relative to state-of-the-art engines because the assumed turbine inlet temperature of 3000°R is quite high. With the engine sized to meet the A_{4max}/A_i constraint, sea level static thrusts are 48,200 lbf and 36,500 lbf for operation with and without the afterburner, respectively.

The value of I_{AB} = 418 lbf/lbm for the ducted rocket at U = 0 is slightly in excess of a hydrogen-oxygen rocket. Two branches are shown between U = 1100 and 2904 ft/s. The lower branch corresponds to large $\dot{w}_P/\dot{w}_a$ and the highest C_T that the ducted rocket could produce following the prescribed trajectory without exceeding a 60 deg. climb angle. On the higher branch the engine is operated at a propellant flow rate and mixture ratio that results in minimum propellant usage for the high vehicle drag case assumed in this study. Both engines have sufficient thrust to accelerate the vehicle through the transonic "pinch" without requiring rocket augmentation.

The values of I_{AB} for the ram/scramjet correspond to an ER schedule that results in a minimum fuel requirement for acceleration to orbit. Minor differences in I_{AB} occur at M_0 > 16 for the three values of M_D examined, but they are imperceptible at the scale shown. At M_0 > 16, oxygen is added to the hydrogen prior to injection to obtain maximum I_{AB}. In reality, the ram/scramjet operates as a ducted rocket at very high speeds, since the fuel injectors are just small rocket "ejectors."

Specific impulse for the rocket increases with velocity in accordance with the change in ambient pressure along the flight path. At sea level the I_{SP} is 392 lbf/lbm and it increases to 474.7 lbf/lbm at Z = 175,000 ft.

Thrust coefficients for the propulsion cycles are shown in Figs. 45 and 46. For clarity, portions of the curves at C_T > 3 are not shown. The relatively low values of C_T for the turbojet cycle at low speeds (Fig. 45) are

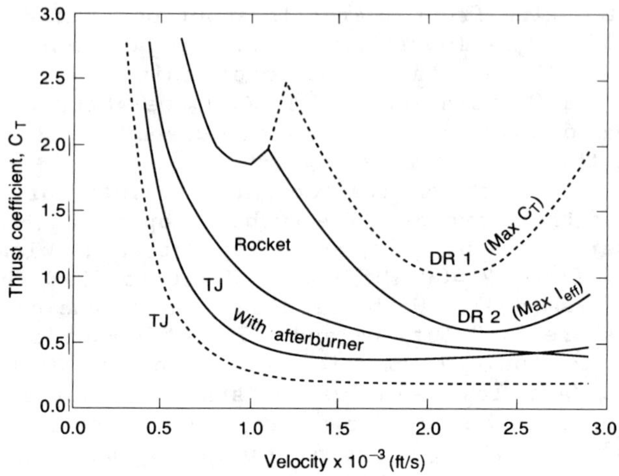

Fig. 45 Engine thrust coefficient

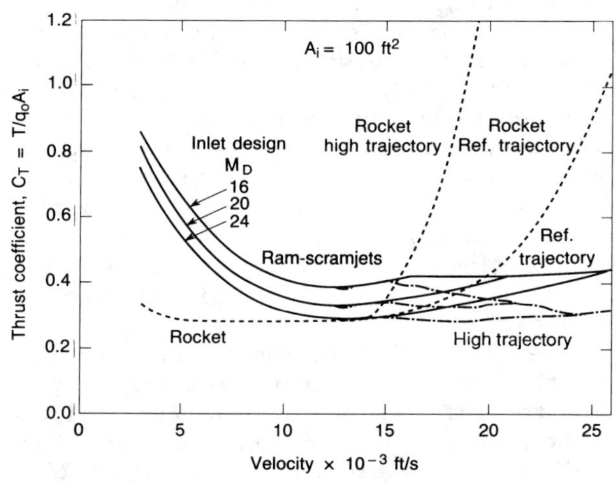

Fig. 46 Engine thrust coefficients: A_i = 100 ft^2.

apparent. The range of C_T available in the ducted rocket cycle is quite broad, which permits the opportunity to provide engine thrust in accordance with the drag characteristics of the vehicle. Rocket C_T decrease monotonically with U since Z increases with U. Had a larger rocket been chosen, C_T values would have been proportionately larger. Figure 46 compares the C_T values for the ram/scramjet and the rocket. For the lower speeds, the differences in C_T for the different M_D lead to differences in the propellant required to accelerate. At U > 16,000-18,000 ft/s, the addition of oxygen in the ram/scramjet can be beneficial. The amount that is added and, therefore, the resultant C_T are dependent on the trajectory and the vehicle drag. The rocket C_T curve rises rapidly at U > 14,000 ft/s because q_0 is rapidly decreasing while the thrust is increasing slightly.

Trajectory Analysis

With C_T, C_D, the specific impulse of the air-breathing cycle I_{AB}, and the specifc impulse of the orbit I_R defined, it is now possible to integrate Eq. (12) to obtain the propellant mass fraction needed to accelerate to orbit. A simple test can be made to determine whether it is beneficial to augment the air-breathing propulsion cycle with a rocket.

$$\text{If} \quad \frac{C_D}{C_{TAB}} > 1 - \frac{I_R}{I_{AB}} \tag{24}$$

then the rocket should be turned on, and the denominator of Eq. (12) becomes

$$g\frac{T}{\dot{w}_p}\left(1+\frac{D}{T}\right) = gI_{AB}\left(\frac{C_{T_{AB}} + C_{T_R} - C_D}{C_{T_{AB}} + C_{T_R}\, I_{AB}/I_R}\right) \tag{25}$$

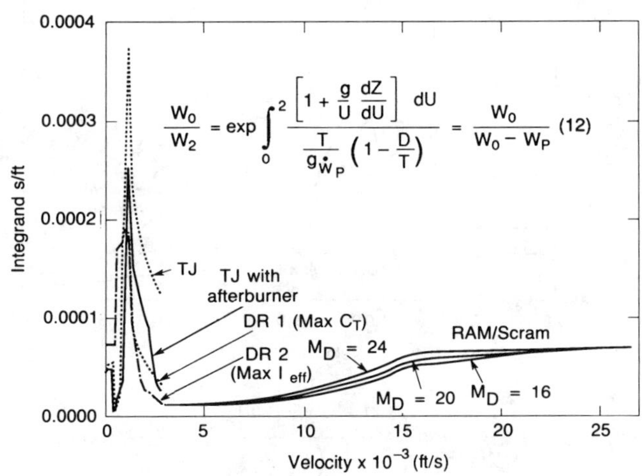

Fig. 47 Values of integrnad of Eq. (12).

Equation (25) also holds for rocket assist prior to takeoff.

Plots of the integrands of Eq. (12) for several engine combinations and the nominal drag are shown in Fig. 47. The propellant required to accelerate to orbit is proportional to the area under the curve. It is apparent from these curves that engine/vehicle performance at transonic and high-hypersonic-speed ranges dictates the propellant requirement. Curves are shown for the three inlet design Mach numbers. Lowering the design Mach number of the inlet leads to saving fuel. The curves for the lower-speed range have been enlarged in Fig. 48.

On the runway, rocket augmentation of the turbojet engine nearly negates the large specific impulse advantage a velocity of 500 ft/s in 10,000 ft, the rocket must be on until velocities of 330 and 432 ft/s are reached for the turbojet with and without the afterburner, respectively. After the rocket acceleration is no longer needed and until it must be turned on transonically, the high specific impulse of the turbojet provides the desired benefit of lower fuel expenditure. At U = 900 ft/s with the basic turbojet, the rocket is turned back on in accordance with the criterion of Eq. (24). It remains on until the switch to the ram/scramjet. With the afterburner, the rocket is on from U = 1100 ft/s until U = 2200 ft/s. Rocket augmentation

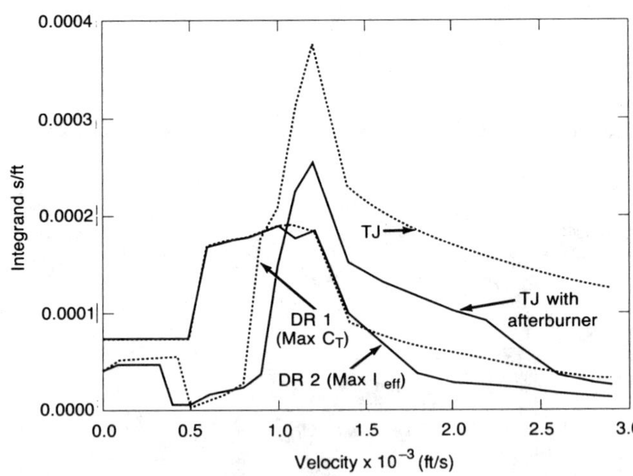

$$\frac{W_0}{W_2} = \exp \int_0^2 \frac{\left[1 + \frac{g}{U}\frac{dZ}{dU}\right] dU}{g \frac{T}{\dot{W}_P}\left(1 - \frac{D}{T}\right)} = \frac{W_0}{W_0 - W_P} \quad (12)$$

Fig. 48 Values of integrand of Eq. (12).

is not needed with the ducted rocket cycles. In fact, the ducted rocket can produce up to 175,000 lbf of thrust at sea level static. If desired, the 160,000-lb airplane, powered with a ducted rocket, could be airborne 3500 ft down the runway. Since there is no rocket augmentation, the ducted rocket having maximum I_{eff} has a lower value of the integrand than the maximum C_T ducted rocket and therefore uses less propellant to accelerate.

Table 9 shows the propellant required to accelerate to Mach 3, U = 2904 ft/s for the four engine cycles. For the nominal drag case, the ducted rocket operating at maximum I_{eff} consumes slightly less propellant than the afterburning turbojet. Operating the ducted rocket at maximum C_T requires more fuel than operating at maximum I_{eff}, but the difference is not too great. The basic turbojet cycle uses about 50% more propellant, which provides a rather convincing argument for avoiding the use of low- thrust, high-specific impulse cycles for SSTO missions. If the base drag could be canceled or, moreover, if a very low drag vehicle were designed, the saving in propellant would be significant, and the use of a higher-specific-impulse air-breathing engine would become more attractive. On the other hand, it can be argued that the ducted rocket is an ideal engine choice because of its relative insensitivity to vehicle drag and high thrust to weight.

When the propellant weight fractions to orbit are compared, the differences in propellant fraction are smaller, as indicated in Table 10. Values are shown for the two cycles of interest and for the three inlet design Mach

Table 9 Fuel required to accelerate to Mach 3

Weight of fuel expended/gross takeoff weight

Engine Cycle	Nominal	Nominal Without Base	Low Without Base
Turbojet	0.341	0.290	0.236
Turbojet with afterburner	0.212	0.172	0.103
Ducted rocket maximum C_T	0.228	0.220	0.210
Ducted rocket maximum I_{eff}	0.202	0.192	0.180

Table 10 Single-stage-to-orbit vehicles: fuel required to accelerate

Fuel weight/gross takeoff weight

Low-Speed Engine Cycle	Inlet design M_D	Vehicle Drag		
		Nominal	Nominal w/o base	Low w/o base
Turbojet	16	0.756	0.737	0.682
	20	0.767	0.748	0.693
	24	0.786	0.767	0.712
Turbojet with afterburner	16	0.708	0.693	0.626
	20	0.722	0.707	0.640
	24	0.744	0.729	0.662
Ducted rocket maximum I_{eff}	16	0.704	0.694	0.659
	20	0.718	0.708	0.673
	24	0.741	0.731	0.696

numbers that were examined. Decreasing M_D from 24 to 16 saves propellant equal to 3-4% of gross takeoff weight (GTOW). From a performance perspective, the choice between the afterburning turbojet and the ducted rocket for the SSTO mission could ultimately rest on the amount of propellant reserved for landing. If an operational requirement dictated that the vehicle must have the capability on a landing approach to "touch and go," "fly around" and then land, the high specific impulse of the turbojet at low speed at reduced throttle would result in large fuel savings relative to the ducted rocket. Conversely, if powered flyaround is not a requirement, the ducted rocket would be the better choice.

With the mass fraction of propellant for the SSTO mission defined, the performance capability of the same vehicle on a hypersonic cruise mission can be calculated if the lift-to-drag ratio is specified. The acceleration phase is ended at the desired cruise velocity, and the remainder of the propellant that would have been used to accelerate further is used instead to cruise. The same propellant reserve that is presumed to be adequate for the SSTO mission is assumed in the cruise mission. In general, the altitude at the end of the acceleration phase on the reference trajectory would not be the optimal altitude for cruise, and so the first segment of the cruise trajectory would be an altitude change at constant velocity. For this assessment the small difference in cruise range from that which would accrue from the same propellant expenditure at near-constant altitude is neglected. Indeed, the comparison of cruise performance will avoid the details of the aerodynamic characteristics of the vehicles by simply specifying a constant lift-to-drag ratio L/D, independent of velocity. With this simplification Eq. (13) can be used to obtain the cruise range.

Figure 49 shows the cruise range for a L/D = 3 vehicle powered by the ducted rocket ram/scramjet engine cycle. Drag during the acceleration phase is the nominal case with base drag included and $M_D = 16$. From Table 10 the total propellant mass fraction is 0.704. The cruise range curve has some interesting features. For cruise velocities U_c between 16,000 and 25,000 ft/s, the effect of centrifugal lift and higher U_c more than compensate for the smaller propellant fraction and lower specific impulse. Thus, R decreases monotonically with U_c, reaching a minimum of 9044 miles for cruises at 16,000 ft/s. For $4000 \leq U_c \leq 16{,}000$ ft/s, the balance among the four terms is such that a maximum $R_c = 15{,}880$ miles is reached at about $U_c = 9000$ ft/s.

The results show that the SSTO vehicle has a very formidable capability as a cruise vehicle. Indeed, for many possible missions, increased payload capability could be traded for propellant and a corresponding decrease in R.

It would be apparent from these results that highly efficient low-speed engine cycles have limited utility in the SSTO mission. The simple reason is that the vehicle

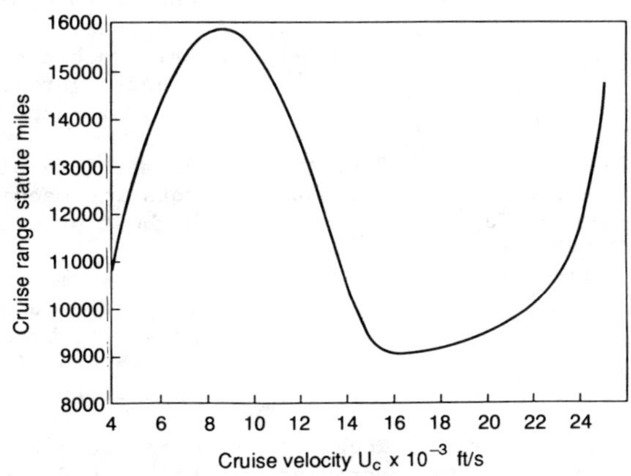

Fig. 49 Cruise range capability for SSTO vehicles: L/D = 3; $W_p/W_{GTOW} = 0.704$ ducted rocket ram/scramjet.

design has to provide an engine flow path that will permit the ram/scramjet to operate efficiently over a very broad Mach number range. The consequence is the very small air capture at low speeds; i.e., the engine is small compared to the typical supersonic airplane.

The alternative that avoids the constraint is the two-stage-to-orbit (TSTO) system. For TSTO vehicles the first stage can be tailored to exploit the best features of the more efficient lower-speed propulsion systems. Moreover, alternate propellants for the first stage could relieve the packaging problem due to low density that is inherent in hydrogen-fueled systems. Table 11 lists some of the candidate systems and staging velocities for TSTO systems. Staging at Mach 3 would permit the use of conventional turbojets. If the propellants were cryogenic, then their cooling capacity could be used to increase the cycle efficiency and reduce the size of the turbomachinery. In-

Table 11 Candidate propulsion cycles for two-stage transatmospheric accelerators

Mach no. at Staging	Propulsion cycles 1st stage	Propulsion cycles 2nd stage	Reasons
3	1st stage turbojet (or) cryocooled (TJ)	Rocket (or) ram/scramjet	Eliminates penalty of carrying weight of highest efficiency, low-speed engine to orbit
			Uses rocket or more efficient lightweight air-breathing 2nd stage
6	Air turboramjet, cryo-cooled TJ/ramjet or TJ/ramjet	Rocket (or) scramjet	Permits use of efficient rotating machinery up to highest practical speed
			Simplifies design of 2nd stage air-breather or allows use of more optimal rocket
12	TJ/Ram/scramjet, cryo-cooled TJ/RJ-SJ (or) ATR scramjet	Rocket (or) scramjet	More optimal staging velocity for rocket or permits use of fixed-geometry 2nd stage scramjet

creasing the staging Mach number to 6 would bring into consideration the cycles that avoid the maximum turbine inlet temperature inherent in the basic turbojet cycle.

Staging at Mach 12 would require integration of the lower-speed propulation system with the ram/scramjet, but the constraints imposed in a SSTO design would be relaxed. This staging velocity also permits consideration of a fixed-geometry scramjet as the second stage. Some loss in performance from that of the variable-geometry configuration would have to be accepted, but the optimal point geometries do not vary significantly at $M_0 > 12$.

The complete analysis of optimal cycles for TSTO vehicles is beyond the scope of the paper. However, the techniques that permit that analysis to be made are contained herein. It should be apparent that intuition, in the absence of a thorough examination, can be deceptive in the selection of the optimal engine cycle for a prescribed application.

Acknowledgment

The author greatly appreciates the information provided by Mr. Hank Lopez of the Garrett Division of Allied Signal Aerospace Company that was the basis for Figures 23, 31 and 32.

Likewise, thanks are due to Messrs. Joseph Bendot and Tom Piercy of The Marquardt Corporation of ISC Defense and Space Company for Figures 17 and 30.

References

[1]Sutton, G.P., Rocket Propulsion Elements, 3rd ed., Wiley, New York, 1949.

[2]Billig, F.S., Waltrup, P.J., and Stockbridge, R.D., "Integral-Rocket Dual-Combustion Ramjets: A New Propulsion Concept," Journal of Spacecraft and Rockets, Vol. 17, Sept.-Oct. 1980, p. 416.

[3]Billig, F.S., "Ramjets with Supersonic Combustion," AGARD-NATO PEP Lecture, Series No. 136, Ramjet and Ramrocket Propulsion Systems for Missiles, Sept. 1984.

[4]Billig, F.S., "Combustion Processes in Supersonic Flow," Journal of Propulsion and Power, Vol. 4, May-June 1988, pp. 209-216.

[5]Burnette, T.D., "Dual Mode Scramjet," The Marquardt Corporation, AFAPL-TR-67-132, June 1968.

[6]Harshman, D.L., "Design and Test of a Mach 7-8 Supersonic Combustion Ramjet Engine," AIAA Propulsion Specialist Meeting, July 1967, Washington, DC.

[7]Waltrup, P.J., Dugger, G.L., Billig, F.S., and Orth, R.C., "Direct-Connect Tests of Hydrogen-Fueled Supersonic Combustors," Proceedings of the XVIth International Symposium on Combustion, The Combustion Institute, Pittsburgh, PA, 1977, pp. 1619-1629.

[8]Waltrup, P.J., Anderson, G.Y., and Stull, F.D., "Supersonic Combustion Ramjet (SCRAMJET) Engine Development in the United States," The 3rd International Symposium on Airbreathing Engines, The Johns Hopkins University Applied Physics Laboratory, Laurel, MD, Preprint 76-042.

[9]Heins, A.E., Jr., Reed, G.J., and Woodgrift, K.E., "Hydrocarbon Scramjet Feasibility Program Part III. Free Jet Engine Design and Performance," AFAPL TR-70-4, Jan. 1971.

[10]Kay, I.W., MeVey, J.B., Kepler, C.E., and Chiappetta, L., "Hydrocarbon Fueled Scramjet, Vol. VIII, Piloting and Flame Propagation Investigation," AFAPL TR 68-146, May 1971.

[11]Stockbridge, R.D., "Combustor Performance Parameters for Shelldyne-H/Air," The Johns Hopkins University Applied Physics Laboratory, Laurel, MD, Rep. BBP-79-37, May 1979.

[12]Oates, G.C., Aerothermodynamics of Gas Turbine and Rocket Propulsion, AIAA Education Series, AIAA, New York, 1984.

[13]Kerrebrock, J.L., Aircraft Engines and Gas Turbines, The Alpine Press Inc., 1977.

[14]Bendot, J. and Piercy, T., "Composite Engine Concepts," Working Symposium on Liquid\Hydrogen Fueled Aircraft, NASA Langley Research Center, May 1973.

[15]Ribaud, Y., "Inverse Cycle Engine for Hypersonic Air-Breathing Propulsion," Ninth International Symposium on Air Breathing Engines, Sept. 3-8, 1989, Athens, Vol. 2, pp. 1044-1050, AIAA, New York, 1989.

[16]Pandolfini, P.P., "Instructions for Using Ramjet Performance Analysis (RJPA) IBM-PC Version 1.0," JHU/APL NASP-86-2, Nov. 1986.

[17]Schetz, J.A., Billig, F.S., and Favin, S., " Modular Analysis of Scramjet Flowfields," Journal of Propulsion and Power, Vol. 5, March-April, 1989, pp 172-180.

[18]Schetz, J.A., Billig, F.S., and Favin, S., "Numerical Solutions of Scramjet Nozzle Flows," Journal of Propulsion and Power, Vol. 3, Sept.-Oct. 1988, pp 440-447.

[19]"Assessment of Cryogenic Hydrogen-Induced Air Liquefaction Technologies," Astronautics Technology Corporation, Madison, WI, Final Report, Sept. 1986.

[20]Billig, F.S., Waltrup, P.J., Gilreath, H.E., White, M.E., Van Wie, D.M., and Pandolfini, P.P., "Proposed Supplement to Propulsion System Management Support Plan," JHU/APL-NASP-86-1.

[21]Small, W.J., Weidner, J.P., and Johnston, P.J., "Scramjet Nozzle Design and Analysis as Applied to a Highly Integrated Hypersonic Research Airplane," Nov. 1976.

Chapter 2

Propulsion System Performance and Integration for High Mach Air Breathing Flight

F. A. Hewitt* and M. C. Johnson†
Rolls Royce plc, Bristol, England, United Kingdom

ABSTRACT

High speed flight vehicles are dominated by the need to capture, process and exhaust streamtubes of air that are extremely large in relation to vehicle size and in comparison to more familiar aircraft. This processing must be carried out with the minimum overall loss and at very high temperatures, the requirements then combining to make extreme demands on the design of very large propulsion systems. The situation is further complicated by the limits of maximum Mach number associated with familiar aircraft propulsion systems. This paper examines some of the interactions between the intakes, nozzles and powerplant, and the vehicles they are propelling.

I. INTRODUCTION

Although the field of propulsion for high-speed flight is extremely complex when observed in fine detail, a number of observations can be made that enable a clear understanding of the limits of the achievable. First, combustion processes: dissociation limits the temperature which is achievable by fuel combustion, even hydrogen combustion, to around 3000 K, when air is the oxidizer. Second, the stagnation temperature ratio of the intake air increases proportional to the square of the flight velocity, reaching 2000+ K by Mach 7 which, in conjunction with the first point, progressively reduces the work that can be added to the stream and, hence, the momentum change and specific thrust. The result of these effects is to limit the

*Senior Project Engineer, Advanced Design Engineering.
†Specialist Engineer (Propulsion Systems) Advanced Design Engineering.

useful maximum flight velocity of any air-breathing system in which the internal airflow is subsonic to Mach 6 - 7. If the intended air-breathing velocity is beyond this limit, then some form of supersonic combustion is required. Supersonic combustion ramjets should permit operation to very high Mach numbers, but specific thrust and specific impulse become very low, reducing the advantage relative to rocket systems at the highest velocities. Some of the detail aspects contributing to overall behaviour are discussed, with emphasis placed on subsonic combustion systems.

II. VEHICLE DYNAMIC PRESSURE AND TRAJECTORY

Freestream dynamic pressure based on $0.5\rho v^2$ is often used as a measure of the air loads associated with a particular trajectory. As Mach number increases, this becomes of

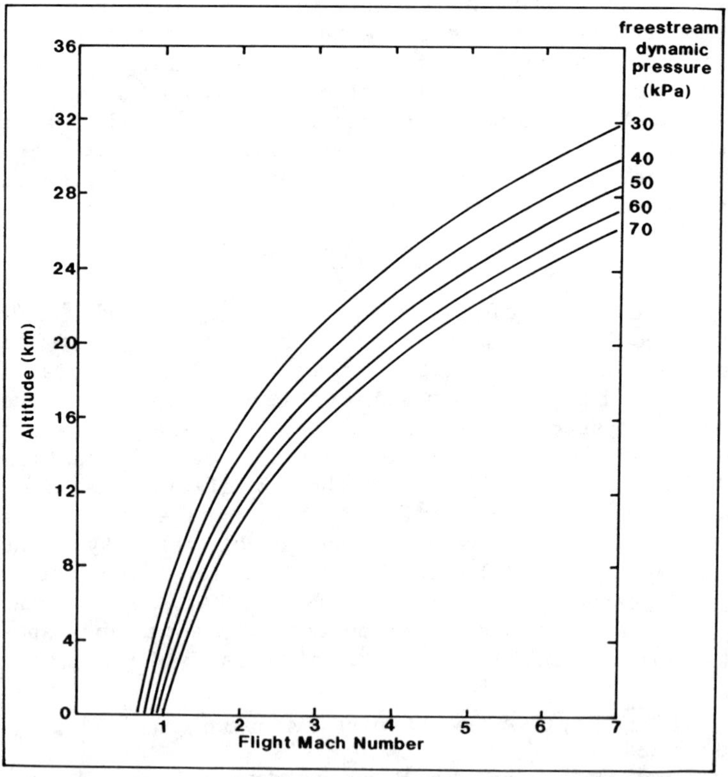

Fig. 1 Trajectories at constant dynamic pressure; altitude vs flight Mach Number.

reducing validity, since compressibility effects modify actual pressures and forces. The effect is greatest where flow is stagnated. Recovered pressures are then greatly in excess of freestream static $+0.5\rho v^2$. It is therefore important to bear this effect in mind when considering trajectories defined in these terms. Fig. 1 shows a range of trajectories in terms of Mach, Altitude and dynamic pressure.

In supersonic flight, lift coefficient at fixed incidence tends to reduce with rising Mach number due to the effect of shock losses and, unless the progressively reducing gross vehicle mass due to fuel usage balances the effect, the result is to require progressively increasing incidence to maintain total lift. The effect of such changes of incidence in compressing a greater freestream cross-section of air into an undersurface intake is shown in Fig. 2. This fortunately tends to offset the effect of falling air mass flow per second on an increasing Mach, "constant dynamic pressure" trajectory, as shown in Fig. 3, allowing a good mass flow to be maintained, and helping to combat the effect of falling specific thrust at high Mach.

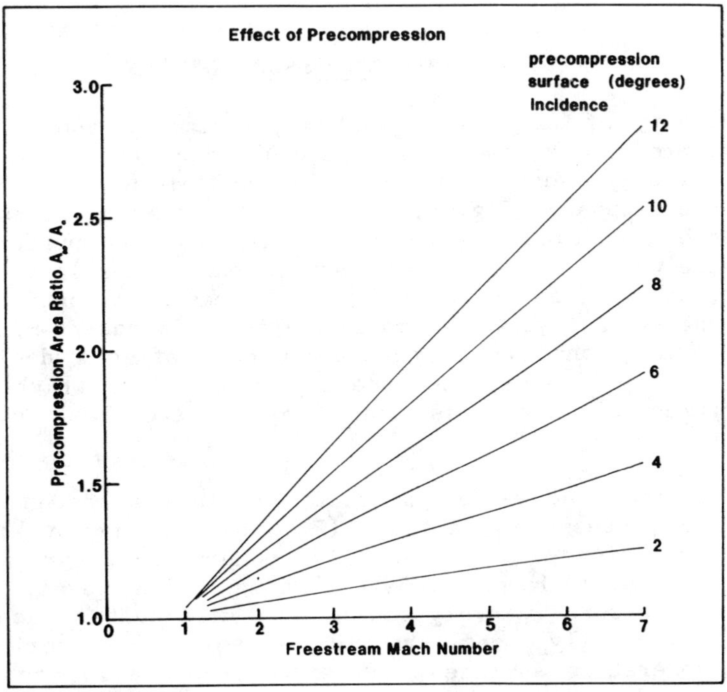

Fig. 2 Precompressive effect of undersurface incidence angle.

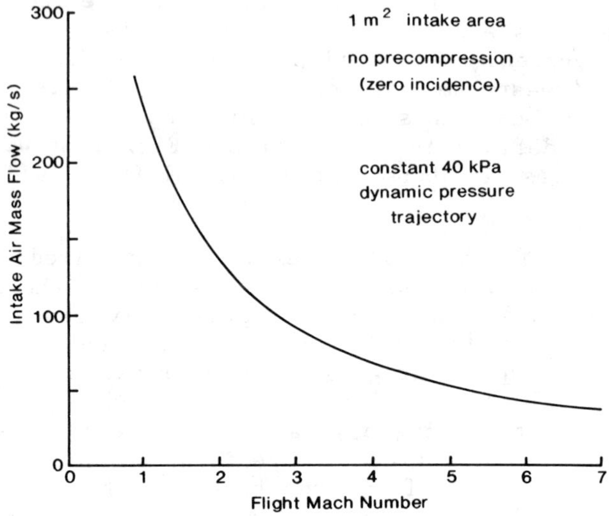

Fig. 3 Decline of captured air flow with increasing Mach number on a constant dynamic pressure trajectory.

III. AIR INTAKE PRESSURE RECOVERY

Sensitivity of propulsion system performance and mass to intake pressure recovery is dependent on cycle. The ramjet is obviously most sensitive in performance to intake pressure recovery (Fig. 4), as it lacks other mechanisms to increase pressure. There are limits to the performance increase which can be achieved by increasing nozzle pressure ratio, however, as shown in Fig.5, whereas mass continues to increase steadily. For mass critical launcher missions where ramjet and rocket powerplant mass effects dominate, low recovery intakes, the modest pressures from which allow structures of low mass, may produce the best overall performance.

The actual recovered pressure obviously depends on isentropic pressure available (Fig. 6) and intake kinetic energy efficiency. All-isentropic compression, turning the flow through a large number of compression waves, gives maximum pressure recovery but is very difficult to achieve. It is generally more practicable to use a series of shock-generating surfaces. Pressure recovery in each case then depends on the relative and absolute strengths of the shocks. Recovery increases with an increased number of shocks and matching of shock strengths (Fig. 7).

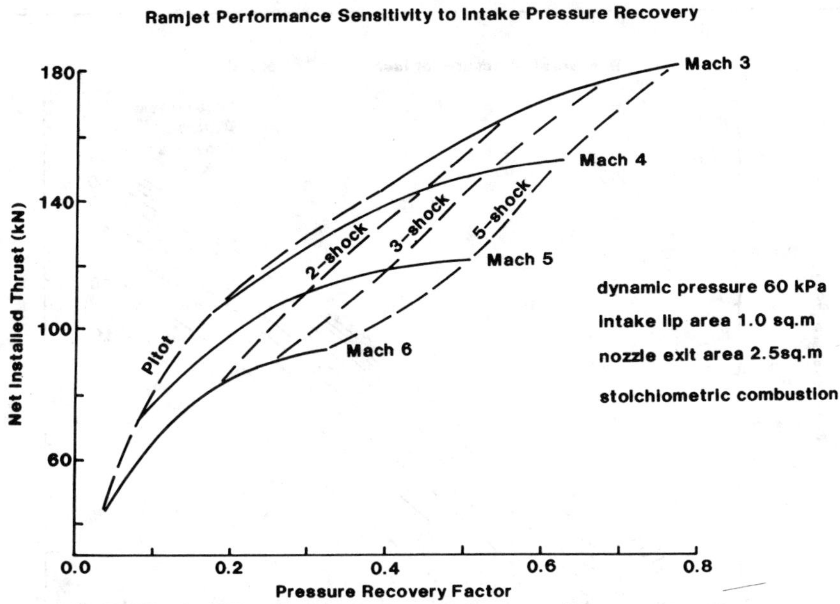

Fig. 4 Sensitivity of ramjet engine performance to intake pressure recovery factor showing the influence of increasing aerodynamic complexity.

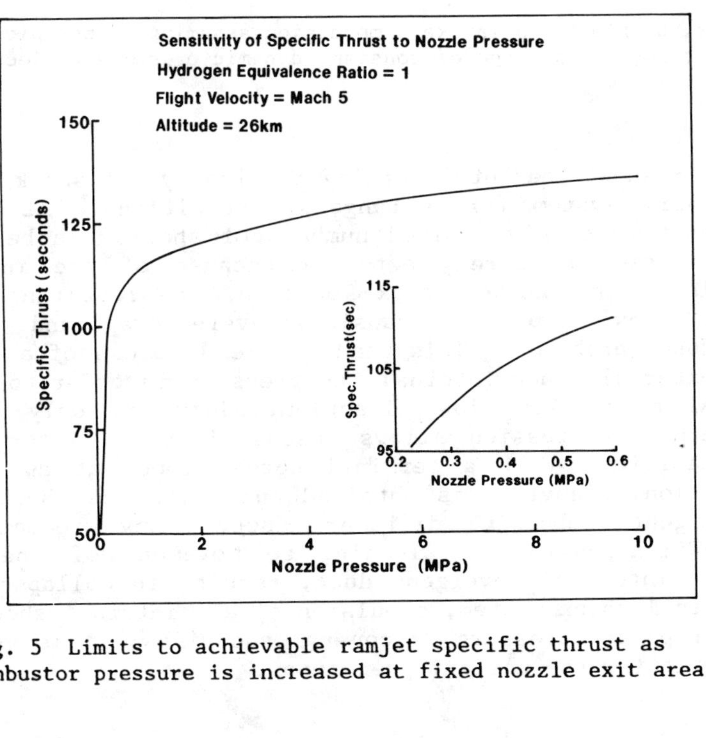

Fig. 5 Limits to achievable ramjet specific thrust as combustor pressure is increased at fixed nozzle exit area.

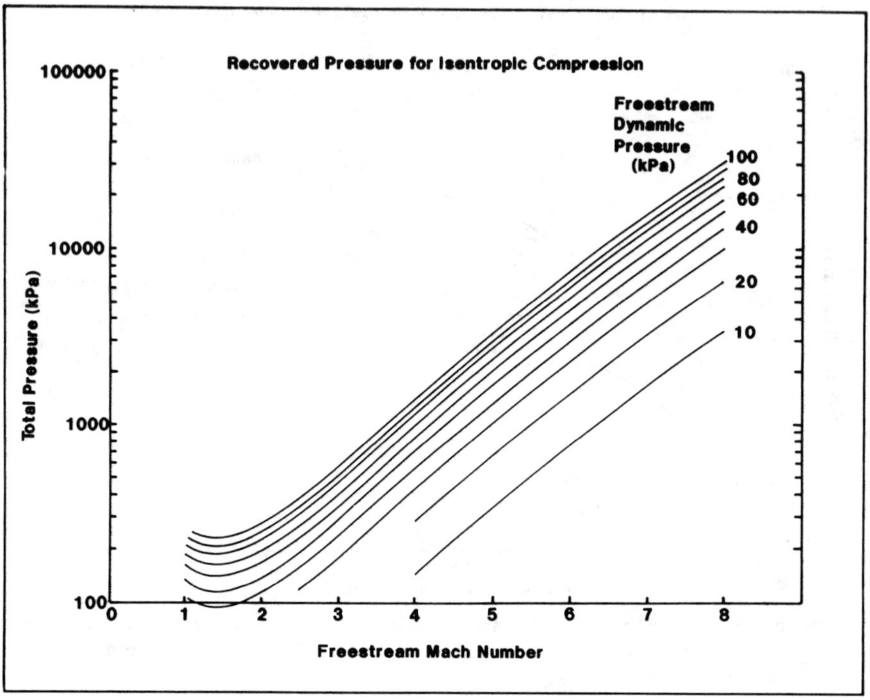

Fig. 6 Limits to intake compression assuming isentropic process, for a range of constant dynamic pressure trajectories.

Fig. 8 shows the total turning implied by the shock pressure recovery system for a range of conditions. It is then evident that only limited numbers of shocks can be used if compression is purely external because of the increasing angle at the normal shock location, the resulting "rolled over" cowl lip then causing severe drag and subsonic diffuser problems. This tends to lead to use of a sequence of external, then internal, compression in order to allow a final near-axial flow direction. However, only entirely external compression allows insensitivity to internal flow instability. With a terminal normal shock at an internal location, stability is obtained only with the shock in the divergent subsonic diffuser duct. Any reduction in non-dimensional flow, leading to movement of the normal shock into the convergent duct, results in collapse of the internal shock system, expulsion of all internal shocks, and a drop in pressure recovery and flow; this sequence constitutes the unstart phenomenon.

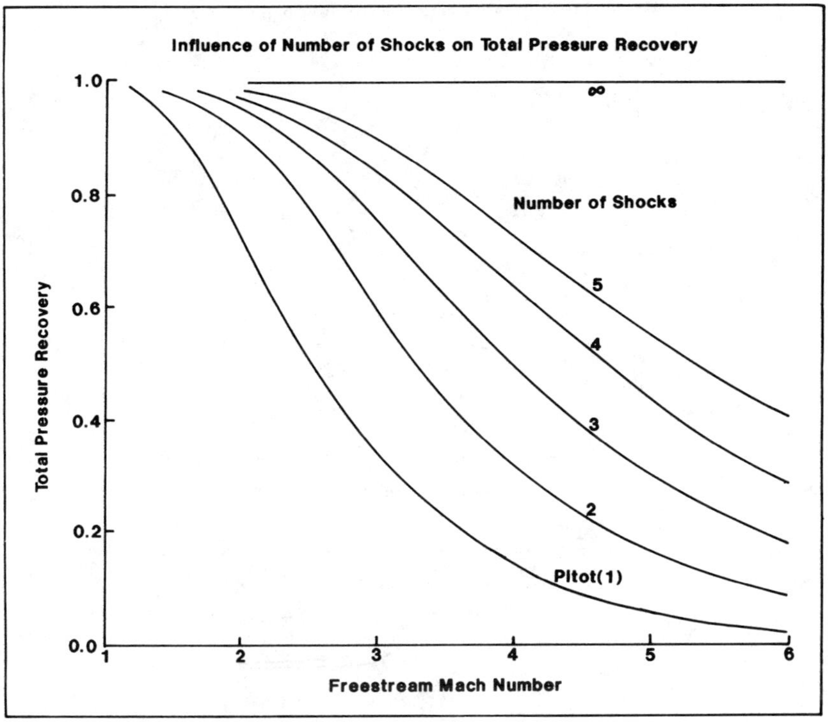

Fig. 7 Pressure recovery as a fraction of isentropic: sensitivity to shock multiplicity.

All intakes with controlled rather than dump subsonic diffusion require control of the boundary layer at the diffuser entry. An excessively deep boundary layer leads to flow separation in the subsonic diffuser. The boundary layer can be partially removed or cooled or can be re-energised by high pressure air co-flow injection. Removal is usually simplest; the benefits are shown in Fig. 9. Removal of boundary layer leads to a cycle penalty, but the air can be used productively, e.g. by nozzle cooling, and re-expanded to reduce losses. A typical bleed schedule is shown in Fig. 10, indicating the need for increasing bleed fraction as the difference in energy between boundary layer and mainstream increases.

The near constant lift/drag performance and lift requirement on the vehicle leads to a larger intake area requirement at high Mach to capture sufficient air than at low Mach. Ramjets for single-stage-to-orbit (SSTO) launchers may then require throttled operation below approximately Mach 3 due to low intake pressure ratio, leading to severe spillage

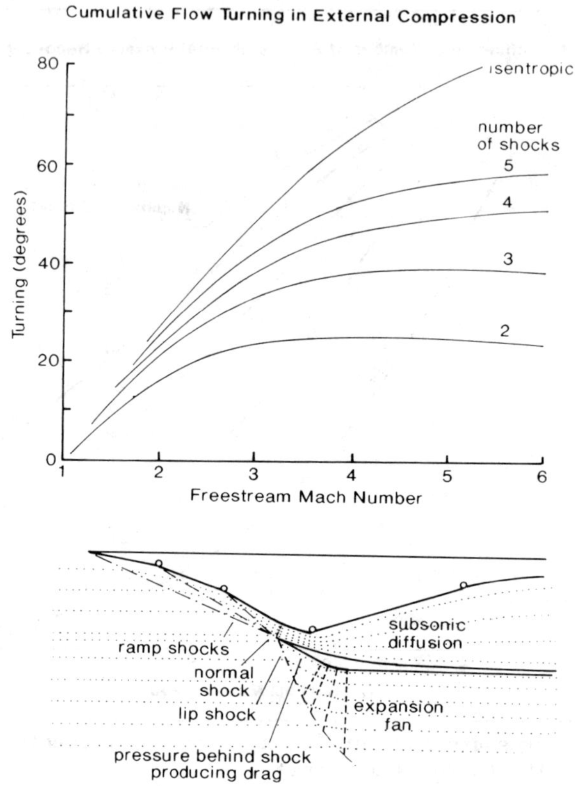

Fig. 8 Flow turning in all-external compression.

loss. Similarly, a turbomachine to give optimum overall performance at low Mach may require <0.5 of the intake air delivery capability. Excess air at low to intermediate Mach must either be swallowed and ejected or spilled/diverted before entry. Swallowing the excess air generally leads to minimum installed performance penalty, but also to a major mass and complexity increase, in ejector nozzle, bypass ducts, etc.

The two main design categories of high Mach intake to deliver subsonic flow are based on rectangular or circular cross-section. Rectangular designs generally use some combination of fixed - and variable-angle linked ramps and curved surfaces to provide a combination of shock and isentropic compression. Axisymmetric or partial axisymmetric designs typically use conical or half-cone surfaces, with relative axial translation of cone or cowl to control shock position and flow capture.

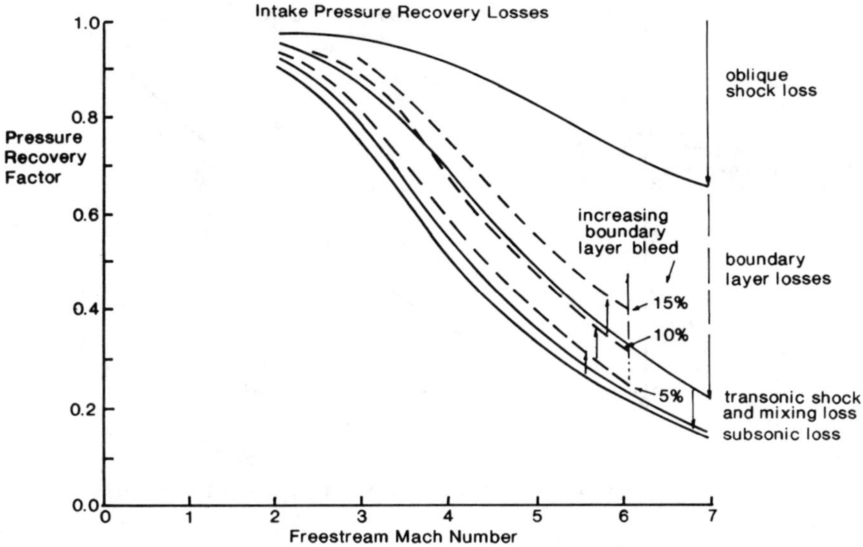

Fig. 9 Intake losses and the influence of boundary-layer bleed.

Collapsible cone structures to provide a variable cone angle have been designed and tested (Fig. 11) but tend to be heavy.

Variations in the basic aerodynamic design of the intake not only control the efficiency of the intake in pressure recovery terms but may also have a major effect on spillage or pre-entry drag. In general terms, intakes with a larger number of controlled-angle shocks can also reduce the penalty of spillage by reducing the shock compression work done before the air is diverted (Fig. 12).

Structural techniques and materials employed in intake designs can vary widely depending on, inter alia, maximum flight Mach number, duration of exposure to high temperature, cooling capacity available, and material capabilities. Up to about flight Mach 4, uncooled designs in titanium should be viable. Beyond this speed, the combination of the installation size and low strength/density of metals at high temperature requires either cooled structures or ceramic materials for prolonged flight. In the case of vehicle design for short periods at elevated Mach, it may well be possible to develop adequate designs in insulated low-temperature materials. Fig. 13 shows the results of a calculation of temperature rise in high Mach flight of a mineral-insulated metal structure.

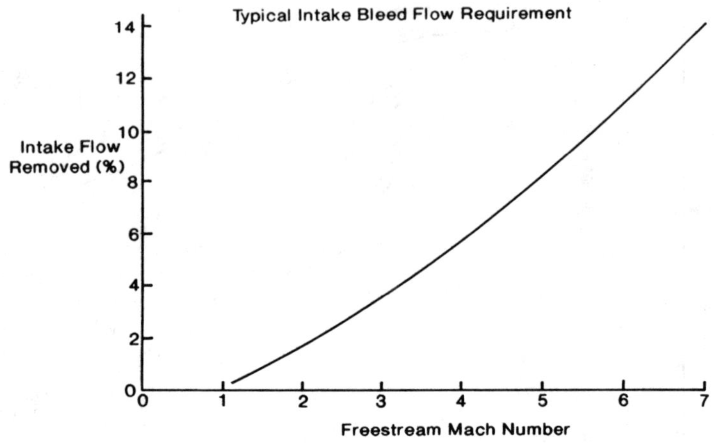

Fig. 10 A typical intake boundary layer bleed schedule.

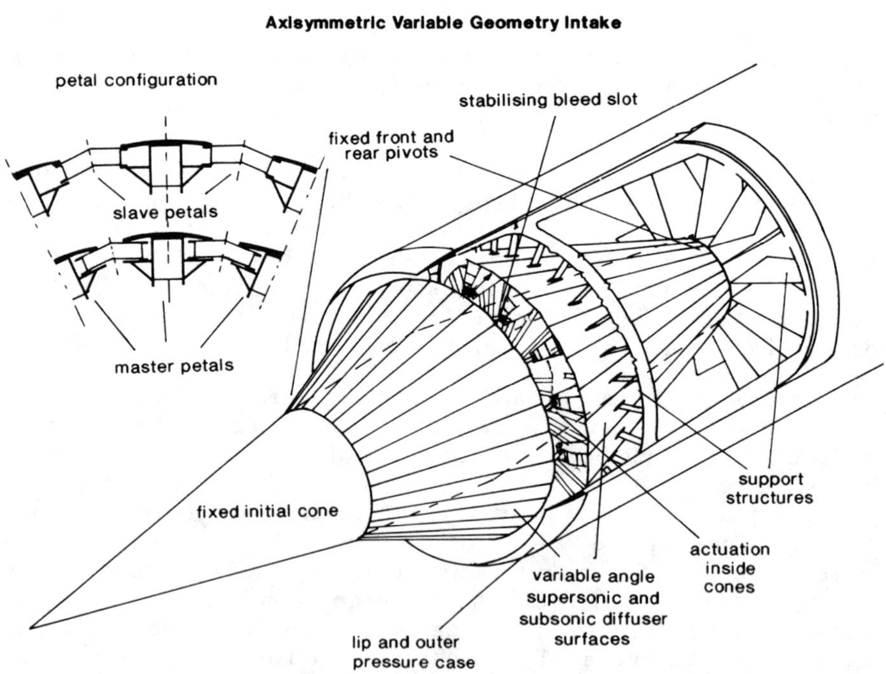

Fig. 11 A design of "collapsing spike" variable-geometry intake.

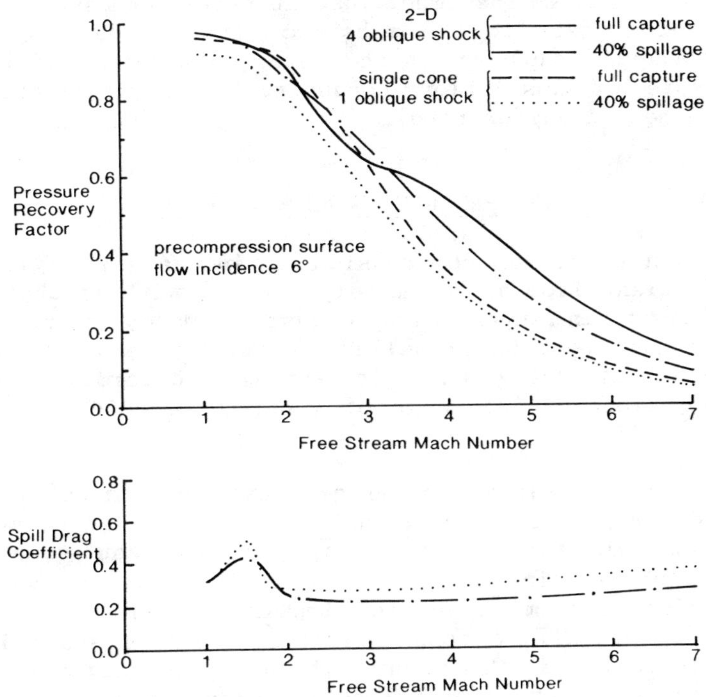

Fig. 12 Sensitivity of intake pressure recovery factor and spill drag to major intake geometry variation.

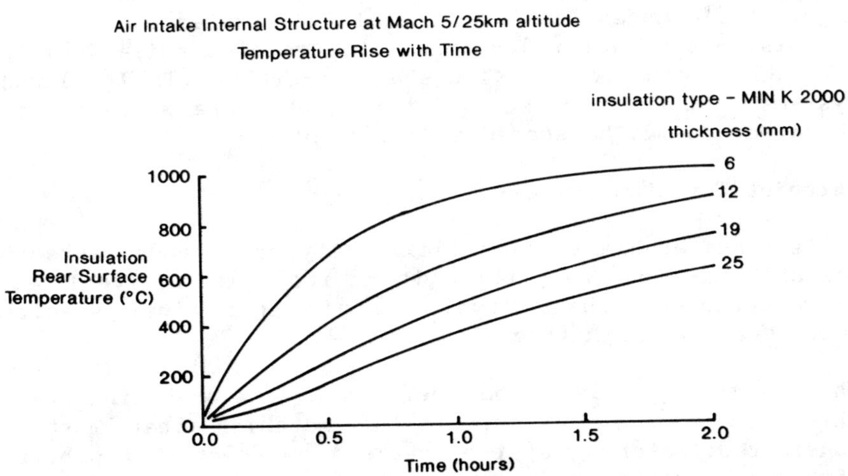

Fig. 13 Control of intake structure temperature rise using surface insulation.

(This calculation assumed instantaneous exposure of the insulated surface to the subsonic diffuser delivery flow, and no thermal capacity in the metal substrate, and is thus a measure of insulation conductivity and characteristic surface heat transfer rate).

IV. PROPULSION FROM TAKE OFF

Propulsion using ram compression of intake air is feasible and efficient from approximately Mach 2, with (probably) no upper limit in supersonic combustion mode; however, the vehicle must then be propelled to ramjet speed. Choice of the means of doing this is strongly determined by the mission, since a wide range of adequate technical solutions exist.

Mission inputs which can be critical determinants of the low-Mach propulsion system include the range which must be flown at this flight condition, and any requirements for loiter, pre-landing hold, or go-around. These, with the propulsion system specific impulse, will then define potential propellant consumption in these phases. For an SSTO mission, the vehicle definition is severely constrained by achievable overall propellant impulse and minimum vehicle structure, to the point where achieving any useful payload becomes the dominant problem with even a minimal (takeoff, climb/accel, orbit, descend, land) mission, and propulsion system mass tends to become more limiting than fuel burn during the brief low Mach flight phases. There may thus be major differences in the resulting preferred low-Mach propulsion system for long range hypersonic cruise vehicles and SSTO launchers. Two-stage-to-orbit (TSTO) launch systems form an interesting intermediate case where mission requirements may be shared in different ways.

Turbojet/Turbofan Systems

Where considerable mission flexibility or extended subsonic operation is required, there is no real alternative to use of a turbojet/turbofan cycle to give good fuel specific impulse at low Mach (Fig. 14).

Where a turbojet is to be operated over a wide Mach range there are two major aspects of matching that must be considered: matching of the intake flow delivery capability with the maximum engine flow capability, and matching of vehicle drag and engine thrust profile. Severe mismatch of the engine and intake is more or less inevitable at low Mach

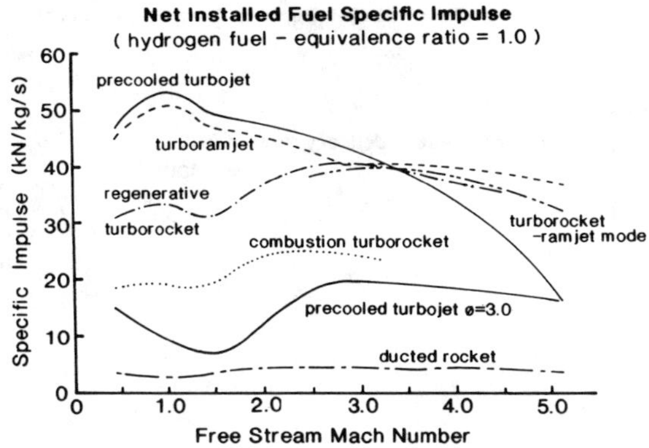

Fig. 14 Installed specific impulse variation of a range of propulsion cycles.

number. This is due to the large intake area required to provide the necessary high Mach airflow and to the inefficiently large mass and unusable high performance of the turbomachinery required to swallow the airflow which the large intake can provide in the transonic flight regime. The main sizing constraint on the low Mach turbomachinery in several recent studies, has been found to be in the regime around changeover to ram operation near Mach 3. The desirability of operation to this speed to minimize ramjet burner and nozzle throat areas has already been noted, but there are consequences for the turbomachinery. The rising recovered temperature leads to falling non-dimensional speed (and this effect may be made worse by a need to reduce rotationally induced stress as the compressor delivery and hence turbine cooling air temperature rises). The effect of this is to reduce flow and compression ratio as flight Mach rises (or operate in an artificially derated condition at lower speed, thus not using the full potential of the machinery). The overall effect is thus loss of thrust in a flight regime where drag may still be high, with a resultant overall penalty in fuel burn, and low usage of the available airflow.

The value and distribution of the mismatch that results, then depends on the design pressure ratio of the engine. Fig. 15 shows the effect of this parameter on the mismatch, where the area ratio is relative to full power engine demand at fixed turbine entry temperature. This should be compared with Fig. 16, where an on-trajectory comparison of available thrust is shown. Clearly, a compromise solution is necessary, based on calculation of the cost of excess flow

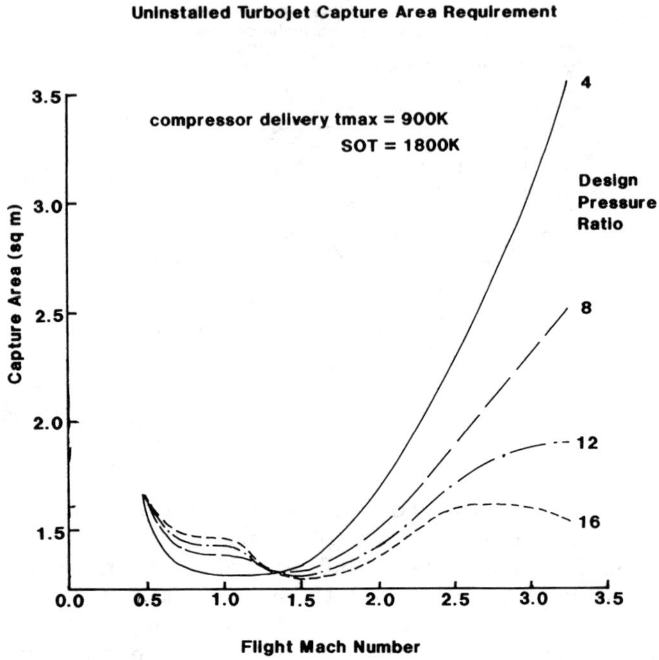

Fig. 15 Variation of turbojet capture area/Mach number profile with design pressure ratio.

in either spillage drag or additional engine or bypass mass to handle the excess flow.

Turborocket (ATR) Systems

Where reduced flexibility, but still some sustained subsonic/low supersonic performance, are required, and where turbojet solutions are excessively heavy, one of the turborocket (often known as Air Turboramjet (ATR)) variants (Fig. 17) may be more suitable, especially in the ability to provide high thrust in the supersonic regime around Mach 3, where the drags of complex launch vehicles can be high.

This class of engine has the advantage that it can be operated to higher Mach numbers than a conventional turbojet/turbofan. Compressor pressure ratio for a given nozzle pressure ratio is much lower than an equivalent turbojet. At sea level static (SLS), compressor pressure ratio is only slightly higher than the nozzle pressure ratio, whereas a typical turbojet might have a nozzle pressure ratio of 3-4 associated with a compressor pressure

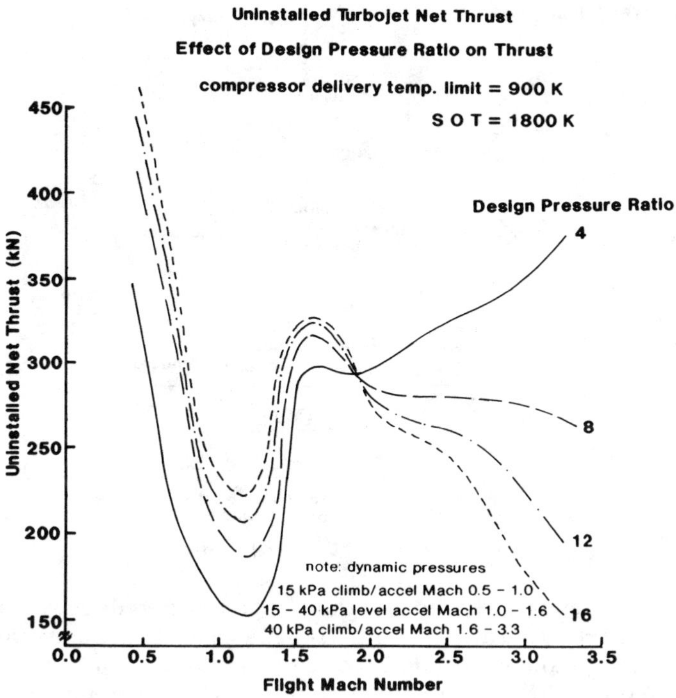

Fig. 16 Variation of turbojet thrust/Mach number profile with design pressure ratio.

ratio of 10-15. Thus compressor delivery temperature will be much lower for a turborocket than for a turbojet. The lower pressure ratio also means that there is sufficient area to pass all the engine flow at high Mach number without the turbomachinery operating, therefore there is no need to supply a separate duct (as is the case with a turbojet), so long as the material of the compressor is adequate to withstand stagnation temperatures. Since the source of the turbine working fluid is independent of the mainstream flow, it is not directly affected by increases in stagnation temperature. Turbine power can be increased to match the compressor requirement by increasing the mass flow rather than the temperature.

Where the turborocket is fueled by hydrogen heated either regeneratively or by partial combustion there is an inherent mismatch between optimum compressor speeds and optimum turbine speeds, which results partly from the high sound velocity and heat capacity of hydrogen. A gearbox between

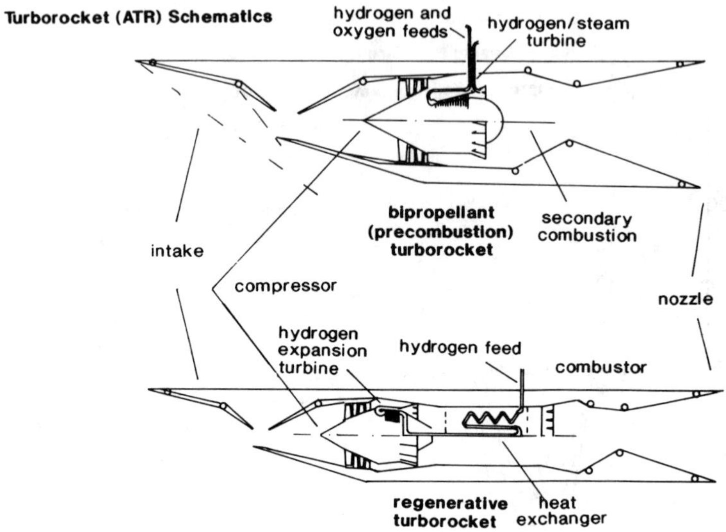

Fig. 17 Turborocket (ATR) concepts.

the turbine and compressor has been considered, but the technical risk at large power levels and the resultant high mass of the gearbox tended to eliminate the advantage over the turbojet. The alternative of a direct coupled turbine was found to be a better solution, even though many stages are required because of the large turbine specific work resulting from the small turbine flow implied by the (preferred) stoichiometric overall fuel-air ratio in the reheat combustor.

Use of stored oxidant in the partial combustion turborocket leads to low specific impulse and reduces the utility of this mode for prolonged low-Mach operation, although the lightness and compactness of the unit can make it attractive where only short periods of operation are required. The regenerative unit would be expected to provide higher specific impulse, but the range of throttled operation is small and maximum efficiency lower than a gas turbine because of the low overall pressure ratio in the air side of the cycle. Again, a mission stressing high power high-Mach operation can favour such an approach.

The logical limit to cycles of this type is shown in Fig.18, in which the variable intake and nozzle are combined with a ducted rocket engine. This configuration can provide very high takeoff and climb performance, but mechanical problems due to mixing of the high-energy stream in the duct would be formidable. This form of ejector mixing is also an

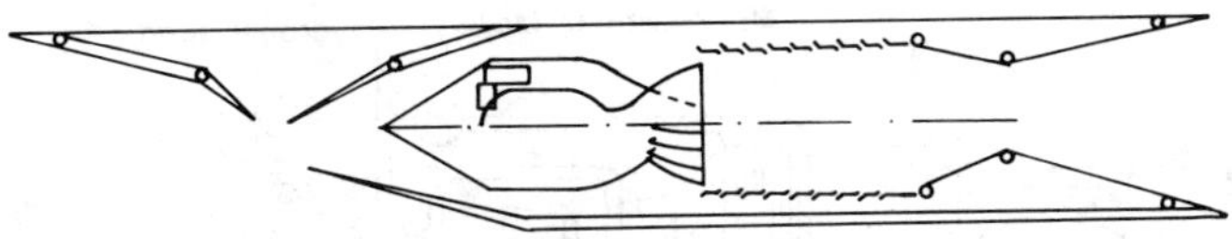

Fig. 18 Integrated ram-rocket

inefficient means for producing pressure rise in the air stream and overall thermodynamic performance is not very good.

Air Heat Exchange Cycles

Use of the cooling potential of the cryogenic fuels, the most effective fuel in this sense being hydrogen, can lead to major rearrangements of engine cycles for wide Mach operation. The most obvious application lies in reducing the temperature of inlet air, although the second feature of hydrogen, the high specific heat, and hence work, is of nearly equal importance. Consider first the case of a precooled simple turbojet cycle: if the final fuel-air ratio overall is stoichiometric, then the potential exists for cooling the inlet air ~300 K with fuel available at ~50 K. This is roughly equivalent to reducing Mach 4 entry temperatures (~900 K) to Mach 3 values (~600 K).

There are several ways of operating a precooled engine. For example, the engine can be operated without heat exchange at low altitude and Mach number, with pre-cooling progressively increased with rising Mach number above the tropopause, where humidity is low and icing is no longer a problem. This permits a lower rate of rise of compressor entry temperature than the stagnation temperature, allowing operation to higher Mach number. If this principle is applied to a turbojet of conventional design e.g. 16:1 pressure ratio, 2000 K (3600°R) TET, with stoichiometric overall fuelling (including afterburning) and all the fuel eventually passing through a high effectiveness heat exchanger, the operating limit rises from Mach 3 to Mach 4. If the mission of the vehicle can accept the increased fuel burn, it is also possible to fly faster still, with overfuelling. For example, at Mach 5, two times stoichiometric fuel flow is required to maintain Mach 3 engine entry temperature. The low molecular weight of the additional exhaust gas (with hydrogen fuel) offsets the reduced nozzle entry temperature

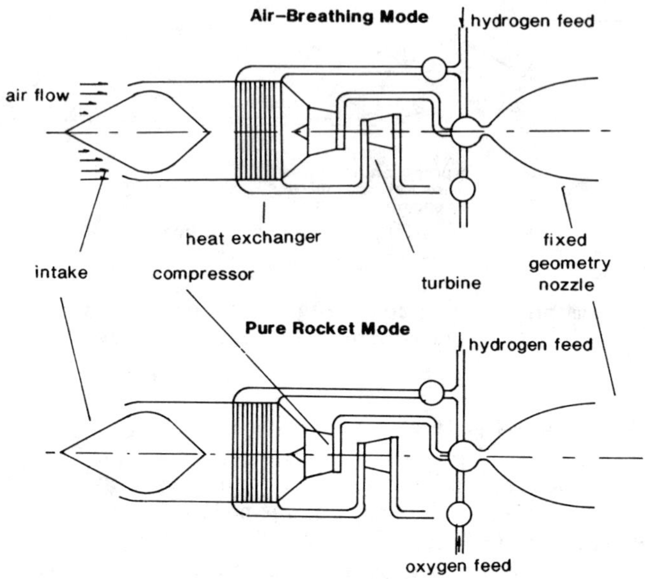

Fig. 19 Air-breathing rocket engine.

with a net contribution to thrust. Such a system could provide single mode propulsion from Mach 0 to Mach 5.

There is a secondary effect of overfueling, that also influences choice of nozzle design; this is the resultant increase of nozzle area required relative to the stoichiometric condition. For an engine propelling a vehicle on a trajectory of constant dynamic pressure, with a moderately efficient intake, nozzle throat area may be required to vary over a range of greater than 5 to one between take off and Mach 5, with the minimum area at Mach 5. Overfueling can bring the minimum area back within the area range achievable with a conventional military petalled nozzle.

Several of these propulsion systems are capable, with appropriate materials, of operation to flight Mach 4-5, permitting overall propulsion concepts for launch vehicles, in which an initial propulsion system provides thrust to rocket takeover, without a pure ramjet phase. Use of precooling with cryogenic fuel may permit both operation to higher Mach number than otherwise possible and definition of dual-mode air-breathing and rocket engines with single combustion chambers operating at high pressure in both modes (Fig.19).

Clearly the process of overfueling deserves some further examination. There are two major aspects here, design and performance. By application of a sufficiently large fuel-air ratio, the air can be liquefied in the heat exchanger, allowing efficient compression (pumping) of the air to very high pressure - forming the basis of the Liquefied Air Cycle Engine (LACE). The fuel-air ratio must however be greater than four and leads to specific impulse below 1000s unless combined with a ramjet to use the excess fuel. If the air is simply very heavily precooled, many of the advantages can be obtained at equivalence ratio ~3. With an air delivery temperature below 100K for a pressure loss of perhaps 30%, the streamtube contracts dramatically, and for given rotational Mach numbers, absolute compressor rotational velocity is very low, leading to low diameter with low rotational stress levels and light construction. As much of such a heat exchanger can be fabricated in low-temperature light alloy, the large mass savings on the small light engine can provide for a large heat exchanger before the savings are consumed. Indeed overall mass for such a system can be very attractive, particularly where high pressure ratios have been employed, minimising combustion and nozzle area requirements. Typical precooled turbojet core engine mass savings for an airflow are shown in Table 1 for progressive reduction of maximum entry temperature. Further mass savings are possible if advantage is taken of the very low combustion entry temperature to reduce turbine entry temperature to a level at which uncooled operation e.g. with ceramic blading, is possible. A comparison of the overall masses of several partially-optimised engine systems for a launcher application shows the major advantage to be derived from

Table 1 Sensitivity of precooled turbojet mass to heat exchanger exit air temperature at fixed airflow.

1800K stator outlet temperature
20:1 overall pressure ratio
315 kg/s air mass flow

Temperature K	Engine mass kg
250	3680
150	2500
80	1660

such a precooled system, Table 2, although the associated Fig. 20 indicates the complexity of the necessary vehicle based comparative study.

In principle, the use of cooling can also be applied in an inverse turbojet cycle (Fig. 21). Here the airflow is initially expanded through a turbine, cooled by heat exchange before compression, heated by combustion, and expanded through a nozzle. However, the use of the heat exchangers in the expanded flow leads to a large heat-exchange cross-sectional area and large compressor diameter, the effect of which, especially when considered in conjunction with the additional preburners needed to achieve adequate low Mach performance, is to reduce the advantage relative to the simple precooled turbojet.

Equally, a wide range of composite heat exchange cycles is possible, e.g. Fig. 22. However, the full analysis and

Table 2 A comparison of the mass of SSTO powerplant airbreathing components[a] (All masses in kg)

Component \ System	gas turbine-turboramjet	turborocket-turboramjet	precooled-turbojet	rocket-ramjet
Intake	980	980	890	980
Heat exchange	-	-	620	-
Turbomachine	3670	2970	320	-
Bypass duct/ reheat	940	940	900	70
Nozzle	1170	1170	220	1170
Rocket[b]	-	-	-	1050
Total	6760	6060	2950	3270

a all masses relate to performance shown in fig.20

b the rocket of the rocket-ramjet also provides the thrust from the end of airbreathing at ~Mach 6 to orbit. A similar mass must be assigned to each of the first three systems in reaching an overall propulsion system comparison design tends to result in such systems being very bulky and heavy for a given thrust.

Furthermore, definition of control systems for powerplant of this degree of complexity poses serious problems in obtaining stable operation.

Combination Cycles

It is possible to resolve the engine/intake mismatch, as already discussed, by provision of bypass ducts, ejector nozzles, etc., but it is also possible to make more effective use of the excess air when the maximum possible thrust is required.

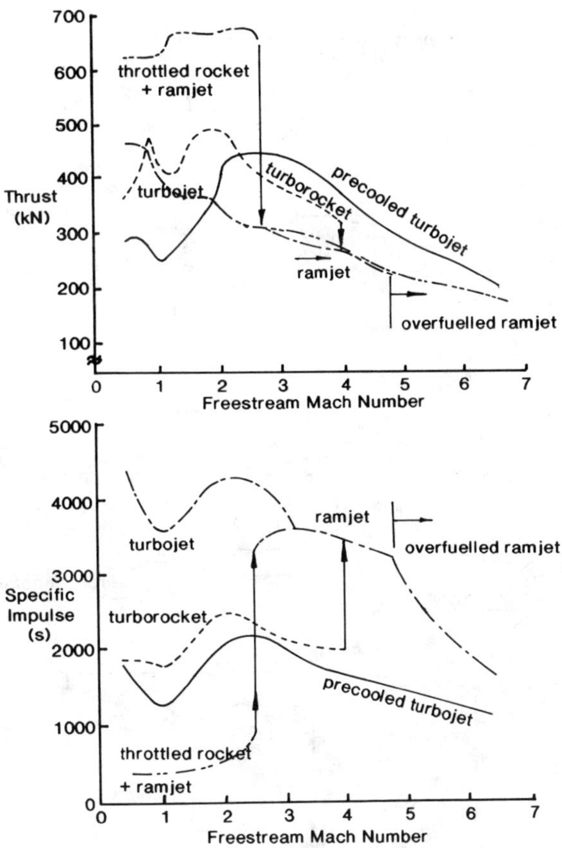

Fig. 20 Full-power performance comparison on a constant dynamic pressure trajectory (600 knots E.A.S. kPa).

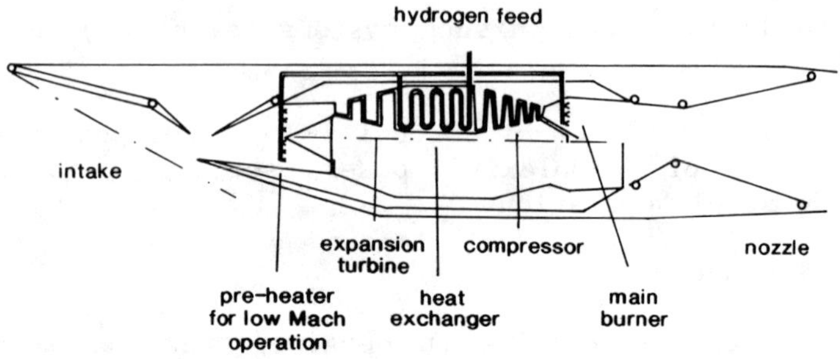

Fig. 21 Inverted turbojet heat-exchange cycle.

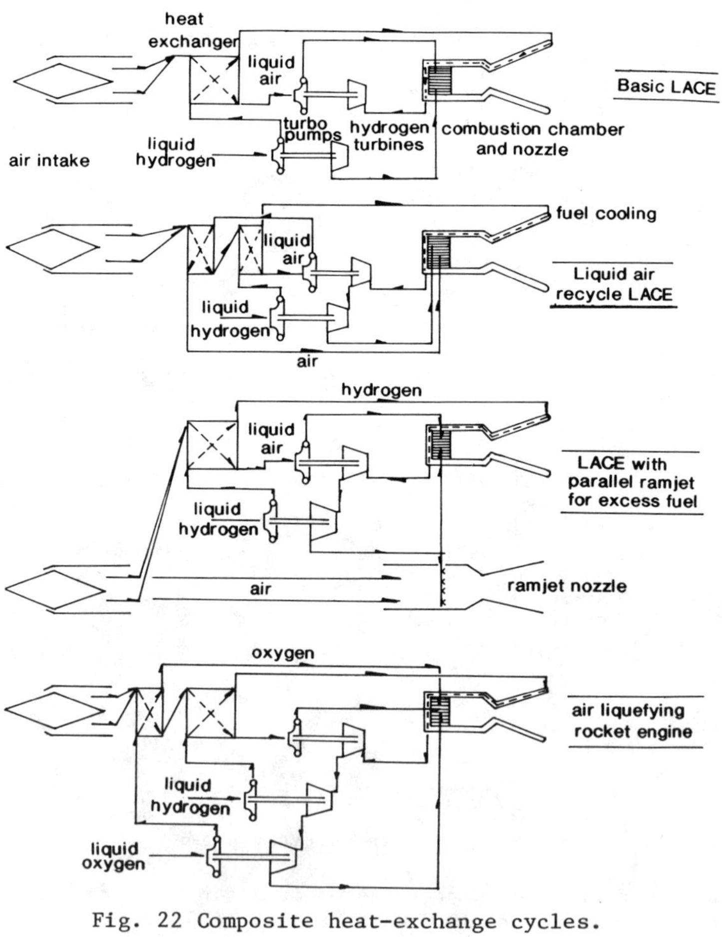

Fig. 22 Composite heat-exchange cycles.

This involves use of the additional air in a "parallel" cycle, typically of ramjet format. A particular example of this type of system was recently considered by Rolls Royce, for propulsion of an SSTO launcher. The system utilized a heavily precooled turbojet, operating at an overall equivalence ratio of 3 with a parallel subsonic combustion ramjet cycle. The ramjet was constrained to a nominal cross-section and nozzle area which limited the airflow at stoichiometric fueling. Optimization of a system of this complexity in conjunction with the vehicle is an immensely complex task: there is no reason, a priori, to suppose that the ramjet should be sized to swallow all excess air for equivalence ratio 1 combustion. Even considering only the propulsion system, a wide range of relative scales of the ramjet and precooled turbomachine is possible, and a similarly wide range of achievable thrust/weight and specific impulse vs. velocity. It is equally possible to investigate variation in cycle and scale within each powerplant: the influence of nozzle area and area variability on ramjet mass is very large. The gains in performance that are the goal of this activity are demonstrated by Fig. 23, in which the achieved fractional thrust increment in the typically high-drag Mach 1-3 regime is impressive (precooled engine and intake are matched at

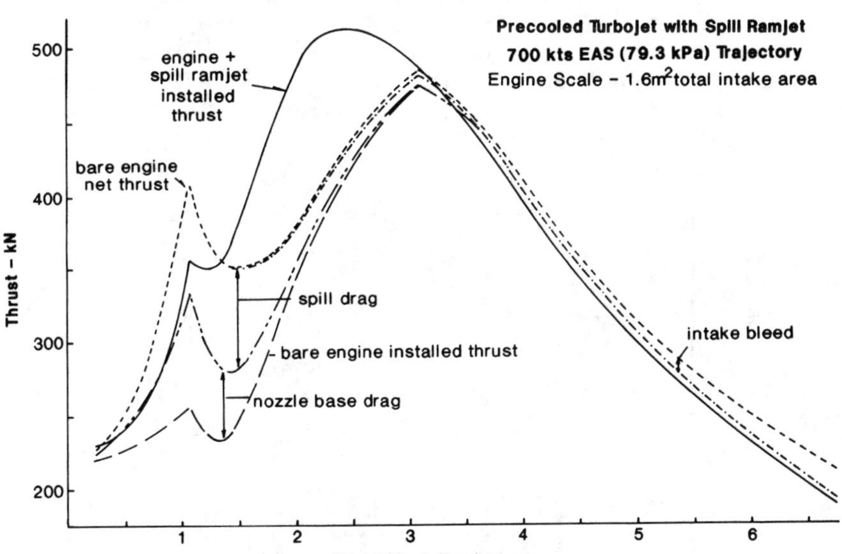

Fig. 23 Summary of installation loss variation with Mach Number for a precooled turbojet and improvement using a spill-air ramjet.

Mach 3, beyond which the engine has to throttle back progressively).

It will often be possible to integrate these "bypass ramjet" systems with boundary-layer removal systems for the airframe/intake interface, making a modest virtue out of the necessity to feed a clean flow to the intake if it is to operate effectively at high Mach number. In this context it is worth noting that one advantage of having precooled cycles operating to high Mach number is that the options for stabilising intake flows are increased, it being possible to use compressor bleed to re-energize the intake boundary layer by co-flow injection.

V. BURNER AND NOZZLE RELATIVE SCALE EFFECTS

As the flight velocity increases beyond about Mach 3, the net specific thrust of all subsonic-combustion systems tends to reduce, due to rising ram drag at the near-constant exit velocity which results from constructional pressure limits and combustion temperature limits. The smaller margin between gross thrust and ram drag then leads to exhaust losses becoming increasingly important. A typical distribution of such losses is shown in Fig. 24. The implication is of a need to obtain most efficient nozzle performance at high Mach, implying large area ratio. Typically, however, this improvement is by no means linear, and the overall effect is shown in Fig. 25. This plot does not include friction loss, inclusion of which further

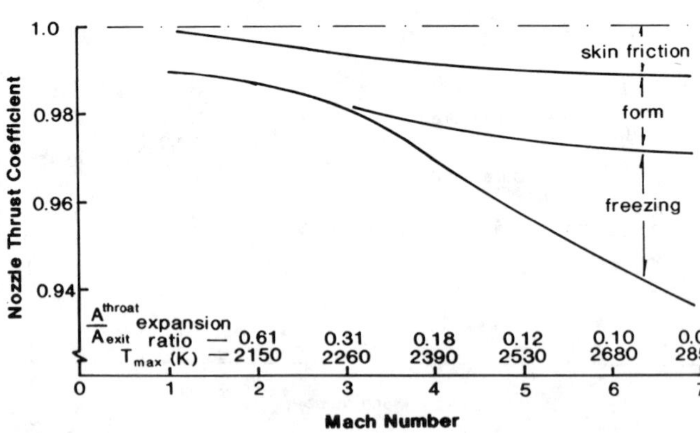

Fig. 24 Relative variation of contributory elements to overall nozzle loss vs flight Mach number.

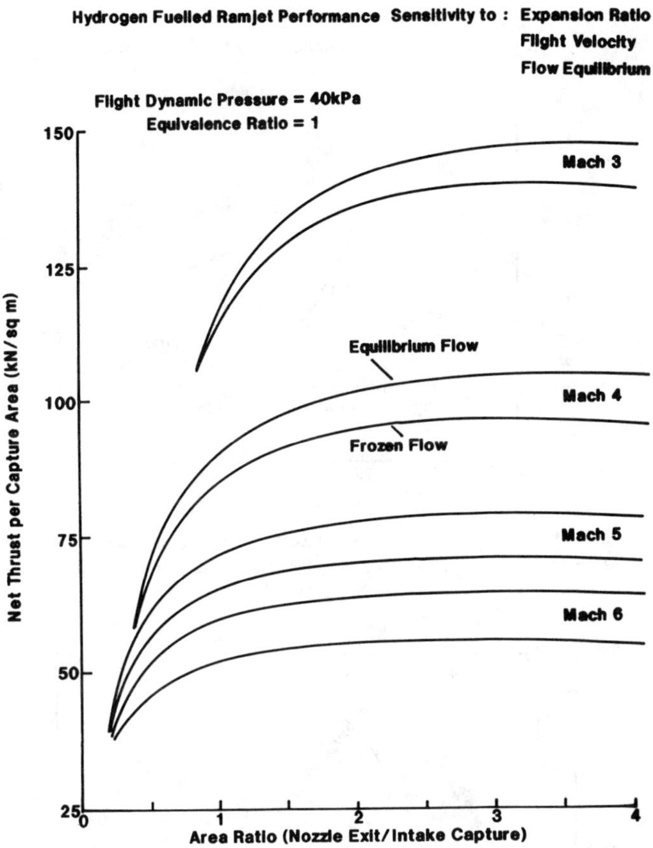

Fig. 25 Ramjet thrust as a function of nozzle to intake capture area ratio.

reduces the advantage of long nozzles with large exit areas. The capture area shown will be larger than the actual inlet by the effect of precompression shown in Fig. 2. The relative scale of nozzle to inlet is thus somewhat greater than shown on the horizontal axis of Fig. 25, typically by ~1.5 at incidence values of ~5 degrees at Mach 5-6. It is thus clear that nozzle exit areas two to three times physical inlet areas are likely to be desirable at this speed.

Nozzle throat and burner/duct areas are obviously a strong function of the pressure ratio of the system; for a subsonic combustion ramjet, where only the intake is available to provide compression, the effect is shown in Fig. 26. From

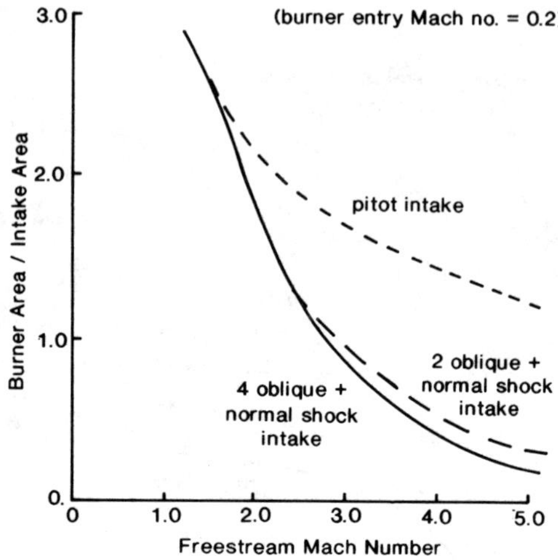

Fig. 26 Combustor duct cross-sectional area required relative to intake area - variation with flight Mach number and intake design.

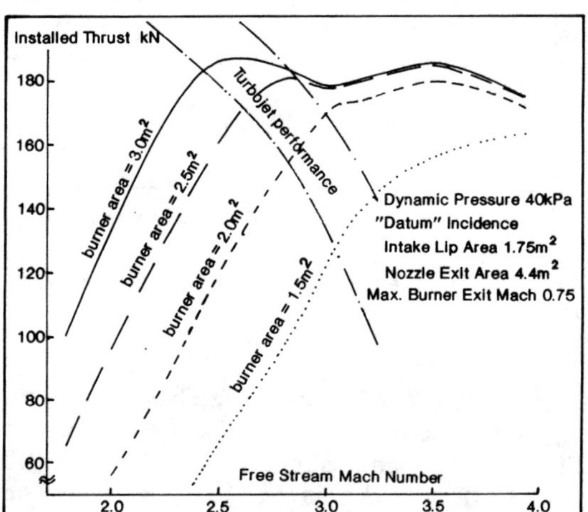

Fig. 27 Sensitivity of ramjet thrust at lower Mach numbers to burner cross-sectional area.

this figure it is clear that, with reasonable compression intakes, burner areas and hence nozzle throat areas are consistent with intake areas beyond ~Mach 2.5, allowing efficient packing of multiple units provided that the accelerating unit does not cause excessive cross-section area blockage. The effect of combustion cross section area on ramjet performance is shown in rather more detail in Fig. 27, where the effect of inadequate area in reducing both combustion flow and thrust is very evident.

If it is found possible to pack accelerator and burner assemblies behind the intake, then it may be possible to achieve efficient side-by-side packing of multiple engine modules under the airframe wing/body compression surface, using an asymmetric final expansion area. A problem that will require prior resolution is the mechanical means of providing the necessary range of nozzle throat area variability.

VI. NOZZLE AREA VARIABILITY

The combination of high mass flows at low Mach, and low mass flows at high Mach, in conjunction with combustion pressures which can often be considerably higher at high Mach as a result of intake compression, leads to a requirement for a large nozzle throat area at low Mach and a low area at high Mach. Compared with typical military engines, with a throat area range of typically ~2:1, the high-Mach unit may require up to 10:1.

Convergent-Divergent Petalled Nozzles and Plug Variants

A throat area variation of up to 10:1 is far beyond the capability of typical overlapping - petal axisymmetric nozzles, unless a plug is inserted at the convergence, (Fig. 28), in which case an increase in total cross-sectional area at this plane is required. It is possible in principle to use three layers of petals rather than two, but the area range is still marginal, and the mechanical complexity, mass and frictional losses may become excessive. With a plug in the nozzle, the cooling problems are clearly greater and the overall cross section may present an increased installational problem. However, a study of variable throat area plug nozzles found that the overall performance of the "plugged" petal-type nozzle was superior to that of a plug nozzle with translating cowl area variation.

Two-Dimensional Rectangular Ramp Nozzles

In order to achieve the necessary area ratio range, the alternative is a rectangular nozzle with moving flat ramps, working between fixed sidewalls. Fig. 29 shows an ejector variant of this type of nozzle. It is possible to vary either throat and exit areas, or only the throat. The main

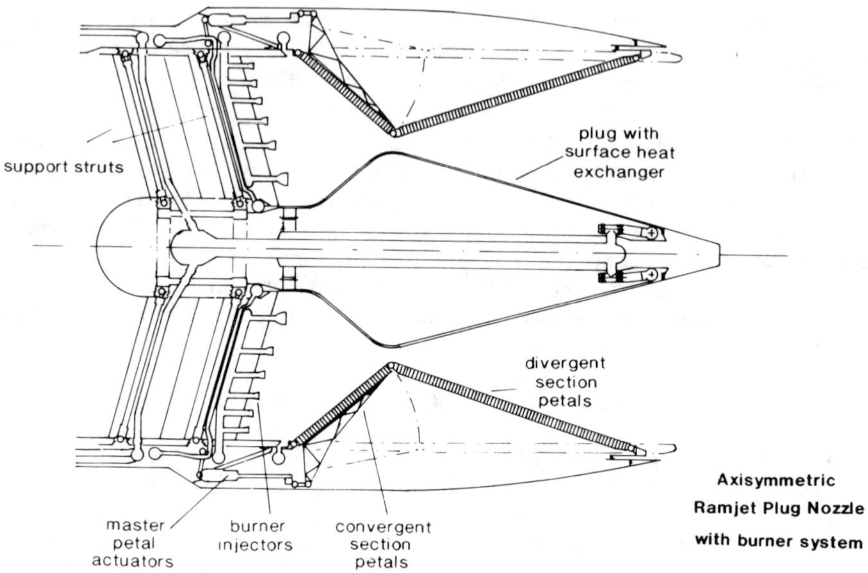

Fig. 28 Axisymmetric convergent-divergent plug nozzle.

problem is the length of the ramps required, to reduce divergence angles and hence losses while achieving large area ratios. Integrated pressure loads on the ramps and hence on the actuators can then be very large, with high ramp and sidewall masses tending to result from the need to meet minimum stiffness criteria in achieving adequate edge seals at sidewalls. It is clearly not possible to pressure-balance these ramps by pressurising the cavity behind the ramps, as can be done for the intake ramps, because the static pressure varies continuously in the flow direction. If the alternative of force-balancing the ramps is adopted, then there will be a possibility of leakage into the cavity from regions in the hot flow at higher pressure. Where the ramps are cooled with air and there is a possibility of overfuelled combustion and, hence, fuel-rich exhaust gas, there is an obvious explosion hazard to be considered. The large throat area range is accompanied by a considerable range in the desirable nozzle exit area.

Nozzle Exit Area

Even where an ideal exit area is not attempted for the high Mach case, our prior discussion has concluded that an exit area at least twice that of the intake is desirable at high Mach to achieve adequate net performance. At low and

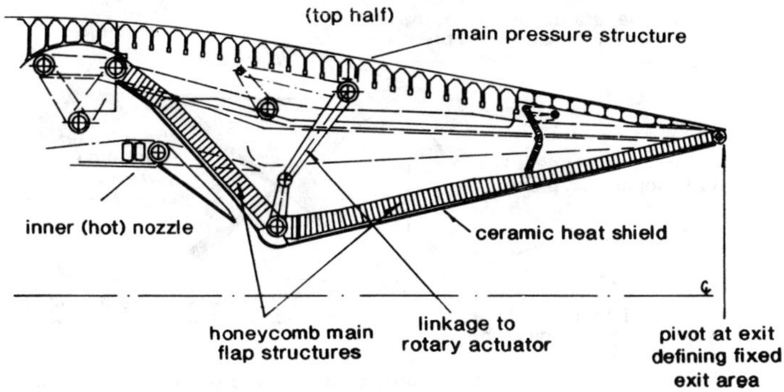

Fig. 29 Convergent-divergent ejector nozzle of rectangular cross-section.

transonic flight conditions, however, the ideal exit area for matched ambient and nozzle static pressure will probably be considerably below the intake area. Varying the exit area to this extent would obviously be possible in principle, but it is not the only solution.

The nozzle exit area could be fixed at the large high Mach value. This allows relatively simple design and operation of all of the nozzle types, but leads to overexpansion and possibly internal flow separation at low altitudes and low to transonic flight velocity. The resultant performance deficit calculated as in Fig. 30, may be worse than the combined effect of a fully variable nozzle with base drag on the resultant unfilled base area at transonic conditions. The overall vehicle performance penalty may, however, be small, or offset entirely by mass savings in the simpler system, particularly in accelerating roles where the vehicle spends little time at flight velocities where severe base drag or overexpansion losses occur.

The nozzle could be ventilated to atmosphere to cause separation at better-matched conditions. This could allow performance close to that of the fully variable system, but at considerable mass saving.

The nozzle film cooling system could be used to induce flow separation.

Operation with separated internal flow may cause problems with flow stability and nozzle area control. This will probably require some further investigation, although limited available data in this area suggests that the problem may not be serious in nozzles in which area changes rapidly with length.

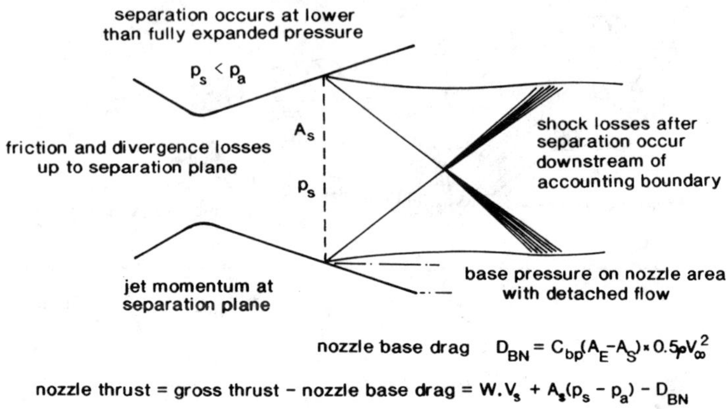

Fig. 30 Force accounting in nozzles with flow separation.

Single Expansion Ramp Nozzles

One approach that could help reduce nozzle mass is to use fixed vehicle surfaces for some part of the nozzle expansion. The most common approach has been to assume that the "engine" nozzle would have a variable area throat, the main part of any further expansion taking place against a curved ramp forming the rear undersurface of the aircraft fuselage. Clearly this surface would require insulation, and possibly cooling, although considerable radiation cooling would occur. The major performance challenge in obtaining the best results from such a system in transonic flight would be to persuade the low volume flow, which will not fill the vehicle base, to remain attached to the ramp and the external flow to turn smoothly with the nozzle flow. If this is not achieved, then base drag losses will still be high.

At high Mach numbers, there is a further consideration. Typically, on hypersonic vehicles, the center of lift moves forward as Mach number rises. This is due in large part to an increase in required incidence for lift. This tends to lead to a strong nose-up pitch moment, which must be trimmed out either by centre of gravity shift (e.g., fuel transfer or selective fuel tank usage) or by aerodynamic control adjustment. Use of the SERN half-nozzle when running full tends to counter this moment, requiring less c.g. or aerodynamic control, but the coupling between engine thrust and trim then becomes very powerful, further complicating control system operation. At lower Mach numbers, with flow overexpansion, flow attachment to, or detachment from, the expansion surface could be a source of serious pitch instability. Clearly, these points and associated control considerations will be relevant in selecting a nozzle of this type.

Flame Tube and Nozzle Cooling

Cooling of the reheat combustor and nozzle of modern military engines relies entirely on use of unburned fan delivery air, which is never, even at Mach 2, at very elevated temperature. For hypersonic propulsion, unless a precooled system is used, stagnation temperature air is generally too hot for cooling purposes (Fig. 31). It may also be desirable to use all of the air in the cycle to achieve maximum performance from the powerplant which will already be very large in relation to the vehicle. Where the cycle is not precooled, it should be possible to use fuel as a heat sink, although the total cooling capacity is clearly shared with the airframe. This may be a critically limiting feature with hydrocarbon liquid fuels, especially in a prolonged hypersonic flight. Accelerator missions may be

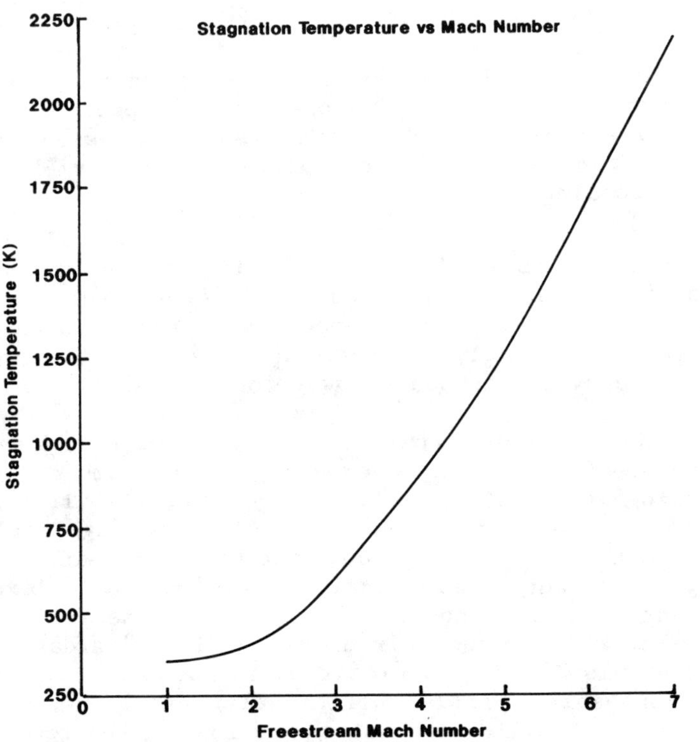

Fig. 31 Rise in air stagnation temperature with increasing freestream Mach number.

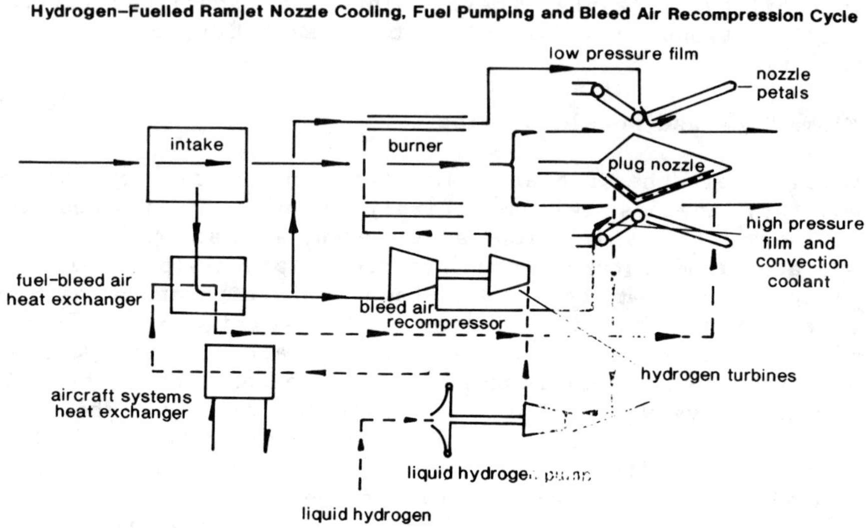

Fig. 32 Schematic of fuel cooling of hot parts and use of the heat in fuel pumping and auxiliary power generation.

more tolerant in this respect. Use of extreme cryogenic fuels such as methane or hydrogen, should provide generous heat sink capability for airframe and propulsion systems, and both could probably be used for either direct or indirect cooling.

Where hydrogen is used as a fuel, the further possibility exists of combining cooling duty and power supply as in the regenerative rocket engine concepts by, for instance, using the high work capacity of heated hydrogen fuel to drive fuel pumps and provide secondary power for the vehicle.

There is an obvious correlation between the rising level of cooling required with increasing Mach number and the increasing level of intake boundary-layer stabilising bleed required. An effective overall cooling package could well use a heat exchanger to cool intake bleed air, with the hydrogen coolant subsequently used to cool fixed hot components, such as nozzle plugs or two-dimensional nozzle sidewalls, and the cool air used for "leaky" areas, such as nozzle petals. Low air velocities in much of the flame tube would probably yield sufficiently low heat transfer coefficients to allow convection cooling on the reverse side with air in transit from intake bleed air heat exchanger to nozzle cooling system. (see the schematic, Fig 32).

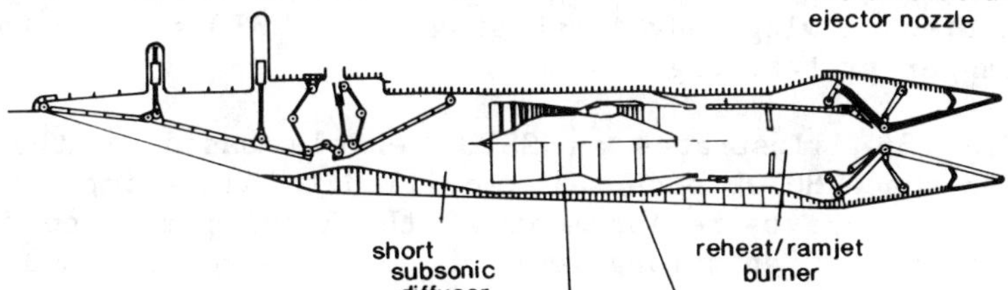

Fig. 33 Bypass ejector turboramjet configuration.

Installation and Installed Performance

Installations of total propulsion systems for hypersonic vehicles will be large and very complex: large for reasons of low specific thrust at high Mach, already described; complex as a result of the need for multiple cycles, engines, or multi-mode systems with, for example, variable heat exchange.

Some typical examples of such installations are shown in the following sequence of figures. Fig. 33 shows an individual turboramjet installation optimized for use in a Mach 5 cruise aircraft. The installation has a mixed compression intake with wide bleed slot, and a relatively conventional pure turbojet, insulated and capable of being closed by inlet vanes against hot flows in high Mach ramjet mode. A bypass duct around the engine carries ramjet flows into the reheat/ram-burner and excess air flows into the two-dimensional ejector nozzle. The relatively compact nature and near-constant cross-section was a result, in this case, of relatively low transonic and supersonic airframe drag. This made possible a modest core engine size, and the absence of a need for very rapid high Mach acceleration minimized the ramjet mode total flow area and, hence, core/intake mismatch.

By contrast, Fig. 34 illustrates the effect of a requirement for high thrust in the Mach 2.5-3.0 regime, where the specific thrust of both turbojet engines and ramjets is low, and also for operation to Mach 6+, requiring large ramjet area. Here the relative size of the turbomachine led to cross-sectional areas aft of the intake being considerably larger than the intake, the combined effect of larger size and a circular pressure vessel around the engine and bypass duct leading to a requirement for a increasing area behind a single line of intakes. This installation was proposed for a blended wing/body first stage of a TSTO similar to the MBB Sanger system, Fig. 35.

Fig. 36 illustrates an SSTO vehicle based on the British Aerospace HOTOL proposal, modified to allow improved ascent cross range by replacement of the RB545 precooled dual-mode engine with an arrangement of separate rockets and ramjets. Fig. 37 shows the back-to-back intake configuration of the ramjet engines, allowing a large flow area to be kept within the limited vehicle/wing compression field. Fig. 34 and Fig. 37 illustrate the use of an axisymmetric con-di nozzle with plug centre. Typical installed performances for powerplant of this class have been illustrated in Figure 20. In these plots installed performance includes all of the losses such as intake pre-entry and spillage, bleed loss and recovery, individual nozzle base drag effects and over/underexpansion losses. All of the engines employed a four-shock, mixed compression intake, with suitably matched turbojet or turborocket cores and the fixed-exit-area axisymmetric con-di plug nozzle.

Additional losses due to the multiple installations, for example additional base drag due to base areas between nozzles is not included in these plots, but can be a major

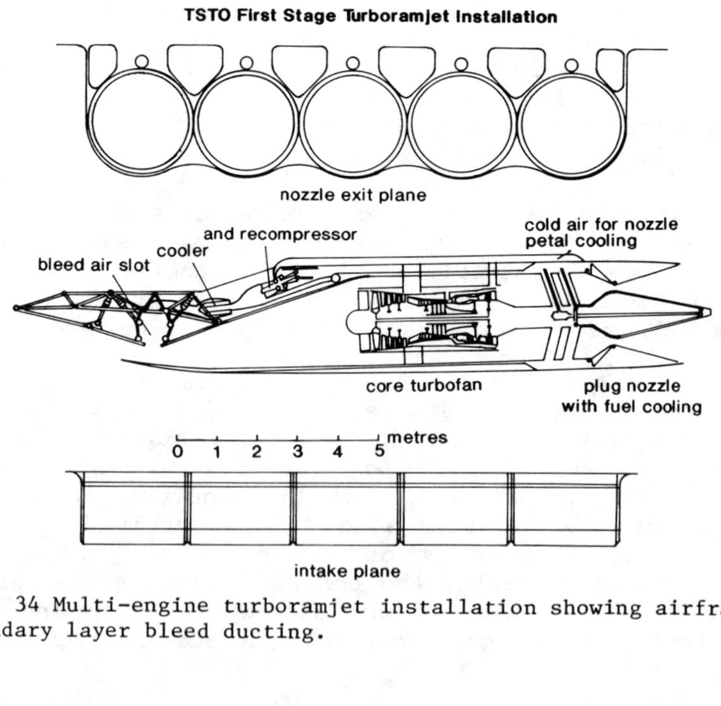

Fig. 34 Multi-engine turboramjet installation showing airframe boundary layer bleed ducting.

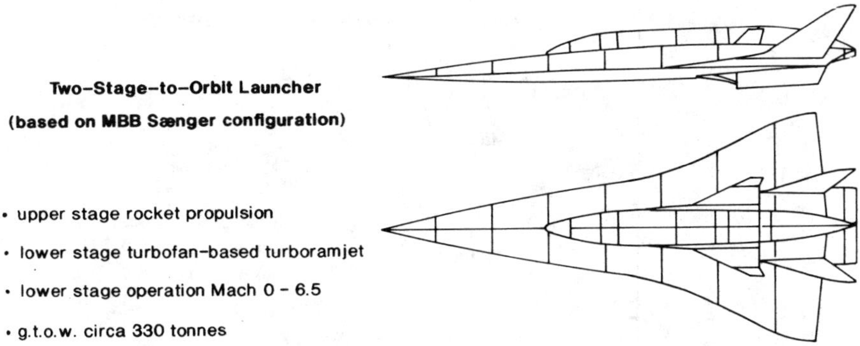

Fig. 35 Two-stage-to-orbit vehicle configuration with turboramjet-powered first stage.

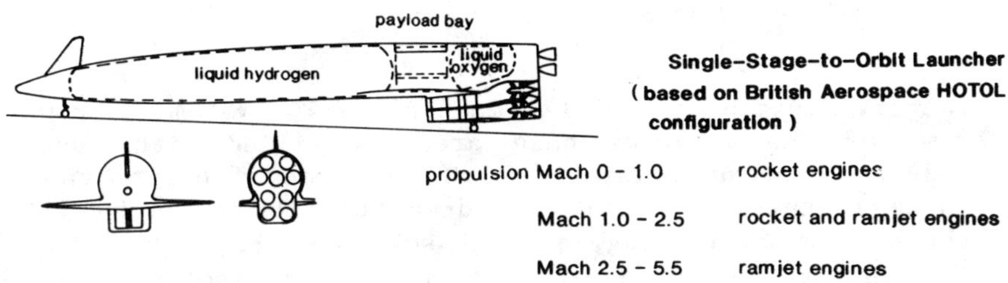

Fig. 36 Single-stage-to-orbit vehicle with rocket and ramjet engines.

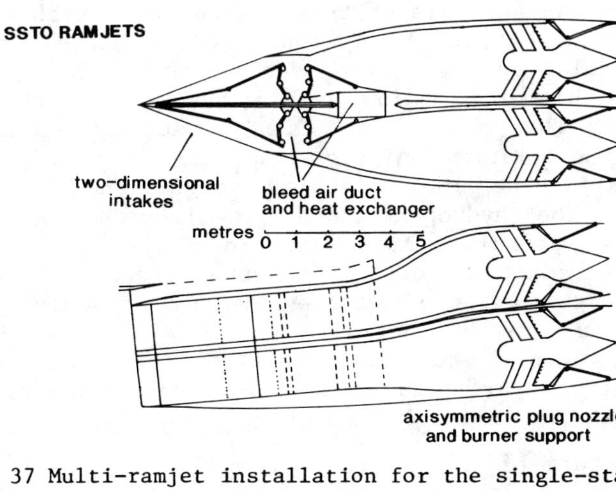

Fig. 37 Multi-ramjet installation for the single-stage-to-orbit vehicle.

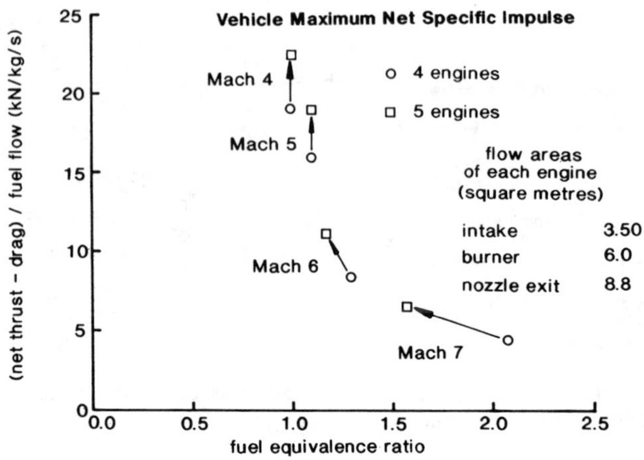

Fig. 38 Reduction of performance benefit of over-fuelling at high Mach number as total engine scale on the vehicle increases.

factor. In the case of the multiple TSTO system, even where some of the included base area is filled with fuselage underbody boundary layer (which was removed before the air intake), the additional base drag contributed up to 30% of the total vehicle drag, at high subsonic flight condition. One feature of performance which is worth noting, even in the ramjet case, is the effect of equivalence ratio on overall vehicle performance. For an accelerating vehicle, the net vehicle specific impulse [defined as (propulsion system net thrust-vehicle drag)/propellent flow] is the term which must be maximised to achieve maximum fuel efficiency. Fig. 38 shows an analysis of this nature for a turboramjet-powered vehicle. Here the optimum fuelling level for this hydrogen-fuelled system can be seen to transition gradually from substoichiometric below Mach 5 through to overfuelled by Mach 7. The reason for this effect is essentially similar to the optimum fuel-rich condition of a hydrogen-oxygen rocket: the low molecular weight and high sonic velocity of the hydrogen added outweighs the effect of reduced working fluid temperature (in conjunction with the simple mass addition effect to the change of airflow momentum). Fig. 38 also shows the effect of a vehicle/propulsion rematch, by increasing the number of engines of fixed size from 8 to 10. As might be expected, the requirement for overfuelling to achieve optimum performance is reduced at the higher Mach numbers.

A major feature of all of these propulsion systems is the extent to which propulsion mass is dominated by intake and nozzle. In our calculations, even where the intake, nozzle

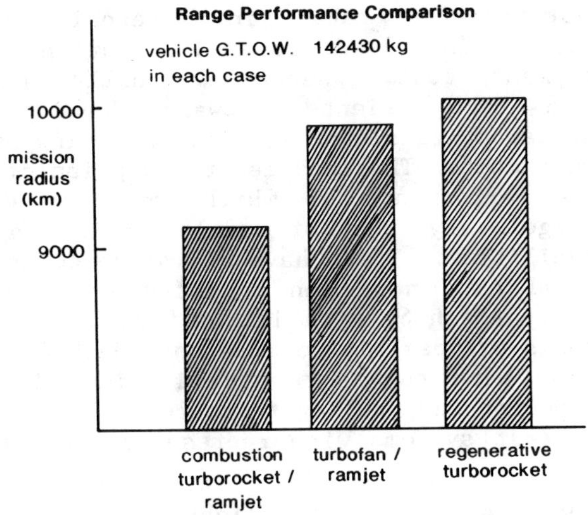

Fig. 39 Modest influence of low-Mach propulsion cycle choice on ramjet-cruise vehicle range for a fixed-format mission.

and intermediate pressure vessel structures have been assumed to be constructed in advanced ceramic composite material with ultimate tensile strength around 600 MPa for a density of 2300 kg per cubic metre, and core engines have been weighed in conventional materials, the core engines have typically contributed less than half of the mass of the powerplant. A result of this has been that, where very long range cruise vehicles were compared, with cruise in the ramjet mode of several different powerplant, the differences in range performance for fixed vehicle total mass were very small. (Fig. 39).

VIII. SUPERSONIC COMBUSTION

No mention has so far been made of supersonic combustion: this is essentially because the Rolls Royce and indeed British and European studies over the last 5-6 years have, after careful initial consideration, deliberately avoided the area. Several factors were active in this decision, some commercial, some technical.

The scramjet is the quintessential aerothermodynamic device; it is very difficult to conceive of a convincing performance check other than full-scale, full-speed model tests, with obvious implications for test facilities and flight test requirements, although CFD modeling, if sufficiently full, may provide useful guidance. Even testing of cooling systems could be quite demanding, given uncertainties in surface heat-transfer rates in internal inhomogeneous supersonic flow. This leads to an implication of extremely high development cost, whereas virtually all European efforts have been oriented toward development of a minimum-cost launch/retrieval system, including amortization of development cost. This has led to adoption of propulsion systems, the performance of which could be defined and achieved largely by use of scale tests on currently available facilities. These have all proposed only subsonic combustion and have not been expected to operate beyond flight velocity ~Mach 6, with intake flow velocities behind airframe shocks of possibly no more than Mach 4-5, for which adequate scale wind tunnel test facilities already exist. Equally, component and cooling tests could be carried out in conventional test systems with representatively modest flow velocities.

The benefits of air breathing flight beyond about Mach 5-6 have been assessed in varying ways by the several European groups working in the area. Two stage systems, examined in a simplistic manner, will deliver a higher payload with a given upper stage as separation velocity increases. Realistically, however, the mass of the lower stage will rise as this is done, probably with a step increment when supersonic combustion is required. For the reasons already outlined, and the limited additional advantage of a faster first stage, in conjunction with the view of a common technology "set" for a first stage launcher and global transport, German groups have preferred the Sanger TSTO system with its turboramjet subsonic combustion powerplant.

In Britain, the HOTOL project study focused on a SSTO system as a minimum operating cost launcher. Launch trajectories were minimal, based on optimization for equatorial launch eastbound into low equatorial orbit. After a major study effort involving considerable detail design, typical real payload fractions of 2.5% GTOW delivered to orbit were predicted. There was believed to be very little residual uncertainty left in this estimate, and materials were not generally exotic, metals being used in the wing and major load bearing structures. Transition from air-breathing operation with subsonic combustion occurred at around Mach 5.5, hydrogen/oxygen rocket operation completing the ascent to orbit. A major conclusion of this work was the costliness, in vehicle performance terms, of hydrogen tankage for air-breathing flight. More than half of the hydrogen loaded was used in the airbreathing ascent to Mach 5. The hydrogen tanks weighed, typically, 12-15% of their contained fuel mass, compared with ~3% for LOX tankage, and the drag increment due to the large volume required was considerable. A relatively elementary analysis showed that an orbital solution would not be possible with typical scramjet specific impulse and thrust beyond Mach 6 and the lift/drag and structural factors characteristic of the vehicle. In conjunction with all of the economic and test arguments, there was thus little enthusiasm for the supersonic combustion approach.

When high-speed transport, rather than launch activities, was considered, it was clear that, for global, i.e., ~20000 km, range, if 4-5 hour flights continued to be considered acceptable, then flight velocities of order Mach 4-5 were adequate. Such velocities are marginally more efficiently achieved with subsonic than supersonic combustion.

Furthermore, if cryogenic fuel was not found to be economically feasible for a high Mach fleet, then Mach 4 was likely to be an upper limit to hydrocarbon fueled operation, albeit over shorter ranges.

Assuming dual-mode scramjet operation to be feasible from ~Mach 3, and assuming also that a high-Mach transport would require an efficient hold/loiter system with acceptable take off/approach noise levels, then the total propulsion machinery loading on the airframe becomes alarming. For Mach 3 a full variable geometry turbo fan engine with variable intake/nozzle is required. This would be similar in scale and mass to the Mach 4/5 turboramjet.

Additionally, a scramjet duct of larger capture and exit area would be required, with a major mass increase.

IX. Conclusions and Directions

High Mach propulsion presents some formidable technical challenges, at the same time offering great scope for innovation in concept and design. Choice of optimum concept and cycle is strongly influenced by mission and vehicle considerations. As well as being able to perform satisfactorily in new flight regimes, the systems must have a very wide operating range, and this generally implies more than one operating mode. This can mean that whole sections of the system are inoperative over significant portions of the mission, emphasizing the importance of minimizing mass.

In high Mach flight, specific thrusts are low because of air momentum and the effects of dissociation in limiting the effective heat release. Large flows are necessary, implying proportionately larger intakes and nozzles than are required for lower-speed aircraft. Achieving acceptable masses for these components depends on meeting new materials goals.

Aerodynamic performance is equally critical: the thrust loss in a ramjet due to inlet-stabilising bleed, for example, becomes an increasing penalty as Mach number and bleed fraction rise. (The loss in net thrust due to a 12% inlet bleed at Mach 7 is around 30%, even where a good proportion of the bleed momentum is recovered).

Sizing minimum-mass powerplant for high ramjet flows at high Mach number with adequate takeoff and mid-supersonic thrust from the turbomachine leads almost inevitably to severe transonic flow mismatch, the intake potentially delivering large excess flows. Several means of dealing with this problem have been assessed, from deflecting (spilling) the excess flow at minimum pre-entry drag penalty through swallowing it and using it as the secondary flow of an ejector, to using it in a parallel cycle. The choice is conditioned by both the mission and the Mach range. Missions requiring minimum mass and moderate maximum Mach favor spillage, whereas missions requiring good performance over a wide Mach range and high maximum Mach at the expense of some mass penalty favor the parallel cycle option. Ejector designs can be useful where efficient subsonic operation is required.

Selection of cycle combinations is also conditioned most strongly by mission and mass constraints. Efficient subsonic cruise, loiter and ferry are feasible only with a relatively conventional gas turbine component, and a severe mass penalty is paid. Reducing the requirement to acceleration or short-period-only subsonic operation permits interesting trade-offs between the mass of rocket (or turborocket) oxidant and tankage and turbomachinery (and expander cycle heat exchangers). The entire class of pre cooled turbomachinery allows a spectrum of optimised solutions from acceleration-only machines with low mass and high fuel flow, through heavier systems capable of economic operation across the speed range up to about Mach 4.

Joint propulsion-airframe studies have led to the general conclusion that all the most likely hypersonic missions, including transport aircraft up to Mach 6+, SSTO and TSTO launch/retrieval systems are feasible using appropriately optimized combinations of subsonic combustion air-breathing propulsion and rockets. There is scope for valuable further study to compare the cost/benefits of such systems with the technically more demanding scramjet-based combinations at a common technology level.

Acknowledgments

The authors gratefully acknowledge the many contributions to the progress of the work reported here which were made by their colleagues both inside Rolls Royce, and from many of the European aerospace companies, and the support of the European Space Agency and other agencies in some parts of the work.

Bibliography

This work upon which this paper is based has been reported in several major reports, and relevant extracts have also appeared in a number of conference papers. The list below includes all of the items which will be publicly available.

Hewitt, F.A., Mattinson, R.M., and Ward, B.D. "High Speed Propulsion Assessment - Volumes 1 and 2", AFWAL-TR-88-2005, May 1988.

Hewitt, F.A., et al., "Evaluation of Advanced Air Breathing Propulsion Concepts", Report on European Space Agency Study Contract 6822/86/F/HEW(SC).

Grallert, H., et al., "Report on the Winged Launcher Configuration Study", European Space Agency Study 7379/87, February 1989.

Schoettle, U.M., Grallert, H., and Hewitt, F.A., "Advanced Air-breathing Propulsion Concepts for Winged Launch Vehicles", Proceedings of the 39th International Astronautical Federation Congress (IAF-88-248), Bangalore, India, October 1988.

Hewitt, F.A., and Schwab, R.R., "Parametric Assessment of Propulsion System Mass for Airbreathing Launcher Configurations", Proceedings of the 2nd European Aerospace Conference Bonn-Bad Godesberg, May 1989 ESA SP-293.

Schwab, R.R., and Hewitt, F.A., "Optimisation of Hybrid Propulsion System", Proceedings of the European Symposium on Future Supersonic Hypersonic Transportation Systems, Strasbourg, France, November 1989, paper IV.6.2.

Schwab, R.R., Andrei, G., and Hewitt, F.A., "Installation Technologies for Winged Launcher Configurations "Proceedings of the 41st International Astronautical Federation Congress, (Paper 90-260), Dresden, Germany, October 1990.

Hewitt, F.A. and Ward, B.D., "Airbreathing Propulsion for Supersonic and Hypersonic Vehicles", AIAA Joint Propulsion Conference, Boston 1988.

Hewitt, F.A., and Ward, B.D., "Advanced Airbreathing Powerplant for Hypersonic Vehicles", Proceedings of the 9th International Symposium on Air Breathing Engines, Athens, Greece, September 1989.

Holmes, A.T., and Varvill, R.A., "Design Considerations of a High Expansion Ratio Hydrogen Turbine", AGARD Conference Proceedings No. 479, 75th Symposium of the Propulsion and Energetics Panel, Madrid, Spain 28 May to 1 June 1990.

Hewitt, F.A., and Hodges, G.S., "Intake/Engine Matching for High Mach Number Propulsion Systems", European Space Agency, 1st European Industrial Workshop on Advanced Space Transportation Technology, Rome, Italy, March 1987.

Hewitt, F.A., and Ward, B.D., "Airbreathing Propulsion for Supersonic and Hypersonic Vehicles", 34th ASME International Gas Turbine and Aeroengine Congress and Exposition, Toronto, Canada, 1989.

Chapter 3

Energy Analysis of High-Speed Flight Systems

P. Czysz*
McDonnell Douglas Corporation, St. Louis, Missouri
and
S. N. B. Murthy†
Purdue University, West Lafayette, Indiana

Nomenclature

A = parameter in eq. (B.18)
B = parameter in eq. (40)
c = specific heat
C = Q_C / H_0, eq. (16).
D = drag
e = $[(V^2/2) + gh]$
E = energy
f = fuel
F = factor
g = gravitational constant
h = height
H = enthalpy
I = internal energy; specific impulse
j = stream
J = conversion factor from work to heat units
k = component; efficiency factor; fraction
K = $V_0^2 / 2gJH_0$, eq. (17)
n = number of moles of species
N = number

---

*Currently President, HyperTech Concepts, St. Louis Missouri.
†Senior Researcher.

o = oxygen
p = potential
P = pressure
Q = heat energy; heat generated per unit mass of reactants
R = gas constant, ratio
S = entropy; wing surface area
t = time
T = temperature, thrust
U = internal energy
V = velocity
$\mathscr{V}$ = volume
$\mathcal{V}$ = vehicle volume
W = weight
X = parameter, (e / e_f)
Y = fixed weight to scalable weight ratio
z = altitude
Z = gravitational potential energy; $(V_1 / V_0)^2$

Greek

α = angle of attack
χ = external bond coefficients
δ = element
Δ = element of change
ε = energy availability
η = efficiency
γ = ratio of specific heats
μ = chemical potential
ϕ = equivalence ratio
ψ = enthalpy ratio
Ψ = rational efficiency
σ = coefficient of structural bond

Subscripts

A = aerodynamic; pertaining to area
A = air
ABS = absolute
ch = chemical
C = crew; combustion; thermodynamic cycle
CC = combustion chamber

CE = fixed part excluding payload related
CV = control volume
D = diffusion
E = empty; effective
EN = engine
f = final; fuel
F = fuel; fixed
F = fuel
FE = fixed part unrelated to payload and crew
FG = fixed-to-gross weight
G = gross
GF = gross-to-fixed weight
H = heat transfer; heat; higher value
H = heat
HS = stagnation enthalpy
i = initial; parameter
ie = internal
j = species; stream
j = species
k = kinetic; component
KE = kinetic energy
L = lower value
LC = losses in combustor
MAX = maximum
MAX = maximum
n = species number
OPT = optimum
p = potential; constant pressure
pc = phase change
P = payload
PN = planform
PR = propulsive
S = scalable
SE = scalable equipment
ST = structure
T = total
Σ = overall
TC = total crew and crew installation
TE = total fuel-contributed energy

TH = thermal

0 = initial; take-off; overall propulsive

V = velocity

$\mathscr{V}$ = vehicle volume

W = weight

WT = wetted

x = direction x

Superscript

$(\dot{\ })$ = time rate

$(\ )^*$ = per unit weight of fuel

1.0 Introduction

A hypersonic vehicle in which the propulsion system includes, as an essential element, an engine that operates utilizing atmospheric air as a working fluid is a complex aerothermodynamics system, whether intended primarily for cruise or for acceleration of a payload to orbital velocities with small cruise segments. In order to be reasonably effective in meeting mission and operational requirements, it is generally agreed that the performance of such a vehicle depends primarily on how fully it is integrated with respect to the vehicle frame (which provides volume, lift, and friction-generated heat), propulsion, control, and given payload. One can expect in the design of such a flight vehicle only small margins to become available in the selection of several critical parameters, taking into account the nature of constraints in mission, geography of takeoff and landing (or recovery), weights, volumes, and their distributions, and energy available. Regarding the latter, the amount of energy that can be generated in the vehicle for overcoming drag, providing the desired acceleration capability, and generating control actions is invariably limited. This is a central concern in vehicle development, regardless of the three-dimensional trajectory, how the propulsion is compounded, or how effectively a thermal management system and control system are incorporated.

When atmospheric air and and specific fuels are introduced in propulsion as major means of increasing effective specific impulse (based on thrust minus drag rather than thrust) and decreasing takeoff weight for a given payload and mission, there is still a wide choice of engines and their combinations that can be considered. For example, the propulsion engines may consist entirely of engines breathing external, atmospheric air or partly of such engines and others that depend entirely on the use of propellants carried onboard. Engines that utilize a substance (for example, atmospheric air) from outside the vehicle for propulsion are referred to as "extrinsic engines"; engines that utilize only propellants onboard the vehicle for propulsion are called "intrinsic engines." An open-cycle gas turbine or a ramjet is an extrinsic engine, and a rocket is an example of an intrinsic engine. It is then possible to develop a substantial morphology for an extrinsic engine in itself or in combination with intrinsic engines.[1] However, an extrinsic engine must collect air from the atmosphere, wherein the density decreases with altitude, for direct use or for processing and subsequent use. This dictates the frontal cross-sectional area and also, in part, the volume of the vehicle and its disposition in a vehicle configuration. It is also of interest to note that the mass of air taken in from the atmosphere has energy that is directly proportional to flow (i.e., (flight velocity)^2). Therefore, the amount of fuel combustion energy required locally depends on how well the energy in the inlet air and any recovered energy from internal and external heated components are utilized, especially at higher flight speeds. The amount of energy that can be obtained by combustion and heat recovery using a fuel is of interest with respect to both its weight and volume; therefore, the density of a fuel becomes a major factor in the determination of vehicle volume for a given mission and chosen flight path. The volume required for fuel is of concern in all engines, whether wholly extrinsic, intrinsic, or a combination of the two.

Next, the takeoff weight of a vehicle is a combination of the so-called fixed and scalable (vehicle) weights, fuel weight, crew weight, and payload weight. The fixed weight of a vehicle is, by definition, the sum of the weights of the parts of the vehicle that do not scale with the vehicle size. It is assumed to be the sum of the weights of the following: payload and its installation structure, crew, crew equipment and crew-unique installation, and other fixed equipment. The scalable weight of a vehicle is again, by definition, the weight that depends on the size that the vehicle may become for the given payload and mission, based primarily on the level of technology at which the flight vehicle is designed. Therefore, it may be considered as the sum of the following weights: 1) the structure that scales with size, 2) the engine and 3) the scalable equipment. Fig. 1 provides a schematic illustration of the breakdown of weights for two types of vehicles. It is of interest in any option for the flight system to examine how the fixed weight and the volume parameter [volume/(planform area)$^{3/2}$] change with variations of size and gross takeoff weight. Thus, the problems of choosing materials and of structural design and packaging also become involved in the evolution of aerodynamic and propulsion design and in the choice of the trajectory for a given mission.

Last, it is clear from the previous discussion that the control system cannot be superposed on the system but needs to be integrated into the evolution of the flight system. The control must be effective in providing a vehicle that can fly all parts of the mission in a stable and controllable manner with the least expenditure of energy. As various aerodynamic and propulsion devices are introduced to regulate forces, moments, and energy in the vehicle, the method of incorporating centralized control, distributed sensors and responses, and a level of artificial intelligence will naturally demand a totally integrated system not only in concept but also in the functions performed by various subsystems with the given energy in the vehicle.

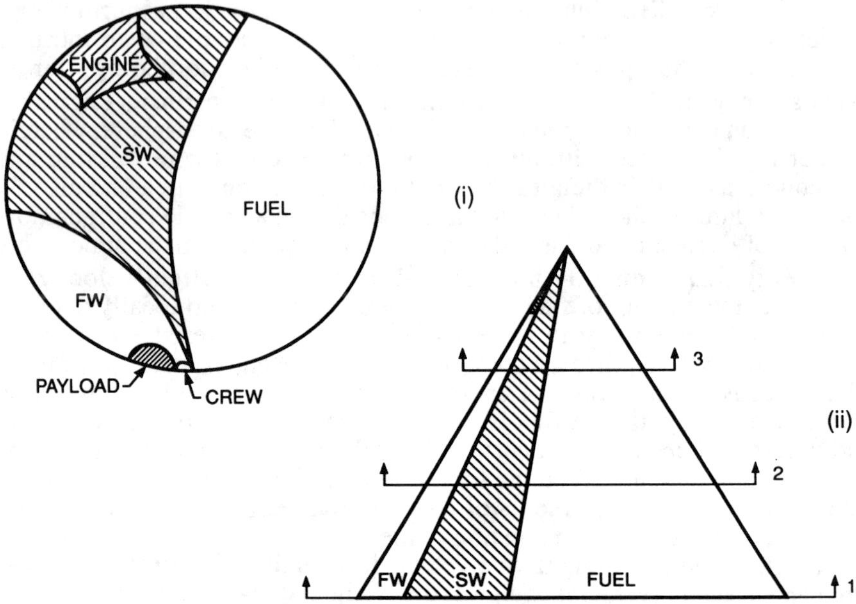

1. Schematic illustration of break-down of weights for (i) a SSTO and (ii) a TSTO vehicle.

The needs and opportunities in the analysis of such a flight vehicle that may be viewed as a flight system can be seen in Figs. 2 and 3. The first noticeable feature in those diagrams is the extensive morphology or the large number of combinations that can be derived from the given set of components. A second and more striking feature is that both combinations and individual components tend to be fuzzy in definition if one considers the input and output (of work, energy, or both, as relevant) as the two main variables. The key words may be said to be "*stretch*" , "*switch*", and "*stage* " in defining a subsystem or a system with respect to a chosen baseline combination or configuration. It may be useful to point out that stretching may be desirable in the use of a component or a large part of the system such as an engine. Switching from the use of one part of a system to another, for example, from one engine to another, or from one fuel to another, does not imply the absence of all commonality in the components of the system. Finally, staging is illustrated in Fig. 1 with a single-stage-to-orbit (SSTO) and a two-stage-to-orbit (TSTO) system, and other configurations can be developed in a similar manner. If one wishes to establish some order and visibility to the possible set of combination systems, mathematical models are available or may be devised as necessary.

The optimization of such a complex system can be carried out, in principle, with several different points of central concern and in various ways. However, it becomes quickly realized that a high speed flight system, particularly when it is centered around the utilization of ambient air

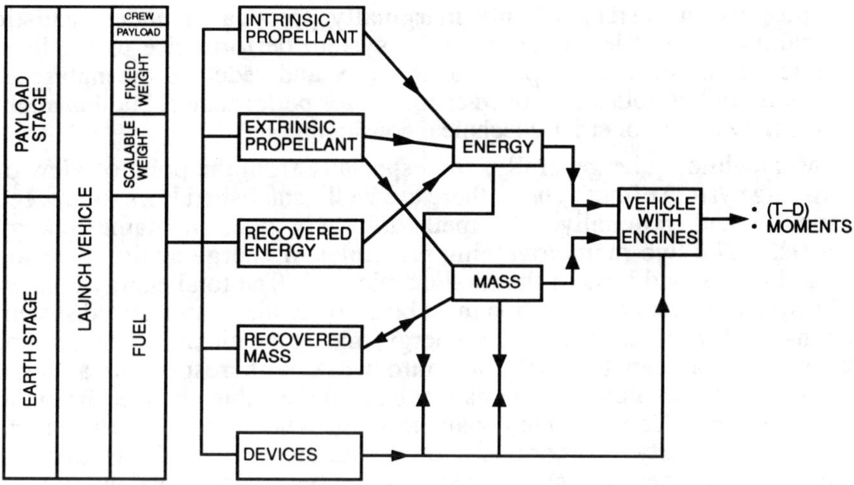

2. General high speed vehicle system.

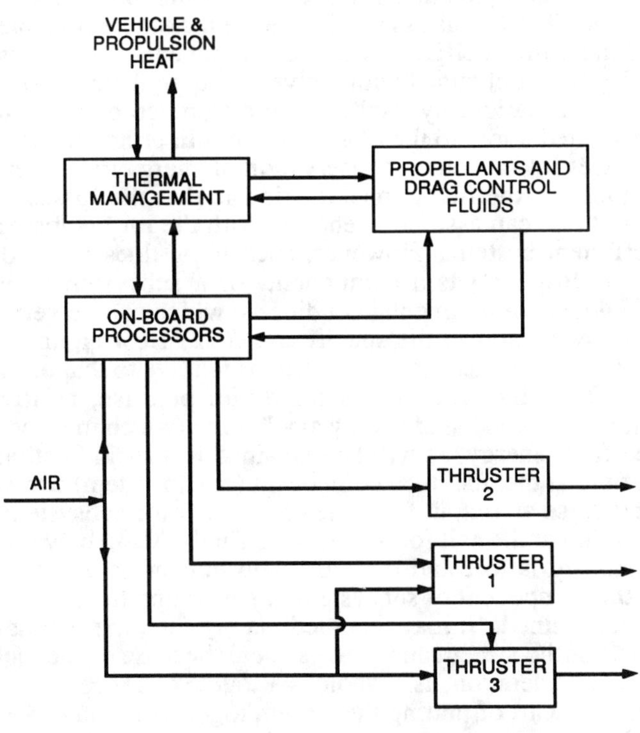

3. Propulsion system morphology.

in a propulsion system, is only marginally practicable under realistic constraints as one tries to improve the system performance to the limit feasible. It is, of course, possible to relax and redefine the nature of constraints and establish improved estimates for performance, but this might involve moving out of strictly technical considerations.

Meanwhile, quite generally, and especially from the point of view of improving system effectiveness, there are well-established laws to proceed reasonably and rationally. The methodology is given the name "energy analysis." The two main governing principles of energy analysis are the first and the second laws of thermodynamics: 1) The total energy content of the system under consideration must be accountable as an isolated system obeying the law of conservation of energy; and 2) the maximum fraction of the energy that can be converted into work with respect to a given environment by whatever means cannot exceed the value obtained from the well-established Carnot cycle considerations. The latter is referred to as "energy availability" in thermodynamics literature. It follows that the analysis of any thermodynamic system must be based on accounting for both the total energy and, in general more importantly, the maximum availability of energy for conversion to work with respect to the environment.

In optimizing a system it is invariably the common practice to obtain the best out of every component (and every combination thereof) over its operating regime. The best is assumed to be the maximum possible value of the so-called mechanical efficiency of the component. This efficiency is a measure of the output obtained with a given input to the component under consideration in isolation, by itself. The difference between input and output is considered a loss to the closed system. In principle, taking input, output, and losses into account, conservation of energy must prevail in the system. In practice the main consideration is that the losses should be minimized, and one can assume, then, that with the lowest losses one has an energy-efficient system. However, such logic does not address two questions: 1) How well is a component (or a subsystem) performing relative to the local environmental conditions, which can be very different from the input condition to the specific part? 2) How significant is the performance of a component (or a subsystem) relative to that of the system as a whole? The first question is important because, relative to the environment, there is a value of energy availability (which must be carefully distinguished from energy, as will be explained further in Section 2.0) for the working fluid at the inlet to a component (or subsystem) corresponding to which there arise an output from the component (or subsystem), and an energy availability at the exit for the working fluid. Also, how much of the energy availability is lost and therefore how much the energy becomes degraded in the component or subsystem in generating the output obtained needs to be determined. It may be noted that, in fact, the output obtained both in the part of the system and in the system, because of the inclusion of the part under consideration, as a whole is a source of concern. The second question is at the heart of putting the system together because one is again interested in knowing how much the component is significant in comparison with others and, if, in fact, it is truly and maximally effective where it is included or, more generally, wherever it is introduced into the system.

Energy analysis addresses those two questions, in addition to including, necessarily, the determination and implications of the mechanical efficiency of each component, subsystem, and total system. Energy analysis can be undertaken at various levels of sophistication utilizing the available knowledge base. In the absence of reliable data, parametric analysis may be needed based on a rigorous application of physical laws to the governing processes involved. This will become evident in the illustrative examples given later. However, since each stage of development of a flight vehicle system depends on prior accomplishments, energy analysis becomes an essential and continuing activity throughout the preliminary design and testing stages.

The energy analysis of a flight vehicle must, in general, include consideration of geography of launch and landing (or recovery) sites and of the flight trajectory. In addition, the analysis of weights and the volumes for the components and the system can also be included. Therefore, optimization of the flight system may be said to consist of 1) recognizing the implications of the given mission; 2) recognizing the constraints imposed by geography in relation to flight profiles; 3) selecting various applicable processes and components for the system and the levels of technology associated with them, including performance, weight, and controllability; 4) performing an energy analysis, and 5) interactively establishing a combination of trajectory, system, and operational schedule to meet mission needs. The resulting system details should then provide the following: 1) the inherent value of all components and recognizable subsystems, their optimal synthesis in the system, and the needs for their improvement; 2) the controllability of the system and 3) the ranges of parameters over which testing of different types is desired. Thus, energy analysis is a comprehensive tool in the development of a flight vehicle system.

1.1 Objectives and Outline of Presentation

The objectives of this Chapter are twofold: 1) to describe and illustrate the methods of energy analysis applied to high-speed propulsion systems and 2) to delineate the inter-relations in the analysis of the trajectory, energy utilization, and weights. Energy analysis is shown to have a central role in flight vehicle optimization, for which there is no substitute.

Energy analysis requires a variety of data related to the performance of each of the components and various processes involved in them. Such data are to be generated from analysis, diagnostics, testing, computational experiments, and prediction schemes. Various aspects of methods are discussed in the current volume. We have drawn upon those as well as cited references.

Energy analysis, including basis, methodology, and illustration of applications to simple aerothermodynamic processes and the basic Brayton cycle, is discussed in Section 2.0. Examples of analyzing various types of practical engine cycles with reference to energy and energy availability are presented in Section 3.0. Considering flight vehicle systems, two additional building blocks of analysis are 1) the geography of the launch and recovery sites and 2) the weight fractions. These topics are discussed in Sections 4.0 and 5.0, respectively. Finally, Section 6.0 is devoted to an outline of the methodology for optimizing propulsion for a given flight vehicle system.

2.0 Energy Analysis

The energy analysis of a flight system is the thermodynamic means of accounting for the energy state of the system. The energy state can be described in terms of 1) the energy supplied to the flight vehicle with intrinsic and extrinsic propellants and energy, 2) the energy converted into useful work in the vehicle utilizing energy conversion and propulsive work generation units, and 3) the losses incurred in the performance of useful work. Energy analysis is of interest everywhere along the trajectory for several reasons: 1) Critical deficiencies in performance may arise at several initially unknown locations along the flight path. 2) The performance over one part of the trajectory may affect the performance over another part, both preceding and following the part under consideration. 3) Several design and operational features depend on the integrated performance of the vehicle over the entire trajectory. The accounting must be accurate and unambiguous and provide adequate visibility in regard to need, performance and physical disposition of all of the identifiable components and processes in the system and the system as a whole.

In a high-speed flight system incorporating an air-breathing engine as an essential element of the propulsion system, energy is supplied to the flight vehicle by the air ingested, the intrinsic fuels, the thermal energy recovered from various parts of the vehicle and the propulsion system, and the mass processed onboard the vehicle (e.g., water or oxygen) and utilized for propulsive work generation. The useful work obtained from that energy is the work that overcomes or reduces drag force on the vehicle and also provides the needed excess power for acceleration and maneuvering of the vehicle. There are obviously losses and resulting entropy increases incurred in the performance of useful work. They are generally sought to be expressed in the form of the efficiency of individual or groups of components, processes and the system as a whole. The efficiencies eventually enter into the system optimization analysis. Therefore, it is critical that they reflect the performance of each element of the system in an unambiguous and comprehensive manner.

The system may be considered as consisting of various elements (any component, part, or process in the system being referred to as an "element"), each governed by a set of parameters, and there are several input streams to the system. In the optimization of the system, then, one is interested in at least three aspects of the element performance: 1) the inherent performance with respect to various input parameters, 2) the role in the system, and 3) the influence of the element on the system with respect to different input streams. A combination of the measures pertaining to the three types of performance is significant in optimization studies.

This section is devoted to a discussion of the formalism and methodology available for such an energy analysis scheme. Reference may be made to standard texts in thermodynamics for details of concepts, formulation, and calculation procedures; some useful references are included here as Refs. 2-7. A note on thermodynamic analysis is presented in Appendix A. It includes a brief discussion on energy availability calculations and illustrations.

2.1 Thermodynamic Analysis

The feasibility and performance of an aerothermal system can and, invariably, *must* be assessed to determine whether the system meets the requirements of the first and second laws of thermodynamics. The first law of thermodynamics can be paraphrased to demand conservation of total energy of any isolated system. Energy is the property of the system that can be defined so that its value between two states is equal to the work in an adiabatic process that has those two states as the end states. In the case of a chemically reactive fluid in flow, its total energy may be identified as the sum of potential, kinetic (including turbulence when required), chemical, and ionization components.

The second law of thermodynamics can be stated in different ways. The Kelvin-Planck statement is as follows: No cyclic process is possible whose sole result is the absorption of heat from a reservoir and the conversion of this heat into work. The Caratheodory statement is as follows: In the neighborhood of each equilibrium state of a thermodynamic system states exist that are inaccessible by adiabatic processes. One can also introduce the concept of entropy, which is an extensive property and therefore does not depend on whether the system is in equilibrium. Thus, the following statement can be derived: The entropy of an isolated system can never decrease; a constant entropy, or isentropic, process is reversible in nature. This provides an operational form for the second law.

Another useful concept in this context is that of energy availability, or "exergy," as it is sometimes called (ex is the Latin prefix for *from* or *out of* and *ergon* is Greek for *work*; hence *exergy* is a word neologized from those). Exergy pertains to the quality, or potential, of energy, the ability to perform work, or, more generally, the capacity to cause a thermodynamic state change relative to an environment. If energy is separated into ordered and disordered forms, it is clear that the quality of energy pertains to the disordered energy component of the system, since ordered energy is fully convertible. The quality is measured by means of a standard, namely, the maximum work that can be obtained from a given amount of energy using the local environment as the reference state. The environment is any convenient large medium in a state of perfect thermodynamic equilibrium. In view of the definition just given for energy availability, it is clear that no conservation law can be formulated for it. However, accounting for energy availability can yield its degradation or, more generally, the change in it. Any degradation is a result of irreversible processes in the system under consideration and therefore of irretrievable loss of energy. It is also clear that availability cannot be negative, but it can be zero for a system that is in equilibrium with the environment.

In discussion of concepts such as entropy and energy availability, the problem arises of determining absolute values (independent of environment) for them for a given thermodynamic system. Regarding entropy, an absolute value can be determined based on quantum mechanical considerations. However, within the framework of classical thermodynamics, based on the third law of thermodynamics, one can set up a datum relative to which the entropy of any system can be given in absolute terms: The entropy of a pure, crystalline substance tends to zero as the absolute zero of temperature is approached, and an assumed value of zero at zero of temperature can therefore be utilized as the datum. Regarding an absolute value for energy availability, the local environment of the system under consideration can be replaced by the state of absolute zero of temperature as the reference state for each of the substances involved. However, the local environment is of immediate interest in the case of flight vehicles.

A definition of energy availability may be written as follows with reference to an environment denoted by subscript *0*:

$$\varepsilon = \underset{(1)}{(I - I_0)} - \underset{(2)}{T_0(S - S_0)} + \underset{(3)}{\frac{P_0}{J}(\mathscr{V} - \mathscr{V}_0)} + \underset{(4)}{\frac{V^2 - V_0^2}{2gJ}} + \underset{(5)}{(Z - Z_0)\frac{g}{g_0 J}} + \underset{(6)}{\sum_c (\mu_c - \mu_0) N_c} \qquad (1)$$

Here the availability is shown as the sum of internal energy (1), entropy (2), work (3), kinetic energy (4), gravitational potential energy (5), and chemical potential energy (6). The expression for availability applies to a general substance but accounts for only gravitational potential and chemical potential fields. In a specific case one or other of the terms in Eq. (1) may be identically zero when the state or field is in equilibrium with the environment. On the other hand, in other cases it may be necessary to include new considerations such as radiative transfer or electromagnetic fields; it is then necessary to account for availability with respect to state parameters associated with those considerations.

It is often of interest in energy analysis to consider the energy availability rate with respect to time for a substance that crosses the boundary of a control volume that is set up to enclose a (closed) process or a (closed) system, or, quite simply, a control boundary. One can then consider the change in irreversibility rate, for example, across a control volume or between two control boundary surfaces. The rate concepts for energy availability and irreversibility are found to be necessary in the analysis of the thermodynamic cycles of engines, the components of propulsion systems, and aerothermodynamic processes.

2.2 Implications of Energy Availability

There is often considerable doubt and skepticism concerning the need for the concept of energy availability and its determination and accounting. The determination of mechanical efficiency and its improvement through reducing losses is considered adequate. However, there are several ways of demonstrating the need and, often, the necessity of accounting for energy availability throughout the development of the flight vehicle. The energy availability of a system at any given state is the work that can be obtained from the system when it is brought from its current state to equilibrium in all details with the environment surrounding the system. The energy content of the system, on the other hand, can include all of the energy components and can again be determined with reference to chosen zero-value states with respect to mechanical, gravitational, chemical, and thermal fields; however, the value for energy content merely denotes the total energy that can exist in different forms and not the part of energy that can be utilized with respect to a given environment. Thus, there is less information in the energy value than in the energy availability value.

In the following, two basic approaches are described: 1) introducing the concept of rational (or, as it sometimes called, the "true thermodynamic") efficiency, based on consideration of energy availability and irreversibility in the components, processes, and the overall system and 2) evolving sensitivity parameters with respect to isolated, individual components and processes and with respect to (external) inputs to the system.

2.2.1 *Rational Efficiency*

One method of establishing the primary significance of energy availability in practical engineering is to reconsider the concept of efficiency of a system or of a component of a system. Accounting for loss of energy, one can calculate efficiency, referred to as mechanical efficiency or overall efficiency of the system under consideration. It may be important to note that mechanical efficiency includes consideration of all of the losses in the system under consideration: friction, wave, heat, mass, aerodynamics, chemical action and so on. Correspondingly, based on energy availability and irreversibilities in the system, one can calculate two parameters: 1) the rational efficiency, which is a measure of the change in availability across the system or between endpoints of a process; and 2) the effectiveness, which is a measure of the output obtained for a given change in availability, again across the system or between the endpoints of a process. The formalism associated with these is also given in Appendix A.

Ambiguities in identifying losses in relation to irreversibilities may arise in certain cases. Losses may not necessarily be irreversible, although they may arise in isentropic processes, but no practical process can be wholly reversible. Two consequences follow: On one hand, in an apparently isentropic process, when losses occur, it is necessary to clarify the distinction between the system associated with the isentropic process and the adjoint environment to which "lost" energy has been rejected. On the other hand, one may consider the following example: A chemical reaction, such as combustion of a fuel with an oxidant, may or may not involve losses (with respect to overall energy) in a given reactor. However, as long as the chemical reaction is associated with a finite rate, irreversibilities and an increase in entropy arise.

It is found that, in general, the mechanical efficiency differs from the rational efficiency since the latter accounts for all of the internal irreversibilities, whereas the former only accounts for energy losses. The internal irreversibilities can be expressed in terms of a net entropy gain. Therefore, in simple cases one can relate mechanical and rational efficiencies to each other. However, the mechanical efficiency alone cannot provide a measure of all aspects of the performance of a component or a system.

In principle, losses must be reduced and mechanical efficiency increased as far as possible in any system. However, overcoming losses may leave the energy availability at the outlet of the system unaffected if the total irreversibilities remain nearly the same; therefore, there may be no significant gain in rational efficiency or, equivalently, in the effectiveness of the system. In other words, the capability of the working fluid at the exit of the component or system to yield maximum work (based on its energy potential) may only become marginally improved. A diffusing duct with friction, heat, and shock-wave losses can be improved with respect to mechanical or rational efficiency or both by lowering, as appropriate, the overall loss or the individual losses relative to the other two. If one is interested in the capability of the outlet fluid from the diffuser to perform work, it is obviously important to improve its potential for performing work; therefore, one should reduce both losses and irreversibilities. Reducing only losses may improve a chosen parameter of duct performance but may in fact have reduced or left unaltered energy availability at the outlet. The irreversibility in the duct is due to friction, the presence of shockwaves, and the change in enthalpy of the fluid along the duct during the course of diffusion. It is clear that in practice, if the diffusion is small, the presence of shockwaves probably causes the greatest degradation in energy availability. Another example which shows that irreversibilities may be more significant than losses is a flow expander in which there is momentum and heat loss. As long as the fluid enthalpy reduces continuously in the course of expansion in the nozzle, friction and heat transfer can be expected to have a decreasing effect on the thrust generated, for example, as the expansion ratio increases to high values. Finally, an attempt is made to introduce isothermal combustion in some systems on the basis of the same reasoning when energy addition is by combustion. However, one should note that the system then must include a means of continuously extracting heat or work and thus becomes complex, although the combustor performance is itself independent of the efficiencies associated with heat or work extraction.

It should be clear from the foregoing that 1) losses and irreversibilities must be considered together in the performance of a system; 2) loss of energy availability results in reduction of capability to perform work, by definition; and 3) mechanical efficiency must be distinguished from rational efficiency, with the latter providing an assessment of total thermodynamic performance.

2.2.2 *Coefficients of Structural and External Bonds*

In the Introduction, attention was drawn to the main consideration in the inclusion of a component in a multicomponent system with several alternative input streams: the significance (nature and magnitude) of the effectiveness of the component to the performance of the overall system when the system is operated with a fixed input stream or several variable input streams. Depending on the effectiveness, one can proceed to determine 1) whether the component should be developed further to yield major improvements in its performance; 2) whether the component performance should be optimized in order to obtain the best system performance; or 3) whether the component should be displaced in or removed from the system. In many respects such questions are of the same type as those raised in conventional sensitivity analysis. However, in the current analysis the emphasis is on energy availability and its loss for reasons stated earlier. Thus, the sensitivities being considered are more general, significant, and comprehensive than the conventional sensitivities.

A basis to and a method of establishing, in quantitative terms, two measures for the current sensitivity parameters have been provided in Refs. 8 and 9. The application of these measures to flight systems has been discussed in Ref. 10. The measures are based on an assessment of irreversibilities in the system. Before a formalism for the determination of the measures is given, it is useful to discuss the general approach.

Approach to defining sensitivity measures. The nature of these measures may be illustrated with reference to the example of a ramjet engine. In that engine there are two input streams to the engine system, namely, air and fuel, which may be referred to as external inputs to the engine. Each of those is characterized by its state and flow properties. There are several components in the engine, each of which is again characterized by its input and performance parameters. Considering the combustor, for example, the temperature of the air entering the combustor is an input parameter, and the pressure loss in the combustor is an output parameter.

In order to determine the sensitivity of engine overall performance with respect to changes in a performance parameter of any component, it is necessary to note that an input parameter to a component can suffer a change due to two different types of circumstances as follows: 1) Under fixed external inputs to the engine, a change may arise in an input parameter to a component due to some internal (to the engine system) reason; and 2) a change may occur in a parameter of the external input streams. In each case a sensitivity parameter can be defined. These parameters must be distinguished from each other. At the same time, for a comprehensive analysis of sensitivity it is essential to examine both of those types of circumstances, since changes can occur in practice due to both internal reasons and external input changes. Both of those types of sensitivity are discussed in the following.

In the case of the ramjet engine, when the combustor is considered, for example, the first type of sensitivity may be illustrated as follows. Suppose that the external input conditions are fixed and, for some internal (to the engine system) reason, a change arises in the temperature of air entering the combustor, which causes, in turn, a change in the pressure loss across the combustor. Then one can expect a change to arise in the engine overall performance. Since the external input conditions are fixed, the engine system can be treated in isolation. The change in engine overall performance with respect to the change in a performance parameter of the combustor is related to the manner in which the combustor has been designed to function in relation to the other components of the engine system. In the case being considered the sensitivity of the engine performance to the specific change in combustor performance (pressure loss, in the example) is due to causes solely internal to the system. In other words, such sensitivity is dependent on what may be referred to as the "architecture" or "structure" of the system. This type of sensitivity is referred to as the *"structural bond coefficient."* This coefficient provides a measure of the extent to which the overall performance of a system changes due to the change in a chosen input parameter of a specific component when the external input parameters to the system are held constant. The first type of circumstance in which sensitivity is of interest thus leads to the concept of the structural bond coefficient.

Next, regarding the second type of circumstance in which sensitivity is of interest, there can arise changes in the input streams of the engine system. A change in an external stream, that is an input to the system, can given rise to three types of effects: 1) a change at entry to various components of the system; 2) a modification in performance of various components; and 3) a change in the performance of the overall system. In the case of the ramjet engine a change may arise in the temperature of air or fuel. This may cause a change in the temperature of the input (air or fuel) to the combustor, which may in turn lead to a change in the pressure loss across the combustor and therefore in engine performance. All of the changes in the engine are a consequence of the change in the external input stream. Hence, the sensitivity to such changes is referred as the *"external bond coefficient"*. This coefficient is a measure of the effect on system performance due to a change in performance of a component when an external input parameter is changed. The second type of circumstance in which sensitivity is of interest leads to the concept of the external bond coefficient.

It is of interest to discuss here another aspect of the concept of external bond coefficient. This aspect pertains to a component of a system that is necessary for the functioning of the system but whose performance is unaffected by variations in the state or flow properties of an input stream to the system. In that case, regarding that input stream, the component may be said to be connected to the system by a "rigid bond". A component with a rigid bond can have any value of irreversibility. However, the irreversibility has no effect on the irreversibility of the system.

The sensitivity of a system to various changes is thus determined by a combination of 1) the structural bond coefficient and 2) the external bond coefficient. It is important to note that in both cases a complex system involves, in general, several external input streams, components, and internal inputs and outputs of components, each of which is characterized by several parameters. Thus, for a given complex system a series of values of structural and external bond coefficients become of interest in practice, reflecting a spectrum of sensitivities with respect to sets of parameters related to input streams and components. The determination and assessment of these values becomes a very large task. In preliminary design and, subsequently, during various development cycles for the system, such a detailed evaluation of sensitivities may prove to be feasible only when a single, comprehensive parameter is utilized to denote the state of the working fluid at all locations in the system.

As discussed earlier in Section 2.1, energy availability is a comprehensive energy state parameter. Changes in it are related to changes in thermodynamic state and flow properties anywhere throughout the system. Therefore, for a flight vehicle propulsion system, energy availability can be used to reflect, throughout the system, a single parameter, the current state of the working fluids. At the same time, losses in energy availability denote irreversibilities across any component or process or the system as a whole. Therefore, it is possible to proceed as follows in determining system sensitivities:

1) establish relations between each of the two bond coefficients and irreversibility rates in individual components, processes, and the system;
2) establish relations between irreversibility rates and pertinent variables in inputs to the system and each of the components; and
3) utilize the relations as obtained in points 1 and 2 as necessary to evaluate specific components and variables.

The foregoing provides a more general, comprehensive method and yields a simpler method of establishing and evaluating sensitivities with respect to a catalog of input parameters for each component and process.

Methodology for determining bond coefficients. A system is considered to operate with one or more of the external input streams 1, ..., j, ..., ℓ and has components or elements 1 , ..., k, ..., n. Variables or parameters are denoted by X, with the subscripts 1, ..., i, ..., m. Irreversibility rates in the system are represented by $\dot{I}$, with the subscripts j and k as appropriate for an external input stream and a component. The two bond coefficients may then be defined using the method of parameter optimization.

The structural bond coefficient, σ, may then be written as follows.

$$\sigma_{k,i} = \left(\frac{\partial \dot{I}_T}{\partial X_i}\right) \Big/ \left(\frac{\partial \dot{I}_k}{\partial X_i}\right) = \left(\frac{\partial \dot{I}_T}{\partial \dot{I}_k}\right)_{X_i} \tag{2}$$

where the subscript T denotes the total system. The coefficient relates the irreversibility rate in the system to that of the component, with irreversibility itself being the loss of energy availability. Therefore, the coefficient of structural bonds for a component of a system, establishes the significance of the loss of energy availability in the component to the performance of the system as a whole. The coefficient depends on 1) the thermodynamic process in the component operation and 2) the structural relation of the component to the system as a whole.

One can then draw the following conclusions from the value of $\sigma_{k,i}$:

1) If $\sigma_{k,i} > 1$, the component has a significant influence.on the system and it is possible to optimize k in order to obtain the least I_T or the best performance from the system.

2) If $\sigma_{k,i} = 0$, there is nothing to be gained by working on element k.

3) If $0 < \sigma_{k,i} < 1$, this indicates that the effect $\Delta \dot{I}_T$ is less than the cause $\Delta \dot{I}_k$. It is then necessary to be cautious in modifying the element k in itself or in the structure of the system.

4) If $\sigma_{k,i} < 0$, the structure of the system obviously needs a change concerning insofar as the element k .

The preceding conclusions are given in a general form. In practice, it is necessary to identify which variable is likely to be the most significant regarding a particular component and then proceed to the evaluation of σ_k with respect to it.

Next, one can proceed to examine the influence of inlet streams for the system in terms of external bond coefficients. The coefficient relates the change in $\dot{\varepsilon}$ at any location in the system to a change in $\dot{\varepsilon}$ in an input stream due to a change in a parameter of that stream. Considering a change in parameter X_i in a stream j, the external bond coefficient $\chi_{j,i}$ may be defined as follows:

$$\chi_{j,i} = \left(\frac{\partial \dot{\varepsilon}}{\partial X_i}\right) \Big/ \left(\frac{\partial \dot{\varepsilon}_j}{\partial X_i}\right) \tag{3}$$

$$= \left(\frac{\partial \dot{\varepsilon}}{\partial \dot{\varepsilon}_j}\right) \tag{4}$$

The coefficient may be determined at entry to the system and at any location downstream along the path of the working fluids in the system.

A change in X_i of j affects the values of $\dot{\varepsilon}_j$ and $\Sigma\dot{\varepsilon}_j$, the summation being over all of the streams at that location. When such a change occurs at entry to the system, two changes could arise as a consequence in the following:

1) The value of $\dot{\varepsilon}$ at exit of the system, or, equivalently, $\dot{I}$ across the system; and

2) the values of $\dot{\varepsilon}$ in input, IN, or output, OUT, of various components, or, equivalently, $\dot{I}$ across one or more components.

It can be easily realized that such possibilities are the reason for introducing the concept of the external bond coefficient for a system and its components.

In general the input streams at entry to a system may interact in various ways within the system. Such interactions, when they arise, make the calculation of external bond coefficients extremely complex. When the input streams for the system are independent of one another throughout the system, then the values of $\dot{\varepsilon}$ for various streams can be simply added together if necessary, or can be considered individually at any location. In most of the flight vehicles, there may, however, be only one input stream for the propulsion system. In such cases the numerical value of χ_i becomes one at entry to the system; the input and output values of χ_i for various components, it must be carefully noted, can be any number equal to or greater than zero.

In order to assess the significance of the external bond coefficient, it is useful to relate that coefficient to the structural bond coefficient. Considering compound k and assuming a single input stream j for the system, the structural bond coefficient may be written as follows for the case in which a change arises in X_i.

$$\sigma_{k,i} = \frac{\left.\dfrac{\partial \dot{\varepsilon}_k}{\partial X_i}\right)_{IN} - \left.\dfrac{\partial \dot{\varepsilon}_k}{\partial X_i}\right)_{OUT}}{\left.\dfrac{\partial \dot{\varepsilon}}{\partial X_i}\right)_{IN} - \left.\dfrac{\partial \dot{\varepsilon}}{\partial X_i}\right)_{OUT}} \tag{5}$$

where the terms in the denominator apply to the system as a whole. If it is assumed that the output state of the system is in equilibrium with the environment, ε_{OUT} is equal to zero. It follows then that

$$\frac{1}{\sigma_{k,i}} = \left.\frac{1}{\chi_{k,i}}\right)_{IN} - \left.\frac{1}{\chi_{k,i}}\right)_{OUT} \tag{6}$$

Utilizing Eq. (6) the value of $\sigma_{k,i}$ can be determined corresponding to the values of χ_i in a given case. The implications of various values for $\sigma_{k,i}$ have been discussed earlier (following Eq. (2)).

One special case of interest is that which $\sigma_{k,i}$ becomes zero due to either $\chi_{k,i})_{IN}$ or $\chi_{k,i})_{OUT}$ being zero. In that case it follows that I_k is independent of the entry conditions to the system. The component k is then said to be in rigid bond, as stated earlier. When a component is in rigid bond in the system, the input conditions to the system should have no effect on either the input or the output conditions of the component. Here, case 3 naturally reflects a rigid bond and the futility of working on improving element *k* in that case.

2.2.3 Structure diagrams. In order to appreciate the significance of structural and external bond coefficients, it is useful to construct a diagram for the system under consideration that schematically represents the structure or architecture of the system. This diagram presents the connections among the various components of the system. Along with a working fluid path diagram and the energy availability flow diagram (referred to as the Grassmann diagram, Ref. 4), a structure diagram serves to illustrate the interactions among components of the system.

Considering a ramjet engine, which was utilized earlier for illustration in the subsection on the approach to defining sensitivity measures of bond coefficients, Fig. 4 presents a block diagram of the engine as well as a structure diagram. The block diagram illustrates the components and the path of the working fluid, and the structure diagram provides the interactions between components. A block diagram and a structure diagram are presented for a simple gas turbine in Fig. 5 as another example of the construction of such diagrams. One can observe the interactions between components at common boundaries in the structure diagram. Referring to Fig. 4, it can be seen that the combustor of the ramjet engine interfaces with the inlet, the fuel storage, and the nozzle. Therefore, compared to other components, the combustor can be expected to have a considerable influence on overall engine performance. Noting that air and fuel may be considered as two external input streams, one can calculate both structural and external bond coefficients with respect to changes in irreversibility rates for the combustor, caused by changes in selected variables. Although details are not presented here, it can be shown that energy availability in the fuel stream at the entry to the combustor has a major influence on ramjet

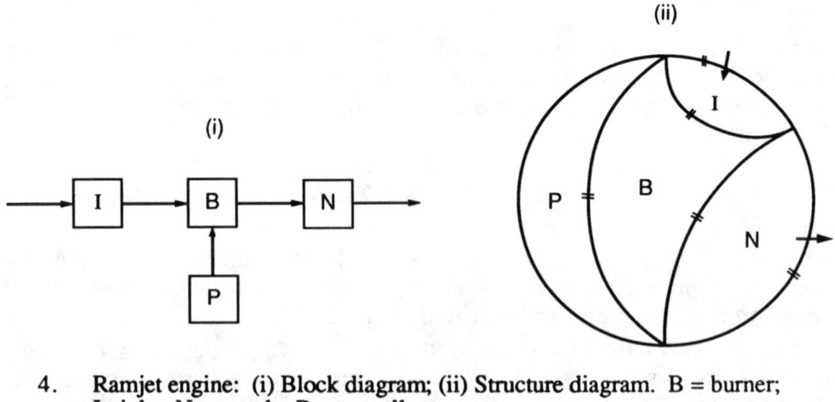

4. Ramjet engine: (i) Block diagram; (ii) Structure diagram. B = burner; I= inlet; N = nozzle; P = propellant.

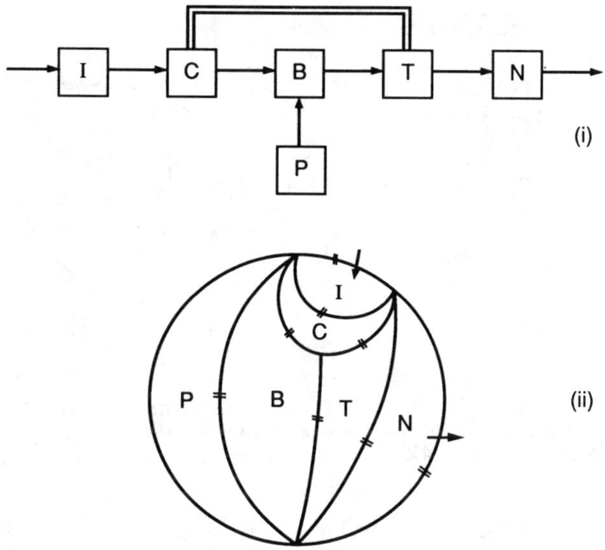

5. Simple gas turbine engine: (i) Block diagram; (ii) Structure diagram. C = compressor; T = turbine. Other symbols as in Fig. 4.

engine performance. Energy availability of the fuel depends, in part, on the internal and kinetic energy contents of the fuel: therefore, one may wish to examine the influence of increasing fuel enthalpy and kinetic energy on structural and external bond coefficients.

Referring again to Fig. 4, one can next examine the influence of nozzle performance on overall engine performance. It is clear that the nozzle interfaces only with the combustor; also, there must be a substantial amount of unused energy availability at its exit plane in the exhaust gas. One may consider recovering part of the energy through the nozzle wall by heat-transfer to fuel, which is recirculated over the nozzle wall before admission to the combustor. A block diagram and the corresponding structure diagram in this case are illustrated in Fig. 6.

In Section 3.0 more complicated propulsion systems are presented along with structure diagrams. In all cases one considers each component of the system from the points of view of the following: 1) the total number of interfaces, 2) the energy availability change across the component with a given input and process, and 3) the external input streams. Structural and external bond coefficients are calculated to establish 1) whether a component needs to be and can be improved and 2) what the response of the system is to a change in the component performance due to modifications in external input stream parameters.

The two coefficients provide powerful, clear means of establishing the role of a component in a system operating with a given structure and one or more input streams. Based on the concept of energy availability and its

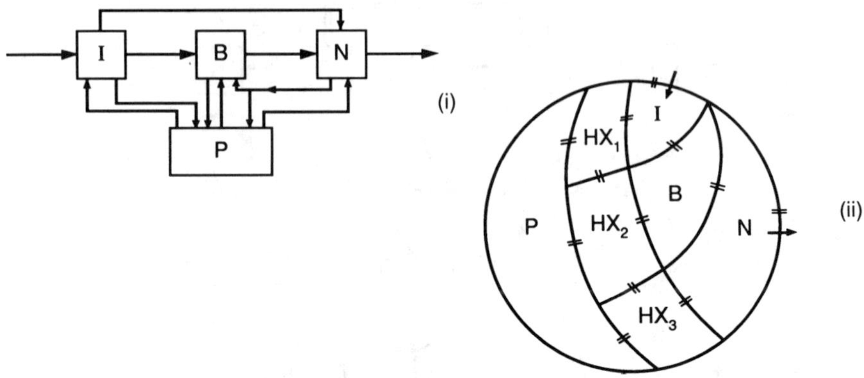

6. Ramjet engine with cooling of inlet, burner, and nozzle, and recirculation of fuel with heat exchangers: (i) Block diagram; (ii) Structure diagram. 4X = heat exchanger. Other symbols as in Fig. 4.

change in the component under consideration, the coefficients are universally applicable to any system. In practice, considering a complex system, it is necessary to identify the variables of greatest influence in order to reduce the total amount of work involved. However, in order to obtain the best results from research and development on components or system configurations, an energy availability analysis for determining and using these coefficients is probably the most effective, rational method.

In summary, energy analysis of a propulsion system consists in accounting for energy and energy availability, utilizing the laws of thermodynamics, throughout the system by identifying it in terms of a series of variously connected elements. An element, as stated earlier, is a component or a process that can be identified by means of a control volume. The performance of an element or a part of a system is assessed in terms of 1) mechanical efficiency, 2) rational efficiency, 3) coefficient of structural bonds, and 4) coefficient of external bonds. Mechanical and rational efficiencies can also be determined for a system as a whole.

3.0 Applications Of Energy Availability Analysis

The applications of energy availability analysis arise in a variety of contexts, and a large body of literature has been generated in the last two decades. The problems addressed include those dealing with economics such as the cost of energy generation, the optimization of plant design and operation for minimal cost, and the management of resources and investments. This Chapter emphasizes the applications to high-speed flight propulsion systems.

In the following an account is given of the manner in which energy availability concepts can be applied to 1) working fluids such as air and fuel, 2) various aerothermodynamic processes of interest in propulsion and 3) thermodynamic cycles of propulsion engines. After a discussion of the application to the Brayton cycle in Section 3.3, the supersonic ramjet engine is analyzed in Section 3.4 along with its application to a high-speed vehicle. In Sections 3.4-3.7 a discussion is then given of several cycles suitable for application to 1) the lower-speed (at flight Mach numbers below about 6) part of propulsion requirements for high-speed vehicles or 2) cruise-type vehicles (again at flight Mach numbers below about 6).

In performing an availability analysis it is obviously necessary to take into consideration the thermophysical properties of working fluids in detail. The properties of interest are the state and the transport properties. In particular, it is extremely important to include effects of dissociation and departures from equilibrium at high temperatures of the working fluids in analyzing several components of the propulsion system and the flight vehicle. Both the performance of such components and, therefore, of the system as a whole may depend critically on the occurrence of such high temperature phenomena. In the case of a scramjet (ramjet with combustion under conditions of supersonic flow) engines, for example, as the vehicle speed increases all of the components can be expected to be affected by departures from equilibrium state of the working fluid; furthermore, the effects in various components become coupled nonlinearly along the flowpath in the engine. An availability analysis must account for local thermophysical properties. However, in the illustrative cases of the current Chapter, no attempt is made to take into consideration the detailed thermophysical properties. Rather, mean values of equilibrium properties have been utilized over the relevant ranges of interest in various parts of the systems considered. This is, in part, due to the complexities and ambiguities involved in determining and taking into account accurately the state and transport properties[11]. More significantly, it is felt that so many significant aspects of the relative performance of various configurations of a system can be evaluated based on assumptions of thermal equilibrium; the emphasis is on the importance of utilizing energy availability and its changes as the basis of analysis.

3.1 Energy Availability in Air and Fuel

In an air-breathing engine the working fluids are comprised of air, its constituents, various fuels, and other gases or liquids. (Air is a working fluid both as a generator of aerodynamic lift and as the oxidant for combustion of fuel.) Oxygen may be carried onboard at take-off for possible use in a chemical rocket over a part of the flight path, or it may be processed from air collected during flight. Nitrogen may also be similarly processed and utilized as a propellant after it is suitably energized. The fuels may be divided into two groups: hydrocarbons and hydrogen. Other gases and liquids may be carried on board at takeoff for use as propellants or for flow control in the air inlet or elsewhere; water, in particular, may also be processed during flight and utilized for similar purposes.

In the following the methods of determining the energy availability of air and fuel are presented. Similar methods may be employed for other gases and also various inert and chemically active substances in different phases.

3.1.1 *Atmospheric Air*

The composition and properties of air in the atmosphere and the Mollier diagram for air can be found in Refs. 12 and 13 . The ranges of pressure and temperature along typical flight paths of high-speed vehicles include the rarefied gas state as well as the dissociated and the ionized gas states. Three parameters of interest along a flight trajectory (which is, in general, three dimensional in space) are 1) the mass flow per unit area, 2) the structural mechanical load, and 3) the thermal load. The mass flow of air per unit area is proportional to density and the mean velocity of air normal to the area under consideration. The structural mechanical load on a part exposed to oncoming air depends, to some extent, on the dynamic pressure of air, which varies directly proportional to density and the square of velocity. The thermal load similarly depends on the enthalpy equivalent of the kinetic energy of air. It may be pointed out here that the static enthalpy of air is about 100 Btu/lb and is nearly constant over the isothermal altitudes above the Earth.

Considering energy availability, the following applies to a stream of any flowing matter in the absence of electromagnetic fields and free surfaces and hence to air.

The energy availability of a stream of matter can be expressed by the following time-rate relation

$$\dot{\varepsilon} = \dot{\varepsilon}_p + \dot{\varepsilon}_k + \dot{\varepsilon}_{ph} + \dot{\varepsilon}_{ch} \tag{7}$$

where the four terms on the right hand side represent potential, kinetic, physical, and chemical components, respectively. The rate pertains to flux. The variables ε_p and ε_k are the same as E_p and E_k, respectively, the energy components, each relative to the environment

The physical component is the maximum amount of work obtainable from the stream of air when it is brought to the thermodynamic state of the given environment involving only physical processes and thermal interaction with the environment. For a gas at the state given by (P, H), the physical exergy with reference to an environment at (P_0, H_0) is given by the following when the environment temperature is T_0.

$$(H_1 - T_0 S_i) - (H_0 - T_0 S_0) \tag{8}$$

Considering air as a perfect gas with constant specific heats, the flux of availability per unit mass becomes

$$\dot{\varepsilon} = c_p (T - T_0) - T_0 [c_p \ln (T / T_0) - (R / \gamma) \ln (P / P_0)] \tag{9}$$

Although the necessary properties in the foregoing discussion are commonly available over certain ranges, special data sources may have to be found to cover the entire ranges of ambient conditions for high-speed flight[14,15].

The chemical component of energy availability is the maximum amount of work obtainable from the substance when it is brought to the given environmental state by means of heat transfer and exchange of substances only with the environment. It is of interest to note here that chemical energy availability may also be considered as being equal to the amount of work required to obtain, in the environmental state, the substance by means of processes involving only heat transfer and substances present in the environment. Thus, the chemical component of a stream of air with respect to an environment of identical composition and temperature is zero.

A common occurrence in atmospheric air is the presence of humidity and wetness. It is often convenient to treat moist air as a mixture. The energy availability of moist air may then be expressed with reference to energy availability of dry air. This is the basis on which the well-known psychrometric charts are constructed.

Finally, it may be necessary to consider air in the liquid state as in engines wherein air is condensed into liquid form; also, heat exchangers may have to be operated at the triple point, at critical points, or in the vicinity of saturation lines. Energy availability in such states can be determined on the basis of a knowledge of properties under equilibrium conditions and the departure from equilibrium of the state under consideration.

3.1.2 *Fuels*

The fuels of interest may be characterized by physico-chemical and combustion characteristics. The physico-chemical properties include density, heat capacity, changes in properties during phase change, transport properties, rate of chemical reaction and controllability of reaction. The combustion characteristics are the stoichiometric fuel/oxidant ratio, enthalpy generation, ignition temperature, and various aspects of flame chemistry and stability. In many cases combustion characteristics are further affected by the aero-thermal environment, including turbulence and shock wave or expansion processes in a combustor.

The general characteristics of hydrogen, in liquid, slush and gaseous forms, and various hydrocarbons can be found in Refs. 16-17. The implications of utilizing hydrogen in the cryogenic state are described in the first Chapter in this volume.

From the point of view of overall flight vehicle system analysis, the main considerations in regard to fuels are the following: 1) density, 2) heat capacity and latent heats, 3) spontaneous ignition temperature, 4) stoichiometric ratio with chosen oxidizers and corresponding heat-release rates, and 5) minimum pressure for combustion.

The following applies to any fuel for which there exists, in the given environment, all of the elements in the fuel either directly or in a chemical compound form. Thus, it applies to a hydrocarbon or hydrogen fuel provided carbon dioxide and water are present in the environment. If matter is present in the fuel for which no "reference" substance is present in the environment, it is necessary to proceed by either 1) assuming the "foreign" substance to be present both in the fuel and in the environment, or 2) appealing to the heat of formations of all substances involved using quantum mechanical considerations. Regarding the environment and reference substances, it is necessary to utilize only substances that are in chemically saturated, stable, and equilibrium states.

It is assumed that the composition, the chemical reactions involved with the oxidizer, and the calorific value of the combusting fuel are known in adequate detail. Furthermore, the following data must be available: composition and properties of the oxidizer, processes involved prior to and during combustion, and extent of heating, if any, of fuel or oxidizer.

The fuel and the oxidizer involved may be in nongaseous phase. Several types of chemical action and combustion may then be visualized. Reactions may occur between substances of the same or different phases. However, except for rockets, air-breathing engines for high-speed flight involve, in most cases, the use of gaseous reactants, especially when the fuel is already being utilized for cooling and thermal management and for subsequent recirculation into the fuel storage tanks. In any case, in all practical engines the reactions occur only in the gaseous state, and the products of combustion are also desired in the gaseous state for use in thrust-producing nozzles. Thus, the energy availability of a (solid, slush, liquid, or gaseous) fuel that is of interest is in the given state and the final, gaseous, product state with specified products, both with respect to the given environment.

As a general case, a fuel may be considered that is not in the gaseous state, is available in the moving vehicle intrinsically at a point along the trajectory, and yields specified products and a specified quantity of heat when reacted with a given oxidizer. The available energy of the fuel is then given by the following.

$$\dot{\varepsilon}_F = (\dot{\varepsilon}_{ie} + \dot{\varepsilon}_p + \dot{\varepsilon}_k + \dot{\varepsilon}_h + \dot{\varepsilon}_{pc})_F - (\dot{\varepsilon}_{ie} + \dot{\varepsilon}_p + \dot{\varepsilon}_k + \dot{\varepsilon}_h + \dot{\varepsilon}_{pc})_0 + \Delta Q_C^* \tag{10}$$

If the oxidant is nothing but the environment, $(\dot{\varepsilon}_p + \dot{\varepsilon}_k + \dot{\varepsilon}_h + \dot{\varepsilon}_{pc})_0$ becomes zero.

It is of interest to point out that ΔQ_C may be positive or negative depending on whether the fuel reaction with the given oxidizer is exothermic or endothermic, respectively. In the case of exothermic fuel reactions the chemical potential of reactants is always greater than that of products. In the case of endothermic reactions it is necessary to note that they automatically terminate when heat is not available to sustain the reaction. For both exothermic and endothermic reactions, finite reaction rates give rise to irreversibilities.

3.2 Aerothermodynamic Processes in a Flight Propulsion System

A brilliant account of the possible application of energy and entropy analysis to aerospace systems has been provided by J. Ackeret in 1963[18]. Ackeret considers a series of elementary processes and, after calculating entropy change in each case, draws important conclusions of aerospace field interest. Some of these processes and others of direct interest to high-speed flight systems are discussed in the following.

1) *A shock wave*: This is a known, predictable and controllable flow process that can be utilized for increasing static pressure and static temperature of a gas or, more generally, a compressible fluid. For example, a waverider[19] and a shock-supported combustion burner with supersonic flow[20] both utilize a shock wave. The shock tube is a simple test facility that is also dependent on a shock-wave formation. It is possible that the total energy content of the fluid under consideration remains constant across a shock wave in most cases. However, there is always a loss of energy availability in view of the irreversible processes occurring in the shock wave. Oswatisch[21] shows how a finite magnitude of compression utilizing shock waves is carried out best with a series of equal strength shock waves rather than with a single shock wave, on the basis of energy availability becoming maximized downstream of the train of shock waves. It is also of interest to note that the thickness of a shock wave remains nearly the minimum value above a certain value of an upstream Mach number in air, for example, of about 6. The availability loss across a shock wave depends on the changes in state properties occurring over its thickness and also in the region downstream that may be required for the gas state to come to equilibrium with respect to its internal energy.

2) *A boundary layer*: Molecular viscosity and turbulence in wall-bounded flows invariably lead to loss of energy availability of the fluid over a length of flow, even when the flowfield can be treated as adiabatic. When the process in the boundary region is nonadiabatic, a loss of energy and a further loss of energy availability across the control volume occur. In high-speed flows there are several further considerations: 1) When shock waves are present, additional losses in energy and energy availability can result from shock-wave/boundary layer interaction processes. Those processes can occur over a region extending over a length of several times the momentum thickness of the boundary layer both upstream and downstream of the shock-wave location. The control volume for the calculation of losses must therefore be extended accordingly. 2) It is possible to recognize a region within the boundary layer wherein the largest change in entropy occurs, the so-called entropy layer[22] in hypersonic flows. 3) The fluid is, in general, not in equilibrium, and this in itself can cause an energy availability loss. Flow past a body, such as a wing, gives rise to a wake behind the body. A direct consequence of molecular viscosity and turbulence of a fluid in flow past a boundary, the formation of the wake leads to viscous drag of a flight vehicle. If the energy availability in the wake of a flight vehicle can be increased to the initial

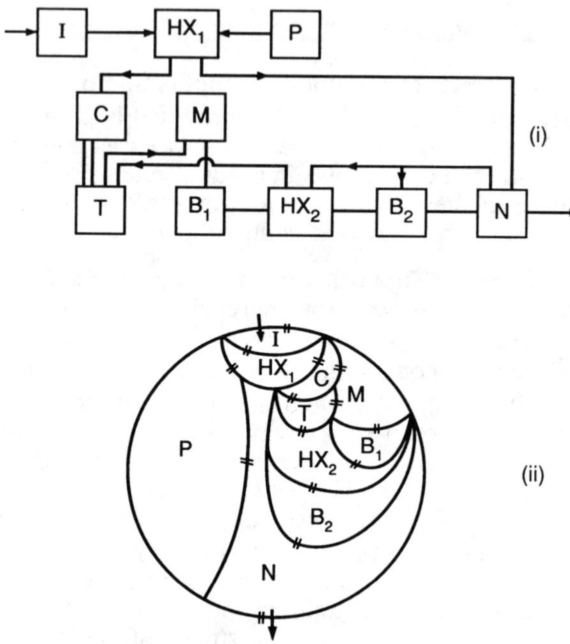

7. Air turbo-ramjet expander cycle engine (ATREX): (i) Block diagram; (ii) Structure diagram. M = mixer. Other symbols as in Figs. 4,5, and 6.

value in the air upstream of the vehicle, there is no need for propulsion power for overcoming viscous drag.

3) *An expansion process in a nozzle*: The expansion ratio from a constant-pressure reservoir becomes infinite when the pressure in the atmosphere into which a fixed area nozzle discharges approaches zero. The highest velocity that is feasible in an expanding gas at a given stagnation enthalpy is proportional to the square root of the enthalpy value; this occurs when the external temperature is near absolute zero. A further interesting consideration in a practical nozzle arises when the temperature of the gas adjoining the boundary wall becomes nearly equal to that of the wall permitting no heat transfer through the wall. The following question then arises: Are there possibilities for maintaining energy availability of the expanding gas nearly constant or for varying it in a specified manner during an expansion process?

In any thrust nozzle an energy availability loss arises because of the state and kinetic energy of the nozzle exhaust flow relative to the environment. Considering the vehicle as a whole, the propulsive efficiency of any extrinsic system becomes higher, but the specific thrust reduces, as the difference between the velocities of the working fluid at entry and exit of the vehicle becomes smaller. Regarding the loss of heat in the exhaust, one can ask the following question: Are there ways of reducing it in practice?

More generally, one can seek to accomplish the following in a flight system: 1) Maintain the energy availability of the working fluid streams at values close to the highest value possible; and 2) make the best use of energy availability in all components or elements of the system. A few examples of such schemes with respect to high-speed flight systems[23,24] are shown schematically in Figs. 7 and 8.

4) *The mixing process*: Mixing of two bodies of fluids results in an entropy gain unless the fluids involved are identical, a trivial case, or the process is accompanied by heat addition and performance of work, which may not be practicable. The reference mixed state that is usually chosen corresponds to uniform, molecular-level mixing. The initial states of the two bodies of fluid as well as the final mixed state may be in a chemical nonequilibrium state. A body of gas may be in a chemical nonequilibrium state or a frozen chemical state that is also a form of a chemical nonequilibrium state. A fluid can be characterized by its absolute energy availability, although this requires a knowledge of the precise composition and state of the fluid. The problem becomes complicated when the fluids in a mixing process are initially in chemical nonequilibrium and also involve reaction, combustion, or a different state of chemical nonequilibrium in the course of mixing. No other parameter can, in such cases, provide the same amount of information as energy availability for the initial or the final states.

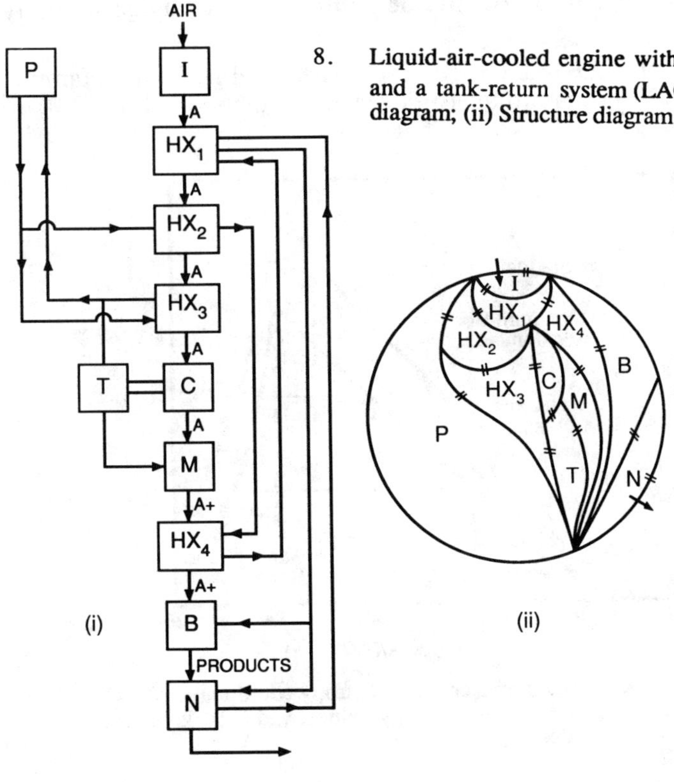

8. Liquid-air-cooled engine with an air compressor and a tank-return system (LACTR): (i) Block diagram; (ii) Structure diagram.

It must be pointed out here that various length, time, velocity, and chemical reaction scales become significant in the analysis of mixing, combustion, and high-speed gas dynamic processes. (Dilatation and the second coefficient of viscosity also introduce additional scales.) When flows are turbulent, the presence of various scales and of a peakness in the distribution of scales becomes very important. Finally, unsteadiness and waves are characterized by length and time scales. Relaxation from a nonequilibrium state to another state requires time and, in flows, a length of distance. Calculation of entropy or energy availability taking into account such scales is only feasible with an appeal to quantum mechanics, even in the case of relatively simple thermodynamic systems.

Mixing losses are of critical importance in high speed flows, for example, in combustion chambers and ejectors and also in such processes as film cooling and maintaining a slurry or a slush in a homogeneous state.

3.3 Application to Brayton Cycle

The Brayton cycle (Fig. 9) is a commonly adopted engine cycle in aircraft propulsion. Its application in the context of high speed vehicles has been discussed in an important paper by Builder[25]. Several aspects of Builder's analysis are discussed here and in Section 3.4. In the following the application of energy availability analysis to the Brayton cycle is presented utilizing the approach of Builder. Other cycles may be similarly analyzed.

It is common practice in the case of low speed vehicles (for example, vehicles for which the maximum flight Mach number is about 3.0) to

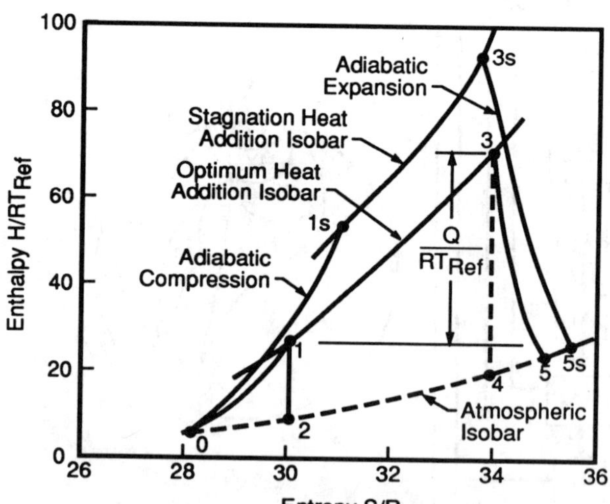

9. Brayton cycle representation is enthalpy-entropy coordinates. 0-1-3-5-4-2-0 path refers to the static enthalpy values, and 0-1s-3s-5s-0 to stagnation enthalpy values.

analyze the Brayton cycle of the propulsion engine in terms of stagnation properties of the working fluid. The stagnation properties permit a proper, direct means of accounting for the internal, kinetic, and combustion or other added energies. However, at higher flight speeds, the kinetic energy of air entering the engine from the atmosphere becomes a substantial fraction of the total energy of the working fluid at the entry to the thrust nozzle; in fact, it becomes considerably larger than the energy added by combustion of an ordinary fuel as the flight speed increases. Therefore, it becomes imperative, as first pointed out by Builder, to consider the internal and the kinetic energies of the working fluid separately and distinctly throughout the cycle. Figure 9 provides both static and stagnation values of enthalpy in the Brayton cycle representation.

Builder's analysis of the basic Brayton cycle is based on the following assumptions: 1) The output thrust is produced by expansion of the working fluid; and 2) the flow velocity is constant across the combustor. It can then be shown that the thrust developed by the engine is at a maximum when the static enthalpy change across the inlet diffuser and the mechanical compressor unit, either separately when only one of them is utilized or together, is given by

$$\psi_A = H_1 / H_0 = [\eta_\Sigma / (1 - \eta_\Sigma)] \cdot (Q_T / H_0) \tag{11}$$

where $\eta_\Sigma = \eta_C \cdot \eta_E$ represents the product of the inlet-compressor efficiency, $\eta_C = (H_1 - H_2)/(H_1 - H_3)$, and the turbine-nozzle efficiency, $\eta_E = (H_3 - H_5)/(H_3 - H_4)$. The energy added per unit mass of air is $Q_T = (H_3 - H_1)$, a given quantity in the optimization. It is clear that, when the air entry speed is high so that adequate diffusion in the inlet yields the desired pressure, there is no need to include a compressor and therefore a turbine to drive it.

Builder has defined a so-called energy conversion coefficient, θ, as follows for the total amount of energy Q_T added per unit mass of air in the cycle:

$$\theta = FV_0 / (gJ\dot{m}Q_T) \tag{12}$$

where thrust work is denoted by FV_0 and $\dot{m}$ represents the mass flux of (only) airflow through the cycle. The energy conversion coefficient is in the nature of overall efficiency, also referred to as overall propulsive efficiency, of the cycle.

The propulsive efficiency η_{PR} (sometimes referred to as jet efficiency) provided by the engine may be defined as follows:

$$\eta_{PR} = 2V_0 / (V_0 + V_5) \tag{13}$$

In order to obtain a high value of η_{PR}, a low value of V_5 is obviously needed. In the simplest case considered by Builder, the following is obtained:

$$V_5^2 / 2gJ = \left(H_1 + V_1^2 / 2gJ\right) + Q_T - H_5 \tag{14}$$

It follows that $(Q_T - H_5)$ should be small with a given performance of the inlet for high η_{PR}. On the other hand, a high value of thrust requires a high value of V_5 and/or a high value of mass flow through the nozzle. It may also be noted that the net production of kinetic energy across the engine $(V_5^2 - V_0^2) / 2gJ)$ is given by the product of Q_T and the thermal efficiency of the cycle η_{th}. Thus, in order to improve propulsive efficiency it is necessary to reduce all three of the quantities Q_T, H_5, and V_5, since Q_T is the heat input in the cycle and H_5 and $V_5^2 / 2gJ$ both constitute losses in the cycle.

It may also be noted that the energy added by the fuel to the airstream is entirely due to combustion in the current simple case. Denoting energy added per unit mass of air by combustion by Q_C and, thus, setting $Q_T = Q_C$, the following expression can be obtained for V_5 from Eqs. (11) and (14):

$$(V_5 / V_0)^2 = (V_1 / V_0)^2 + [(\psi_A - 1) / K] \cdot [\eta_\Sigma (1 + C / \psi_A)] \tag{15}$$

where

$$C = Q_C / H_0 \tag{16}$$

and

$$K = V_0^2 / 2gJH_0 \tag{17}$$

Several modifications may now be considered in the basic Brayton cycle.

First, it is possible to include an intercooler in the diffuser-compressor, especially at low flight speeds when a compressor may be included in the cycle along with an inlet. The optimum ratio H_1 / H_0 for maximum thrust is still given by Eq. (11).

Second, as the mass of fuel added becomes close to the stoichiometric value or higher, it is necessary to take that mass into account in the total mass of working fluid available for turbine operation and thrust production. It may also be useful in some cases to enrich the incoming air with oxygen. This is discussed by F. S. Billig in Chapter 1 in this volume. Denoting the fuel equivalence ratio by ϕ_F, Eq. (12) for the energy conversion coefficient becomes modified as follows, taking into account fuel mass in the total mass of propellant.

$$\theta = \frac{2K}{Q_T/H_0} \left[\frac{V_5}{V_0} (1 + \phi_F \frac{f}{a}) - 1\right] \tag{18}$$

and V_5 is given by the following relation:

$$(1 + \phi_F \frac{f}{a}) V_5^2 = V_0^2 + 2gJ\, H_0 (\psi_A - 1) \cdot \left[\eta_\Sigma \left(1 + \frac{Q_T/H_0}{\psi_A}\right) - 1\right] \tag{19}$$

Here ψ_A and Q_T may be chosen as desired, or one can introduce the optimum value of ψ_A for given values of Q_T and η_Σ. It may be pointed out that the optimum value of ψ_A is given again by Eq. (11).

Third, the total amount of energy added per unit mass of fuel may in general consist of 1) the energy released by combustion, Q_C^*, 2) the energy due to its velocity, Q_V^*, and 3) its internal energy, Q_H^*. The fuel may be energized with heat when it is utilized for cooling of vehicle skin or internal parts and with added momentum. It is clear that the energy required in the burner for heating and accelerating the fuel to air temperature and velocity values can be reduced by means of supplying the fuel at temperature and velocity values close to those of air. The ease of ignition may also demand fuel heating. Denoting the combustion efficiency, or combustion completeness factor by η_{CC} and the sum of losses in the combustor due to mixing between fuel and air, friction, shock waves, and heat transfer by Q_{LC}, the total heat added, Q_T, may be written as follows:

$$Q_T = \eta_{CC} \frac{f}{a} Q_C^* + \phi_F \frac{f}{a} \left(Q_V^* + Q_H^*\right) - Q_{LC} \tag{20}$$

where f/a denotes the stoichiometric fuel-to-air ratio. It is often of interest to examine the influence of Q_{LC}. Denoting the gross amount of heat supplied to the working fluid by Q_T^* that is defined by the following

$$Q_T^* = \eta_{CC} \frac{f}{a} Q_C^* + \phi_F \frac{f}{a} (Q_V^* + Q_H^*) \tag{21}$$

the value of V_5 can be expressed, using Eq. (19), by the following relation, which explicitly shows the effect of Q_{LC}:

$$\begin{aligned}(1 + \phi_f \tfrac{f}{a}) V_5^2 = V_0^2 &+ 2gJ\, H_0 (\psi_A - 1) \cdot \left[\eta_\Sigma \left(1 + \frac{Q_T^*/H_0}{\psi_A}\right) - 1\right] \\ &- 4gJ\, H_0 (Q_{LC}/H_0)\end{aligned} \tag{22}$$

Ignition and completeness of combustion, and mixing, on which the other two depend, all involve various length and time scales. Chemical reaction (and also processes such as compression or diffusion and expansion) must, include a consideration of chemical equilibrium during those processes.

Finally, returning to the consideration of the methods of realizing both the required thrust and a high value of η_{PR}, one can proceed as follows: For given values of Q_C^*, Q_H^*, and Q_V^* and also, η_C and η_E, the values of ψ_A and Q_T are chosen in order to obtain, through an iterative procedure, a desirable, high value of η_{PR}. The desired value of thrust per pound of air is obtained by choosing the required value of ϕ_F (or ϕ_o, the oxidant equivalence ratio). This step also involves an iteration in view of the connection between Q_T and ϕ_F.

It may be pointed out that the foregoing procedure is somewhat similar to the one adopted in improving the efficiency of a high-bypass-ratio engine: The core flow is restricted to a level required for adding the desired amount of fuel energy, and the bypass ratio and the fan pressure ratio are increased to a level required for obtaining the desired thrust. Since the high-speed vehicle engine has to operate under a variety of flight conditions depending on the trajectory and the variation in flight speed chosen, ϕ_F and Q_T have to be varied so that η_{PR} and thrust are high enough everywhere along the flight. Again, one can draw analogy with a high-bypass-ratio engine with variable geometry (and also variable cycle) and mixers for air and combustion products.

Returning to the problem of obtaining the desired value of thrust and η_{PR}, once ψ_A, Q_T and ϕ_F are chosen, the thrust per unit mass of air in the cycle becomes determined; therefore the mass flow of air required to yield the desired thrust is also determined. For the given conditions of the environment and flight velocity (or air entry velocity into the inlet), one can find the inlet cross-sectional area and the nozzle exit area. From considerations of the resulting volume and length of the vehicle, a total of three iterations are required to meet energy and air mass flow needs and configuration constraints. It is useful to note here that, in the evolution of the flight vehicle system, there are yet other considerations pertaining to weight, trajectory, and geography of vehicle takeoff and landing; an approach to the preliminary design of a flight vehicle is discussed in Section 6.0.

3.3.1 Energy Availability Considerations

As seen in Fig. 9, energy availability can be calculated at stations 0, 1, 3, and 5 given the state conditions at those points and the properties and energy of fuel added between 1 and 3. When the working fluid is in a gaseous state, the energy availability rate can be established at each station utilizing Eq. (9). The calculations yield 1) the energy availability rate relative to a chosen environment (generally, most conveniently, the ambient atmosphere) at each station; 2) the change in the energy availability rate over the process paths 0-1, 1-3, and 3-5; 3) the thrust work generated in the engine; 4) the loss of energy availability rate in the exhaust; and 5) the rational (or true thermodynamic) efficiency of any of the individual processes and of the cycle as a whole. Regarding item 2, the change in availability takes into account mechanical efficiency, heat transfer, work input and heat input as they apply over the processes under consideration. Item 4 may be divided into loss resulting from temperature of the exhaust in excess of that of ambient conditions and loss in kinetic energy form due to exhaust velocity.

Regarding items 3 and 5, it is of interest, in practice, to determine and maximize the thrust power output as well as the overall propulsive efficiency of the engine for the amount of fuel and air utilized. The *overall propulsive efficiency* may be defined as follows.

$$\eta_0 = \eta_{TH} \cdot \eta_{PR} \tag{23}$$

$$= \left[\frac{1}{2}(V_5^2 - V_0^2)/Q_T\right] \cdot \left[V_0 (V_5 - V_0)/\frac{1}{2}(V_5^2 - V_0^2)\right] \tag{24}$$

where the subscripts *0* and *TH* indicate overall propulsive, and thermal efficiency values, respectively; overall propulsive efficiency is sometimes referred to as overall efficiency. It can be observed that the energy-conversion coefficient of Builder [Eq. (12)], is merely the following.

$$\theta = V_0 (V_5 - V_0) / Q_T \tag{25}$$

and hence, equal to η_o.

For a given value of Q_T the jet efficiency is maximized, as stated earlier, when $[2V_0 / (V_5 + V_0)]$ is a maximum. At the same time, it is of interest to maximize θ as in Eq. (23). The method of accomplishing these two objectives is to convert the maximum possible fraction of energy availability in air and fuel entering the engine into thrust power. In other words, it is necessary to reduce the loss of energy availability in each of the components and, also, in the nozzle exhaust.

3.4 Supersonic Ramjet Engine

A supersonic flow ramjet engine may be said to operate on the basis of the Brayton cycle, as do the common gas turbine and the subsonic ramjet. Thus, the thermodynamic cycle is the same as that shown schematically in Fig. 9. A brief discussion of the energy availability analysis of the cycle has been presented in the preceding section. In the following a supersonic flow ramjet engine is analyzed on the same basis utilizing Builder's approach. The application of the analysis is illustrated with reference to a flight vehicle that operates with a supersonic flow ramjet engine.

The application of the Brayton cycle to high-speed propulsion systems must take into consideration 1) the increasing kinetic energy available in air entering the engine as flight speed increases and 2) various methods of energizing the working fluid in the engine. Regarding consideration 1, Fig. 10 shows the ratio of kinetic energy of air to the combustion energy that can be added under stoichiometric conditions by burning hydrogen fuel in air. Regarding consideration 2, fuel mass itself is a means of increasing thrust as is the addition of any mixing substance to the propellant stream in the engine. For example, water (which may be generated intrinsically from the products of combustion of hydrogen in air) may be added to the propellant

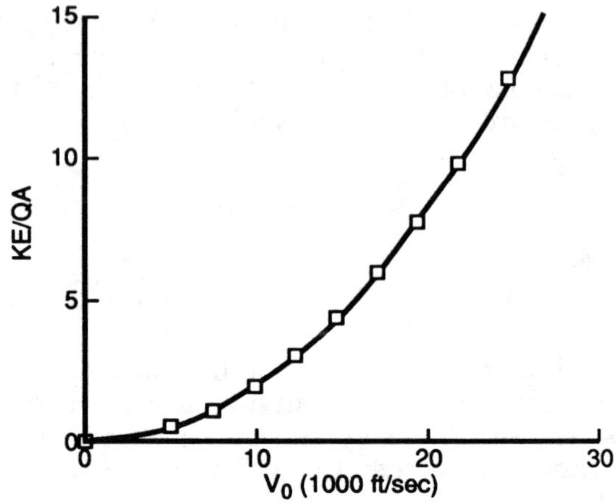

10. Ratio of kinetic energy of air (KE) to the combustion energy of hydrogen under stoichiometric conditions (QA) as a function of flight-speed (V_0).

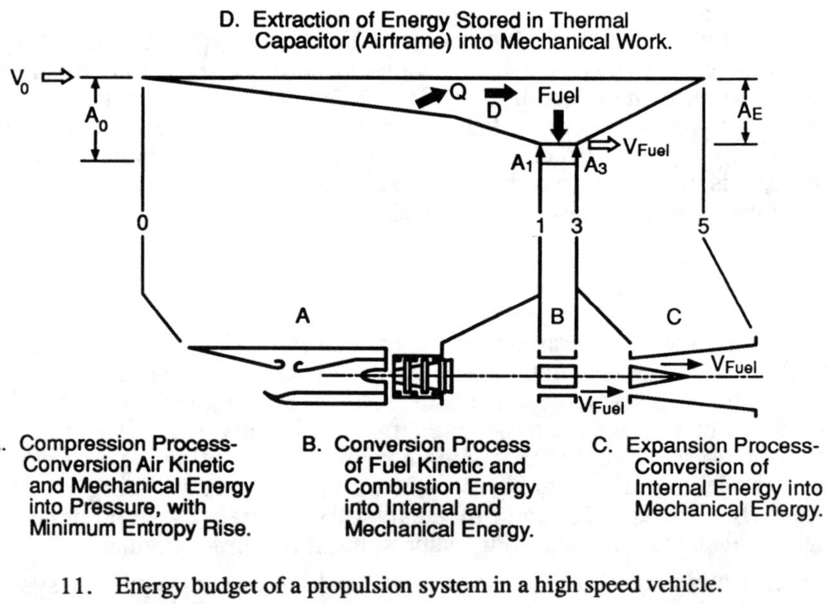

11. Energy budget of a propulsion system in a high speed vehicle.

stream at several useful locations in the engine and provides additional mass for thrust generation. Similarly, the fuel stream may be energized before supply to the combustor by means of heat addition while it is utilized for cooling of the vehicle skin or engine components themselves. The fuel stream may also be provided with additional momentum before entry into the combustor so that the loss of energy in accelerating the fuel to that of air speed can be eliminated or reduced. These possibilities are shown schematically in Fig. 11.

Inlet Performance. The method of treating internal and kinetic energies separately, which has been discussed in Section 3.3, is particularly valuable in establishing inlet performance unambiguously and with the greatest clarity regarding diffusion of momentum, gain in pressure, and loss of heat due to heat transfer. Fig. 12 provides a thermodynamic diagram for the inlet process. Three efficiency parameters may be defined as follows:

$$\eta_{HS} = (H_{1'S} - H_0)/H_{1S} - H_0) \tag{26}$$

$$\eta_C = (H_1 - H_2)/(H_1 - H_0) \tag{27}$$

$$\eta_{KE} = (H_{1S} - H_2)/(H_{1S} - H_0) \tag{28}$$

or, expanding the terms,

$$\eta_{KE} = \left[(H_1 + V_1^2/2gJ) + Q_{LI} - H_2\right]/(V_0^2/2gJ) \tag{29}$$

where Q_{LI} represents the total loss in the inlet due, for example, to friction, shock waves, and heat transfer. The first efficiency parameter refers to the loss of energy due to heat transfer. $H_{1'S}$ is the value of stagnation enthalpy obtained at the exit of the inlet while H_{1S} is the value of stagnation enthalpy at the entry to the inlet. The heat transferred to the inlet walls may be regained by means of fuel circulation, but insofar as the inlet is concerned, it leads to an inefficiency. The second equation relates to the efficiency of compression with respect to static pressure; it reflects the extent to which the enthalpy change is converted into pressure. Finally, Eq. (29) yields the efficiency of conversion of initial kinetic energy into pressure and kinetic energy at the exit of the inlet. Thus, inlet effectiveness involves all three of

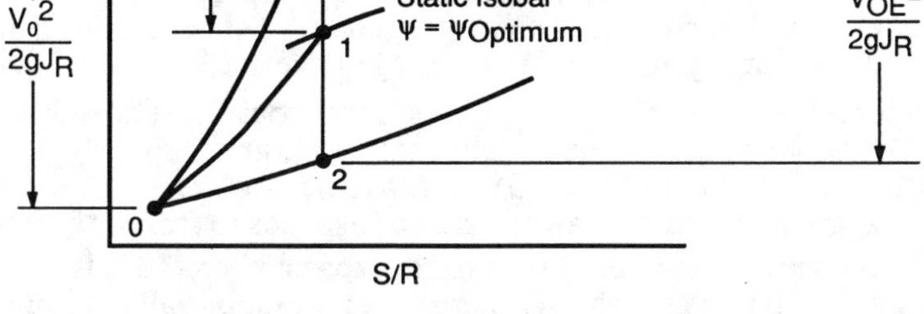

12. Thermodynamic diagram for the ilet process in enthalpy (H/R) entropy (S/R) coordinates.

the efficiencies. It can be noted here that the standard approach to defining inlet effectiveness in terms of actual change in stagnation or static pressure compared with the dynamic head available at inlet does not yield the details provided by the current performance parameters, and can often be misleading.

Another method of examining the losses and efficiency parameters is to write the following.

$$\eta_{KE} = k_D + (1 - k_D)(V_1 / V_0)^2 \tag{30}$$

Thus, on rewriting, the factor k_D becomes

$$k_D = \left[\eta_{KE} - (V_1 / V_0)^2\right] / \left[1 - (V_1 / V_0)^2\right] \tag{31}$$

Here the ratio (V_1/V_0) can be chosen at any value between zero and unity, noting that

$$(V_1 / V_0)^2_{OPT} = 1 - [(\psi_{OPT} - 1) / K] \tag{32}$$

An inlet provides a diffusion of air momentum and thus an increase in static pressure. In Section 3.3 expression (11) has been given for the optimum enthalpy ratio across the inlet for obtaining the maximum value of thrust with given heat energy input and losses in the system. The optimum enthalpy ratio, along with inlet loss, yields a static pressure ratio across the inlet. It may be pointed out that the pressure rise obtained at entry to the burner must be high enough to permit sustained combustion of given fuel, for example, about 7 psi in the case of hydrogen fuel.

Engine performance parameters. The two main performance parameters are as follows: 1) the thrust output for a given value of (H_1/H_0) and fuel energy input Q_T, both per unit mass of air entering the engine; and 2) the energy- conversion coefficient.

Regarding parameter 1, it was shown by Builder that the thrust can be maximized in an engine with given component efficiencies and fuel energy input rate when ψ is chosen appropriately [viz., Eq. (11)]. The optimum value of ψ, ψ_{OPT}, is independent of flight velocity, provided it is assumed that 1) H_0 is nearly constant with altitude; and 2) heat addition in the combustor can be arranged with no net change in velocity.

In determining the value of ψ_{OPT} several considerations should be noted: 1) Air properties are such that at higher enthalpy levels (corresponding to large flight speed), entropy changes are high for various processes; 2) the air thermodynamic state corresponds to frozen equilibrium as altitude and speed increase; 3) the nozzle expansion process also leads to a frozen equilibrium state; and 4) there is a minimum value of pressure below which a given fuel will not undergo combustion. Based on these considerations and assuming the use of a particular fuel, one can construct a map of ψ_{OPT} as a function of $\eta_\Sigma = \eta_C \cdot \eta_E$, utilizing heat addition by combustion of fuel per unit mass of air as a parameter. Figure 13 provides such a diagram for the case of hydrogen fuel and other parameters as indicated therein.

The manner in which ψ_{OPT} changes for a number of fuels is presented over a wide range of flight speeds in Fig. 14. The range of flight speed also covers low values; at such speeds combustion occurs at subsonic flow velocities. In Fig. 14 it is extremely significant that, above a flight Mach number of about 5.5, only partial diffusion of airflow momentum is required within the engine, with no necessity of mechanical compression. The implication is that the reduction in momentum may, in fact, be such that airflow into the combustor may remain supersonic optimally. Thus, in general, whether combustion in the burner should occur under conditions of supersonic or subsonic airflow velocity depends entirely on the extent to which air momentum should be diffused for best performance. This fact is also illustrated in Figs. 15 and 16.

A similarly significant fact concerning the low-speed region in Fig. 14, where both diffusion and mechanical compression may be needed, is that it includes supersonic flight speeds toward the higher speed region. One can then examine whether it is worth introducing diffusion to subsonic velocities in the inlet before using a subsonic fan for mechanical compression, or whether using a supersonic through-flow fan with little prior diffusion in the inlet may be more advantageous. A discussion on using a supersonic through-flow fan is included in Section 3.5.

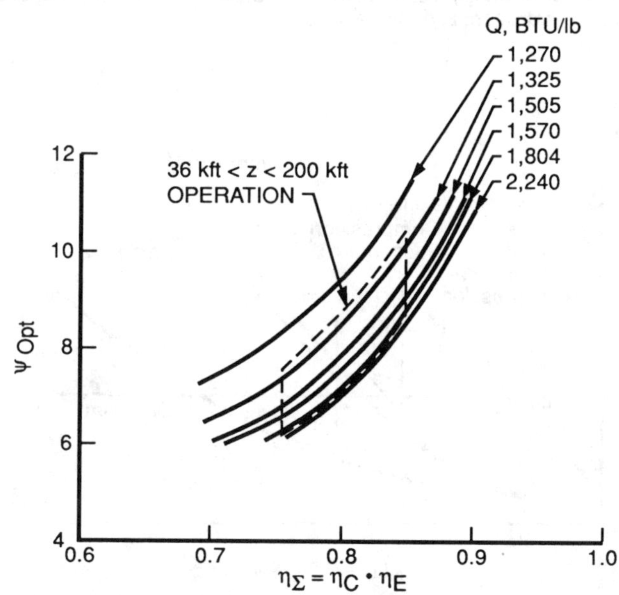

13. Optimum enthalpy ratio in inlet (ψ_{OPT}) as a function of overall efficiency ($\eta_\Sigma = \eta_c \cdot \eta_E$) utilizing heat addition by combustion (Q) as a parameter. Fuel is hydrogen, and flight altitude range 36,000 - 200,000 ft.

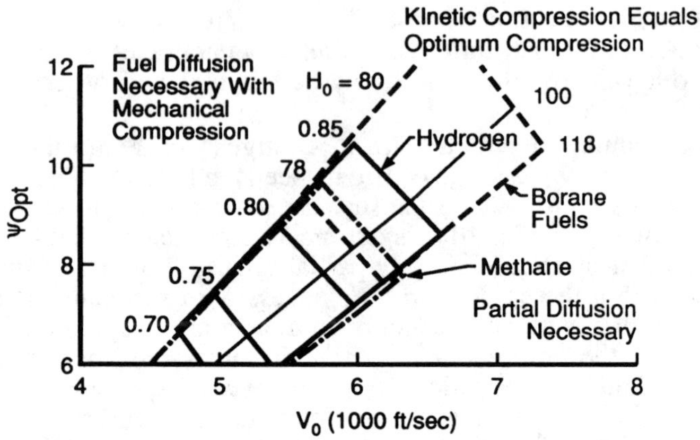

Minimum Pressure (7 psia)				Maximum Pressure (150 psia)			
H_1/H_0	η	$(S/R)_0$	$z_0/1000$	H_1/H_0	η	$(S/R)_0$	$z_0/1000$
6.17	0.716	28.2	111	5.80	0.692	25.1	59
7.40	0.785	29.0	127	7.53	0.790	26.1	70
8.64	0.833	29.8	142	9.74	0.863	27.4	96
12.34	0.910	31.5	182	12.59	0.913	28.7	122

14. Optimum enthalpy ratio in inlet as a function of flight speed for borane, hydrogen, and methane fuels.

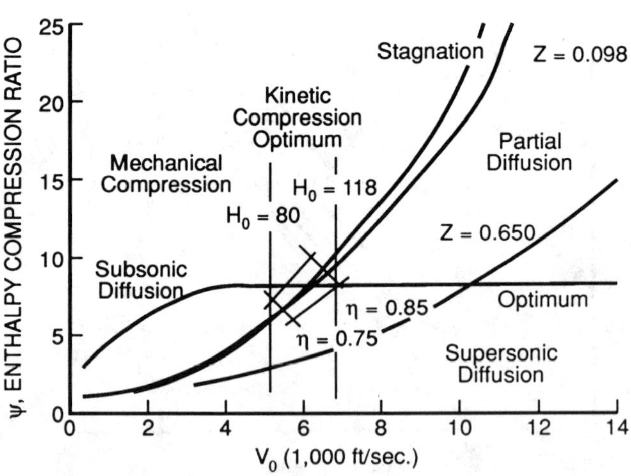

15. Enthalpy (compression) ration in inlet as a function of flight speed. $Z = (V_1 / V_0)^2$.

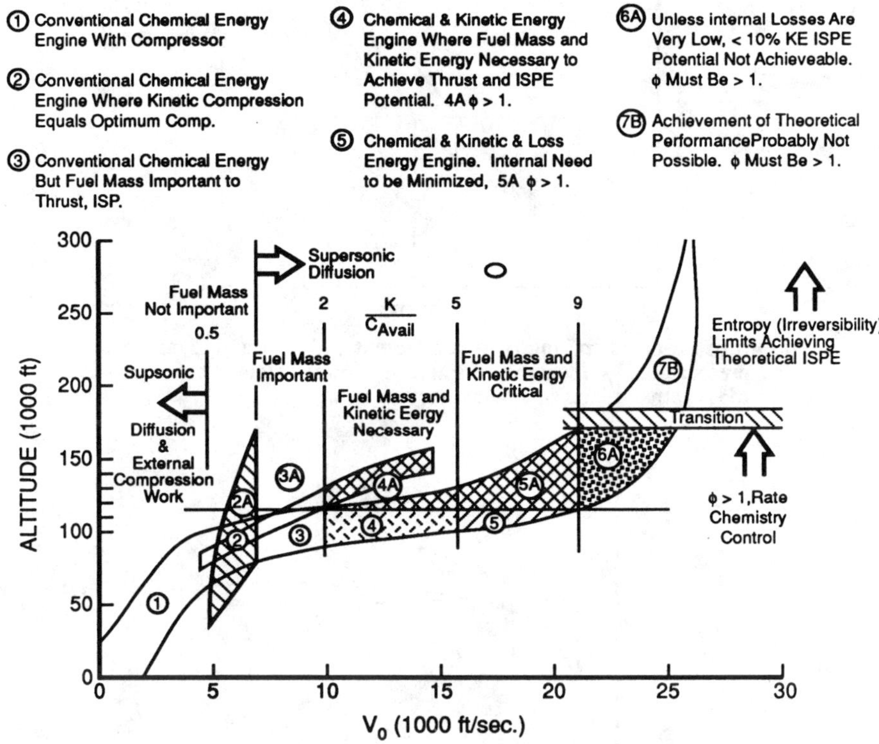

16. Energy and mass addition along the flight trajectory in altitude-flight velocity coordinates.

Other considerations in the Brayton cycle. Both the optimal compression for maximum thrust and the energy conversion coefficient are functions of several other parameters: 1) mixing loss in pressure or energy in the combustor arising because of the need to mix fuel and air to the extent (molecular level) that chemical reactions become feasible; 2) mass addition due to addition of fuel (at stoichiometric or another value) or additional propellant (for example, water, as mentioned earlier.); 3) momentum addition to the fuel stream; and 4) heat addition to the fuel stream.

A brief discussion of the mixing of streams of fluids was given in Section 3.2. A diagram that illustrates the nature of mixing loss is given in Fig. 17. In the diagram the manner in which entropy gain occurs in mixing can be observed. Some aspects of estimating mixing loss and its effect on engine performance are discussed by J. Swithenbank, and P. E. Dimotakis and J. P. Drummond in their respective Chapters in this volume.

The effects of parameters 1-4 (in the foregoing list) are illustrated most usefully in terms of the changes in energy-conversion coefficient in various cases.

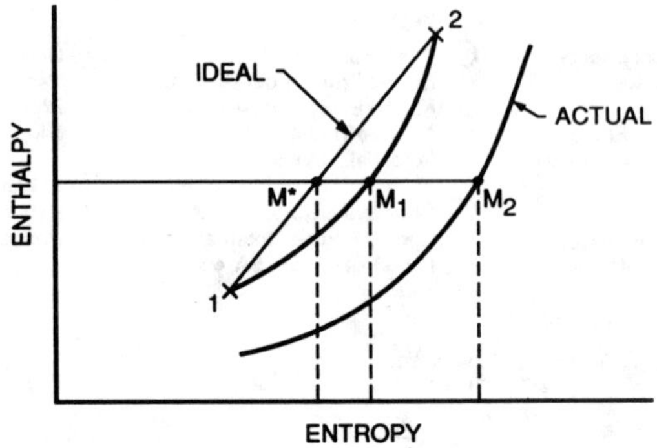

17. **Illustration of mixing of two different fluids, 1 and 2, at the same pressure. M^* = ideal mixed state; M_1= mixed state at same pressure; M_2 = actual mixed state with pressure loss.**

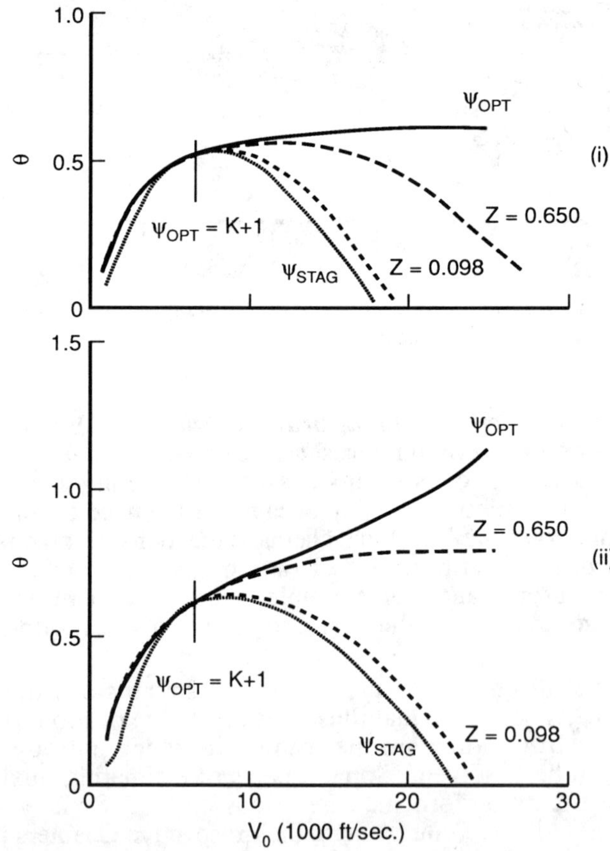

18. **Energy conversion coefficient as a function of flight speed (i) without and (ii) with fuel mass addition included, utilizing ψ and Z as coordinates.**

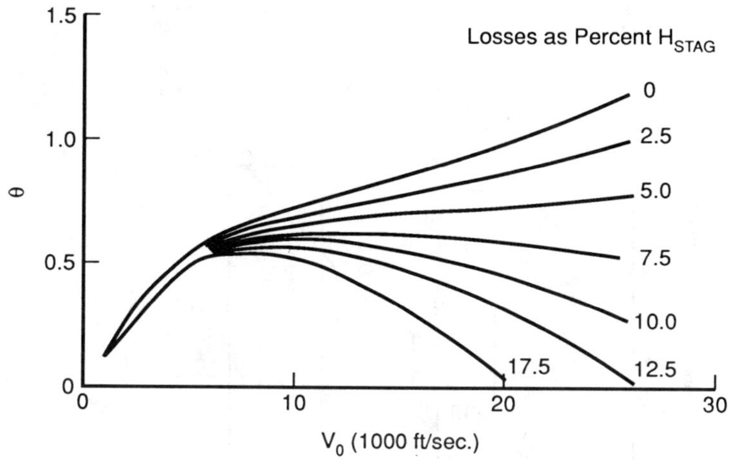

19. Effect of combustor losses (friction, mixing and combustion) on energy conversion coefficient as a function of flight velocity.

The effect of mass addition on energy-conversion coefficient is presented in Fig. 18. In Fig. 19 the effect of combustor losses, expressed as a percent value of stagnation enthalpy of the working fluid, is also accounted for in presenting energy conversion coefficient as a function of flight velocity. It may be noted that, in Fig. 19, $\psi = \psi_{OPT}$. The combustor losses have a substantial effect on energy-conversion coefficient, as may be observed further in Fig. 20. Here the change in the energy-conversion coefficient for a percent increase in losses in the combustor is presented as a function of flight velocity; the ratio of kinetic energy (corresponding to the flight speed) to fuel combustion energy is also shown on the diagram as a matter of additional interest.

In summary, the effect of various parameters on energy conversion coefficient is illustrated in Fig. 21.

The application of a supersonic flow ramjet engine to a flight vehicle is discussed in Section 6.0.

The following examples of engines relate to the low-speed flight of a vehicle, generally less than about 8000 ft. However, these engines may be utilized in cruise-type vehicles or over the low-speed part for an accelerator vehicle.

3.5 Air-Turbo-Rocket with Supersonic Through-Flow Fan

This is an example of an engine for a flight speed below about 8000 ft. The air-turbo-rocket scheme suggested in Ref. 26 is based on a cycle illustrated in Fig. 22. The supersonic through-flow fan (STFF), with no diffusion in an inlet, is driven by a turbine operated with fuel-rich exhaust gas from a rocket. The turbine exhaust is then burned in a second

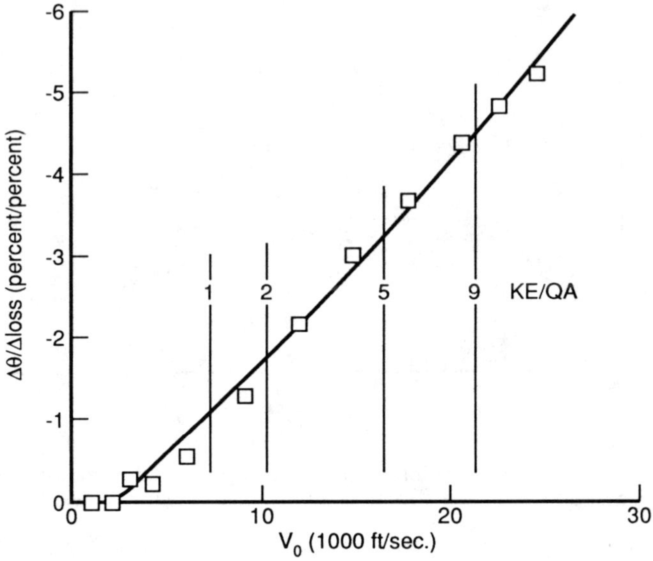

20. **Change in energy conversion coefficient per unit change in combustor losses as a function of flight speed.**

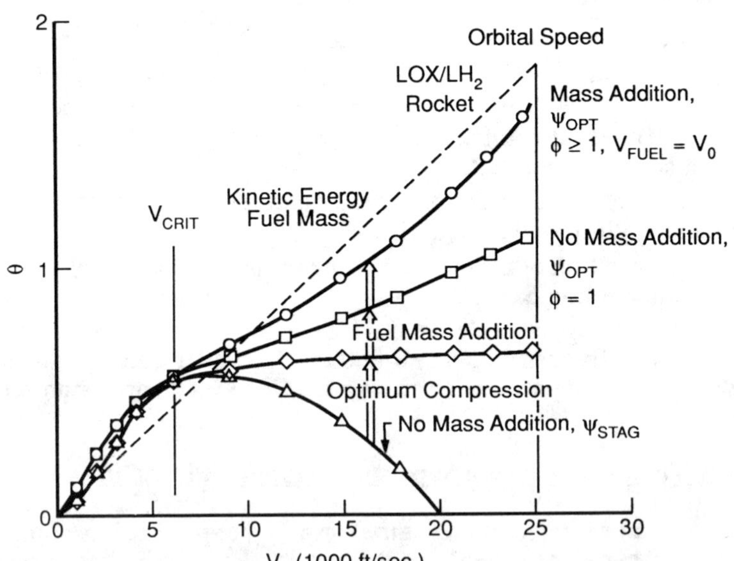

21. **Summary of effects of various parameters on energy conversion coefficient as a function of flight speed.**

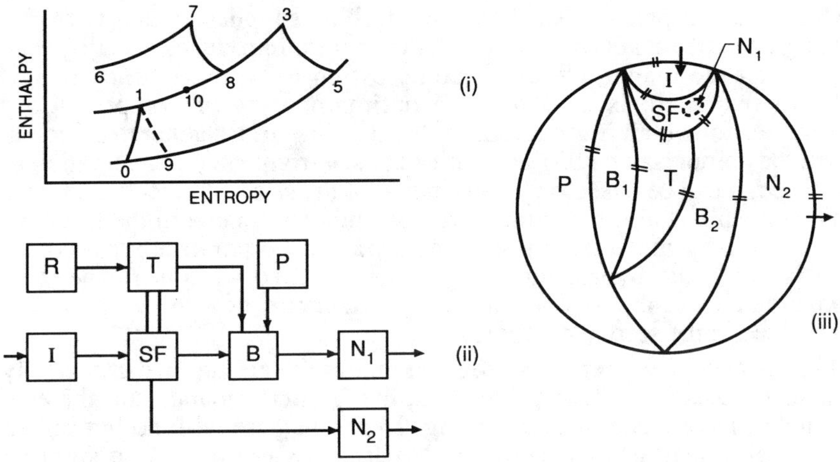

22. Air-turbo-rocket with supersonic through-flow-fan. (i) Cycle diagram; (ii) Block diagram; (iii) Structure diagram.

Table 1

CYCLE DATA UTILIZED FOR
SUPERSONIC THROUGH FLOW FAN-AIR TURBO-ROCKET

Altitude, kft.	0	20	50	70
Flight Mach Number	0.1	1.5	3.0	5.0
Fan Pressure Ratio	3.8	3.5	1.8	0.6
Turbine Temperature, R	2,500	2,500	2,000	-
Turbine Expansion Ratio	9.6	10.0	6.7	-
Nozzle Temperature, R	4,360	4,420	4,630	5,050
Nozzle Pressure Ratio	3.1	10.9	44.8	180
Net Thrust lb/lb air/sec	115.5	97.5	123.0	103.0

combustor that utilizes compressed air from the fan, along with additional fuel, if necessary, at constant pressure, and the products are expanded in a thruster nozzle. Therefore, the cycle is a variant of the Brayton cycle in which the compressor is driven by a turbine supplied with gas from the (independent) rocket motor. The latter feature introduces a major change regarding the magnitude of air compression to be used since it can be shown that there is no longer an optimum value of compression, for example, for obtaining maximum thrust. The thrust generated increases with the compression ratio and can be chosen from other considerations; for instance, it can be based on the limitation of the size of the rocket motor that is practicable and the maximum temperature acceptable in the turbine. In Ref. 33 it is assumed that the engine may also incorporate a bypass of air at the fan exit, the bypass air being either expelled without further use or expanded to ambient conditions in a separate nozzle, without being energized in the second combustor.

The cycle parameters as provided in Ref. 26 are summarized briefly in Table 1. Based on those parameters, the fan performance and the engine performance are summarized in Fig. 23 for the case with no bypass flow. Although not illustrated here, the bypass does not provide a significant improvement in engine performance. The performance also does not vary dramatically through adjustments of the fuel equivalence ratio in the rocket motor and the air-breathing combustor, except for a minor mass flow effect, noting that the air mass flow is naturally high at low altitudes. It is clear that the system is component-efficient and material-intensive. However, there

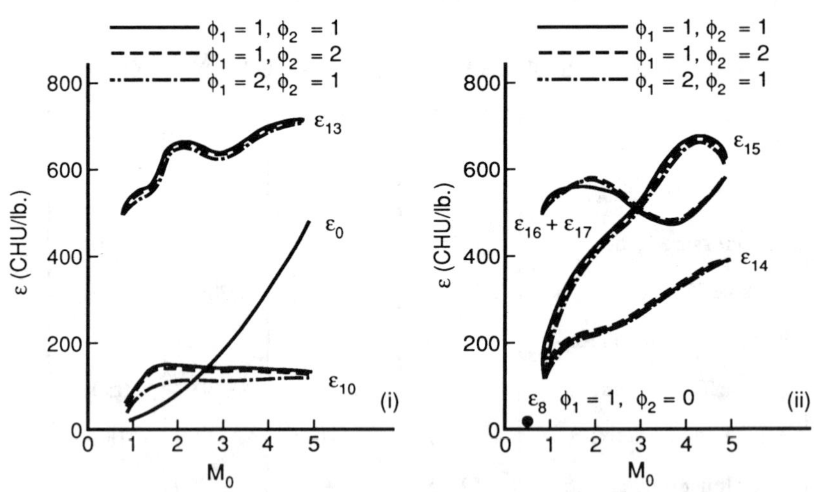

23. Performance of air-turbo-rocket with supersonic through-flow-fan: energy availability as a function of flight Mach number at (i) entry (0), state 10, and change from state 10 to state 3 (13), and (ii) change from state 3 to state 5 (14), propulsive energy (15), and total losses in exhaust stream (16 + 17). $\phi_{1,2}$ = equivalence ratio in combustors 6-7 and 3-5.

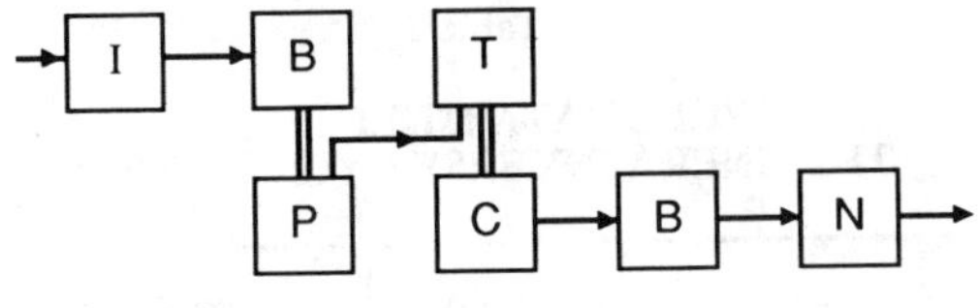

24. ONERA Inverse cycle engine: Block diagram.

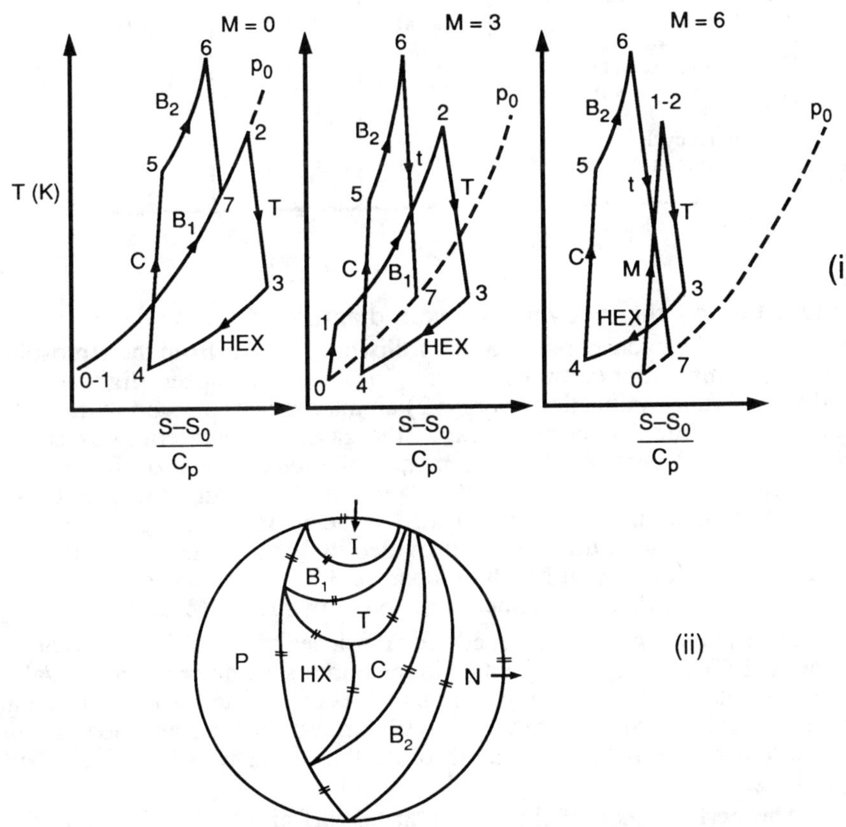

25. ONERA Inverse cycle engine: (i) Thermodynamic cycle diagram; (ii) Structure diagram.

seem to be other important gains in the cycle that can be enhanced with certain modifications, but they are not discussed here.

3.6 ONERA Inverse Cycle

As in the previous case, the ONERA inverse cycle engine is also intended for flight speeds below about 8000 ft/s. Figs. 24 and 25 provide

Table 2

CYCLE DATA UTILIZED FOR
THE ONERA INVERSE CYCLE ENGINE

Altitude, kft.	0	59.5	98.5
Flight Mach Number	0	3.0	6.5
Temperature at end of initial compression, k	2,000	2,000	2,000
Temperature at end of expansion and cooling, k	300	300	300
Maximum cycle temperature, k	2,500	2,500	2,500

schematic and thermodynamic cycle diagrams for a so-called inverse engine[27]. The cycle consists of the following: 1) Air from the atmosphere drives a turbine after compression and, in certain low-speed flight regimes, addition of fuel combustion energy; 2) an intercooler operated with the fuel cools the turbine discharge gases; 3) the gases are then compressed by a compressor; 4) heat is added by means of fuel combustion in the air at constant pressure; and 5) thrust is generated utilizing the products of combustion in a thruster nozzle. Data for the cycle operating conditions at various flight Mach numbers are provided in Table 2; the thermodynamic cycle diagram for a flight Mach number of 3.0 may be chosen as reference for comparison with performance at the other two flight Mach numbers.

Several assumptions are made for purposes of analysis: 1) Hydrogen is the fuel; 2) the equivalence ratios in the burners are any reasonable values; and 3) turbine and nozzle entry temperatures are limited from material and other considerations. In the cycles illustrated, turbine and nozzle entry temperatures are held constant through the entire range of flight Mach numbers.

The performance of the engine at takeoff and at the other two flight Mach numbers is given in Fig. 26.

The controllable parameters of the cycle are the fuel equivalence ratios in the rocket motor and the other combustor and four temperatures, namely, at the turbine entry, of the fuel at the intercooler entry, of air at the intercooler exit, and at the nozzle entry. It is interesting to compare a Brayton cycle engine with an intercooler and the inverse cycle engine. The intercooler in the inverse cycle engine needs to be more effective than that in the Brayton cycle engine. Also, although the Brayton cycle engine can be optimized for a given amount of fuel energy input, it can be shown that there is no optimum in the case of the inverse cycle engine. Meanwhile, the

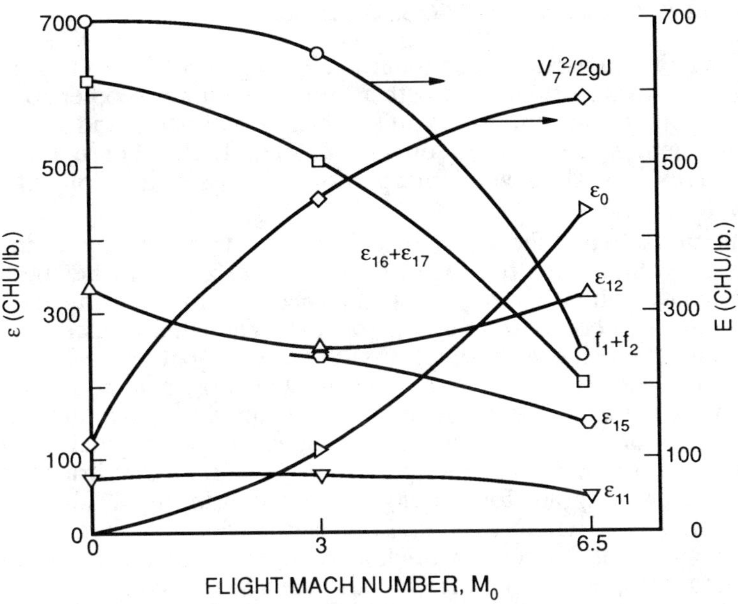

26. Performance of ONERA Inverse cycle engine under conditions given in Table 2: energy availability as a function of flight Mach number. 0, 11, 12, 15 and (16 + 17) indicate entry condition, change from state 3 to start 4, change from state 4 to state 5, propulsive energy, and losses in the exhaust stream, respectively.

inverse cycle engine can be modified to yield better utilization of fuel and air energy availabilities than those for the scheme as shown. Such possible modifications are not discussed here.

3.7 Two Other Engine Schemes

The simple Brayton cycle engine discussed in Section 3.3 can be modified in various ways for utilization at low flight speeds, as shown in the two preceding sections. In the next three Chapters of this volume several other modifications are discussed. Generally these schemes involve various means of improving the cycle by modifications to the treatment and use of air, fuel and other propellant substances, and introduction of several engines in variously compounded forms. Such improvements are considered for SSTO, TSTO, and multiple-stage-to-orbit (MSTO) applications as well as predominantly cruise missions at relatively low hypersonic flight Mach numbers. Without presenting results of numerical calculations, we briefly discuss two other engines that are not included elsewhere and that can serve to illustrate certain possibilities and weaknesses in apparently interesting schemes.

3.7.1 *Air turbo-ramjet expander cycle engine*

The air turbo-ramjet expander cycle engine (ATREX), proposed in Ref. 23, consists of an air-breather thruster with a precooler for a fan, a staged compressor incorporating a heat exchanger, and a thermal management system based on recirculated fuel. These elements in themselves provide several morphological possibilities of engineering interest.

In the scheme shown in Fig. 7 the fan is driven by a tip turbine operated with heated hydrogen fuel. The engine can be modified in operation over the flight regime with respect to air processing and input of mass, heat, and combustion energy of fuel. The latter is accomplished by correlating the fuel energy supplied with the energy of local airflow and the power derived from the turbine. Thus, the energy availability of air and fuel can be made to change the minimum required for obtaining the highest values of both the thrust generation factor and the propulsive efficiency while ensuring the lowest energy availability loss in the thruster exhaust. The structure diagram for the engine, included in Fig. 7, shows that the coefficients of external structural bonds of various components are high. The conclusion is that it is worthwhile to improve the mechanical efficiency of several components and thereby raise the true thermodynamic efficiency to a significantly high value. In other words, the system architecture yields a high effectiveness of energy utilization and improvements in component performance are directly beneficial to the system.

3.7.2 *Liquid-Air-Cooled Engine with an Air Compressor and a Tank-Return System*.

Considering a generic liquid-air-cooled engine (LACE), a series of modifications have been considered in Ref. 24, culminating in a LACE engine with a tank return and an air compressor (LACTR). A schematic of the engine is illustrated in Fig. 8

A significant point made in Chapter 3 of this volume is that, in the use of a fuel such as hydrogen for cooling air, depending on the properties of air and fuel, the amount of fuel required for cooling may exceed that corresponding to the desirable fuel equivalence ratio in the combustor. A part of the coolant fuel may then be recirculated into the fuel storage tank after it has gained heat elsewhere. In this connection it may be pointed out that an increase in air pressure increases the condensation temperature. Thus, there is a basis for introducing a compressor and optimizing the pressure ratio of compression of air relative to hydrogen recirculation parameters. The LACTR engine includes several such features and thus provides various options for different parts of a trajectory. A structure diagram is also included in Fig. 8. It shows how individual components are expected to perform with inputs to them and also to the system as a whole from external streams. The engine scheme obtains the best out of each component and, can therefore be expected to yield a high value of effectiveness in energy utilization. A study of the structure diagram seems to indicate several ways in which the basic system can be improved further based on energy availability considerations; these methods are not discussed specifically here.

The processing of air, including collection, separation of oxygen from nitrogen, and storage, provides a number of options over the flight Mach number range of 0-10, for example, a scheme such as that shown in Ref. 28. However, such schemes need a thorough analysis based on energy availability considerations in order to avoid expensive developments on a basically ineffective system.

In the foregoing the illustrative cases deal predominantly with propulsion selection. The method of extending the analysis to system selection has been mentioned in Section 3.4.1 but is discussed further in Chapter 6.0. However, it may be pointed out that, even in propulsion selection analysis, it is often necessary to take into account flight vehicle design. For example, the selection of an engine with commonality in elements over the entire speed range cannot be separated from vehicle considerations. Similarly, external and base combustion for drag reduction in, say, the transonic regime can be developed only in the context of flight vehicle geometry[29,30].

Before a procedure is given for optimizing propulsion systems for a flight vehicle (in Section 6.0), it is necessary to discuss two other fundamental aspects of a flight vehicle: (1) mission and geography of takeoff and landing (or recovery) and 2) weight and weight distributions (or fractions) for the vehicle. These subjects are rather complex and are discussed briefly in Sections 4.0 and 5.0.

4.0 Missions and Geographical Considerations

The two main missions of interest, excluding weapons and other defense needs from consideration, are 1) hypersonic transport in which cruise is a dominant mode and 2) orbital launch vehicles. The high-speed vehicle obviously has to take off from Earth, and either the vehicle or some part of it should land on Earth. Thus, the missions include the entire speed range from take off to cruise and to landing or from launch to cruise and to orbit as desired in different vehicles. Then, an important question in the case of an accelerator vehicle for orbital launch is whether it could be a SSTO or must be a TSTO or MSTO system. This question is examined in terms of two factors: 1) energy availability utilization and the technological needs and 2) geographical constraints. The first factor is included in Sections 3.0 and discussed further in Section 6.0. The second factor follows immediately in Section 4.1.

4.1 Geographical Considerations

It is well known that a typical velocity for an orbital launch vehicle is about 7-8 km, and the geostationary transfer orbital (GTO) plane is about 7

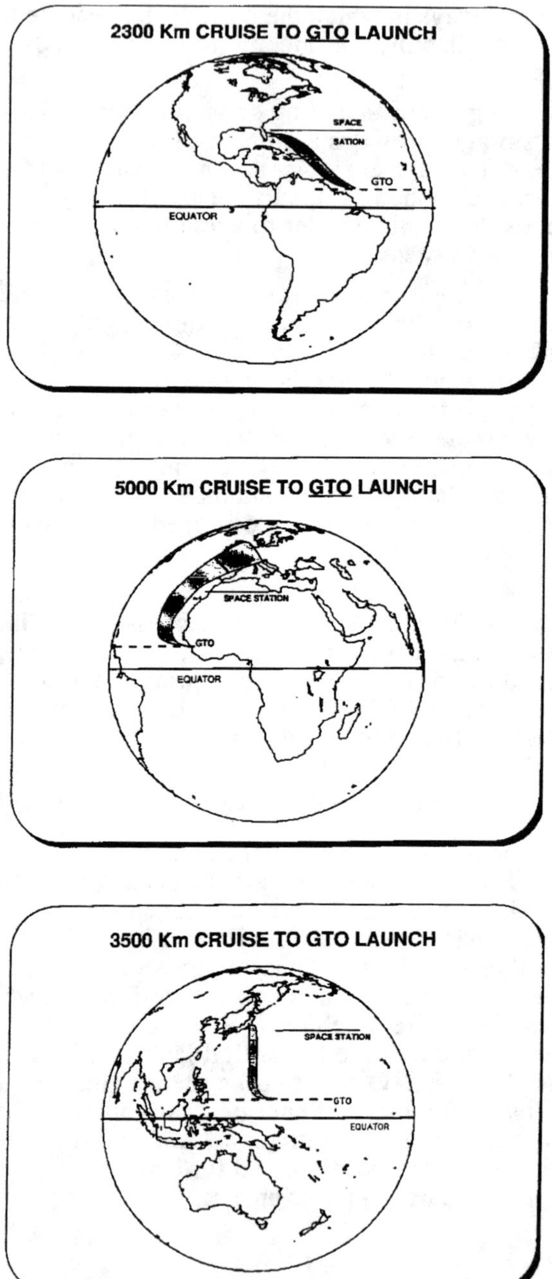

27. **Launch trajectory:** (i) United States, (ii) Europe, and (iii) Japan and Asia.

deg. off of the equator. In determining whether the required velocity is to be reached with one or more stages, the relation between a desired launch site and the GTO plane must be taken into account. In addition, several other considerations may be significant: 1) whether a horizontal or a vertical launch is desired; 2) what type of landing is desired, for example, conventional aircraft-type landing; 3) whether the vehicle is required to place a spacecraft, for example, at an altitude that is suitable for rendezvous with an already available orbiter and to provide a significant increment in velocity or altitude; and 4) the other uses to which the first stage of a multistage vehicle can be adapted, for example, a cruise-type hypersonic vehicle in a lower-Mach-number range.

One can examine the implications of those considerations for four typical geographical units on Earth: 1) a western European country, 2) the Soviet Union, 3) Japan, and 4) the United States. It may be pointed out that the extent of land in the Soviet Union is the largest among those. Also, China, India, and Indonesia are located favorably with respect to the GTO plane, with the latter two countries actually including land at 7 deg. N latitude (Fig. 27). Heuristic reasoning then yields the following conclusions, based on allowing a flight of about 3000 km-range between the launch site and the location of the GTO plane:

1) A TSTO configuration may prove advantageous to European nations desiring horizontal launch and conventional landing capability.

2) In the case of the United States and the cited Asian nations, either a SSTO or a TSTO system is practicable.

3) a multi-stage-to-orbit system provides no additional advantages compared to a TSTO system based on geographical considerations.

5.0 Weights Analysis

Based on the Jones-Donaldson analysis[31], described in part in Appendix B, a comparison can be made of a rocket and an air breather. The parameters of interest are the specific impulse, the efficiency of energy utilization in the vehicle, and the weight ratio. The specific impulse of a rocket fuel is a fixed value, but the specific impulse of a fuel used in an air-breathing engine is dependent on the state properties of local atmospheric air, specifically, the temperature[32].

The efficiency of energy utilization has been discussed in Appendix B and can be expressed by the following relation.

$$\eta = \frac{e_f}{gJ\,(R_W - 1)Q}\left[1 + X_i R_W + \int_i^f \frac{W/W_f}{(T/D) - 1}\,dx\right] \tag{33}$$

where $X = e / e_f$, $e = [(V^2/2) + gh]$, $Q = Q_C / [1 + (o/f)]$, and

$$\frac{W}{W_f} = (R_W - 1)\frac{(1 - X)}{(1 - X_i)} + 1 \tag{34}$$

W is the weight of the vehicle at any location along the trajectory, and the subscripts i and f denote initial (launch) value and final value, respectively.

R_W is defined as the ratio of gross weight and empty weight. The empty weight differs from the gross weight by the weight of propellant taken onboard at launch and is defined as follows:

$$W_E = W_F + W_S \tag{35}$$

where W_F denotes fixed weight, defined by

$$W_F = W_P + F_P \cdot W_P + W_C \cdot N_C + W_C \cdot F_C \cdot N_C + W_{FE} \tag{36}$$

where, on the right hand side, the first term is the payload; the second is the product of the payload with the payload installation factor; the third is the product of the weight of crew and crew's equipment, W_C, and the number of crew; the fourth, is the product of the crew-unique installation factor, and the number of the crew; and the fifth is the fixed equipment weight unrelated to the crew or the payload. Also, W_S in Eq. (35) denotes the scalable weight, defined by

$$W_S = W_{ST} + W_{EN} + W_{SE} \tag{37}$$

where, on the right-hand side, the first term is the structural weight that scales with size; the second term is the engine weight, which again scales with vehicle size and geometry; and the third term is the scalable equipment weight.

Thus, weight-ratio may be written, as follows:

$$R_W = W_G / (W_F + W_S) \tag{38}$$

$$= W_i / W_f \tag{39}$$

since the final weight of vehicle is equal to $(W_F + W_S)$, except for any weight of propellant that is left unused at the end.

The so-called rocket equation may be written as follows.

$$W_i / W_f = R_W = \exp B \tag{40}$$

where

$$B = (V_i - V_f) / (g \cdot I_E) \tag{41}$$

where I_E is the effective specific impulse that is related to the specific impulse in the following manner.

$$I_E = I \cdot (T/D - 1) / (T/D) \tag{42}$$

Thus, the effective specific impulse is based on the local value of the difference between thrust and drag. As stated earlier, the specific impulse of a fuel in an air-breathing engine is a function of local ambient conditions.

In summary, Eqs. (40-42) provide relations between the weight-ratio, the effective specific impulse, the thrust-to-drag ratio, and the efficiency of energy utilization. At any location along a given trajectory of the vehicle, it is also possible to determine the local value of (W/W_f) and η for a given vehicle.

A rocket and an air-breather have been compared in Fig. 28 with respect to the values that can be realized for η and R_W, utilizing I_E in one case and I_E and η in the other as parameters. Two conclusions from the two cases are that 1) a rocket cannot be competitive with a "good" air-breather; and 2) an air breather approaches its limit value of R_W at a much lower value of T/D ratio or $(T - D)/T$ ratio than a rocket does for the same value of η in the two cases.

5.1 Extension of Jones-Donaldson Analysis

In Ref. 33 an interesting approach has been discussed for assessing the potential of a given type and level of technology to meet the goal of launching into orbit a desired amount of payload. It is argued that the

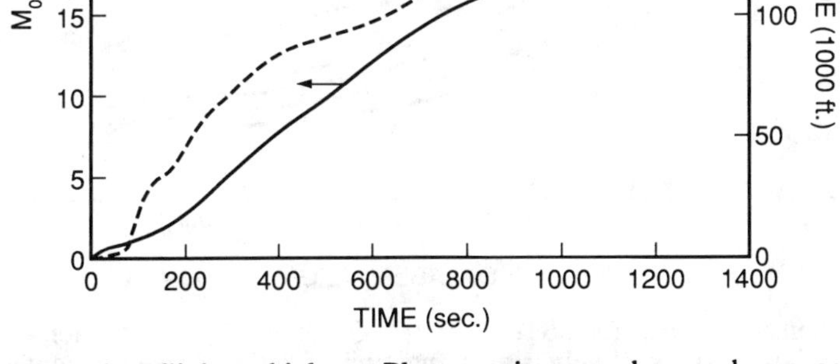

28. Illustrative flight vehicle: Planar trajectory chosen shown as variations of flight Mach number and altitude as a function of flight time.

measure for acceptability of a level of technology must be the vehicle launch weight and the sensitivities pertaining to weights. The gross take off weight is directly affected by the empty weight of the vehicle and the fuel weight. The empty weight of the vehicle is, according to Eq. (35), the sum of the fixed and scalable weights. Thus, based on Eq. (40), one can write the following.

$$F_{FG} = \exp(-B) - F_{SG} \tag{43}$$

where F_{FG} and F_{SG} are fixed-weight growth factor and scalable-weight growth factor, respectively;

$$F_{FG} = W_F / W_G \tag{44}$$

and

$$F_{SG} = W_S / W_G \tag{45}$$

F_{FG}, as expressed by Eq. (43), is not a practical sensitivity factor since F_{SG} for a given value of F_{FG} becomes a function of R_W as defined in Eq. (38). In the following, an alternative formulation is presented by means of which F_{FG} becomes expressed by known design values pertaining to the vehicle; this alternative formulation is based on an extension of the Jones-Donaldson analysis. The details of the extension may be found in Appendix B.

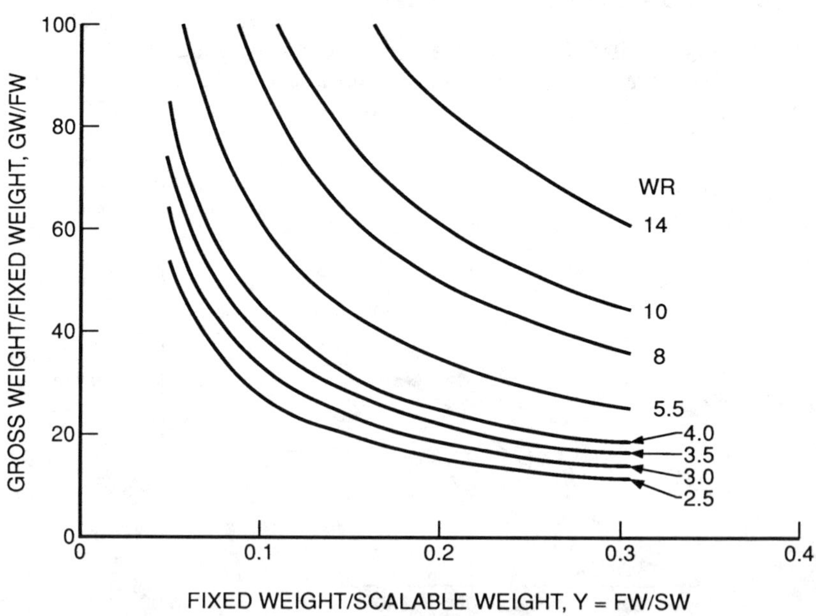

29. Sensitivity map for SSTO vehicle: Gross weight growth factor in relation to fixed-weight as a function of ratio of fixed-weight to scalable-weight utilizing weight ratio as a parameter.

Quite simply, starting with the definition of the fixed-weight growth factor, one can write the following utilizing Eq. (40):

$$F_{FG} = \exp(-B)\,[Y / (1+Y)] \tag{46}$$

It is clear that Eq. (46) applies to a SSTO vehicle. When dealing with a TSTO or MSTO vehicle, each stage, other than the first, is a part or the total of the payload for each of its preceding stages, as shown in Fig. 1. Based on that premise, the sensitivity factors, namely, fixed weight growth-factors, can be deduced for TSTO and for any MSTO vehicle. Fig. 29 provides the solution map for a SSTO air-breathing vehicle. In Fig. 30 a comparison is provided between SSTO, TSTO, and MSTO; the figure also includes an estimate for a rocket vehicle.

It may be pointed out that both F_{FG} and its inverse, denoted by $F_{GF} = W_G/W_F$, are useful sensitivity parameters since they provide the growth of one with respect to the other for W_F and W_G.

The manner in which weights analysis enters into vehicle optimization analysis can be seen from the foregoing. In fact, the weights analysis is related to energy analysis. The scalable weight includes engine weight, which, in general, may be taken as also encompassing the thermal management system. Also, one has to account for fuel weight, which is

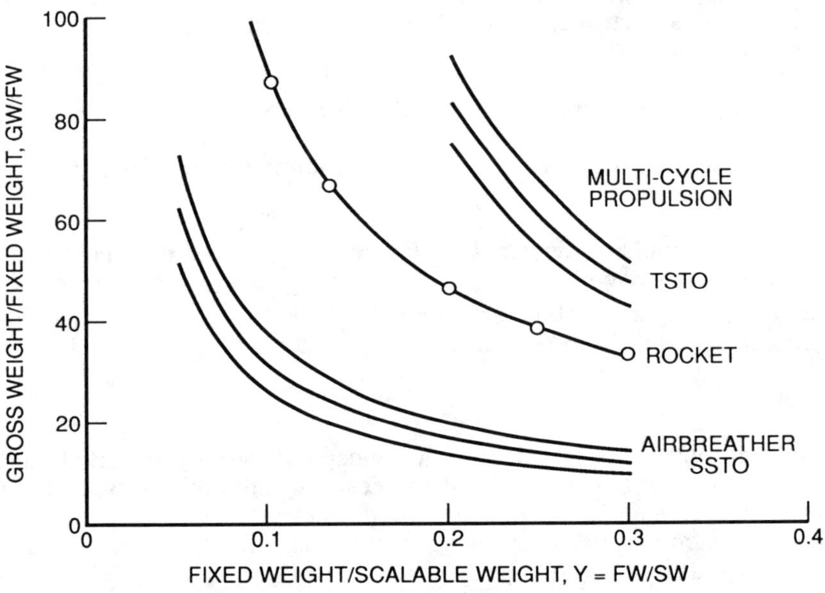

30. Sensitivity map for SSTO, TSTO, and rocket: Gross weight growth factor in relation to fixed-weight as a function of ratio of fixed-weight to scalable-weight utilizing weight-ratio as a parameter.

equal to the range along the flight path multiplied by the fuel utilized per unit range, locally along the trajectory. The latter is one of the outputs from energy analysis. The engine and fuel weights affect vehicle structure and packaging, severally and simultaneously.

6.0 An Approach to Preliminary System Selection

The flight vehicle system under consideration is an accelerator vehicle for orbital launch with cruise capability as needed. The vehicle system may be divided, for convenience, in two ways as follows: 1) for aerothermodynamic performance analysis, a) vehicle frame and payload, b) propulsion, and fuels or, more generally, propellants and c) control; and 2) for weights analysis, a) fixed-weight, b) scaleable weight, and c) fuel weight. The vehicle system may be a SSTO, TSTO, or MSTO. In all cases it is assumed that an extrinsic airbreathing propulsion system is to be utilized over a part of any trajectory chosen, and at least a part of the vehicle is to be recovered in each flight and reused.

The system parameters may be chosen as the following:

1) A baseline flight trajectory including flight speed;

2) a baseline vehicle frame that meets requirements with respect to a) predictable lift, lift-to-drag ratio, stability and controllability; b) volume and volume disposition ("packaging"), taking into account fuels and payload; and c) cross-sectional and other geometrical features for thermal management and other needs;

3) fuels, with their physical and chemical properties;

4) a thermal management scheme combining both vehicle frame and propulsion needs;

5) a propulsion system with different, selected engines compounded for a) operation with stretching, switching and staging capabilities, including external combustion, b) air and fuels supply, c) commonality of components, d) thrust production, and e) moment production; and

6) a control system.

The constraints in selecting a conceptual system design should be limited to as few as possible for the widest range of options to be obtained. For example, one may choose just two constraints as follows: 1) materials and manufacturing or, better, weight fractions, and 2) launch and landing schemes and sites.

There is perhaps little meaning to the generation of an optimal design in a strictly mathematical sense at this stage of the development of necessary technologies and design tools. However the selection of the system parameters may be based on realizing the best matching collectively, at various critical locations along a trajectory, with respect to the following: 1) generating a thrust minus drag value that is adequate to meet the trajectory speed map and associated lift, drag and moment variations; 2) obtaining the highest utilization of energy and energy availability (or the lowest losses and inherent irreversibilities) in the energy resources and the working fluids in all parts of the system; and 3) realizing a sufficiently low growth factor of fixed weight to changes in gross weight as the vehicle system is scaled in parts or as a whole.

In connection with the first factor there is a distinct possibility that more than one type of "pinch" point will occur in system operation and performance. The first Chapter in this volume presents some details on those points and also approaches to meeting such difficulties. Here it is the objective to deal with only the gross outlines of generating a design.

A procedure for flight vehicle selection has been outlined in Appendix C. The procedure is centered around the selection of a propulsion system, which, it is generally agreed, dictates the integrated design of the overall system. It may be pointed out that, although a thermal management system is not explicitly discussed, it is assumed to be included in the propulsion system. An example of the application of the procedure to a given flight vehicle for launching a payload to orbit is presented in the following. The analysis deals with the flight regime beginning with a flight speed of 6000 ft/s. No account is taken of the requirement for propulsion below that speed, nor of its interaction with the propulsion over the higher speed range. The propulsion system for the speed range of 6000 ft/s to orbital speed is assumed to be one of the following: 1) a scramjet engine over the entire speed range and 2) a scramjet engine over a part of the trajectory, followed by a rocket toward the end of the trajectory. The fuel is assumed to be hydrogen in all cases. The oxidant for the rocket motor is assumed to be liquid oxygen carried onboard at takeoff.

6.1 Flight Vehicle

The planar trajectory of the vehicle is chosen, as illustrated in Fig. 30, in the form of a map of altitude and speed as a function of time elapsed from takeoff. Assuming a payload of 3.0% of W_G and a ratio of W_E/W_G of about 0.29, a base vehicle is determined for which the total thrust required is given in Fig. 31; the total thrust includes the part of thrust required for the desired acceleration of the vehicle.

6.2 Scramjet Engine

As stated earlier, a scramjet engine may be utilized either by itself or in combination with a conventional liquid hydrogen-liquid oxygen rocket. Further details on the scramjet engine are provided in the following.

Several features may be noted regarding the Brayton cycle utilized in the scramjet engine. First, in the range of flight speeds under consideration,

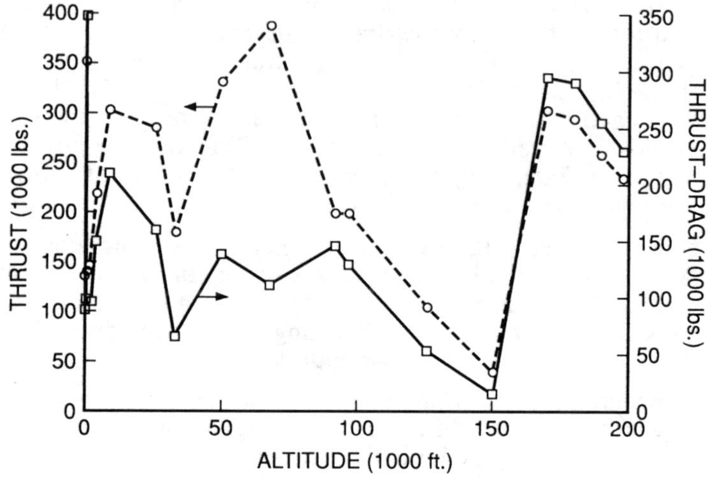

31. **Illustrative flight vehicle: Thrust and thrust-minus-drag required as a function of altitude.**

there is no necessity for mechanical compression; this can be verified in Figs. 15 and 16. A certain amount of momentum diffusion is required in the inlet in order to meet the requirements of minimum pressure for combustion of hydrogen fuel, pressure loss in the burner, and thrust generation. Second, fuel can be energized, before supply to the burner, both by the addition of heat recovered from the hot sections of the engine and the vehicle as well as by the addition of kinetic energy. Finally, the fuel equivalence ratio can be increased as desired in order to provide additional mass flow in the thrust nozzle.

The following assumptions are introduced in carrying out energy availability analysis:

1) The inlet effectiveness k_D is either constant at a value of 0.5, or it varies with respect to flight speed as shown in Fig. 32.

2) The fuel is heated to a temperature close to the self-ignition temperature of hydrogen, about 900 K (Ref. 16), and its kinetic energy is increased by increasing its velocity very nearly to that of air entering the burner.

3) The burner has a mixing loss of 8.0% and a combustion inefficiency of about 10%; both of these are accounted for in terms of a reduction in energy supplied to the air.

Regarding assumption 1, it is of interest to determine a reference value for the inlet optimum enthalpy ratio in the cycle diagram of Fig. 9, as a function of flight speed. Fig. 33 presents such reference values under the assumptions of a constant value of k_D equal to 0.5, stoichiometric combustion of hydrogen, fuel equivalence ratio of unity, and adiabatic efficiency of thrust nozzle equal to 0.98.

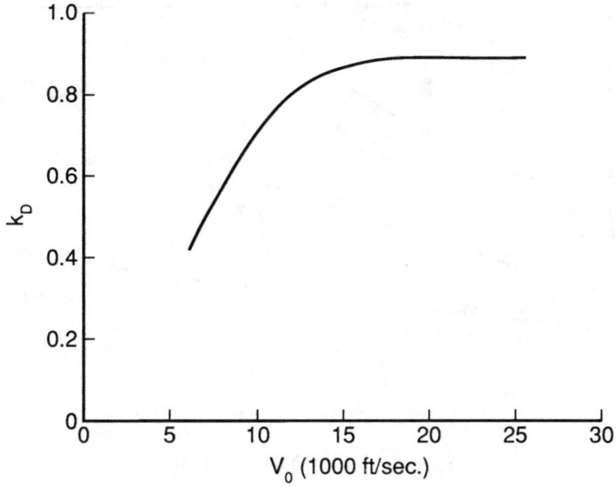

32. Inlet effectiveness k_D chosen as a function of flight speed.

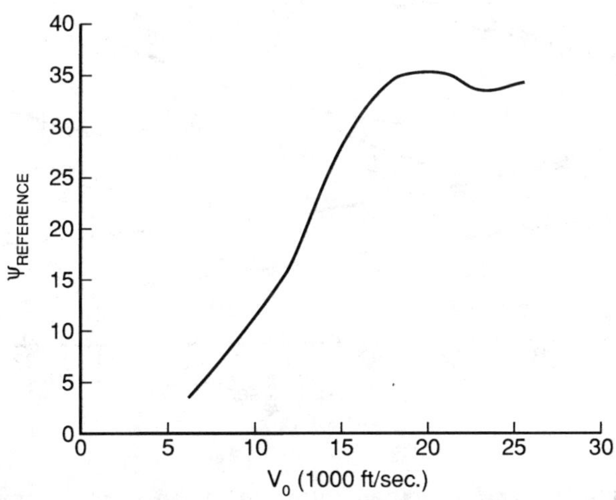

33. Reference inlet enthalpy (compression) ratio as a function of flight speed.

6.2.1 Energy Availability Analysis

The procedure for energy availability analysis is presented in Section 3.3 and Appendix A.

Briefly, as seen in Fig. 9, with chosen values of (H_1/H_0), k_D, fuel equivalence ratio, heat and kinetic energy added to the fuel before entry into the burner, mixing and combustion losses in the burner, flow velocity ratio

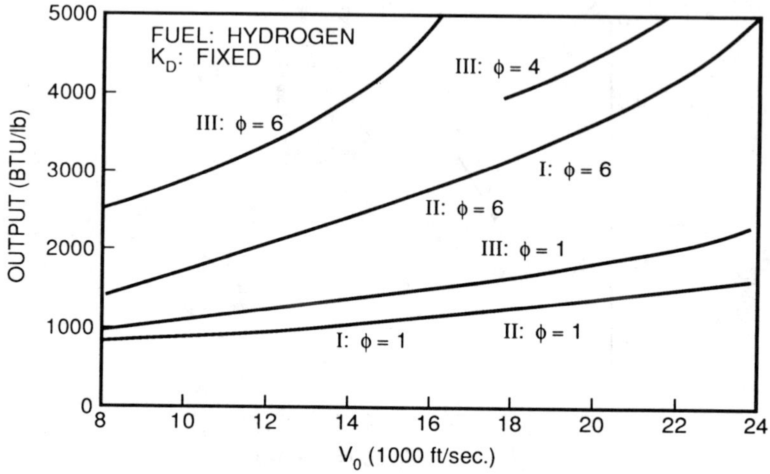

(i)

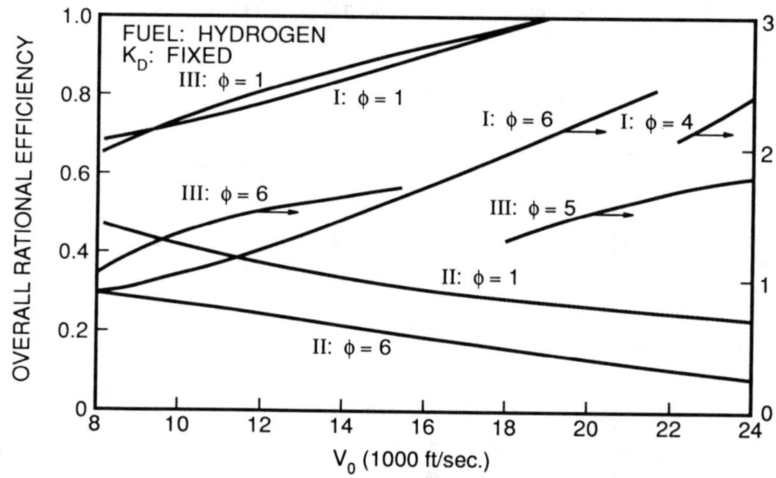

(ii)

across the burner, and an expansion ratio in the nozzle, the following are determined: Thrust obtained per unit mass of air; specific impulse of fuel; thermal, jet, propulsive and rational efficiencies; energy availability values at conditions 0, 1, 3, and 5 in the cycle; and losses in the nozzle exhaust in excess enthalpy and kinetic energy relative to the local environment conditions. The output, the losses in the combustor, the losses in the nozzle and the overall rational efficiency are shown as a function of flight velocity in Figs. 34 and 35; in Fig. 34, k_D is fixed and in Fig. 35, it is varied according to Fig. 32.

It is clear that a large number of parameters influence the performance of the scramjet engine. However, in order to obtain simultaneously the desired high values of thrust, specific impulse, and jet, propulsive and

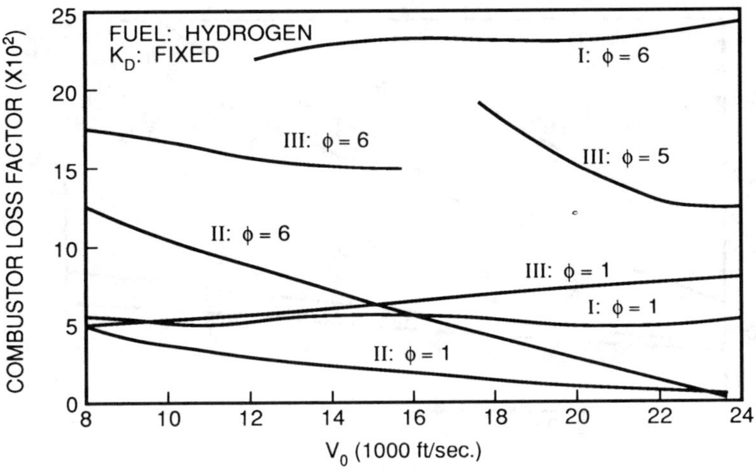

(iii)

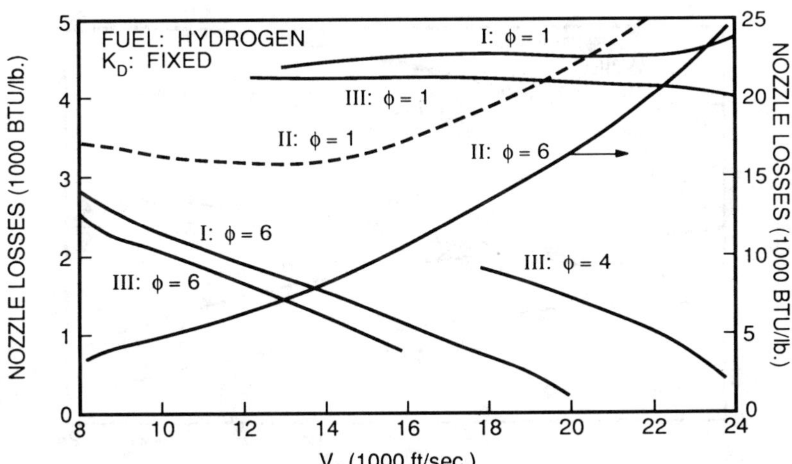

(iv)

34. **Output, availability losses, and rational efficiency as a function of flight velocity for a scramjet engine utilizing hydrogen fuel equivalence ratio as a parameter. k_D is fixed at 0.5 throughout. Combustor loss factor = Loss of energy availability / energy availability at combustor inlet.**

rational efficiencies, and also, low values of energy availability losses in individual components, it is practicable only to adjust (H_1/H_0), ϕ_F, and Q_T. In general, it is not possible to visualize substantial improvements in component efficiencies.

For the illustrative case currently under consideration, the values of (H_1/H_0), ϕ_F, and Q_T, chosen following substantial trial-and-error attempts, are shown in Figs. 36-38. The resulting engine performance is given in the following figures: Fig. 39 shows the thrust work generated per unit mass

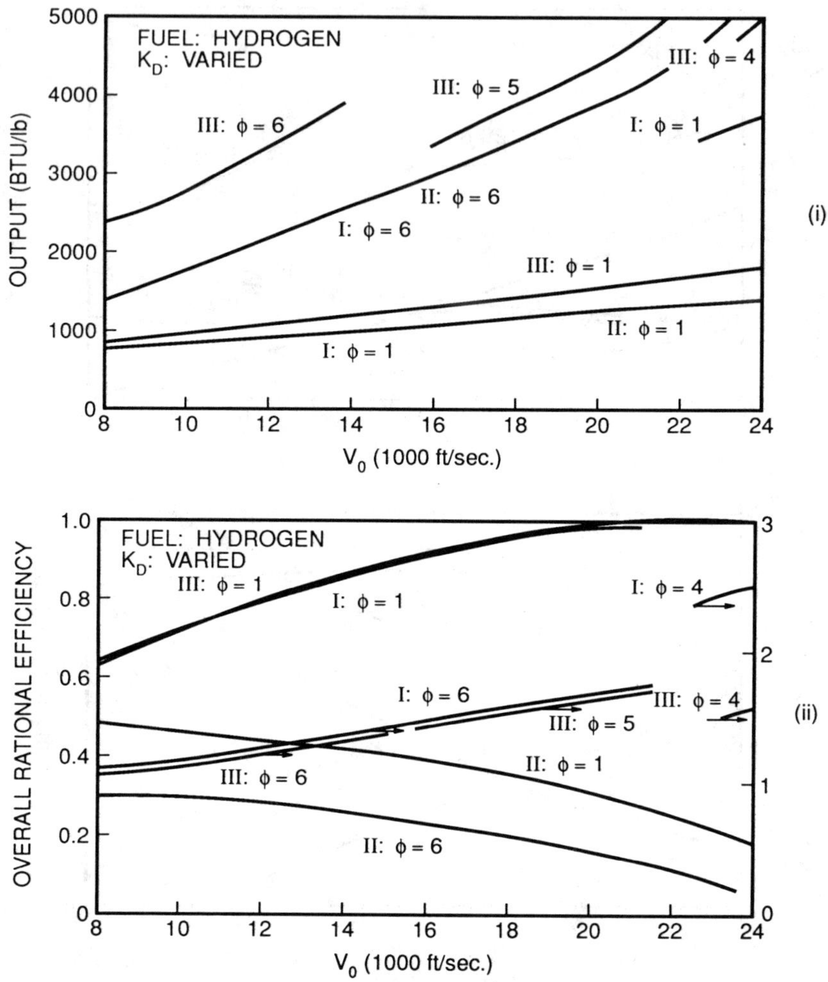

of air. Fig. 40 shows the propulsive efficiency; and Fig. 41 shows the two major losses of enthalpy and kinetic energy in the thrust nozzle. In all cases the individual performance parameters are shown as functions of vehicle speed. In each of Figs. 39-41, the performance parameter is given both for the case of ϕ_F being chosen equal to unity and that of ϕ_F being varied as in Fig. 37. The latter case, along with the choice of other parameters, provides a scramjet engine that has low losses of energy availability both in the thrust nozzle as well as in the system as a whole, although it cannot be claimed that the engine has the lowest energy availability loss (or the highest effectiveness) of all possible scramjet engines. On the other hand, the engine provides simultaneously high values of thrust and propulsive efficiency.

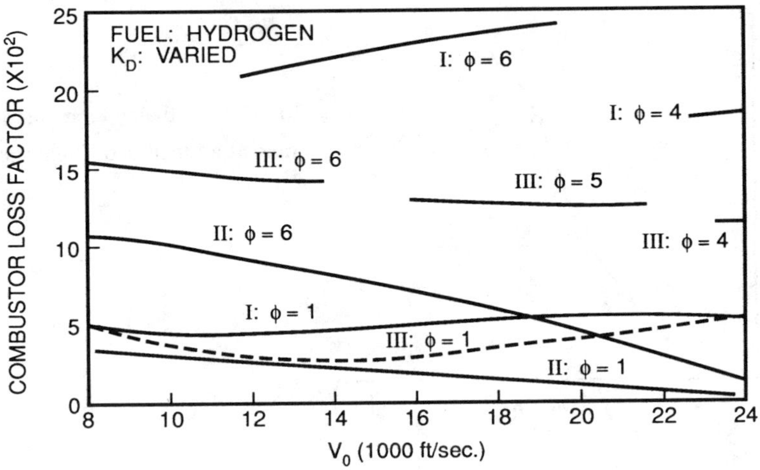

(iii)

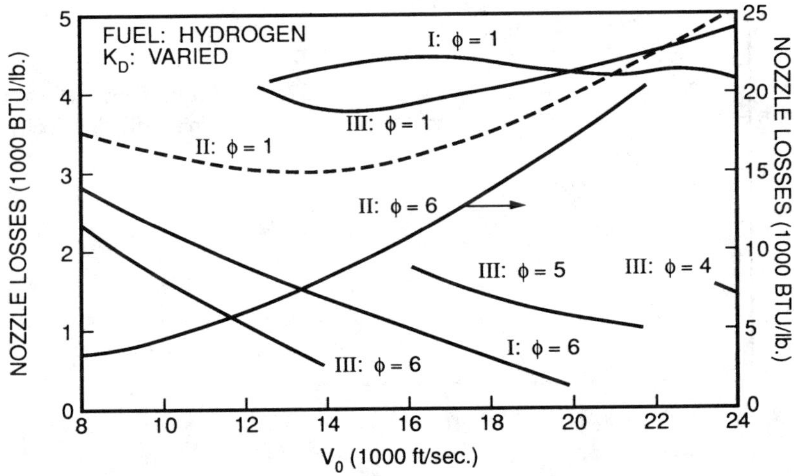

(iv)

35. **Output, availability losses, and rational efficiency as a function of flight velocity for a scramjet engine utilizing fuel equivalence ratio as a parameter. k_D is varied as shown in Fig. 32.**

The performance parameters given in Figs. 39-41 apply at specific locations along the trajectory where calculations have been performed. It is obviously necessary to include all points along the trajectory, where a change in speed or altitude arises, and also all of the points at which any critical conditions are likely to arise with respect to the engine. In particular, it is of interest to determine the integrated values of fuel consumed, thrust power generated, and energy availability input and losses for the system as a whole, and also individual components such as the thrust nozzle. In addition, it is necessary to establish the integrated values of performance parameters over the trajectory as a whole.

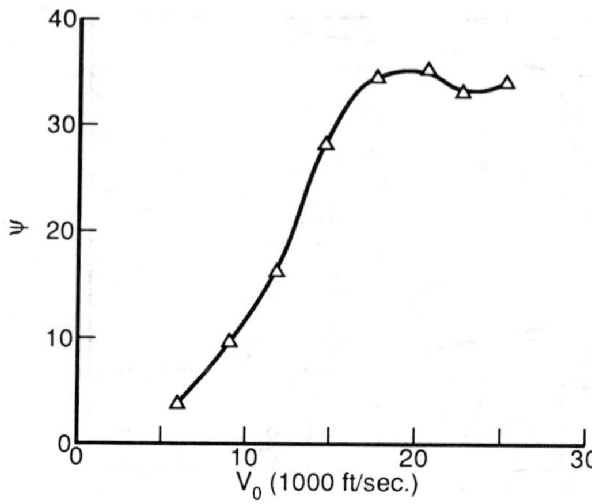

36. Inlet enthalpy (compression) ratio as a function of flight speed.

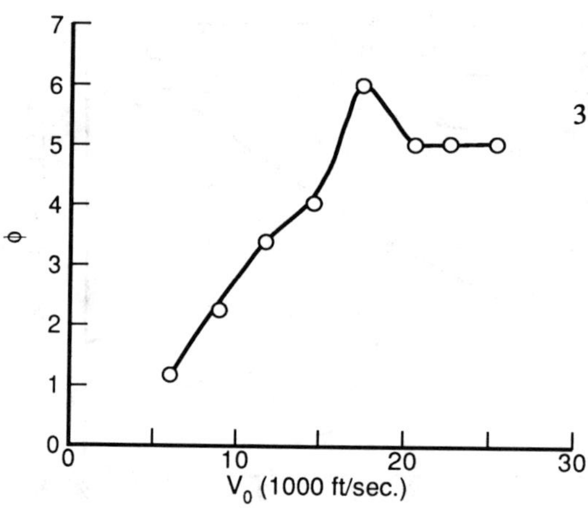

37. Fuel equivalence ratio as a function of flight velocity.

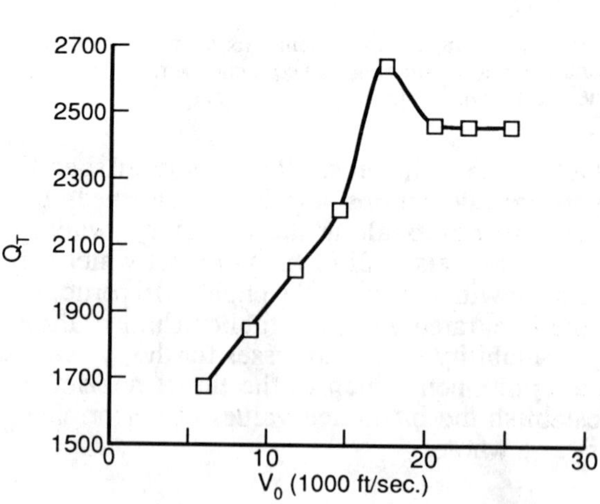

38. Total heat added (velocity energy, enthalpy and combustion) through fuel, Q_T, as a function of flight velocity.

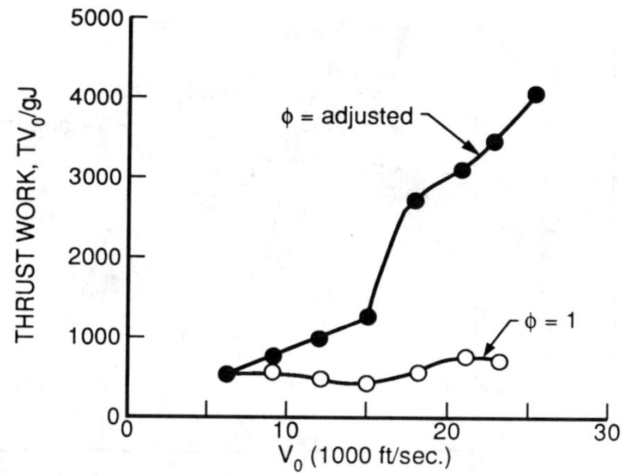

39. Thrust work output as a function of flight velocity with fuel equivalence ratio as a parameter.

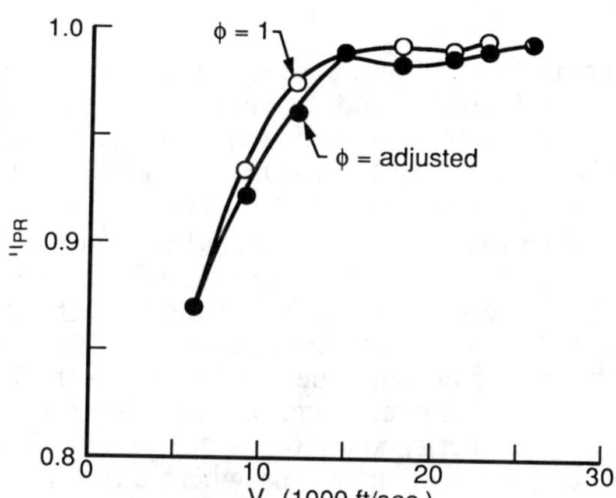

40. Propulsive efficiency as a function of flight velocity with fuel equivalence ratio as a parameter.

6.2.2 Relation Between Engine Performance and Airframe Parameters

Utilizing the performance parameters for the engine, one can determine the total amounts of air and fuel necessary to meet the requirements of the flight vehicle given in Figs. 32 and 33. They may be compared with the values assumed for the base vehicle in determining inlet cross-sectional area, vehicle volume and surface area, drag and weight. If the comparison is favorable, one may consider refinements in optimizing engine

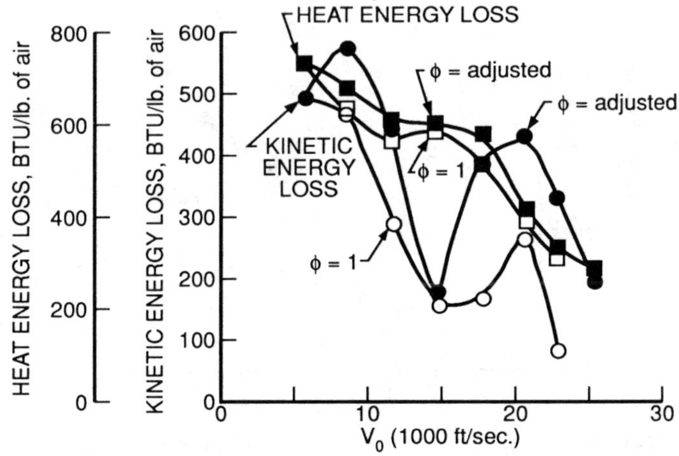

41. Losses in kinetic energy and enthalpy in the nozzle exhaust stream as a function of flight velocity with fuel equivalence ratio as a parameter.

performance parameters. If, on the other hand, the comparison reveals a mismatch between the vehicle and the engine, the vehicle may have to be resized. This may, in turn, affect the engine parameters, especially Q_T. Thus, a series of iterations may become necessary before the engine and the airframe converge to give rise to an effective flight vehicle.

In the preceding discussion, the match between the airframe and the propulsion system was considered with reference to the flight trajectory given in Fig. 32. Obviously, the flight trajectory itself may be treated as a set of parameters. In other words, trajectories other than the one shown in Fig. 32 also have to be considered before the flight vehicle system may be considered to be optimal or even matched effectively for the given mission. As stated earlier, the energy availability analysis must be conducted at various points, including all locations where any form of criticality in energy availability is suspected along the trajectory, as well as over the trajectory as a whole.

6.3 Propulsion System with Scramjet and Rocket Engines

In Section 6.2.1 a scramjet engine has been considered for use over the entire flight trajectory beginning with a flight speed of about 6000 ft/s. Now the performance of that engine will be compared with one in which the same scramjet engine is utilized up to a lower flight Mach number of about 20; then a liquid hydrogen-liquid oxygen rocket motor is utilized over the balance of the trajectory as stated at the beginning of Section 6.0.

An attempt is made to compare the two propulsion systems with respect to two chosen trajectories and with reference to a particular performance parameter. The two trajectories chosen are 1) the trajectory illustrated in Fig. 32 and 2) a trajectory that causes a minimum loss of

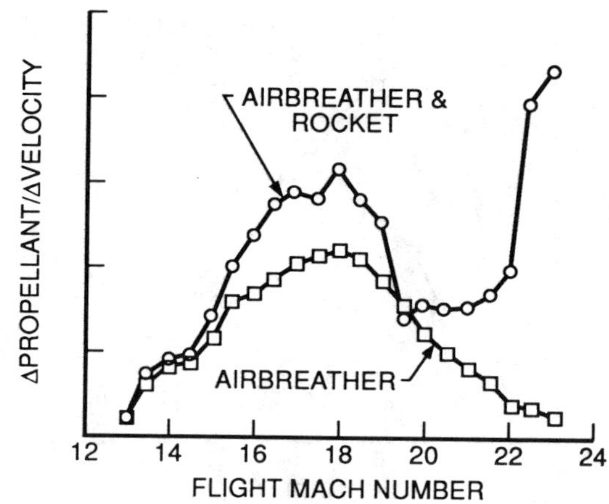

42. Comparison between air breather and air breather-plus-rocket with reference to the trajectory given in Fig. 32.

energy availability, or gain in entropy. The performance parameter chosen for comparison is the amount of propellant required for obtaining an increment of unity in the velocity of flight. This is an indirect measure of the effectiveness in the use of fuel in the production of thrust.

In Fig. 42 the two propulsion systems are compared with reference to the trajectory given in Fig. 32, and in Fig. 43 they are compared with reference to the minimum entropy gain trajectory. In the first case it is clear that the airbreather yields a noticeable reduction in propellant usage for unit gain in vehicle velocity. The airbreather in this case is optimized for minimum loss of energy availability. In the second case also it is found that the airbreather can provide practically the same performance as the airbreather plus rocket. Thus, in principle, an airbreather with combustion in supersonic flow is feasible as an accelerator up to orbital launch speed starting from about 6,000 ft/s.

The performance parameter chosen here (Δ propellant required/Δ increment in flight velocity) is useful when lift, thrust and specific thrust, requirements are specified for the flight system. When the vehicle, propulsion, thermal management, and control are also free to be chosen with only the mission, materials, and weight ratios given, other performance parameters may obviously prove effective. The same also becomes true in scaling a vehicle to a different "size." For example, a parameter such as (Δ propellant/Δ effective thrust work per unit time) may be considered for use, as an indicator of overall effectiveness of the flight vehicle.

* * *

A methodology for analysis of a flight vehicle system and its propulsion has been developed based on considerations of energy

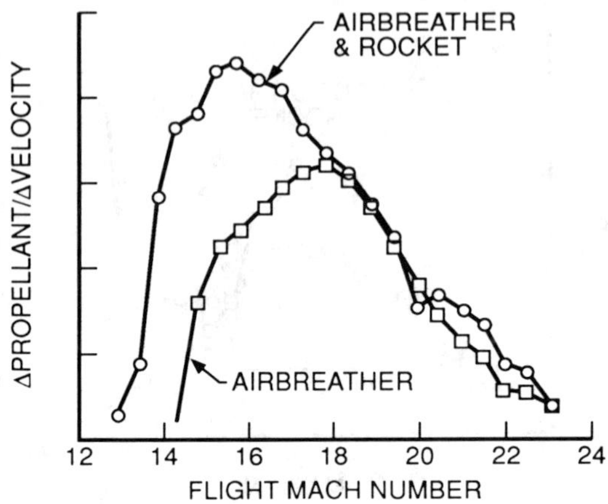

43. Comparison between air breather and air breather-plus-rocket with reference to the minimum entropy gain trajectory.

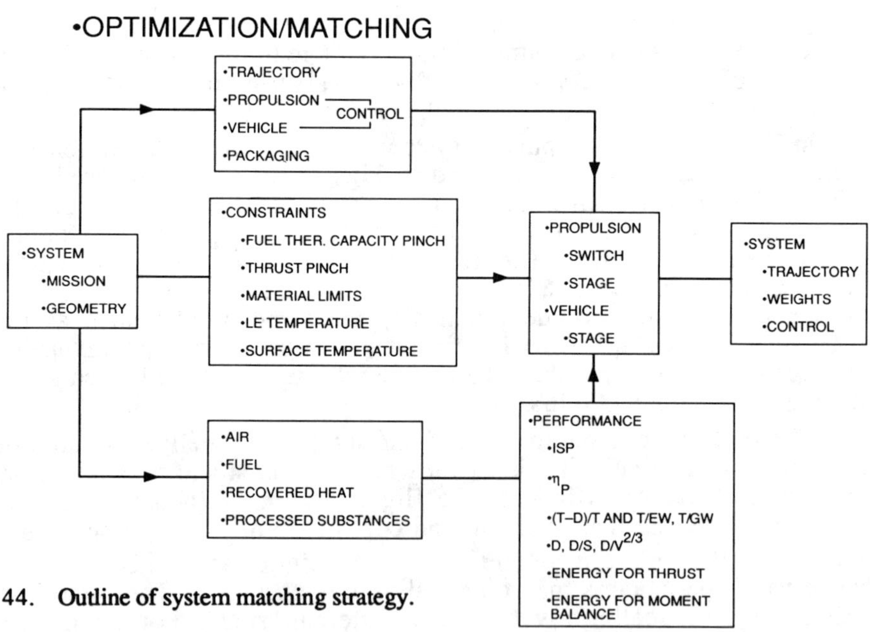

44. Outline of system matching strategy.

availability and its effective utilization. The methodology is summarized in Fig. 44. The results of the analysis serve two important goals: 1) Establishment of "true" thermodynamic performance of all parts of the system, individually and collectively; and 2) determination of the most significant contributors to system performance and their possible improvement in the given system composition and architecture.

Acknowledgment

The research project at Purdue University has been supported by McDonnell Douglas Corporation, R.V. Masek, Technical Monitor.

Appendix A: Thermodynamic Analysis

Basic Principles

1) The first law of thermodynamics: Heat and work can be transformed into each other in a fixed ratio.

2) The second law of thermodynamics: Thermal energy cannot of itself pass from a lower to a higher temperature (Clausius); the entropy, an intrinsic property (for example, a measure of the unavailability of internal energy) of an isolated system, can never decrease (Stodola).

3) The third law of thermodynamics: The entropy at absolute zero of temperature is a universal constant that can be taken to be zero (Nernst).

4) The change in entropy according to quantum statistics is the change in accessible phase volume (Lorentz).

5) A close analogy exists between the change in statistical entropy and the change in the uncertainty function. (Shannon[34]).

Basic Concepts and Definitions

Environment, as a concept, is characterized by a perfect state of equilibrium with respect to all properties that together describe the environment. Equilibrium in part with the environment can be visualized for any substance. Unrestricted equilibrium is referred to as the "dead state", which, it is noteworthy, includes chemical composition.

1) *Energy*: Energy, which is an extensive property, remains constant for an isolated system regardless of whether the system is in equilibrium. Energy can exist in many forms: internal, motion, potential in gravitational field, and chemical potential. The latter is based on the assumption of reversible reactions. The energy of a system and its surroundings is constant, and this fact leads to the law of conservation of energy. Energy cannot be destroyed.

2) *Energy Availability*: Energy availability, which is an extensive property, can be destroyed for any system and is, in fact, destroyed during any and every practical process in which the system is involved. It is often referred to as "exergy." Energy availability is the ability of a system to perform work. It has a maximum value if all processes involved with the system are reversible. It is not conserved for the system. Energy availability loss is due to irreversibilities in the processes involved with the system.

Energy availability can be calculated with reference to any environment state. When calculated with reference to absolute zero of temperature its value is referred to as absolute energy availability under the assumption that the absolute energy availability is zero at zero of temperature. Therefore, energy availability can be assigned to a finite mass of any substance by treating it as a thermodynamic system. For a process, a heat-transfer device, an energy conversion device, or a machine performing work, the change in energy availability of the substance involved can be calculated similarly. Energy availability can never be zero for any substance, except, by assumption, at zero of temperature.

It is sometimes useful to separate energy availability into physical and chemical parts. Physical energy availability is the maximum amount of work that can be obtained from the substance under consideration when it is brought from its initial state to the environmental state by physical processes that involve only thermal interaction with the environment. In considering chemical energy availability, it is necessary to consider the attainment of thermal, mechanical and chemical equilibrium (the state of equilibrium referred to as the dead state) by the substance through processes that involve heat transfer and mass exchange only with the environment. In the calculation of chemical energy availability one can utilize one of two approaches: 1) the energy availability of the substance when it is brought from its current composition to the same as that which exist in the environment or 2) the minimum amount of work required to synthesize and deliver the substance under consideration in the environmental state starting with environmental substances and utilizing processes involving heat transfer and exchange of substances only with the environment.

The energy availability flow in a system starting from the inputs to the system and ending with the outputs from it, while accounting for the various losses in the system, can be represented diagrammatically. Such a diagram is called the "Grassman diagram". The width of bands in the diagram is made proportional to the total value of the energy availability flux.

3) *Efficiency* : The efficiency of a process, a component, or a system with many components can be established by comparison of the output obtained with a given input under the given conditions of operation. The input and the output may be expressed in terms of energy or energy availability. When based on energy, the efficiency is referred to as "mechanical efficiency," and, when based on energy availability, it is called "rational efficiency," "true thermodynamic efficiency" or "energy availability efficiency." In general, both types of efficiency are useful in assessing a thermodynamic system. On the other hand, only the rational efficiency provides a true picture of what a process, component or system of components accomplishes and what is lost irrecoverably.

The difference between the input and output of energy in any identified thermodynamic element constitutes energy loss. However, energy is a conserved quantity, and the input must balance exactly with the sum of output and losses.

The difference between the input and output of energy availability in the element under consideration constitutes energy availability loss through irreversibilities. Irreversibility is always accompanied by an entropy increase in the isolated system (the so-called universe that includes the identified element and its surroundings). The entropy increase may be due to dissipation of work into internal energy and spontaneous, unrestricted change from nonequilibrium state, either individually or together.

Two other parameters of interest in connection with rational efficiency are 1) *effectiveness,* which is the ratio of the output to the change in energy availability; and 2) *efficiency defect*, which is the ratio of irreversibilities to the energy availability input.

It may be pointed out that, in thermodynamic systems involving flow or rate processes, mechanical and rational efficiencies, effectiveness, and efficiency defect are defined with respect to fluxes of energy and energy availability and irreversibility rates.

4) *Bond coefficients*: The structure of a system that is made up of several components may be specified in terms of a) the components and processes, b) the flow of energy and energy availability through the system, c) the input streams to the system, and d) the architecture of the system or the makeup of the system with the components.

In thermodynamics the system is generally considered along with the environment. The latter may be changing with respect to time or the location of the system (as in the case of a flight vehicle), although at each instant of time or location the environment in considered to be in equilibrium and also sufficiently large. In assessing the performance of a system or of any of its components, the reference must include 1) an inertial frame of reference for mechanical energy and energy availability, and 2) the environment or the zero of temperature or, in certain cases, both.

One can then proceed to introduce the concept of structural effectiveness in terms of component effectiveness. Structural effectiveness may be understood as follows: the extent to which the effectiveness of each of the components (or groups thereof) affects the effectiveness of the system as a whole when operating with given input streams in view of the given architecture of the system. Then one can denote structural effectiveness in terms of a combination of two parameters as follows. First, the time rate of change of irreversibility of the total system is obtained with respect to the time rate of change of irreversibility in individual components while the input streams are held constant. Thus, the change in this case in the components is due to reasons entirely internal to the system and dependent on the structure or architecture of the system. The effectiveness of the system in this case is then called the "*coefficient of structural bonds*." Second, the time rate of change of irreversibility of a component or the total system is obtained with respect to the time rate of change of the input stream parameters. In this case the changes are caused by parameters external to the system under consideration. The effectiveness of the system is then called the "*coefficient of external bonds*."

It may be pointed out that, in a system consisting of several components and involving various parameters, it becomes necessary to determine the two coefficients with respect to each of the parameters involved in each of the components and input streams. Thus, a matrix of coefficients has to be assessed for the system in order to gain an understanding of its effectiveness from all points of view.

The foregoing concept, due to J. Beyer (Refs. 8-9) and recently applied to a flight system by Y. Gogus (Ref. 10), is of great value in the analysis of complex systems with several components and processes. Based on the coefficients of structural bonds and external bonds, it is possible to address a series of questions as follows: 1) Does an individual component affect the overall system performance? 2) If it does, should it be optimized? 3) If it does, are there circumstances when improvement of that component may, in fact, affect adversely the overall system performance? 4) Finally, if it does, are there conditions under which it is necessary to modify the structure by displacing or removing the component?

5) *Brayton cycle*: In the thermodynamic analysis of engines it is common practice to introduce the notion of cyclical operation with various parts of the cyclical processes associated with different components. The analysis of the cycle may be carried out based on energy and loss or energy availability and irreversibilities in the engine.

In dealing with energy it is again common practice to consider the total energy content of the working fluid, the sum of internal, potential, and kinetic components of energy. Of course, the total energy must remain constant in an isolated system, or across a control volume, unless, as in any engine, there are losses or processes such as combustion, heat transfer, or performance of work involving energy exchange or conversion processes. Considering an air-breathing engine operating on the well-known Brayton cycle as an example, the input from the environment to the engine consists of air with internal and kinetic energy components. For an engine providing propulsion in a vehicle in flight, the kinetic energy of air at entry to the engine (or, more correctly, to a suitably chosen control volume of the engine) is proportional to the square of the flight velocity. When the flight velocities involved are small and therefore the kinetic energy is relatively small compared with the internal energy of air or energy added to the air (by combustion, say, in the current case) in the engine, it may not be significant to distinguish the kinetic energy from the internal energy of the working fluid. However, as the kinetic energy of entry air increases, changes in that component of energy can have a significant influence on engine performance. This has been pointed out by Builder in an important paper (Ref. 31).

A second point of interest introduced by Builder pertains to the role of energy availability and rational efficiency considerations in engine thermodynamic analysis. The example chosen is the same as the one given earlier, before, an airbreathing engine operating on the Brayton cycle with constant-pressure heat addition by combustion. The argument proceeds as follows.

From several points of view it is found useful in engines to increase the pressure of the working fluid. The increase can be accomplished, in part, by diffusion of momentum and mechanical compression. For given irreversibilities and combustion-generated energy addition, one can then ask the following question: Are there an optimum value of compression and, furthermore, an optimum split between diffusion and mechanical compression so that the maximum output from the engine can be realized? Builder has shown that the answer is, in fact, affirmative, and, furthermore, the optimization of compression split becomes extremely important as the kinetic and, therefore, the available total energy in the entry stream increases. It is obvious that, as the velocity of incoming air increases, there is a decreasing need for mechanical compression. Eventually, with further increases in entry velocity, the proportion of available momentum that needs to be diffused for the desired optimum pressure to be obtained becomes quite small. Thus, the velocity of flow can be maintained with only a modest change from the entry value at the inlet to the value at the entry into the combustor and, also, through the combustor, up to the entry into the thruster nozzle. It may be pointed out that it is, in fact, for this reason that combustion under high-speed, particularly, supersonic, flow conditions for air-breathing engines in high-speed flight vehicles is of interest. Meanwhile, the minimum value of pressure required for combustion must be met with diffusion of the air momentum as the flight altitude may demand.

It is possible to show by yet another argument that the velocity of the flow of the working fluid should, in fact, be maintained at a high value rather than converted to pressure. In the case of the engine for the flight vehicle, the final output of interest is the thrust work produced by the engine. Assuming that thrust is produced through expansion of the working fluid in a thrustor nozzle, there are several considerations: 1) The velocity at the exit of the nozzle is a function of the pressure ratio, provided the nozzle has the required area and the ambient pressure has a non-zero value; 2) the kinetic energy (as well as the thermal energy) in the exhaust is a loss; and 3) any large difference between the exhaust velocity and the air entry velocity adversely affects the propulsive efficiency of the engine. It is clear that, as the flight velocity increases, there is an imperative need for the velocity of the working fluid to remain as close as feasible to the entry value everywhere in the engine, including the combustor and the nozzle exit. It may be of interest to observe here that the method of increasing thrust under high-speed operating conditions in the engine is therefore through increasing the mass flux by mass addition to the working fluid in the engine.

Last, Builder defines an energy conversion efficiency factor as the ratio of the thrust work obtained to the amount of thermal energy added to the working fluid in the engine by means of fuel combustion in order to raise its internal energy. Under constant-pressure combustion conditions, with the velocities across the combustor equal to each other, the energy added by combustion is exactly equal to the change in energy availability across the combustor. Thus, Builder's energy-conversion efficiency or thrust work factor is merely the rational efficiency of the engine or the Brayton cycle with some additional constraints.

The thrust work factor can be shown to be a function of the compression ratio along with some other parameters. Thus, a relation governing the split of compression between the diffusion and mechanical process and the rational efficiency of the engine can be obtained. This is a significant contribution by Builder based on energy availability considerations.

Formalism

A brief summary of the formalism involved in energy analysis is presented in this subsection. Details can be found in several standard texts.

Energy and Entropy

The incremental change in the energy of a substance, d*E,* can be expressed in terms of its internal, kinetic, and potential energy changes by writing

$$dE = dU + VdV + g\,dz \tag{A.1}$$

assuming that the gravitational field is nearly constant over the height dz. It may be noted that internal energy constitutes, as others such as radiation and chemical energy, disordered energy. The potential and the kinetic energy components belong to the ordered form of energy. Whereas the ordered part of energy is fully convertible, the disordered part is never fully convertible.

The enthalpy change in the substance when it is compressible is given by

$$dH = dU + d(P\mathscr{V}) \tag{A.2}$$

The amount of entropy produced in the substance, dσ, in receiving an amount of energy δQ at a boundary which is at temperature *T* can be expressed by the relation

$$d\sigma = dS - \delta Q / T \tag{A.3}$$

where d*S* is the change in entropy. Entropy production is zero in an internally reversible process. The following relations can thus be written for internal energy, enthalpy, and Gibbs free energy, respectively.

$$dU = T\,dS - P\,d\mathscr{V} \tag{A.4}$$

$$dH = T\,dS + \mathscr{V}dP \tag{A.5}$$

$$dG = \mathscr{V}dP - S\,dT \tag{A.6}$$

For a control volume with matter flowing in and out, the time rate of change of energy can be expressed in terms of the time rates of change of energy coming into and leaving the control volume, the time rate of change of work performed by the substance in the control volume, and the time rate of change of energy due to heat interactions. Thus,

$$\frac{dE_{CV}}{dt} = \left(\frac{dE}{dt}\right)_{IN} - \left(\frac{dE}{dt}\right)_{OUT} + \dot{Q} + \dot{W}_{CV} \tag{A.7}$$

Considering a substance with variable chemical composition and denoting the number of moles of species j by n_j, one can define chemical potential by writing

$$\mu_j = \left(\frac{\partial U}{\partial n_j}\right)_{T,P,\, n_l \neq j} \tag{A.8}$$

The chemical potential is a measure of the change in the Gibbs function of the system for a change in the chemical composition of the system.

Another useful relation for the change in internal energy with variable composition is

$$dU = \left(\frac{\partial U}{\partial S}\right)_{\mathscr{V}, n_j} dS + \left(\frac{\partial U}{\partial \mathscr{V}}\right)_{S,\, n_j} d\mathscr{V} + \Sigma \left(\frac{\partial U}{\partial N_j}\right)_{S, \mathscr{V}, n_l \neq j} d\, n_j \tag{A.9}$$

The mechanical efficiency of a heat engine, also called the "thermal efficiency," can be written as follows.

$$\eta_{TH} = W_{NET} / Q_{IN} \tag{A.10}$$

where the net work obtained from the heat delivered to the engine, Q_{in}, is denoted by W_{NET}. The thermal efficiency may also be written as

$$\eta_{TH} = \frac{(E_{IN} - E_{OUT}) + (Q_{IN} - \text{losses})}{Q_{IN}} \tag{A.11}$$

The maximum efficiency of any heat engine operating between two thermal energy reservoirs is the Carnot efficiency, given by

$$\eta_{CARNOT} = (T_H - T_L) / T_H \qquad \text{A.12)}$$

where the subscripts H and L denote high and low values, respectively. It can also be written as

$$\eta_{MAX} = \left(\frac{Q_{IN} - Q_{OUT}}{Q_{IN}}\right)_{REVERSIBLE} \tag{A.12a}$$

It is implied here that there are no losses and the cycle itself is reversible.

From Eq. (A.12), in the case of the Carnot cycle, the unavoidable and irreplaceable loss can be written as follows.

$$\text{Irreplaceable loss} = Q_{IN}\ (1 - \eta_{CARNOT}) = Q_{IN}\ (1 - \eta_{MAX}) \quad \text{(A.13)}$$

It is clear that, if there are any other losses in the system, the thermal efficiency becomes reduced in comparison with η_{max}.

For the case of any heat engine, according to Eq. (A.11),

$$\text{losses} = (E_{IN} - E_{OUT}) + Q_{IN}\ (1 - \eta_{TH}) \quad \text{(A.14)}$$

Here the losses do not reflect the irreplaceable part. Furthermore, if the losses are reduced to zero, it follows that η_{TH} becomes unity: This can be misleading in assessing and improving the system. On the other hand, based on Eq. (A.13), reducing $(T_H - T_L)$ and reducing T_H does provide an improvement in the maximum possible efficiency. There is obviously an approach that is rationally clear whereby improvements in the system can be sought, through reducing both irreversibilities and other losses.

Energy Availability

Energy availability or exergy of a mass of substance that can be treated as a closed system or a control mass in an environment is zero if the substance is in the dead state relative to the environment. If the substance is at a different state from the environment [the state properties of the latter denoted by $(\ )_0$] then the energy availability ε of the substance can be stated as follows:

$$\varepsilon = (E - U_0) + P_0\,(\mathscr{V} - \mathscr{V}_0) - T_0\,(S - S_0) \quad \text{(A.15)}$$

The absolute energy availability with reference to zero of absolute temperature can then be written as

$$\varepsilon_{ABS} = E + P_0\mathscr{V} - T_0\,S \quad \text{(A.16)}$$

since U_0 $\mathscr{V}_0$, and S_0 are zero by definition at that temperature of the environment.

Taking into account heat and work interactions between the system and the environment, the maximum work that the system can yield is given by

$$(W_C)_{MAX} = \varepsilon \quad \text{(A.17)}$$

The time rate of change in energy availability due to heat transfer, $\dot{Q}_A$, from the control mass over an area dA at temperature T is given by

$$d\dot{\varepsilon} = \left(\frac{T - T_0}{T}\right)\dot{Q}_A\, dA \quad \text{(A.18)}$$

For a control volume the rate of change in energy availability over a period of time *dt* can be written as

$$\frac{d\,\varepsilon_{CV}}{dt} = \left.\frac{\varepsilon_{CV}}{dt}\right)_{IN} - \left.\frac{d\varepsilon}{dt}\right)_{OUT} + \int \left(1 - \frac{T_0}{T}\right)\dot{Q}_A\, dA - \left(\dot{W}_{CV} - P_0 \frac{d\mathcal{V}_{CV}}{dt}\right) \quad \text{(A.19)}$$

where the integration is over the control surface.

When the substance under consideration can vary in chemical composition, one can express the additional availability in terms of the chemical potential. Energy availability then can be expressed as follows:

$$\varepsilon_{\text{dead state}} = (E - U_0) + P_0\,(\mathcal{V} - \mathcal{V}_0) - (\Sigma\, n_j\, \mu_{jo} - \Sigma\, n_{jo}\, \mu_{jo}) - T_0\,(S - S_0) \quad \text{(A.20)}$$

Here it is assumed that the environmental state has a different composition compared with that of the system under consideration.

There is often considerable difficulty in establishing chemical availability of a substance with respect to a given environment when the environment does not contain all of the reactants and the products of reaction under consideration. Absolute chemical availability can, of course, be established with respect to the absolute zero of temperature. However, when the main interest is in changes in chemical availability of a substance, it is possible to start with an assumption that all of the reactants and products are present in the environment in sufficiently large quantities.

In the case of a fuel that undergoes combustion, there is often a need for a phase change of the fuel in the burner. This further complicates the problem of determining chemical availability, especially when chemical action accompanies physical, phase change processes. It is also necessary to note the possibilities in practical systems of nonequilibrium states, catalytic action, and quenching of certain reactions. However, these details are of interest in examining specific mechanical components and processes of the system, and in many cases one can establish overall system performance on the basis of simple models for chemical changes.

The destruction of availability or irreversibility can be expressed by the following relation:

$$I = T_0\,\sigma \quad \text{(A.21)}$$

where the entropy produced in the system, σ, is defined by Eq. (A.3). A time rate of change of irreversibility can also be established utilizing the time rate of change of entropy produced. Based on loss of availability, or irreversibility, the so-called rational, or true thermodynamic or second law, efficiency of a process can be defined as follows.

$$\psi = \frac{\Delta \dot{\varepsilon}_{OUT}}{\Delta \dot{\varepsilon}_{IN}} \tag{A.22}$$

$$= 1 - \frac{\dot{I}}{\Delta \dot{\varepsilon}_{IN}} \tag{A.23}$$

where the time rate denotes the flux across a boundary. Other efficiency parameters can be constructed from the foregoing, e.g. an efficiency defect equal to $(1 - \psi)$ and, in the case of a mechanical component that performs work, an effectiveness parameter defined by

$$\varepsilon^* = \frac{\text{Output}}{\text{Change in availability}}$$

$$= \frac{\text{Output}}{\Delta \dot{\varepsilon}_{IN} - \Delta \dot{\varepsilon}_{OUT}} \tag{A.24}$$

It is important to observe that both Ψ and ε^* have maximum values in any given case when the process involved is entirely reversible. Values less than the maximum, which itself can never be unity in practice due to entropy production, arise on account of losses. As losses are reduced, ψ_{MAX} and ε_{MAX} are approached.

In any thermodynamic process, with given changes in state properties, only energy availability and rational efficiency reflect what could have been obtained from the substance undergoing the process and what would be lost irrecoverably from it.

Application of the Energy Availability Analysis to a Brayton Cycle Engine

An engine, such as a ramjet or a scramjet engine, operates according to the Brayton cycle, which is illustrated in Fig. 9. The engine is considered here when it is propelling a flight vehicle moving under given ambient conditions (H_0, P_0) at a speed equal to V_0. Therefore, atmospheric air may be assumed to enter the engine with a value of enthalpy equal to H_0 and of kinetic energy equal to $V_0^2 / 2gJ$. It undergoes momentum diffusion in the engine inlet under adiabatic conditions with an efficiency, say, $\eta_C = (H_1 - H_2)/(H_1 - H_0)$; the pressure of air increases to P_1. Fuel, which has internal, kinetic and combustion (under ideal stoichiometric conditions) energies equal to Q_H^*, Q_V^*, and Q_C^*, respectively, is assumed to be added at 1 with a fuel equivalence ratio of ϕ_F. As a result of the combustion of fuel at constant pressure, taking into account various losses in the combustor, it is assumed that an amount of heat equal to Q_T becomes added to unit mass of air flowing through the burner. It follows that

$$(H_3 - \psi H_0) = (V_1^2 - V_3^2) + Q_T \tag{A.25}$$

Assuming a value for the ratio V_1/V_3, such as unity, across the burner, one can determine the value of $(H_3 + V_3^2 / 2gJ)$. If the products of combustion are assumed to expand, under adiabatic conditions, in a thrust nozzle to atmospheric pressure between 3 and 5, with an efficiency $\eta_e = (H_3 - H_5)/(H_3 - H_4)$, the nozzle jet velocity V_5 can be determined. Then obtain the value of H_5 at the nozzle exit can also be determined.

The thrust generated per unit mass of airflow is given by the following:

$$T = (1 + \phi_F f)V_5 - V_0 \qquad \text{(A.26)}$$

The fuel energy expended in generating the thrust is given by the following per unit mass of airflow.

$$Q_{TE} = (Q_H^* + Q_V^* + Q_C^*)f\,\phi_F \qquad \text{(A.27)}$$

The energy conversion coefficient may therefore be written as follows.

$$\theta = [(1 + \phi_F f)V_5 - V_0]\,V_0 / (gJQ_{TE}) \qquad \text{(A.28)}$$

It follows that $\theta = \eta_0$ the overall propulsive efficiency.

The rational or true thermodynamic efficiency of the engine becomes the following, based on Eq. (A.22):

$$\psi = [(1 + \phi_F f)\,V_5 - V_0]\,V_0 / (gJ)\,(Q_{TE} + V_0^2 / 2gJ) \qquad \text{(A.29)}$$

The thermal or mechanical efficiency of the engine is given by the following:

$$\eta_{TH} = \left[\left(1 + \phi_F f\right) V_5^2 - V_0^2\right] / (2gJQ_{TE}) \qquad \text{(A.30)}$$

Relative to the ambient conditions, the energy availability of air entering the engine per unit time is given by the following:

$$\dot{\varepsilon}_0 = V_0^2 / 2gJ \qquad \text{(A.31)}$$

The energy availability rate of the nozzle exhaust gas is given by the following:

$$\dot{\varepsilon}_5 = \left[(V_5^2 / 2gJ) + (H_5 - H_0)\right] / (1 + \phi_F f) \qquad \text{(A.32)}$$

The energy availability rate lost in the burner irrecoverably is given by the following:

$$\dot{\varepsilon}_{LOSS\ BURNER} = Q_{TE} - Q_T + \left(V_3^2 - V_1^2\right) / 2gJ \qquad \text{(A.33)}$$

where $Q_T = (H_3 - H_1)$, provided it is assumed that Q_C^* represents the chemical availability of fuel relative to the atmosphere.

Appendix B: Weights Analysis

Jones-Donaldson Analysis

The so-called Jones-Donaldson analysis[31] is an important contribution to the evolution of a high-speed vehicle. It leads to sensitivity parameters, based essentially on energy analysis, that can be utilized effectively to assess given designs and levels of technology for a high speed flight vehicle.

The two main assumptions of the analysis are the following: 1) the gain in total (potential and kinetic) energy of a flight vehicle is proportional to the amount of input energy converted into mechanical work, and 2) it is also proportional to the change in the mass of the vehicle.

The element of change in total energy of the vehicle, ΔE, between two locations, i and f, along the flight trajectory may be expressed in terms of the following: 1) the change in the sum of potential and kinetic energies, and 2) the work done in overcoming the drag of the vehicle that is converted into aerodynamic heat. The two may be written as follows:

$$\Delta(E_k - E_p) = \frac{W_f}{gJ}\left(\frac{V_f^2}{2} + gh_f\right) - \frac{W_i}{gJ}\left(\frac{V_i^2}{2} + gh_i\right) \tag{B.1}$$

and

$$H_A = \int_i^f \frac{W}{[(T/D) - 1]\,gJ}\, d\left(\frac{V^2}{2} + gh\right) \tag{B.2}$$

where W is the local weight of the vehicle, and the subscripts i and f refer to the two locations i and f.

Writing $R_W = (W_i / W_f)$, $e = \left[(V^2/2) + gh\right]$, and $X = (e/e_f)$, the energy ratio, one can write ΔE as follows.

$$\Delta E = \Delta(E_k - E_p) + H_A \tag{B.3}$$

$$= \frac{W_f e_f}{gJ}\left[1 - R_W \cdot X_i + \int_i^f \frac{W/W_f}{(T/D) - 1}\, dx\right] \tag{B.4}$$

If the input energy to the working fluid, Q, is proportional to the heat generated by combustion of fuel, one may write

$$Q = Q_C^*\, \phi / (1 + f) \tag{B.5}$$

It follows that

$$\Delta E = \eta\, Q\, W_f (R_W - 1) \tag{B.6}$$

since the mass lost is given by $(W_i - W_f)$. Here η represents the efficiency of input energy utilization, or

$$\eta = \Delta E(\text{output}) / \Delta E(\text{input}) \tag{B.7}$$

Equating Eqs. (B.4) and (B.6), it is possible to obtain an expression for η.

However, in order to carry out the integration in Eq. (B.4), it is necessary to proceed as follows. From the second basic assumption of the analysis, it follows that

$$\frac{W - W_f}{W_i - W_f} = \frac{e - e_i}{e_f - e_i} \tag{B.8}$$

Denoting the takeoff and the final conditions along the trajectory by *i* and *f* respectively, one can express the weight of the vehicle at any point along the trajectory by the following utilizing Eq. (B.8):

$$\frac{W}{W_f} = (R_W - 1)\,\frac{(1 - X)}{(1 - X_i)} + 1 \tag{B.9}$$

From Eqs. (B.3), (B.6), and (B.9), it can be shown, after some algebra, that

$$\eta = \frac{e_f}{gJ(R_W -)1\ Q}\left\{\left(1 - R_W \cdot X_i\right) + \frac{(R_W + 1)}{2\,[(T/D) - 1]}\,(1 - X_i)\right\} \tag{B.10}$$

Here $(R_W - 1)Q$ is the total amount of input energy; the first term within the braces on the right-hand side is the sum of the gain in potential and kinetic energies; and the second term in the braces is the energy left behind in the atmosphere between *i* and *f*.

Equations (B.9) and (B.10) provide two important relations for the flight vehicle: the first relation shows the amount of fuel consumed, and the second relation shows the efficiency of utilization of fuel energy input. It may be of interest to restate Eq. (B.9) in the following form.

$$\frac{W_i - W}{W_f} = (R_W - 1)\left[1 - \frac{1 - X}{1 - X_i}\right] \tag{B.11}$$

The so-called rocket equation is

$$W_i / W_f = R_W = \exp B \tag{B.12}$$

where

$$B = (V_i - V_f) / [g \cdot I_E] \qquad \text{(B.12a)}$$

Therefore, it is possible to apply Eq. (B.9) or (B.11) in terms of either R_W or the velocities of the vehicle and the effective specific impulse of the fuel in the vehicle. Various types of propulsion systems operated with different fuels may be compared with one another on the basis of Eqs. (B.9) and (B.10) for utilization in a flight vehicle of given weight ratio or thrust-to-drag ratio.

An Extension

Flight vehicles for a given mission and payload may also be compared with one another on the basis of the makeup of the initial or gross takeoff weight and of the final weight.

W_G is made up of the empty weight of the vehicle, and the propellant weight. The empty weight may be considered in terms of fixed weight and scalable weight. If the amount of fuel left in the vehicle at the end of the trajectory is considered to be negligibly small, then $W_f = W_E = W_G/R_W$.

The fixed weight is considered to be made up of the payload, the crew, the weight related to payload installation, the weight that is crew-related, and the fixed weight that is not related to the payload and crew. Thus, the following can be written:

$$W_F = W_P + W_P \cdot F_P + \left(W_C \cdot N_C + F_C \cdot N_C\right) + W_{FE} \qquad \text{(B.13)}$$

On the other hand, scalable weight is the weight that scales with the size of the vehicle and consists of structure weight engine weight and scalable equipment weight. Thus, W_S may be written as follows.

$$W_S = W_{ST} + W_{EN} + W_{SE} \qquad \text{(B.14)}$$

In assessing a design for a flight vehicle and its propulsion system, it is therefore of considerable interest to determine how the gross weight changes with changes in fixed weight. This sensitivity factor is designated the fixed-weight growth factor, F_{GF}, which is defined as follows:

$$F_{GF} = W_G / W_F = 1/F_{FG} \qquad \text{(B.15)}$$

where, F_{FG}, the inverse of F_{GF}, may also be considered as a sensitivity factor that represents the change in fixed weight that results from by a change in W_G.

It follows from Eq. (B.11) that

$$F_{GF} = \exp(B) \cdot \left[(1 + Y) / Y\right] \qquad \text{(B.16)}$$

$$= \frac{W_G}{W_C}\left(\frac{1+Y}{Y}\right) \tag{B.17}$$

Thus, in the case of a vehicle for which Y and either the velocities and effective specific impulse or the ratio W_G/W_E are known, one can establish F_{GF}. Alternative vehicles and propulsion systems for a given mission may then be compared utilizing Eq. (B.16) or (B.17).

The foregoing applies to a SSTO vehicle in which the payload is, by definition, a part of the fixed weight. It is possible to extend the concept of the sensitivity parameter to a MSTO vehicle. In a vehicle with N stages, any stage n has a payload consisting of the $(N - n)$ stages following itself. It is necessary to take this fact into account in deducing an expression for the sensitivity factor.

Before proceeding further, one may rewrite the expression (B.13) for the fixed weight as follows.

$$W_F = A \cdot W_P + W_{CE} \tag{B.18}$$

where

$$A = 1 + F_P \tag{B.18a}$$

and

$$W_{CE} = (F_C + W_C) \cdot N_C + W_{FE} = W_{TC} + W_{FE} \tag{B.18b}$$

It can be assumed, in general, that A has a constant value for all stages of a staged vehicle.

Then it can be shown that the following applies to a TSTO system, with the various parameters denoted by $(\ )_{1,2}$ for the total vehicle, and $(\ )_1$ and $(\ )_2$ for the first and the second stage, respectively.

$$W_{G_{1,2}} = W_{E_1} \cdot R_{W_1} + W_{G_2} \tag{B.19}$$

$$R_{W_{1,2}} = W_{G_1} / W_{G_2} + R_{W_2} \tag{B.20}$$

$$W_{E_2} / W_{E_1} = (W_{F_1} / W_{F_2}) \cdot (Y_2 / Y_1) \cdot [(1 + Y_1) / (1 + Y_2)] \tag{B.21}$$

$$W_{F_1} / W_{F_1} = (A - 1)\left(\frac{1+Y_2}{Y_2}\right) R_{W_2} + \frac{W_{CE_2}}{(A \cdot W_P + W_{TC})_2} \tag{B.22}$$

and

$$W_{G_{1,2}} / W_{F_2} = \left[\left(\frac{1+Y_2}{Y_2}\right) R_{W_2}\right] \cdot \left[(A-1)\left(\frac{1+Y_1}{Y_1}\right) R_{W_2} + 1\right]$$

$$+ \frac{W_{CE_1} + \left(\frac{1+Y_1}{Y_1}\right) R_{W_1}}{(A \cdot W_P + W_{TC})_2} \tag{B.23}$$

In the case of a three-stage vehicle a similar analysis yields the following relation, subscript 3 denoting the third stage.

$$W_{G1,2,3} / W_{F3} = \left(\frac{1+Y_3}{Y_3}\right) \left[\frac{W_{E1}}{W_{E2}} R_{W1} + \left(\frac{W_{E2}}{W_{E3}}\right) R_{W2} + R_{W3}\right] \qquad \text{(B.24)}$$

Equation (B.24) can be recast in the same form as Eq. (B.23) if necessary.

It is sometimes of interest to examine the growth in gross weight for a given change in the sum of payload and fuel weights. The necessary relation can be obtained utilizing the relation between gross weight and empty weight.

Other Considerations

Comparison of Air-Breather and Rocket

The weight ratio may be included as a parameter for comparing different vehicles designed for a given mission (for example, an air breather and a rocket) in several ways. One of the ways is based on comparing the efficiency of input energy conversion to mechanical work at any desired location along the flight trajectory. The efficiency is defined by Eq. (B.10).

Assuming that $X_i = 0$, corresponding to stationary conditions at takeoff, one obtains from Eq. (B.10) the following simple expression for η.

$$\eta = \frac{e_f}{gJQ} \cdot \frac{1}{(R_W - 1)} \left\{1 + \frac{R_W + 1}{2\,[(T/D) - 1]}\right\} \qquad \text{(B.25)}$$

A second way is to compare a mean value of equivalent specific impulse over the entire trajectory for the vehicles under consideration.

The mean specific impulse over a trajectory can be defined as follows.

$$\bar{I} = \frac{\int (\theta JQ/V_0)\, dV_0}{\int dV_0} \qquad \text{(B.26)}$$

where θ is the Builder energy conversion factor, Eq. (12).

Heat and Recovery and Utilization

In the case of a vehicle operating with an air-breathing engine, the duration of flight in the atmosphere is, by design, larger than that for a vehicle operating with a rocket. Therefore, there is a substantial amount of H_A that becomes generated, as stated earlier. A part of this heat may be recovered by recirculation of the fuel, which is generally carried in the form of a liquid or slush. The fuel supplied to the burner in the engine may therefore be energized with the recovered heat, either directly by an increase in enthalpy or indirectly by conversion into kinetic energy of fuel. In either case the recovered heat becomes an input to the engine along with the combustion energy of the fuel.

In the case in which the fuel is supplied with heat energy, the additional input of energy, Q_H, may be written as follows.

$$Q_H = \eta_H kH_A \tag{B.27}$$

Here kH_A is the fraction of aerodynamic heat recovered, and η_H is the efficiency of the recovery process; an expression for H_A may be found in Eq. (B.2) with the integration in that equation being carried out.

Then, utilizing the expression for efficiency of energy utilization as in Eq. (B.7), and writing $E = e / gJQ$, after some algebra, it follows that

$$R_W = \frac{\eta + E_f\,(1 - k\eta_H)\left[1 + \frac{(1 - X_i)}{2\,[(T/D) - 1]}\right]}{\eta + E_f\,(1 - k\,\eta_H)\left[X_i - \frac{(1 - X_i)}{2\,[(T/D) - 1]}\right]} \tag{B.28}$$

Taking the recovered aerodynamic heat into account, the total amount of input energy to the engine in this case is equal to the following.

$$(R_W - 1)\left[Q + \frac{\eta_H k H_A}{(R_W - 1)}\right] \tag{B.29}$$

Next, in the case where the recovered energy is in the form of kinetic energy of fuel, the energy added may be written as follows:

$$kE_f = \frac{V_F^2\,(R_W - 1)}{2gJQ} \tag{B.30}$$

Then an expression for R_W is obtained that is similar to Eq. (B.29):

$$R_W = \frac{\eta + E_f\left[1 + \frac{(1 - X_i)}{2\,[(T/D) - 1]}\right] + kE_f}{\eta + E_f\left[X_i - \frac{(1 - X_1)}{2\,[(T/D) - 1]}\right] + kE_f} \tag{B.31}$$

Geometrical Parameters

The total volume of the flight vehicle, $\mathcal{V}_T$, may be written as the sum of the volumes of different parts of the vehicle, including the systems, void, crew and payload, and propellants. The systems include the engine or, more generally, the energy addition section. A volume analysis can be performed, for example, by assuming that the volume of the crew and payload and the volume fraction of the void to the total are constant, and then varying the sum of the volumes of the propellants and systems. The latter vary with energy stored and energy density per unit volume. As did the weights analysis, the volume analysis provides sensitivity parameters.

Next, it is of interest to relate geometrical parameters to weight parameters. Recalling the definitions of R_W and W_E, it is possible to write an expression for structural weight per unit of wetted area in the following manner:

$$\frac{W_{ST}}{S_{WT}} = \left(\frac{\rho_F}{(R_W-1)}\right) \cdot \left[\left(\frac{\mathcal{V}_F}{\mathcal{V}_T}\right) \cdot \left(\frac{\mathcal{V}_T}{S_{PN}}\right) \cdot \left(\frac{S_{PN}}{S_{WT}}\right)\right] \cdot \left(\frac{W_{ST}/W_E}{[1 + (W_P/W_E)]}\right) \tag{B.32}$$

The first and third terms on the right-hand side of the equation are weight parameters and the second consists of geometrical parameters. In particular, the first term on the right-hand side includes various considerations related to propulsion, fuel, and integration technology. Equation (B.32) can be utilized for the analysis of various system configurations and also for the analysis of combinations of air-breathing and rocket engines.

Appendix C: System Selection

A given mission and payload form the starting point for the consideration of a flight vehicle. In the case of a flight vehicle for orbital launch, mission and geographical constraints determine some aspects of vehicle staging and trajectory.

Two main features of a high-speed flight vehicle are the imperative need for total integration of the vehicle and the central status of the propulsion system in the design of the vehicle. The airframe, propulsion, and control integration that is required in all flight vehicles must become enlarged further to include a consideration of the thermal management system and trajectory. The propulsion system must be able to provide the desired thrust power with the most effective use of energy availability in air, fuel and aerodynamic heat generated in the vehicle and engine structure everywhere along the flight trajectory. The flight vehicle may, for various reasons, present a variety of critical conditions at a number of locations along the trajectory, and these conditions may have to be taken into account even in the initial stages of vehicle design.

For a given mission and payload, first, staging and energy sources should be chosen, the latter in terms of energy generated per unit mass as well as volume of fuel. Next, values for the ratio of payload weight to empty weight, the fixed weight, and the fixed weight growth factor are chosen. These lead to an estimate of weights and volumes and their distributions.

At the same time, the flight trajectory can be optimized for minimum requirement of thrust power for meeting vehicle drag and acceleration requirements, noting the constraints on maximum allowable aerodynamic and heat loads on the structure.

At this stage, the propulsion system can be analyzed to yield simultaneously a combination of maximum values for thrust and overall propulsive efficiency; this requires the most effective use of energy availability in fuel, air, and aerodynamically generated heat in the vehicle and the engine and the least loss in energy availability in every component of the propulsion system. Utilizing the results of the analysis, the implications of using compound engines of different types over different parts of the flight trajectory can be estimated. For the air-breathing engine the airflow required can be determined, and for a rocket engine the requirements in intrinsic propellants can be determined.

The propulsion can be analyzed for the sensitivity parameters, namely the structural and the external bond coefficients, utilizing the structure diagram of the system.

The convergence of vehicle design then depends on a synthesis or matching of propulsion, trajectory, and vehicle size. At this stage of development of necessary technologies, including those for testing and for numerical prediction, the methodology for accomplishing convergence cannot take the form of a mathematical method of optimization. Rather, the convergence must be obtained through a series of trials in matching. In the following the basic equations applicable to various aspects of the matching procedure are gathered together and presented in summary form for convenience.

Governing trajectory equations

$$\frac{dV_0}{dt} = g(n_X - \sin\theta) \tag{C.1}$$

$$\frac{dz}{dt} = V_0 \sin\theta \tag{C.2}$$

$$\frac{d\alpha}{dt} = \frac{R_0}{R_0 + z} V_0 \cos\theta \tag{C.3}$$

$$q = q(\text{trajectory}) \tag{C.4}$$

$$\text{Aerothermodynamic parameters} = f(\text{trajectory}) \tag{C.5}$$

Limitations on outer skin, internal wall, and fuel temperatures (C.6)

Limitation on maximum q (C.7)

Flight boundary conditions (C.8)

Lift distribution along trajectory (C.9)

Lift-to-draft distribution along trajectory (C.10)

Acceleration vector distribution along trajectory (C.11)

Vehicle staging equations (C.12)

Propulsion equations

Mechanical efficiency of propulsion components (C.13)

True thermodynamic efficiency of propulsion components (C.14)

Structural bond coefficients for components (C.15)

External bond coefficients for components (C.16)

Jet efficiencies (C.17)

Energy-conversion coefficients (C.18)

Overall propulsive efficiencies (C.19)

Engine-compounding relations (C.20)

Specific impulse values along trajectory (C.21)

Effective specific impulse values along trajectory (C.22)

Engine switching schedules (C.23)

Propellant consumption along trajectory (C.24)

Weights equations:

Payload to empty weight (C.25)

Fixed weight (C.26)

Fixed-weight growth factor (C.27)

Gross weight (C.28)

Propellant weight (C.29)

Volume equations (C.30)

References

1. Zwicky, F., "Future Prospects of Jet Propulsion," in *Jet Propulsion Engines*, (ed) O. E. Lancaster, Princeton Aeronautical Series, Princeton, 1956.

2. Bosnjakovic, F., *Technical Thermodynamics*, Holt, Rinehart and Winston, New York, 1965.

3. Reynolds, W. C., and Perkins, H. C. *Engineering Thermodynamics*, McGraw-Hill, New York, 1977.

4. Ahern, J. E., *The Exergy Method of Energy System Analysis*, Wiley, New York, 1980.

5. Bejan, A., *Advanced Engineering Thermodynamics*, Wiley, New York, 1988.

6. Kotas, T. J., *The Exergy Method*, Butterworth, Boston, 1985.

7. Morgan, M. J., *Availability Analysis*, Prentice-Hall, Englewood Cliffs, 1982.

8. Beyer, J., Structural Investigations - An Essential Part of the Analysis of the Efficiency of Thermal Systems, *Energieanwendung*, Vol. 19, No. 12, Dec. 1970.

9. Beyer, J., Distribution of Primary Energy Costs in Multi-Purpose Processes on the Basis of Structural Analysis, *Energieanwendung*, Vol. 21, No. 6, June 1972.

10. Gogus, A. Y., Methods of Integral Thermo-Economic Evaluation, Proc. First National Conference on Air-Breathing Engines, Istanbul Technical University, Istanbul, Feb. 1990.

11. Park, C., *Nonequilibrium Hypersonic Aerothermodynamics*, John Wiley & Sons, 1990.

12. Sychev, V. V., Vasserman, A. A., Kozlov, A. D., Spiridonov, G. A., and Tsymarny, V. A., *"Thermodynamic Properties of Air"* Hemisphere Publishing Corporation, New York, 1987.

13. Brahinsky, H. S., and Northcutt, D., *"AEDC Mollier Diagram for Equilibrium Air,"* ARO Inc., Pathenon Press, Nashville TN, Feb. 1967.

14. Pinkus, O., "Thermodynamic Properties and Dynamics of Real Air From Subcritical Temperatures to 1,500 K," Republic Aviation Corporation, Farmingdale, Report No. 2716, 1964

15. Jacobson, R.T., McCarty, R.D., and Olien, N.A. "Thermophysical Properties of Air," Center for Chemical Engineering, National Bureau of Standards, Printed by NASA Langley Research Center, NASP TM-1005, March 1987.

16. National Bureau of Standards, *Selected Properties of Hydrogen (Engineering Design Data),* U.S. Department of Commerce, National Technical Information Service PB81-2111492, Feb. 1981.

17. McCarty, R. D., "Hydrogen Technological Survey - Thermo-Physical Properties," Report No. NASA SP-3089, 1986.

18. Ackeret, J., The Role of Entropy in the Aerospace Sciences, *Advances in Aeronautical Sciences*, Vol. 3, Pergamon, New York, 1962.

19. Eggers, A. J., Julian, A. M., and Neice, S. E., A comparative Analysis of the Performance of Long-Range Hypervelocity Vehicles, National Advisory Committee for Aeronautical Report 1382, 1958.

20. Billig, F. S., "Two-Dimensional Model for Thermal Compression," *Journal of Spacecraft*, Vol. 9, Sept. 1972, 702-703.

21. Oswatisch, K., *Gas Dynamics,* Academic Press, New York, 1956.

22. Hayes, W. D., and Probestein, R. F., *Hypersonic Flow Theory,* Academic Press, New York, 1959.

23. Tanatsuga, N., Honda, T., Sagiya, Y, and Higashino, K., Development Study of an Air Turbo-Ramjet for Future Space Plane, International Astronautical Federation Paper 89-311-, Oct. 1989.

24. Ogawara, A., and Nishiwako, T., The Cycle Evaluation of Advanced LACE Performance, International Astronautical Federation Paper 89-313, Oct. 1989.

25. Builder, C. H., On the Thermodynamic Spectrum of Air-Breathing Propulsion, AIAA Paper 64-243, 1st Annual Meeting, Washington D.C., June-July 1964.

26. Kepler, C. E., and Champagne, G. A., Supersonic Through-Flow Fan Assessment, NASA CR-18-2202, 1988.

27. Riband, Y., Inverse Cycle Engine for Hypersonic Air-Breathing Propulsion, *Proc. IX ISABE*, Athens, Greece, AIAA, Washington, D.C. 1989.

28. Gopalaswami, R., Gollakora, S., Venugopalan, P., Nagarathinam, M., and Pillai, A.S., Concept Definition and Design of a SSTO Launch Vehicle Hyperplane, International Astronautical Federation Paper 88-194, Oct. 1988.

29. Townend, L. H., Effects of External Heat Addition on Supersonic Cruise Performance, *The Aeronautical Quarterly*, Vol. XVI, Aug. 1962.

30. Townend, L. H., Base Pressure Control for Air-Breathing Launchers, AIAA Paper 90-1936, July 1990.

31. Jones, R. A., and Donaldson, C. du P., From Earth to Orbit in a Single Stage, *Aerospace America*, August 1987.

32. Kerrebrock, J. L., *Aircraft Engines and Gas Turbines*, The MIT Press, Cambridge, Massachusetts, 1978.

33. Gregory T., Credibility of NASP, *Aerospace America*, Sept. 1989.

34. Shannon, C., and Waver, W., *The Mathematical Theory of Communication*, University of Illinois Press, 1949.

Chapter 4

Waves and Thermodynamics in High Mach Number Propulsive Ducts

R. J. Stalker*
University of Queensland, Brisbane, Australia

Nomenclature

a	=	speed of sound
a_3	=	heat added per unit of stagnation enthalpy
A	=	streamtube cross-sectional area
A	=	$\sqrt{M^2-1}$ / (γM^2)
C_d	=	drag coefficient
d	=	duct height
h	=	enthalpy
M	=	Mach number
p	=	pressure
q	=	heat added per unit mass
Q	=	total rate of heat addition
R_{ex}	=	Reynolds number
T	=	thrust
T	=	temperature
T_δ	=	thrust due to waves from streamtube
u	=	velocity along streamline
W_n, W_s	=	work done normal to and along streamlines respectively

*Professor.

x	=	distance downstream
α	=	leading-edge angle
β	=	flow deflection through transmitted shock
γ	=	ratio of specific heats
δ	=	streamtube thickness
θ	=	flow deflection in clockwise direction
θ_s	=	surface deflection
μ	=	viscosity
ν	=	Prandtl-Meyer function
ρ	=	density

Subscripts

1, c	=	immediately after and before heat addition, respectively
i, r	=	incident and reflected wave, respectively
j	-	jet
s	-	surface
t	-	transmitted wave

I. Introduction

Not only are wave phenomena an important feature of supersonic flight, but they also must be expected to play an important role in the fluid dynamics of supersonic-combustion ramjets. In order to design an efficient propulsive duct with supersonic combustion, it is necessary to understand phenomena such as the drag associated with wave production, and the efficient application of waves to thrust surfaces. As flight speeds are increased, such understanding becomes ever more important, since the flow Mach numbers within the duct are raised from supersonic to hypersonic values, and the penalties of inefficient flow management become more severe.

This chapter is concerned with the propulsive effects of waves in ducts, especially at high Mach numbers. It focuses on drag and thrust and on the conversion of heat into waves which produce thrust. In order to allow reasonably simple approaches to analysis of these effects, a number of simplifying assumptions are made. Thus, real-gas effects are ignored, and the fluid is treated as a perfect gas, with constant ratio of specific heats. Combustion is represented by heat addition, without the addition of mass or change in compressibility factor that normally is associated with combustion. Also, it is assumed that the heat addition is effected in somewhat arbitrarily simplified zones, such as a plane normal to the stream direction. However, it is recognized that the fuel-injector arrays represented by the heat-addition zones could, in practice, be such that fuel-injection rates could be matched to local flow conditions and, therefore, the heat-addition rate is allowed to vary across a heat-addition zone. Only two-dimensional flows are considered, and shock waves are taken to be weak enough to be regarded as isentropic compressions. Except for a short discussion toward the end of the paper, boundary-layer effects are ignored. This implies, of course, that there is no heat exchange between the fluid and the walls.

The chapter begins with the production of thrust at high Mach numbers, demonstrating that essentially all of the work done by an expanding fluid passing through a duct at high Mach number is delivered in the form of waves. It is also shown that duct surface angles exist that are optimum for the production of thrust from a wave.

The effects of wave phenomena on drag and thrust are then considered by extending the concept of a Busemann biplane into that of a "Busemann scramjet," and taking "off-design" performance into account. For this exercise, heat addition is effected, in a somewhat oversimplified "off-design" performance into account. For this exercise, heat addition is effected, in a somewhat oversimplified manner, by assuming a heat-addition zone that is both very thin and is normal to the flow direction.

In the search for a more realistic model, the process of heat addition to a streamtube is discussed, leading to an idealized model of a streamtube with heat addition. This discussion also is extended to include the flow mechanisms involved in generating thrust by the expansion of this streamtube in an exhaust nozzle.

And finally, some boundary-layer effects are briefly considered.

II. Thrust at High Mach Number

A. One-Dimensional Flow

Consider a general cycle for a compressible fluid, as displayed in the p-V diagram of Fig. 1a. Compression of the fluid along path a is followed by heat addition along path b. Thrust may be produced during heat addition and, in subsequent expansion, along path c.

The conversion into work of the internal energy of the fluid or of the heat added to it is shown in Fig. 1b, where an element of fluid of initial length dx is seen to be undergoing expansion in a duct with area increasing in the streamwise direction. In a frame of reference that is fixed in the fluid, the streamwise boundaries separate by an

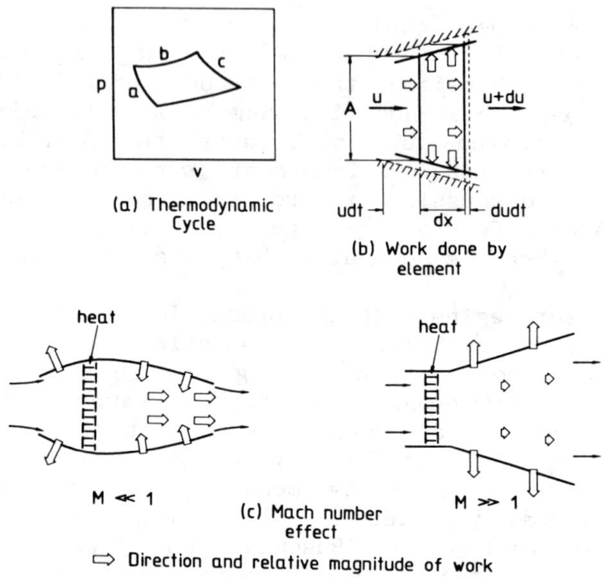

Fig. 1 Thrust generation in one-dimensional flow.

amount du dt in time dt, implying that the fluid is doing an amount of work on adjacent fluid elements that is given by

$$dw_s = p\ A\ du\ dt \tag{1}$$

where the subscript is a reminder that the work is done by movement of the fluid boundaries in a streamwise direction. During the same time interval, the transverse boundaries of the element also separate, because of the apparent upstream movement of the duct within which the fluid is flowing, and the work done by the fluid at these boundaries is given by

$$dw_n = p(dA/dx)\ dx\ u\ dt \tag{2}$$

where the subscript n is a reminder that the work is done in a direction normal to the flow. Combining Eqs. (1) and (2) yields

$$dw_n/dw_s = u(dA/dx)\ /\ A(du/dx) \tag{3}$$

as the ratio of work done normal to the streamlines to work done along the streamlines.

For one dimensional flow, the continuity equation can be written as

$$A^{-1}\ dA/dx + \rho^{-1}\ d\rho/dx + u^{-1}\ du/dx = 0$$

and the momentum equation in the streamwise direction as

$$dp/dx = \rho u\ du/dx \quad .$$

When these two equations are combined, they yield the well-known relation

$$A^{-1}dA/dx=(M^2-1)u^{-1}du/dx \tag{4}$$

where $M^2 = u^2/(dp/d\rho)$. The value of $dp/d\rho$ depends on the thermodynamic processes in the fluid. For isentropic flow, $dp/d\rho = \gamma p/\rho = a^2$, where a is the isentropic speed of sound, and M is the Mach number. For heat addition in a perfect gas of constant specific heats, the equation of state, the first law of thermodynamics, and the streamwise momentum equation lead to

$$dp/d\rho = a^2 [1 - (\gamma - 1)(dq/du)/u]^{-1}$$

which, when substituted into Eq. (4), yields

$$A^{-1}(dA/dx)[1-(\gamma-1)(dq/dA)A/a^2] = (M^2-1)u^{-1} du/dx \tag{5}$$

and when this is substituted into Eq. (3) it is found that

$$dw_n/dw_s = (M^2 - 1)/[1 - (\gamma - 1)(dq/dA) A/a^2] \tag{6}$$

It is obvious that the second term in the denominator can be large when dq/dA is large, corresponding to a large amount of heat addition with small area change. However, such a case represents a nearly constant area combustion duct and, since this is not expected to be a significant source of thrust without further expansion, it need not be considered further. When significant thrust is produced during heat addition, a substantial area change takes place. Taking an area change of a factor of 2 as typical and putting

$$a^2 = \gamma RT = (\gamma - 1) C_pT \sim (\gamma - 1)q$$

it follows that the second term in the denominator on the right side of Eq. (6) typically is of order unity, and

$$|dw_n / dw_s| \geq M^2 - 1 \tag{7}$$

That is, the ratio of work done normal to the streamlines to work done along the streamlines depends on the Mach number.

This Mach number effect is illustrated diagrammatically in Fig. 1c, where two idealized propulsive ducts are shown; one at a low subsonic Mach number, and the other at a high supersonic Mach number. In both cases, heat addition takes place in a parallel part of the duct, where no work can be done on or by the duct walls. This means that thrust generation involves only the converging or diverging parts of the duct where, because $dq/dA = 0$, the equality sign applies in Eq. (7).

It is well known (e.g., Ref. 1) that, for Mach numbers much less than one, the thrust of the propulsive duct is generated through the work done by the fluid on the duct at the upstream end. As shown by Eq. (7), the work done on the duct walls is equal and opposite to the work done in the stream direction by the decelerating flow, signifying that, in that part of the duct, the energy delivered to the walls is taken from the kinetic energy of the stream. If there were no heat addition in the duct, then, for the same pressure at the duct exhaust as at the inlet, the energy taken from the stream would be returned through the work done on the stream by the converging section of the duct, and there would be no net thrust. However, with heat addition, the same exhaust pressure is reached with less energy transfer from the duct walls, and so thrust is generated. The thermodynamic energy associated with expansion (or compression) of the fluid does not play a direct role in this process.

As the Mach number increases, the expansion of the fluid takes on a more significant role, tending to reduce the exchange of energy with the duct walls until, at a Mach number of unity, no energy is exchanged with the duct walls. Then, as the Mach number increases further, in the supersonic regime, the expansion of the fluid provides enough energy to do work on the walls of the duct, as well as increasing the stream kinetic energy. The latter represents an increasingly insignificant fraction of the energy of the fluid expansion until, at high Mach numbers as shown in the diagram, essentially all of the expansion work is done on the walls of the duct.

B. Two-Dimensional Flow

A two-dimensional flow may be considered as an array of thin streamtubes and, since the arguments developed above can be applied to any individual streamtube, it follows that, at high Mach numbers, nearly all of the work done by the expanding fluid is delivered in a direction normal to the stream. Considering a single expanding streamtube, as shown in Fig. 2a, this work is transferred to adjacent streamtubes. If the streamline deflections are not too large and the surrounding flow is homentropic, then the linearized theory of supersonic flow can be applied, according to which, the exact amount of this work is transferred to the next adjacent streamtubes, and to the next, and so on, to produce waves, as shown in the figure. These waves may be modified by the behavior of the adjacent streamtubes, but these modifications do not alter the

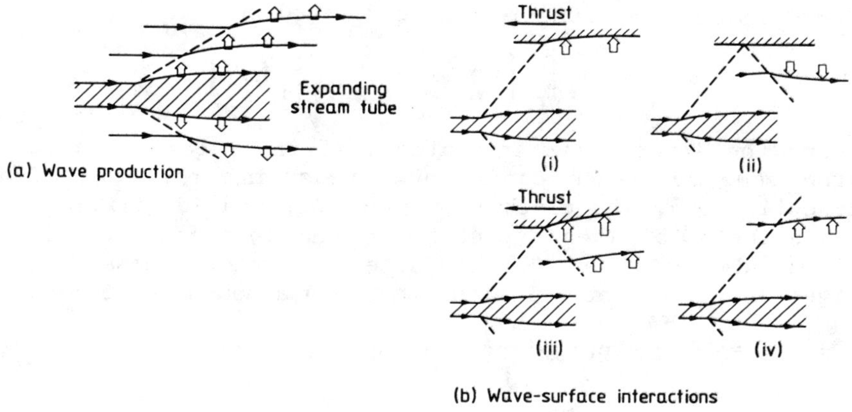

Fig. 2 Waves and thrust generation.

conclusion that, in a high Mach number supersonic flow, the thermodynamic work resulting from heat addition and expansion is essentially all delivered in the form of waves. Thrust is produced by the interaction of these waves with the surfaces of the propulsive duct.

The thrust produced by a single wave transmitted across a uniform flow, such as that shown in Fig. 2a, depends on the contour of the duct surface. If, as shown in Fig. 2b(i), the surface slope matches the distribution of streamline deflection in the wave, then the wave is eliminated, and the transverse flux of energy, which originated as work done by the wave-producing streamtube, is transferred to the surface to produce thrust. However, if the surface slope is reduced to zero, as in Fig. 2b(ii), no work can be transferred to the surface, and the wave must reflect to direct the transverse flux of energy away from the surface. On the other hand, if the surface slope is increased, as in Fig. 2b(iii), then the transverse flux of energy in the initial wave may be increased by expansion of the streamtube adjacent to the surface, which gives rise to a reflected wave in which the transverse flux of energy is directed toward the surface. And, finally, if no surface exists, as in Fig. 2b(iv), then the transverse flux of energy propagates away from the original wave-producing streamtube, and never produces thrust.

These effects can be summarized analytically by changing to a frame of reference that is fixed in the duct, and using the theory of characteristics for supersonic flow.

The Prandtl-Meyer function[2] ν is defined by

$$\nu = -\int_r \left(\sqrt{M^2 - 1} / \gamma M^2\right)\left(dp/p\right) \tag{8}$$

where the lower limit indicates that the integral is taken from some reference condition. Measuring the local flow direction θ in the clockwise direction, and identifying the characteristic lines departing to the left from the local stream direction as (+) characteristics and those to the right as (-) characteristics, then, in a homentropic flow,

$$\nu + \theta \quad = \quad \text{constant along (+) characteristics} \tag{9a}$$

$$\nu - \theta \quad = \quad \text{constant along (-) characteristics} \tag{9b}$$

Using these relations, it can be shown (e.g., Ref. 3) that the increment of flow deflection across a wave remains constant as it propagates through the flowfield. While this is strictly true only for homentropic flows, it can be shown[3] that it also holds true to a close approximation as a wave passes through regions of transverse Mach number gradients, such as those downstream of a region of nonuniform heat addition.

Using this theory, the ideas expressed in Fig. 2 can be generalized to include a wave transmitted across a flow that is not necessarily uniform and arriving at a surface that, upstream of the wave, may be inclined with respect to the incident flow direction at the origin of the wave. If, for example, such a wave is transmitted along (+) characteristics, then, because the flow deflection across a wave remains constant as it propagates, a surface with streamwise change of slope that matches the wave simply replaces the wave streamlines and generates no reflected wave. In that case, the change of ν along the surface also matches that of the wave [by virtue of Eq. (9b)] but, because ν itself may be different, not necessarily all of the original transverse flux of energy in the wave is transmitted to the surface. If the streamwise change of slope differs from that of the wave then a new wave must be created that is a compression or expansion according as the change of surface slope is less or greater than that of the original wave.

C. Optimum Surface Angle for Most Effective Thrust

Scramjet flight vehicles are likely to be very severely weight-limited and, therefore, there is considerable advantage to be gained by minimizing the amount of surface needed to produce a given thrust, i.e., by maximizing the thrust per unit of surface area. The application of waves to produce thrust involves transfer of the transverse energy flux of a wave to a surface that is suitably inclined to the flight direction. There the work done normal to the stream direction at the boundary of an element of fluid adjacent to the surface is experienced by an element of the surface as work done in the flight direction. This is shown in Fig. 3a, using a frame of reference that is fixed in the fluid. In order to increase the thrust per unit length of surface associated with a given surface pressure, it would appear beneficial to increase the surface angle. However, this also involves increasing the transverse kinetic energy of the fluid, as transverse velocities increase to match the slope of the surface, and this will absorb part of the transverse flux of energy of waves incident on the surface. Therefore, there will be an optimum surface angle at which maximum *thrust per unit length* is transmitted to the surface from a given flow.

Following the (-) characteristic from some point, a, in the field, through a homentropic region of the flow to the surface, as shown in Fig. 3b, Eq. (9b) can be written

$$\nu_s - \nu_a = \theta_s - \theta_a$$

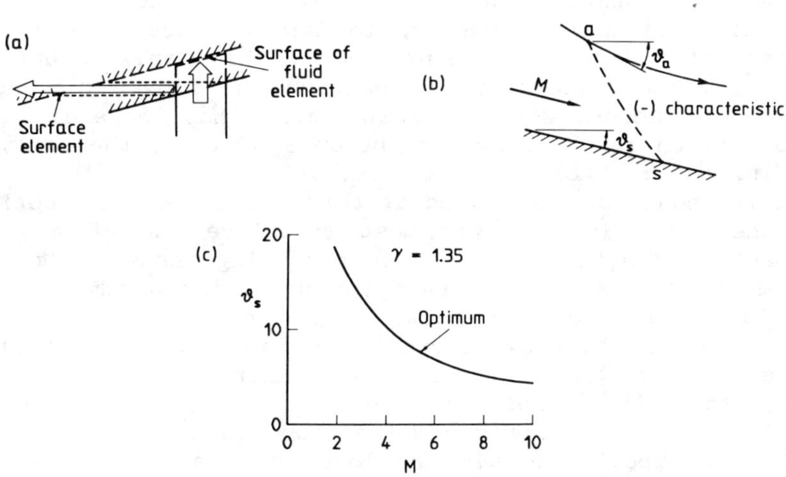

Fig. 3 Optimum surface angle for thrust effectiveness.

where θ_s is the angle made by the surface with the upstream flow direction. Using Eq. (8), and differentiating,

$$dp/d\theta_s = - Ap \tag{10}$$

where

$$A = \gamma M^2 / \sqrt{M^2 - 1}$$

Now, the thrust per unit length of the surface can be written as p $\sin\theta_s$ and, differentiating and using Eq. (10), it follows that

$$d\left(p \sin\theta_s\right) / d\theta_s = p \cos\theta_s \left(1 - A \tan\theta_s\right) \tag{11}$$

which has a zero, corresponding to a maximum in p $\sin\theta_s$. when

$$\tan\theta_s = A^{-1} = \sqrt{M^2 - 1} / \left(\gamma M^2\right) \quad . \tag{12}$$

Thus, with a given wave pattern in the flow, there is an optimum surface angle for local maximum production of thrust, and this angle is dependent only on the Mach number. As shown in Fig. 3c, the optimum angle reduces to quite small values at high Mach numbers.

Of course, the value of the optimum angle may change when the deflection of the thrust surface itself causes waves that generate thrust or drag. An example of this may be seen later, where wave-interface interaction is considered; another example ocurs when the duct is long enough for the waves arising from local deflection of the surface to propagate across the flow and, reflecting from a wall or a plane of symmetry, to impinge once more on the surface at a downstream point. Since this implies that the local surface deflection is changing the flow, such cases are not in accord with the requirement that there is given wave pattern in the flow and, because of this, the analysis leading to Eq. (12) is not applicable.

It should be emphasized at this point that this optimum surface angle is that for most effective use of a given element of length of surface in producing thrust. It does not mean that it is the angle for producing maximum thrust from a given wave. In fact, it can be shown that, if a given wave in the flow is considered, with a fixed distance between the upstream and downstream Mach lines bounding the wave, then the thrust delivered by the wave increases monotonically with surface deflection (at least, until the surface deflection equals the local Mach angle). This is because the length of surface over which the wave delivers energy to the surface increases with surface angle.

III. Thrust and Drag in a Duct

A. The Busemann Biplane.

The phenomenon of the Busemann biplane is well known[4]. When two thin, double-wedge airfoils are placed to form a duct, as in Fig. 4, then, according to the linearized theory of supersonic flow, the wave drag of the combination can be reduced to zero by suitably choosing the lengths of the surfaces of the duct to match the Mach number of the incoming flow.

As shown in Fig. 4a, this requires a duct such that the length upstream of the minimum cross section allows the shock waves generated at the leading edge of the inlet to collide on the duct centerline and to impinge on the surface at the minimum cross section. Also, the length downstream of the minimum cross section allows the expansion waves generated by the corners there to reflect at the centerline and impinge on the surface at the trailing edge. These expansion waves are represented on the figure by broken lines, which may be taken as indicative of the path followed by the center of the expansion fans that will be present. Representing the path of these fans by lines will be a reasonable approximation only if the expansion waves are weak that is, if the flow deflection angles are small.

The surface pressure distribution can be obtained by adding flow deflection angles through the waves and by counting compression waves as positive and expansion waves as negative. Thus, the net flow deflection on the upstream surface is α, whereas the flow deflection on the downstream surface is obtained by adding the deflection through the reflected shock wave and subtracting the same deflection through the corner expansion wave to yield again a value of α. From Eq. (9a) and (9b), it is clear that changes in α

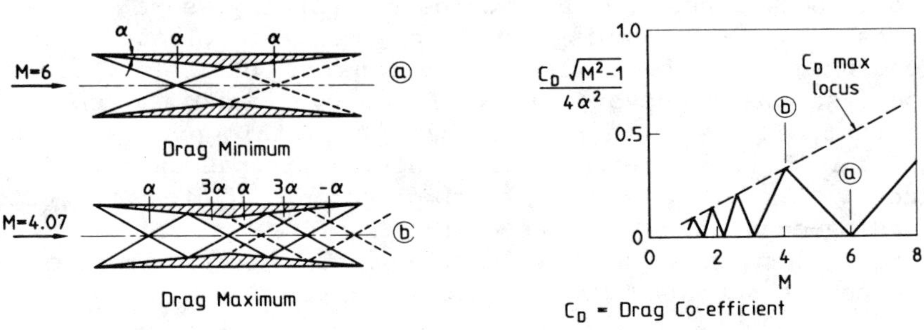

Fig. 4 Busemann biplane.

across the waves will be equal and opposite to changes in ν. Equation (8) then allows one to write, for small changes in α and ν,

$$\Delta p/p \approx \left(\gamma M^2 / \sqrt{M^2 - 1}\right) \alpha \qquad (13)$$

where Δp is the pressure change across the wave. This is the linearized Prandtl-Meyer relation and, when applied to the flow deflection pattern of Fig. 4a, it yields zero drag, provided that the pressure changes are small enough to take the Mach number as constant through the duct.

The flow pattern of Fig. 4a would exist at any inlet Mach number, provided that the lengths of the duct were correctly chosen, but here it is arbitrarily taken to correspond to a Mach number of 6. Thus, in displaying the variation of drag with Mach number[4] at the bottom of Fig. 4, Fig. 4a corresponds to zero drag. As the inlet Mach number falls, the drag increases and comes to a local maximum at a Mach number of 4.07. As can be seen from Fig. 4b, this occurs because the wave patterns within the duct lead to waves escaping downstream, and the transverse flux of energy associated with those waves must be supplied by the duct. Further reduction in Mach number leads to a reduction in drag, until it is zero again at a Mach number of 3.12, when all the waves are once more "trapped" in the duct. These drag maxima and minima continue to occur as the Mach number is reduced, with the magnitude of the maxima continually reducing. This is because the flow is approximating, more and more closely, a one-dimensional flow, for which the drag is zero.

Features of the wave-surface interaction of Sec. II are evident in these flows. Thus, reversal of the traverse energy flux in a wave occurs wherever it reflects from the duct centerline whereas, in Fig. 4b, the increase in surface slope occasioned by the corner occurring some distance downstream of impingement of the reflected inlet leading-edge wave calls for expansion of the fluid to increase the transverse flux of energy toward the surface. In fact, it is worth noting that it is this expansion that gives rise to the drag since it ultimately passes out of the duct. The drag caused by impingement of the reflected leading-edge wave on the inlet surface is recovered as thrust when this wave reflects once again and impinges on the nozzle downstream of the corner.

It will be seen that the behavior of a Busemann biplane can be considerably modified by heat addition.

B. <u>The Busemann Scramjet</u>

1. <u>Heat Addition</u>. The Busemann biplane can be converted to a propulsive duct by heat addition, preferably at the region of maximum pressure, near the minimum cross section. It then may be said to become a "Busemann scramjet." It can be formed by taking the zero drag duct of Fig. 4a and adding a section in which the duct walls are parallel to the upstream flow direction, with heat addition, at the minimum cross section. This section may be regarded as a "combustion chamber" but, until heat is added, it does not affect the inlet or nozzle flow patterns. The heat-addition zone is taken to be very short and normal to the flow, as shown in Fig. 5a. This is simple to analyze, and it could, conceivably, represent an array of injectors in the combustion chamber. Of course, complications in the analysis may occur when waves are incident on the heat-addition zone, as shown in Fig. 5b, and these will be dealt with in Sec. 4.

Analysis of one-dimensional heat addition can be conducted by using a method due to Zierup[1]. Applying the conservation of mass, momentum, and energy across the heat-addition discontinuity, it is found that two solutions are possible when the amount of heat addition is not large in

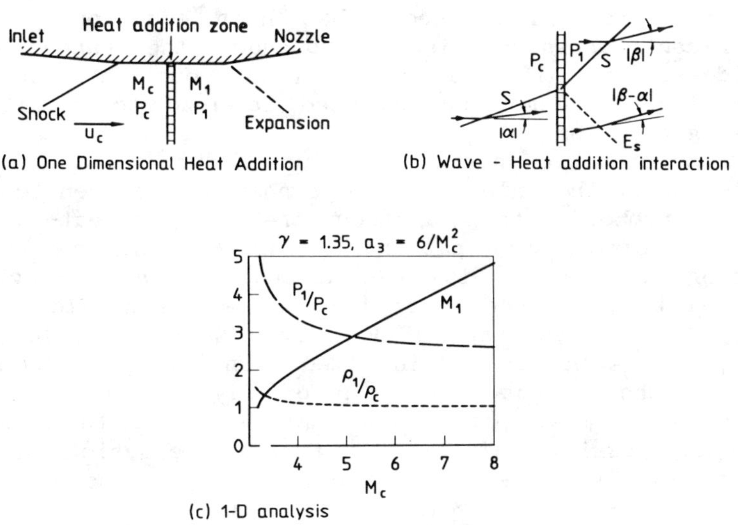

Fig. 5 Busemann scramjet, heat addition.

relation to the stream stagnation enthalpy. One of these approximates a normal shock, which could not be supported by the downstream conditions in the duct and, therefore the other solution is used, yielding

$$p_1/p_c = M_c^2 \left(\gamma + M_c^{-2} - \gamma B\right) / (\gamma + 1) \tag{14a}$$

$$M_1^2 = \left(\gamma + M_c^{-2} + B\right) / \left(\gamma + M_c^{-2} - \gamma B\right) \tag{14b}$$

$$\rho_c/\rho_1 = \left(\gamma + M_c^{-2} + B\right) / (\gamma + 1) \tag{14c}$$

where

$$B = \left\{ \left(\gamma + M_c^{-2}\right)^2 - (\gamma + 1)\left(2M_c^{-2} + \gamma - 1\right)\left(1 + a_3\right) \right\}^{\frac{1}{2}} \tag{15}$$

$$a_3 = q/ \left(0.5u_c^2 + h_c\right)$$

and q is the heat added per unit mass of fluid. If it is assumed that M_c, the combustion chamber Mach number, is one-third of the flight Mach number, then putting

$$a_3 = 6/M_c^2 \tag{16}$$

provides a reasonable approximation to the heat released by stoichiometric combustion of hydrogen. Using this expression, Eqs. (14) may be evaluated to obtain the post-heat-addition variation of flow parameters displayed in Fig. 5c. It can be seen that heat addition produces significant pressure rises throughout the range of M_c considered and, if the waves in the duct are properly controlled, this may be expected to lead to significant thrust generation.

Some possible wave patterns are shown in Fig. 6. When the length of the inlet and of the nozzle is chosen to match the Mach number upstream and downstream of the heat-addition zone, the simple wave pattern of Fig. 6a results. If the incoming Mach number is reduced and the inlet length is adjusted to take account of this while the nozzle remains fixed, the wave pattern of Fig. 6b results. If both the inlet and the nozzle remain fixed when the Mach number is reduced, the the wave pattern of Fig. 6c results. Of course, these wave patterns do not represent the complete range of possibilities. For further reduction of Mach number, more wave reflections will occur in the inlet and the nozzle and, if the incoming Mach number is raised above that of Fig. 6a, then the inlet shock will pass through the heat-addition zone before it impinges on the inlet surface. All these possibilities are represented in the thrust and drag calculations that follow.

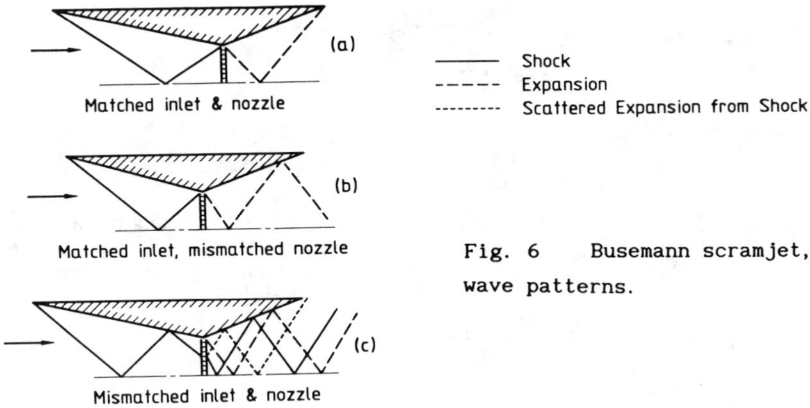

Fig. 6 Busemann scramjet, wave patterns.

All the calculations are done with a duct inlet to minimum cross-section area ratio of two. It will be seen that this makes the calculations somewhat more difficult since it involves abandoning some of the simplifications of linearized theory that are used for the Busemann biplane. However, it has the advantage that a more realistic assessment of wave effects is possible since significant levels of thrust and drag are involved. For example, the difference between thrust and drag at $M_c = 6$ in Fig. 7 would amount to a specific impulse of approximately 600 sec for hydrogen fuel.

2. Matched Inlet and Nozzle. This is the flow pattern of Fig. 6a. For analysis, the linearized Prandtl-Meyer relation of Eq. (13) is abandoned for calculating pressure changes across the waves and, instead, the exact Eqs. (8) and (9) are used, with the reminder that flow deflection across a wave remains constant as it passes through the flowfield. However, in order to obtain the wave paths, the assumption is retained that the Mach number upstream and downstream of the heat-addition zone is constant, at the values M_c and M_1, respectively, and that the expansion waves in the nozzle may be represented as a single wave that follows the Mach lines.

The thrust levels obtained as the Mach number is varied are shown as curve A in Fig. 7. As a test of the accuracy of the assumptions made in calculating the thrust, the results are compared with those obtained for one-dimensional expansion of the heated fluid from pressure P_1, through an area ratio of two. Since both processes are isentropic, and both yield an exhaust flow that is uniform and parallel,

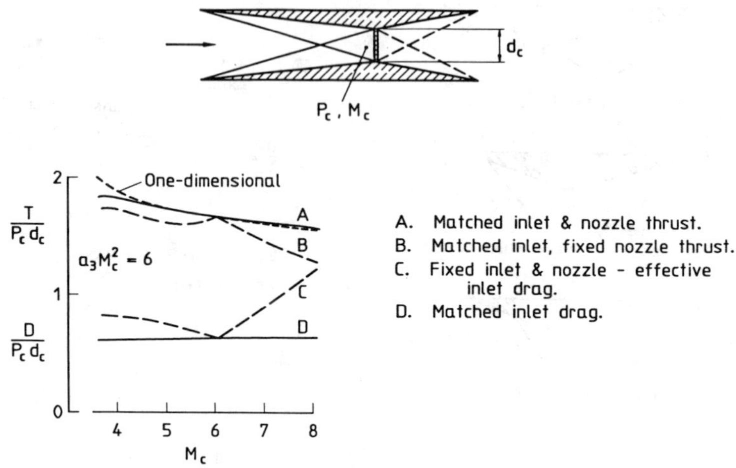

Fig. 7 Busemann scramjet, performance.

they should yield the same thrust. The close agreement between the two methods of calculation indicate that the assumptions made in the wave analysis do not lead to serious errors.

3. Matched Inlet, Fixed Nozzle. Rather than varying the duct configuration to match the Mach number, in practice, it is anticipated that the configuration will remain fixed as the Mach number is varied, thereby changing the wave patterns in both the inlet and the nozzle. In order to separate the effects of the two, it is convenient to begin by considering the effect of allowing the inlet configuration to vary, thereby maintaining a flow in which the waves are matched to the inlet geometry while the geometry of the thrust nozzle remains fixed. To set a value, the fixed geometry is here taken to correspond to $M_c = 6$, but it should be noted that qualitatively similar results would be expected if any other high-Mach-number value of M_c were chosen.

The results of these calculations are shown in curve B of Fig. 7. It can be seen that, by allowing waves to "escape" from the duct and to propagate energy into the surrounding stream, the nozzle mismatch reduces the thrust obtained with a given inlet. This effect is less serious at low than at high Mach numbers since, as for the Busemann biplane, the multiple reflections in the duct in the former case bring it closer to a one-dimensional flow.

Curve D shows the drag of the matched inlet which, when subtracted from curve A or B, will yield the respective net thrust in each case. If the specific impulse available from hydrogen fuel at corresponding flight speeds is taken to be of the order of 2000 s[5], then, as noted earlier, the net thrusts obtained from Fig. 7 represent a significant portion of this specific impulse.

4. Fixed Inlet and Nozzle. When the length of the inlet is not matched to the Mach number of the incoming flow, then, as shown in Fig. 6c, waves are generated that pass into the thrust nozzle. In their passage, these waves must interact with the heat-addition zone. A weak shock wave S is shown impingent on the heat-addition zone in Fig. 5b. Using the approximate Prandtl-Meyer relation of Eq. (13), and noting from Fig. 5c that the density change and, hence, the velocity change are negligible across the heat-addition zone, it can be shown that the original shock produces a transmitted shock S and a scattered expansion wave E_s. If the magnitude of the deflection across the original transmitted shock is $|\alpha|$, and the magnitude of the transmitted shock deflection is $|\beta|$, then,

$$|\alpha|\, M_c^2 \,/\, \sqrt{M_c^2 - 1} \;=\; |\beta|\, M_1^2 \,/\, \sqrt{M^2 - 1} \tag{17}$$

and the magnitude of the scattered wave deflection is $|\beta - \alpha|$. Similarly, an incident expansion wave E produces a transmitted expansion and a scattered shock S_E, with the magnitude of the incident and the transmitted expansion related by Eq. (17) and the magnitude of the deflection through S_E given as $|\beta - \alpha|$.

These waves then may be traced along the Mach lines following the heat-addition zone, as shown in Fig. 6c. When they reflect from the thrust surface, the pressure change is calculated by adding the proportional pressure changes across the incident and reflected waves, as for the Busemann biplane, and referencing the resultant proportional pressure change to the pressure levels at the surface in the absence of inlet mismatch waves. These reference pressure levels are obtained by taking the pressure levels associated with a matched inlet and adjusting for the changed values of P_1 associated with the change in inlet angle.

With a mismatched inlet, a problem arises in choosing an appropriate value of P_c for normalizing the thrust in Fig. 7, since the pressure is no longer uniform upstream of the heat-addition zone. This is resolved by noting that the thrust levels produced by two different configurations should properly be compared when the same amount of heat is added to the flow. Under the assumptions that have been made in this paper, combining Eq. (15) with the mass flow through the heat-addition zone leads to an expression for the total heat added per unit time, Q, as

$$Q = 0.5\, p_c\, u_c\, d_c\, a_3\, \gamma \left\{M_c^2 + 2/(\gamma - 1)\right\} \tag{18}$$

It can be seen that, for given u_c and M_c, the heat added is directly proportional to p_c and, therefore, the mean value of p_c across the heat-addition zone may properly be used as a normalizing value. For these cases, the denominator in the scale of the ordinate in Fig. 7 should be interpreted in such terms.

The effect of the inlet mismatch is obtained by calculating the total thrust of the nozzle in the presence of the mismatch waves and subtracting the nozzle thrust with a matched inlet and fixed nozzle (curve B) in order to obtain the thrust loss due to inlet mismatch. This is added to the mismatched inlet drag to produce curve C for effective inlet drag shown in Fig. 7.

It can be seen that the inlet mismatch produces a strong increase in effective drag as the Mach number increases above 6. In fact, when this drag is subtracted from the reduced thrust due to nozzle mismatch (curve B) at $M_c = 8$, no significant net thrust remains. On the other hand, the effect of reducing the Mach number below 6 is not so serious. For example, the difference between curves B and C at $M_c = 4$ indicates that the net thrust remains at 75% of the matched inlet and nozzle net thrust. This strongly suggests that, if a scramjet is to operate with changing inlet Mach numbers, then it is best to design for the maximum Mach number.

IV. The Streamtube Model.

Up to this point, the use of a constant-area heat-addition zone normal to the stream has been a convenient abstraction for the purposes of analysis. In fact, injection of fuel is likely to take place at a finite number of sources, as shown in Fig. 8. In the figure, injection is seen to occur from both a central strut in the combustion zone, and from the surfaces of the intake. In the former case, mixing and combustion will begin

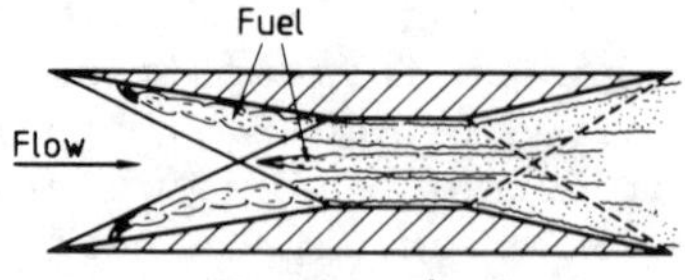

Fig. 8 Scramjet fuel-injection pattern.

immediately upon injection and, in the latter, considerable mixing may take place before combustion is initiated by crossing the shock wave. In both cases, the diffusion and combustion of fuel in adjacent streamtubes will produce a region of heat addition that envelops one streamtube after another until all of the fuel or all of the oxygen in the duct has been consumed. Thrust is produced from the waves generated by the array of streamtubes supplied by a fuel source as they are heated and expand. As indicated in Ref. 3, these waves generally may be expected to be sufficiently weak that characteristics theory will allow their flow deflections to be combined linearly, although interactions between the waves, and between the waves and heated fluid in the streamtubes, may affect the paths along which they propagate in a nonlinear fashion. Thus, to a first approximation, the array of streamtubes influenced by a fuel source may be combined into one, and the basic thrust-producing element becomes a single streamtube with heat addition, within a two-dimensional supersonic flow, as indicated in Fig. 2a.

A. Heat Addition to a Streamtube.

This may be analyzed by assuming one-dimensional flow within the streamtube, which allows the thermodynamic process for an element of fluid as it passes through the streamtube to be represented on a p-v diagram as in Fig. 9a. The fluid enters the heat-addition region at an initial pressure p_c, and the pressure increases as the addition of heat, together with the flow deflection caused by the associated expansion of the cross section of the streamtube, causes a pressure rise in the two-dimensional flow outside the streamtube. As the heat addition ceases, the pressure falls again, returning to the value p_c. This may be seen as an open cycle which, because the pressure changes are induced by interaction with the outside stream, is here referred to as a "self-induced" heat-injection cycle.

The general process path 2 varies according to the rate at which heat is added and may conveniently be idealized by assuming that all of the heat is added at constant volume, as shown by path 1. Because the expansion after heat addition is isentropic by this path and because the path involves the highest fluid temperatures during the heat addition, it is the path of minimum entropy rise and, therefore, it is also the path that allows the fluid to deliver most work to surrounding streamtubes for a given amount of added heat.

For path 1, all of the work done by the expanding fluid in the streamtube is delivered in the isentropic expansion from the peak pressure p_1. If high-Mach-number flow existed throughout the streamtube, then all of this work would be transferred to the surrounding flow in the form of waves. If the resulting transverse flux of energy in the waves is then delivered to a perfectly matched surface, as in Fig. 2b(i), then the resulting thrust can be readily estimated from the work done in the cycle path 1.

However, it will be noted that Fig. 5c shows low supersonic values of M_1, the Mach number immediately after heat addition, at the lower values of M_c, implying that the high-Mach-number approximation will not be valid in that region. Therefore, in order to estimate the thrust resulting from path 1, the streamtube shown in Fig. 9b is considered, and a momentum balance is taken between a station immediately after heat addition, where the pressure is P_1 and the streamtube width is δ_c, and a station sufficiently far downstream for the pressure to return to P_c, where the streamtube Mach number and width are M_j and δ_j, respectively. This yields

$$T_\delta / P_c\delta_c = \left(1 + \gamma M_j^2\right) \delta_j / \delta_c - \left(1 + \gamma M_1^2\right)\left(P_1/P_c\right) \qquad (19)$$

when $a^2 = \gamma p/\rho$ has been used for the speed of sound in the streamtube and T_δ can be interpreted as the thrust delivered to the interface between the streamtube and the surrounding flow. From continuity,

$$\delta_j / \delta_c = \left(M_1/M_j\right)\left(P_1/P_c\right)^{(\gamma+1)/2\gamma}$$

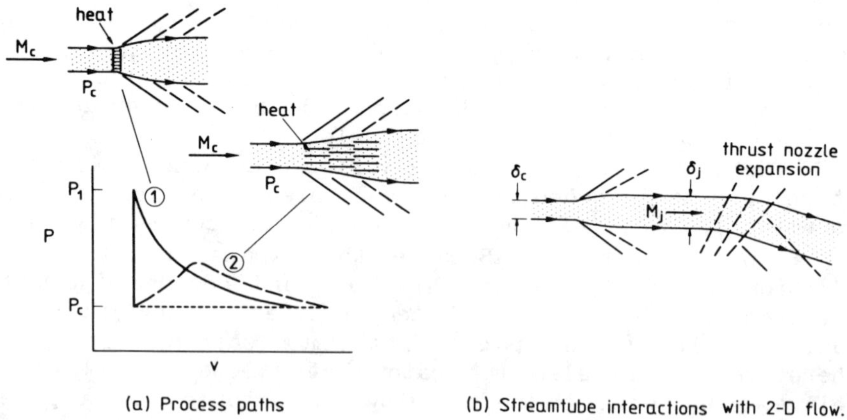

Fig. 9 Heat addition to a streamtube.

and Eq. (19) becomes

$$T_\delta/p_c\delta_c = \left(1+\gamma M_j^2\right)\left(M_1/M_j\right)\left(P_1/P_c\right)^{(\gamma+1)/2\gamma} - \left(1+\gamma M_1^2\right)\left(P_1/P_c\right) \tag{20}$$

The pressure ratio P_1/P_c is given by Eq. (14a), M_1 is given by Eq. (14b), and M_j is obtained by isentropic expansion from M_1, at pressure P_1, to pressure P_c. When the waves produced by the self-induced cycle are delivered to a perfectly matched surface, as in Fig. 2b(i), the thrust T_δ is produced.

Combining Eqs. (16) and (18) with d_c replaced by δ_c in the latter, the rate of heat addition to the streamtube can be obtained for the conditions of Fig. 5c. When this is related to the rate of mass injection of hydrogen fuel required to produce this heat release, the specific impulse contribution associated with the thrust of Eq. (20) can be estimated. For the conditions chosen, this ranges from 1730 s at $M_c = 4.0$ to 670 s at $M_c = 8.0$. These values are sufficiently far below maximum theoretical values to make it worthwhile exploring further expansion of the streamtube after the self-induced cycle in order to see if significant increases in specific impulse can thereby be realized.

For this purpose, the streamtube is once again considered in the context of the simple duct flow of Fig. 8 and thus is represented as in Fig. 9b. The self-induced heat-injection cycle takes place in the nominally parallel part of the duct, but the walls of the duct there are contoured to produce the perfectly matched surface required to accommodate the waves produced by the self-induced cycle and thus to develop the thrust given by Eq. (20). At some distance downstream, after the self-induced cycle is completed, the streamtube experiences the expansion that signals the upstream end of the thrust nozzle, as shown in Fig. 9b. In order to determine the thrust produced by the resulting expansion of the streamtube, it is necessary first to consider the manner in which the streamtube interacts with the thrust nozzle expansion.

B. Wave-Interface Interaction and Thrust by Expansion.

The analysis employed to determine the optimum thrust surface angle may be continued and used to treat the flux of energy through a streamwise interface, across which the Mach number changes. If, as illustrated in Fig. 10a, an incident wave i is transmitted through an interface as the wave j, then an analysis along the two (-) characteristics shown in the figure, following the approach used in obtaining Eq. (11), leads to expressions for the change in transverse energy flux across i and j, respectively, as

$$\Delta(p \sin\theta)_i = p \cos\theta \left(1 - A_i \tan\theta\right) \Delta\theta_1 \tag{21a}$$

and

$$\Delta(p \sin\theta)_j = p \cos\theta \left(1 - A_j \tan\theta\right) \Delta\theta_j \tag{21b}$$

where $\Delta(p \sin\theta)$ and $\Delta\theta$ represent small changes in p sinθ and θ across the wave. Since p cosθ and the transverse flux of energy, p sinθ, must be the same on either side of the interface upstream of the waves i and j, it follows from Eqs. (21) that differing Mach numbers M_i and M_j, and the continuity of energy flux across the downstream interface, imply that $\Delta\theta_i \neq \Delta\theta_j$. Thus, a further reflected wave r must originate at the wave-interface junction. Following a (+) characteristic across this wave and neglecting the small change in a p cosθ and A_1 tanθ across i, it can be shown that, across r,

$$\Delta \left(p \sin\theta\right)_r = p \cos\theta \left(1 + A_1 \tan\theta\right) \Delta\theta_r \tag{22}$$

Then, matching the energy flux across the downstream interface and noting that

$$\Delta\theta_i + \Delta\theta_r = \Delta\theta_j \tag{23}$$

it follows that

$$\Delta\theta_r = \Delta\theta_i \left(A_i - A_j\right) / \left(A_i + A_j\right) \tag{24}$$

and, using these two relations and Eqs. (21), the increase in the transverse energy flux downstream of the primary wave as it passes through the interface is found to be given by

$$\Delta\left(p \sin\theta\right)_j - \Delta\left(p \sin\theta\right)_i = p \cos\theta\left\{1-\left(A_j-A_i\right)\tan\theta\right\}\Delta\theta_r \tag{25}$$

Usually, tanθ and the Mach number difference will be small enough to insure that the term in the brackets on the

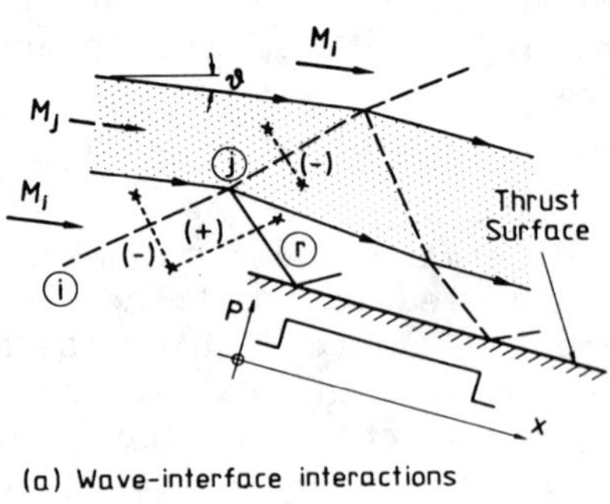

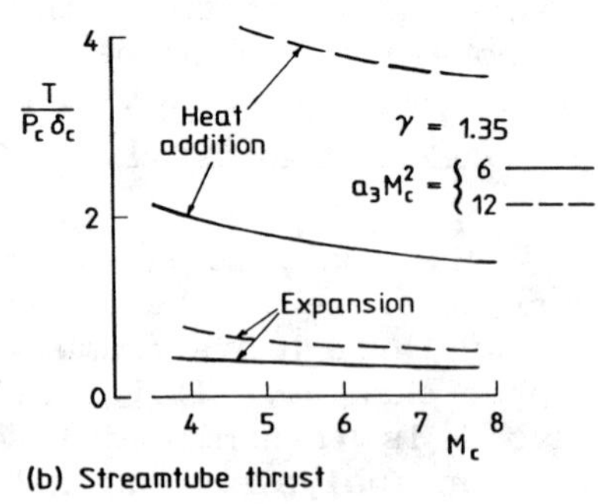

Fig. 10 Expansion-induced waves.

right-hand side of Eq. (25) will be positive. Thus, if the primary wave is an expansion, and the Mach number is reduced in passing from i to j, then $\Delta\theta_r$ will be positive, implying that r is a compression wave, and the transverse flux of energy from j to i increases. Conversely, if the Mach number increases in passing from i to j, then r will be an expansion wave, and the transverse flux of energy from j to i is reduced.

The manner in which this mechanism operates to produce thrust also is shown schematically on the figure. The interface is a boundary of a jet (or wake) that has been formed by heat addition, causing the temperature in the jet to increase and its Mach number to be reduced below that of the surrounding fluid. When a primary expansion (i, say) passes into the low-Mach-number flow of the jet, it generates a compression wave that propagates to the thrust surface. When it passes out of the jet, it generates an expansion wave that also propagates to the thrust surface. This wave cancels the pressure rise at the surface as a result of the compression wave but, because it arrives further downstream, there is a net thrust produced at the surface. The energy required to produce this thrust derives from the expansion of the jet, which is due to the difference between the deflection of the jet boundary as the primary expansion enters the jet and the deflection when it leaves the jet. Of course, the heat-addition jet will, in practice, exhibit a continuous distribution of Mach number but, as shown in Ref. 3, the mechanisms involved and their effects are the same, although the discontinuities in the surface pressure distribution of Fig. 10a are smoothed out.

It might be remarked that this mechanism for producing thrust cannot be represented as a one-dimensional process. The strength of the thrust-producing waves generated by the expansion of the heat-addition jet is limited by the need to provide transverse pressure gradients to bend the streamlines, and these pressure gradients are supplied by the primary wave. Thus, the area ratio, which is a measure of the strength of the expansion in a one-dimensional process, becomes irrelevant in specifying the thermodynamics of the expansion process and is replaced by the strength of the primary wave. This aspect is discussed further in Ref. 6.

Using this expansion-jet interaction model, an approximation to the thrust arising from the streamtube expansion due to the thrust nozzle expansion in Fig. 9b can be obtained. First the streamtube cross section δ_j and its Mach number M_j after the heat-injection cycle are calculated by using an isentropic expansion from the immediate post-heat-addition values to a pressure p_c. Then, assuming that the thrust surface angle will be set at the optimum value of Eq. (12), the flow deflection through the primary expansion is taken from Fig. 3c, using the Mach number obtained by expanding M_c through the optimum angle. Of course, Eq. (12) does not necessarily yield an optimum angle for the generation of thrust in the nozzle expansion, since not only is the wave pattern in the flow influenced by the choice of angle but the nozzle also may be long enough to insure that the wave reflection effects noted in Sec. II can play a role. The angle given by Eq. (12) is chosen for the thrust surface angle here in the belief that vehicle mass considerations will make it impractical to deviate too far from the value that produces maximum thrust per unit length of surface.

To complete the calculation, Eq. (24) is used to obtain the flow deflection through the reflected compression and following expansion and, taking account of the doubling of pressure associated with reflection at the thrust surface, Eq. (13) is used to obtain the pressure change at reflection. The streamwise distance along the surface between the compression and the expansion is approximated as $2\delta_j\sqrt{M_j^2 - 1}$, implying that the geometrical effects of the relatively small flow deflection angles are ignored.

The results of such calculations are presented in Fig. 10b, where the streamtube thrust levels developed for the thrust nozzle expansion are compared with those obtained for the heat-injection cycle, as given by Eq. (20). The contribution of the expansion is modest in comparison to that of heat injection. For example, the combined thrust levels at $M_c = 6$ amount to a specific impulse for hydrogen fuel of 1170 s compared to 970 s for the heat-injection cycle alone. This suggests that only limited gains in specific impulse can be made by further expansion and that the larger of the two specific impulse values quoted might be close to a practical limit for this condition.

Some calculations have also been made to explore the effect of reducing the temperature before heat injection. For $a_3 = 6/M_c^2$, the temperature is approximately 2400 K, and this is reduced to 1200 K for $a_3 = 12/M_c^2$. It can be seen from Fig. 10b that this reduction increases both the heat-injection and the expansion components of the normalized thrust. However, when M_c is related to the flight speed, it is found that the combined specific impulse at a given flight speed is essentially unaffected by the temperature before heat injection, changing by less than 10% as the temperature changes by a factor of 2. The specific impulse varies from 2400 s to 1200 s as the flight speed varies from 3.5 km/s to 6 km/s and, throughout this range of speeds, the thrust arising from the self-induced cycle represents more than 80% of the total thrust.

It might be noted that that the calculations yielded pressure rises due to heat-addition up to seven times P_c, thereby producing strong shock waves in the supersonic flow outside the streamtube. This is not expected to affect significantly the arguments presented in the discussion, provided that the thrust due to the heat-injection cycle is realized on a surface that is sufficiently close to the streamtube to insure that the expansion waves produced after the shock reach the surface before a significant fraction of these waves has overtaken and merged with the shock.

V. Boundary Layers.

Boundary layers make important contributions to the drag of a propulsive duct at high Mach numbers, not only through skin-friction drag but also through the surface pressure increase on inlet surfaces that is caused by the viscous leading-edge interaction. By way of brief comment, two examples of the manner in which these phenomena can interact favorably with duct wave patterns are presented.

To consider first the leading-edge interaction, the matched inlet and nozzle configuration of Fig. 5a is reproduced in Fig. 11a. As shown in the figure, if the leading-edge interaction waves are included, they are seen to reflect from the duct centerline and to pass through the heat-addition zone to the thrust surface. Taking a laminar boundary layer with the product of density and viscosity constant across the boundary layer, Ref. 7 gives a first approximation to the increment of pressure induced at the inlet surface by the viscous leading-edge interaction as

$$0.31\ M^3 \ / \ \sqrt{R_{ex}}$$

where M and R_{ex} are here the Mach number and Reynolds' number, respectively, outside the boundary layer. Putting $R_{ex} = 10^6$, with x the length of the inlet surface, and taking the configuration of Fig. 11a to correspond to $M_c = 6$, it is found that the interaction-induced pressure drag on the inlet is $0.051\ P_c d_c$. Then, following the procedures used in Sec. IIIB4 to calculate the effect of passage of a wave through the heat-addition region, the pressure rise and associated thrust as these waves reflect from the nozzle surface can be estimated, yielding a thrust of $0.167\ p_c d_c$. Thus, because of the energy added to the waves when they cross the heat-addition zone, the viscous leading-edge interaction can produce a net thrust amounting,

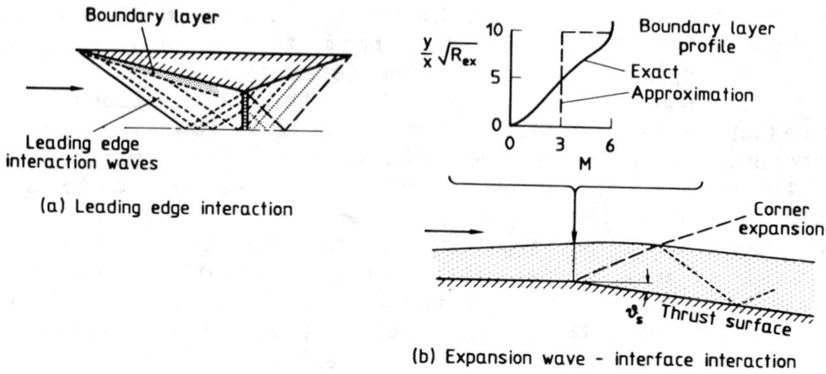

Fig. 11 Boundary-layer wave phenomena.

in this case, to 0.12 $p_c d_c$. Referring to Fig. 7, it can be seen that this represents a gain of 11% in the net thrust produced by the duct. It is worth noting that similar benefits may be expected to flow from the waves as a result of leading-edge blunting.

The skin-friction drag on the inlet can be obtained for the conditions of the preceding paragraph by obtaining an expression for the skin-friction coefficient from Ref. 8 as

$$C_f = 0.664 / \sqrt{R_{ex}}$$

Integrating over the inlet length, using the boundary-layer thickness δ, obtained from the Mach number profile in Fig. 11(b), the inlet skin-friction drag is found to be 1.61 δP_c. However, the skin friction also reduces the Mach number in the boundary layer and, when the boundary layer turns to follow the flow onto the thrust surface, the reduction in pressure through the corner expansion is less in the boundary layer than in the mainstream. Thus, the boundary layer produces an effective thrust over a zone that is terminated by expansion waves transmitted downstream from the interaction of the corner expansion wave with the Mach number gradient in the boundary layer, as shown in Fig. 11b. To estimate the magnitude of this effect, the Mach number in the boundary layer is taken to be uniform at a value equal to half the mainstream value and, following the analysis of the preceding section, it is found that the boundary-layer thrust in this case is 0.175 δP_c, representing only the order of 10% of the inlet skin-friction force.

VI. Conclusion.

A discussion of the flow of an expanding fluid in a duct has been used to point out that, at high Mach numbers, essentially all of the work done by the fluid is delivered transverse to the streamlines and therefore appears in the form of waves. Effective management of these waves is necessary in order to minimize drag and maximize thrust.

Consideration of the thrust produced per unit length of surface from a given wave indicates that there is an optimum angle for this purpose, which reduces as Mach number increases. This suggests that high-Mach-number ducts will tend to be slender ones, not only because of the need to minimize drag but also in order to make most effective use of thrust-generating surfaces.

The role of waves in producing drag and thrust in a simple propulsive duct has been explored by extending the concept of a Busemann biplane to that of a Busemann scramjet. It has been found that the net thrust of a duct configured for optimum operation at a particular design Mach number is reduced as it departs from this Mach number, but the reduction is much more serious for Mach numbers in excess of the design value than for those below it. This suggests that if a propulsive duct is to operate over a range of Mach numbers, then it is best to choose a design Mach number at the maximum end of the range. Again, this implies that a high-Mach-number duct should be a slender one.

Noting that fuel injection will occur at a finite number of points in a scramjet duct, a model for heat injection into a finite streamtube has been developed. It has been found that maximum thermal efficiency is obtained when all the heat is injected into the streamtube at one station and that the cycle representing the overall process involves a self-induced pressure rise, followed by subsequent expansion, which takes place upstream of the thrust nozzle expansion.

Consideration of the flow mechanisms that govern expansion of the heated streamtube in the thrust nozzle indicates that this expansion is limited by the strength of the thrust nozzle expansion, as determined by the nozzle angle, rather than by the area ratio of the nozzle. If the nozzle angle is limited to the optimum thrust value, then only limited thrust can be obtained from the nozzle expansion. Thus, if it is accepted that vehicle mass considerations will demand maximum thrust loadings per unit surface area, then it follows that the upstream, self-induced heat-injection cycle has the potential to produce much more thrust than the nozzle expansion.

Two boundary-layer phenomena have been considered briefly, with differing results. Amplification, at the heat-addition zone, of the waves generated by the leading-edge viscous interaction allows this effect to produce a useful net increase in thrust. By contrast, the thrust produced by expansion of the skin-friction heated boundary layer at the thrust nozzle expansion is only a small fraction of the skin-friction drag on the inlet.

These conclusions apply strictly to two-dimensional flows only, and it is clear that they will be subject to some quantitative modification when three-dimensional effects are taken into account. This will be particularly true for the nozzle expansion thrust. Nevertheless, it may be expected that the qualitative features of these flows will be preserved in three dimensions and, therefore, it is hoped that this two-dimensional treatment may provide a useful guide to the influence of wave phenomena on scramjets in general.

Acknowledgements

This work was supported by the Australian Grants Scheme, and through NASA GRANT NAGW-674.

References

[1]Broadbent, E. G. "Flow with Head Addition." Progress in Aerospace Sciences Vol. 17, Pergamon, London, 1976, pp. 93-103.

[2]Liepmann, H. W., and Roshko, A., Elements of Gas Dynamics, Wiley, New York, 1967, p. 98.

[3]Stalker, R. J., Morgan, R. G., and Netterfield, R.P., "Wave Processes in Scramjet Thrust Generation," Combustion and Flame, Vol. 71, 1988, pp. 63-77.

[4]Liepmann, H. W., and Roshko, A., Elements of Gas Dynamics, Wiley, New York, 1967, pp. 116-117.

[5]Swithenbank, J., "Hypersonic Air-breathing Propulsion," Progress in Aeronautical Sciences Vol. 2, edited by D. Küchenmann, Pergamon, Oxford, 1967, pp. 229-294.

[6]Stalker, R. J., and Morgan, R. G., "Supersonic Hydrogen Combustion with a Short Thrust Nozzle." Combustion and Flame, Vol. 57,1984, pp. 55-70.

[7]Hayes, W. D., and Probstein, R.F., Hypersonic Flow Theory, Academic, New York, 1959, p. 362

[8]Hayes, W. D., and Probstein, R.F., Hypersonic Flow Theory, Academic, New York, 1959, p 296.

Chapter 5

Turbulent Free Shear Layer Mixing and Combustion

P. E. Dimotakis*
California Institute of Technology, Pasadena, California

Abstract

Some experimental data on turbulent free-shear-layer growth, mixing, and chemical reactions are reviewed. The dependence of these phenomena on such fluid and flow parameters as Reynolds number, Schmidt number, and Mach number are discussed, with the aid of some direct consequences deducible from the large-scale organization of the flow as well as from some recent models.

I. Introduction

The mixing of two or more fluids that are entrained into a turbulent region is an important process from both a scientific and an applications vantage point. Mixing in turbulent flows can imply a host of processes and phenomena. Species can be transported by turbulence to produce a more uniform distribution than some initial mean profile. This process is sometimes also referred to as mixing, without regard to whether the transported species are mixed on a molecular scale or not. If the issue of mixing arises in the context of chemical reactions and combustion, however, we recognize that only fluid mixed on a molecular scale can contribute to chemical product formation and associated heat release. The discussion in this paper will be limited to molecular mixing.

*Professor, Aeronautics and Applied Physics.

Molecular mixing by turbulence is important theoretically in that it provides an arena in which models of the behavior at the smallest scales of turbulence can be tested. These smallest scales correspond to a spectral regime that can be treated in a relatively cavalier fashion, if one need only address the *momentum* transport properties of the turbulent region[1], for example, but must be described with some deference to the physics at those scales if molecular mixing is to be accounted for correctly. From an experimental point of view, molecular mixing and chemical reactions in high-Reynolds-number flows provide us with an important probe of diffusion length and time scales that are so small as to otherwise remain beyond the reach of conceivable direct measurement diagnostics. Matters are no better computationally, with the requisite spatial/temporal resolution long recognized to be out of reach[2], a situation that must still be accepted as the case for high Reynolds number flow for the foreseeable future[3,4]. Finally, from a technological vantage point, mixing in a turbulent environment may well dictate the performance of many devices that rely on the details of the turbulent mixing process, such as high fuel efficiency internal combustion engines, chemical lasers, hypersonic propulsion, etc. It is also likely to prove to be an important consideration in other contexts, such the local and global environmental issues involving chemistry in the turbulent atmospheric environment, the dynamics of stellar atmospheres and interiors, etc.

The discussion here will be limited to mixing in turbulent shear layers, formed between two uniform free streams of unequal velocity, not necessarily of equal density, at high Reynolds numbers. In particular, at least for subsonic flow, for values of the local Reynolds number given by

$$Re \equiv \frac{\delta \, \Delta U}{\nu} > 10^4 , \tag{1}$$

where $\delta = \delta(x)$ is the (local) transverse extent of the turbulent shear layer region,

$$\Delta U = U_1 - U_2 \tag{2}$$

is the velocity difference across the shear layer, and ν is some appropriate measure of the kinematic viscosity. In the discussion that follows, issues pertaining to gas-phase mixing will be addressed, for which the Schmidt number,

$$Sc \equiv \frac{\nu}{\mathcal{D}} , \tag{3}$$

with $\mathcal{D}$ the diffusing/mixing species diffusivity, is near unity. Mixing in liquid-phase flows, for which $Sc \gg 1$ (e.g., $Sc_{\mathrm{water}} \approx 600$), will also be discussed. Differences in turbulent mixing between these two fluid phases are important in that they provide important clues to the behavior and the dynamics of the flow at the smallest scales.

Although the two-dimensional, shear-layer flow geometry may not be germane to all the issues alluded to earlier, many of the phenomena that need to be addressed are generic and two-dimensional, turbulent, free-shear-layer flows provide a useful arena in which they can be studied. Additionally, however, the flow within the turbulent region formed between the two bounding free streams is capable of sustaining relatively rapid mixing, and one that can be further enhanced by a variety of flow manipulation means. This is a consequence of the property of shear-driven turbulence which, at least for subsonic flow conditions, can generate interfacial surface area between fluids inducted from each of the two streams at very high rates. At the high Reynolds numbers of interest here, the relatively small (molecular) diffusivity can result in a total diffusive flux across this very large interface that may come close to accommodating the rate at which the free stream fluids are inducted into the turbulent region at the largest scales of the flow. As a consequence, at least for subsonic, high-Reynolds-number, gas-phase flows, one finds that the expected fraction $\delta_{\mathrm{m}}/\delta$ of the turbulent region occupied by molecularly mixed fluid can be significant. It's not called a *mixing* layer for nought!

It is useful to view the molecular mixing process and any associated chemical product formation at a particular station x of the flow as resolved into a sequence of Lagrangian stages in the "life" of entrained fluid elements bearing the chemical reactants. These fluid elements must:

1. be inducted into the mixing zone of relative transverse extent δ/x, then,
2. mix molecularly to occupy some fraction $\delta_{\mathrm{m}}/\delta$, of $\delta(x)$ at x, before
3. reacting to form the chemical product, which represents, in turn, some fraction $\delta_{\mathrm{P}}/\delta_{\mathrm{m}}$ of the mixed fluid.

This suggests expressing the expected chemical product, formed between the virtual origin ($x = 0$) of the shear layer and the station at x, as a product of three factors, i.e.,

$$\frac{\delta_{\mathrm{P}}}{x} = \frac{\delta}{x} \times \frac{\delta_{\mathrm{m}}}{\delta} \times \frac{\delta_{\mathrm{P}}}{\delta_{\mathrm{m}}} . \tag{4}$$

The first factor, δ/x, measures the *growth* of the mixing-layer region; the second, $\delta_{\mathrm{m}}/\delta$, the *mixing* within the shear layer, and the third, $\delta_{\mathrm{P}}/\delta_{\mathrm{m}}$, the *chemical reactions* that can take place within the molecularly mixed fluid in the layer. This partition, at least in the case of high Reynolds numbers, is justified by the fact that the various stages represented by these factors occur in a *succession of Lagrangian times*. This resolution also provides a useful framework within which turbulent mixing and chemical reactions in two-dimensional shear layers can be discussed and reviewed and will be adopted in the discussion that follows.

II. Shear-Layer Growth: δ/x

At the high Reynolds numbers of interest and for Schmidt numbers that, if not large, are not much smaller than unity, the shear-layer growth rate δ/x is an important quantity. It measures the angle of the wedge-shaped turbulent mixing region that confines the mixed fluid and chemical product. As a consequence, the width of the turbulent region, δ/x, represents an upper bound for the amount of mixed fluid in the shear layer, corresponding to a scenario in which the entrained fluids are mixed instantly, on a molecular scale, as soon as they enter the turbulent region within the transverse extent δ.

As we'll discuss below, the normalized transverse extent δ/x of the shear layer is found to depend on several dimensionless parameters of the flow, i.e.,

$$r \equiv \frac{U_2}{U_1} , \qquad s \equiv \frac{\rho_2}{\rho_1} , \tag{5}$$

the freestream velocity and density ratios, respectively, the (convective) Mach numbers of the two streams, i.e.,

$$M_{c1} = \frac{U_1 - U_c}{a_1} \quad \text{and} \quad M_{c2} = \frac{U_c - U_2}{a_2} , \tag{6a}$$

where

$$U_2 < U_c < U_1 \tag{6b}$$

is the convection velocity of the large-scale structures in the shear layer, and $a_{1,2}$ are the speeds of sound in the high- and low-speed freestream, respectively. In the case of combusting flow, it also depends on the relative mean density reduction $\Delta\rho/\rho$ in the interior of the chemically reacting shear layer attributable to heat release. For equal freestream densities ($\rho_1 = \rho_2 = \rho_0$), this can be expressed in terms of the heat release parameter

$$q \equiv \frac{\rho_0 - \bar{\rho}}{\rho_0} \equiv \frac{\Delta\rho}{\rho_0} , \tag{7}$$

where $\bar{\rho}$ is the (reduced) mean density of the flow within the δ/x shear-layer wedge. Finally, the shear layer growth rate is influenced by the presence of streamwise pressure gradients. We note, however, that if $\mathrm{d}p/\mathrm{d}x \neq 0$ (accelerating/decelerating flow), the shear layer will not grow linearly, unless it so happens that the dynamic pressures in the two free streams are matched[5], i.e., if $\rho_1 U_1^2 = \rho_2 U_2^2$. As we'll discuss later, experimental information, as well as some theoretical understanding of the dependence of δ/x on these parameters, is available, even though the picture is as yet far from complete or satisfactory.

By δ, in this discussion, we will denote the local transverse extent of the sheared region that contains the molecularly mixed fluid in a boundary-layer sense, i.e., the distance between the shear layer edges outside which the expected concentration of molecularly mixed fluid is less than some small fraction, say, 1% of its peak value. This definition yields a local width that closely matches the measurements of the "visible" shear layer width δ_{vis}, as would be measured in a schlieren or shadowgraph picture of the layer[1,6]. It is also very close to the 1% width, δ_1, in the case of a chemically reacting layer, defined as the distance between the two points across the layer where the mean product concentration, or temperature rise owing to heat release, has dropped to 1% of its peak value[7,8]. As a result of the similarity properties of this flow at high Reynolds numbers (Eq. 1), we can argue that other transverse scales must simply be proportional to the outer scale δ. By way of example, the vorticity (or maximum slope) thickness δ_ω, defined in terms of the mean streamwise velocity profile $U(y)$ as

$$\frac{1}{\delta_\omega} \equiv \frac{1}{\Delta U}\left[\frac{\mathrm{d}U(y)}{\mathrm{d}y}\right]_{\max} , \tag{8}$$

is found to be roughly half of δ.

Dependence on the Velocity and Density Ratio

Abramowich[9] and Sabin[10] proposed an expression for the shear-layer growth rate given by

$$\frac{\delta}{x} \approx C_\delta \, \frac{1-r}{1+r} , \tag{9}$$

where C_δ is taken as a constant. This is an expression that can be argued for on similarity grounds,† and is found to be in reasonable accord with experimental data of incompressible shear layers with equal freestream densities. The dependence on the freestream density ratio

---

† In a frame moving with the convection velocity U_c, we must have $\delta/t \propto \Delta U$, where $\Delta U = U_1 - U_2$; the only relevant velocity in the problem. Equation 9 then follows if $U_c = (U_1 + U_2)/2$, which is found to be the case for $\rho_1 = \rho_2$, if we transform back to the lab coordinates, i.e., $x = t/U_c$, and normalize all velocities with U_1.

was addressed in the seminal experiments by Brown and Roshko[6,1], originally undertaken to investigate whether the observed reduction in growth rate in supersonic shear layers could be attributed to density ratio effects. Using different gases for each freestream and subsonic freestream velocities, they found that whereas the growth rate depended on the freestream density ratio, compressibility effects could not be identified with density ratio effects. In subsequent experiments, Konrad[11] provided further documentation of the dependence on the density ratio and also noted that a shear layer entrained asymmetrically from each of the freestreams, as will be discussed later.

Brown[12] proposed to account for these phenomena on the basis of similarity arguments that recognized the significance of a Galilean frame translating at the convection velocity U_c of the large structures. Growth and entrainment are to be understood as taking place in this frame, whose convection velocity is a function of both the density and velocity ratio. Brown's theory, which applied to a temporally growing shear layer, was in reasonable accord with the observed dependence of the growth rate on the freestream density ratio, but — as appropriate for a temporal model — predicted no asymmetry in the entrainment ratio for matched freestream densities, contrary to Konrad's observations.

In a subsequent proposal by Dimotakis,[13] the difference between temporal vs. spatial growth of a shear layer was noted and exploited to explain the entrainment ratio asymmetry. These basically geometric and similarity considerations led to an expression for the growth rate of a spatially growing, incompressible ($M \to 0$) shear layer given by

$$\frac{\delta}{x}(r,s) \approx C_\delta \, \frac{(1-r)(1+s^{1/2})}{2(1+s^{1/2}r)} \left\{ 1 - \frac{(1-s^{1/2})/(1+s^{1/2})}{1+2.9\,(1+r)/(1-r)} \right\}, \tag{10}$$

where the coefficient C_δ is taken as independent of the velocity ratio r and the density ratio s. The factor multiplying the braces describes the growth rate of a temporally growing shear layer and is the same as the Brown[12] proposal for shear layer growth. The factor in the braces arises from the fact that the growth is in fact spatial.

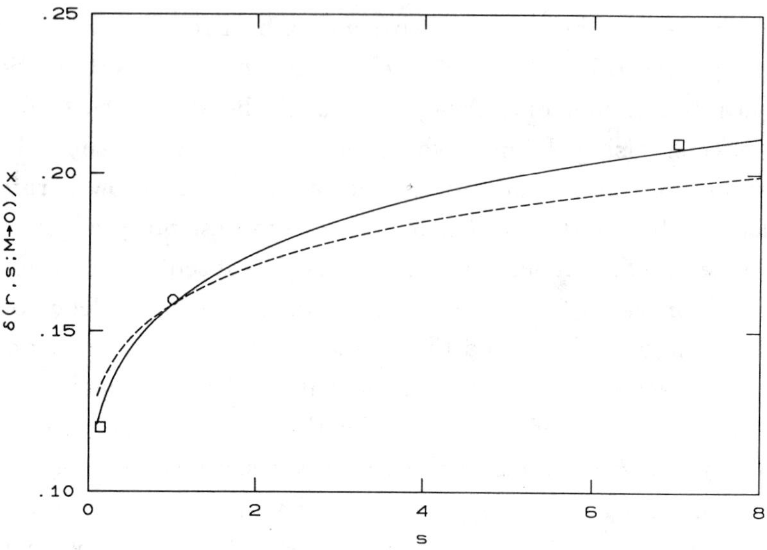

Fig. 1 Incompressible flow growth rate at fixed $r = U_2/U_1 = 0.4$ as a function of the density ratio $s = \rho_2/\rho_1$. Solid line: spatially growing layer (Eq. 10) with $C_\delta = 0.37$. Dashed line: temporally growing shear layer[12]. Squares: $s = 1/7, 7$ (Ref. 1). Circle: $s = 1$ (Ref. 7).

The observed (and predicted) dependence of the growth rate on the density ratio is not weak. It is plotted for density ratios in the range $0.1 < s < 8$ in Fig. 1, computed for a fixed velocity ratio of $r = 0.4$ and the value of $C_\delta = 0.37$, along with the experimental values of Brown and Roshko[1] for $s = 1/7$, 7, and the measurement of Mungal and Dimotakis[7] for $s = 1$. Also plotted for comparison is the Brown[12] prediction for the temporally growing shear layer. As can be seen, the difference between the two predicted growth rates is not large. It vanishes for equal freestream densities ($s = 1$), where the quantity in the braces becomes equal to unity, and where the proposed expression reduces to the Abramowich-Sabin relation (Eq. 9).

For a free-shear layer with no external disturbances,‡ the value of the coefficient C_δ is found to be in the range of

$$0.25 \leq C_\delta \leq 0.45 \tag{11}$$

‡ As opposed to a *forced*, or *driven*, shear layer (e.g., Refs. 14–20).

for the total thickness δ, or roughly half that for the maximum slope thickness δ_ω (cf. Eq. 8). See, for example, Brown and Roshko (Ref. 1, Fig. 10). An understanding, much less an accounting, of this rather large spread of values of the coefficient C_δ, which cannot be attributed to experimental errors, must await further investigations. It is not even clear, at this writing, whether the inequality bounds in Eq. 11 represent actual limiting values, or not. What *is* clear is that this coefficient depends in some way on the initial conditions of the flow. See, for example, data and additional discussions in Refs. 21–26.

It was proposed by Bradshaw[27] that the shear layer is sensitive to its initial conditions. Bradshaw suggests that a minimum of several hundred of the initial momentum thicknesses θ_0 is required for the shear layer to assume its asymptotic behavior. In view of the dynamics and interactions of the large-scale structures in the flow, it can even be argued that this estimate may not be conservative enough.[28] These caveats notwithstanding, sufficient experimental data exist to suggest that a turbulent shear layer will exhibit *linear* growth, even within the Bradshaw x/θ_0 specification, *but that the growth rate may not be a unique function of the freestream density and velocity ratio*. This is illustrated in the schlieren data in Fig. 2 of a shear layer with

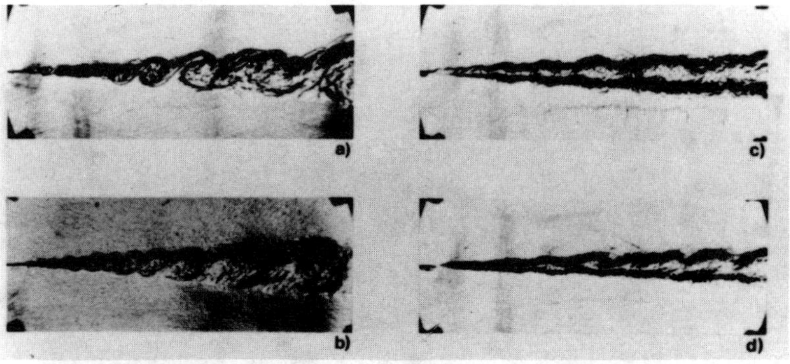

Fig. 2 Shear-layer growth at a fixed velocity ratio (r = 0.4) and equal freestream densities, as a function of flow velocity (field of view ≈ 25 cm). The flow here is made visible by changes in the index of refraction owing to the (small) heat release from a diluted reactant H_2 + F_2 chemical reaction. (a) U_1 = 13 m/s, (b) U_1 = 22 m/s, (c) U_1 = 44 m/s, (d) U_1 = 83 m/s. Unpublished data by Mungal, Hermanson, and Dimotakis.

equal freestream densities ($s = 1$), as the flow velocity was increased, keeping the freestream velocity *ratio* fixed at $r = 0.4$. As can be seen, the reduction in the shear layer growth rate with increasing flow velocity is appreciable.

It is clear that this behavior cannot be attributed to effects scaled by the *local* Reynolds number, for example, which increases linearly with x for this flow (see Eq. 1). Were that the case, or if *decaying* remnants of the effects of the initial conditions were responsible, the shear layer would be growing with curved edges rather than along (straight line) rays emanating from a virtual origin.

This behavior is perhaps better illustrated in the schlieren data in Fig. 3, formed as a composite of two pairs of pictures that cover roughly half a meter of flow, corresponding to an x/θ_0 of several thousand. The reader is invited to sight along the shear-layer edges of Fig. 3a, which is characterized by the better contrast, to ascertain the claim. Note that the highest high-speed stream velocity (U_1) in these data (Fig. 2) was 83 m/s, with the freestream fluid primarily composed of N_2 (diluent) gas. As a consequence, the observed reduction in growth rate cannot be attributed to compressibility (Mach number) effects, which will be discussed later.

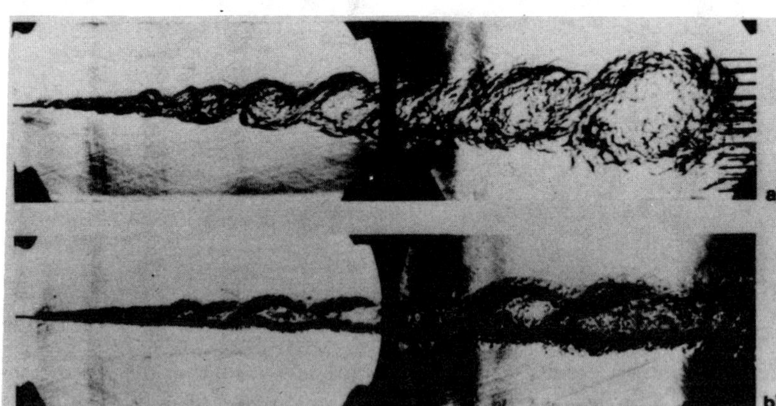

Fig. 3 Composite schlieren data of the same shear layer (probe array at far right at $x = 45$ cm). (a) $U_1 = 14$ m/s, (b) $U_1 = 88$ m/s. Unpublished data by Mungal, Hermanson, and Dimotakis.

In reference to the data in Fig. 2, it should be noted that these were acquired during the early stages of the development of that particular shear-layer facility, under flow conditions corresponding to relatively high acoustic excitation (cf. also Ref. 24); the reason they were not published at the time. Following subsequent improvements in the facility upstream flow management system, the shear-layer growth rates were found to be more typically represented by those in Fig. 3, corresponding to a laminar (Fig. 3a), or turbulent (Fig. 3b) initial high-speed stream boundary layer, respectively (cf. Ref. 106, Table 1), as had also been documented earlier by Browand and Latigo[23]. Nevertheless, the range of growth rates documented in Fig. 2 should not be regarded as generally atypical and was encountered as a result of various initial flow conditions, not necessarily traceable to the laminar or turbulent state of the initial boundary layers, or the absolute value of the flow velocities, in shear layer experiments in water (Ref. 26).

It is intriguing that shear-layer growth, *as far downstream as several thousand of the original splitter-plate boundary-layer momentum thicknesses*, appears dependent on, if not determined by, the initial conditions. It is also intriguing that the large-scale structure spacing appears to be the same in the low- and high-speed flow data in Fig. 3, even as their transverse extent is being reduced; what does appear to be changing is the large-scale structure aspect ratio.

A clue to this behavior may, perhaps, be found in terms of the notions of *convective* and *global* instabilities, first developed in the context of plasma instabilities.[29,30] By that criterion and the results of temporal, linear stability analysis, the fluctuations in a (coflowing) plane shear layer must be classified as the former, as discussed in Huerre and Monkewitz.[31] In that context, it could be argued that the shear layer should be regarded as an amplifier of the externally imposed disturbances, possessing a nonunique growth rate thereby, as opposed to the behavior of an oscillator, which could be characterized by its own growth rate. In my opinion, however, that analysis is a (spatially) local analysis and must be amended to include the contribution to the overall stability of the global effects of the initial development region, which is characterized by the influence of the wake introduced into the flow by the splitter plate (see, for example, discussion in Koch[32]) that Huerre and Monkewitz cite (see also Refs. 26, 33, and 34).

Finally, one would have to incorporate the additional feedback mechanisms present in the real flow in the analysis, such as the ones that act on the initial region owing to the long-range velocity fluctuations induced by the downstream structures, which span a range of *lower* frequencies, or pressure fluctuations feeding back to the splitter plate tip as the last structures leave the shear layer flow domain (test section), as dictated by the facility-dependent outflow boundary conditions, as discussed by Dimotakis and Brown.[28] In addition to the long-coherence-time behavior of the autocorrelation functions in those experiments,[28] additional evidence for this feedback was also available in the form of a flagellation of the initial region of the shear layer initial flow region and the downstream flow with an elliptical region. Its absence in purely supersonic flow deprives shear layer flows of an instability/amplification mechanism that is potentially important to the growth of the turbulent region.

A corollary of the existence of such long-range feedback mechanisms is that *local* descriptions of the dynamics of this flow may be inadequate. Certainly, the important differences between a *temporally* growing layer and the full scale, *spatially* growing shear layers of interest here must also be contended with. This difference was already noted in the context of the shear layer entrainment ratio. In the present context, a local, temporal analysis also fails to represent the long-range coupling of the local behavior to non-local shear layer dynamics. The question of the applicability of *linear* stability analysis to the description of these phenomena aside, the classification of fluid mechanical instabilities into global and convective should also be assessed in this light; it derives from a local, temporal instability analysis. It is difficult to say, at this time, whether a proper accounting of all of these effects would alter the oscillator/amplifier qualitative classification of the behavior that stems from (temporal) convective/global instability analyses.

Compressibility Effects

It has been documented that two-dimensional shear-layer growth decreases as the flow Mach number increases, even after the coupling of the flow Mach number to the freestream density and velocity ratio that would result in a typical flow facility is accounted for. See discussion in Brown and Roshko (Ref. 1, Sec. 7.1). An analysis by Bogdanoff[37] and experimental investigations of compressible shear layers by Papamoschou and Roshko[38] have suggested that, to a large extent, the effects of compressibility can be scaled by the *convective* Mach numbers of the shear-layer large-scale structures with respect to the two streams, which measure the relative freestream Mach numbers as seen from the Galilean frame of these structures (Eq. 6). It is interesting that one can also argue for a similar scaling on the basis of linear stability analysis of compressible shear flow, if the convection velocity U_c is identified with the (real) phase velocity c_r of the unstable mode in the flow.[39–41]

For incompressible flow, the convection velocity U_c can be estimated by recognizing that, in the large-scale structure Galilean convection frame, stagnation points exist in between each adjacent pair of structures.[42] Continuity in the pressure at these points,[13,43] i.e.,

$$p_1 + \frac{1}{2}\rho_1 (U_1 - U_c)^2 \approx p_2 + \frac{1}{2}\rho_2 (U_c - U_2)^2 , \qquad (12)$$

then yields, for $p_1 \approx p_2$,

$$\frac{U_1 - U_c}{U_c - U_2} \approx \sqrt{\frac{\rho_2}{\rho_1}} , \qquad (13)$$

or,

$$\frac{U_c}{U_1} \approx \frac{1 + r\, s^{1/2}}{1 + s^{1/2}} . \qquad (13')$$

This agrees with the differently derived Brown[12] result, the few estimates of this quantity from the (x,t) data in Brown and Roshko,[1] as well as the measurements of Wang[44] in curved shear layers. See also Coles (Ref. 43, Fig. 7) and related discussion.

For compressible flow, the corresponding result can be similarly estimated from the isentropic relation for the total pressure, i.e.,

$$\frac{p_{tj}}{p_j} = \left[1 + \frac{\gamma_j - 1}{2} M_{cj}^{(i)\,2}\right]^{\frac{\gamma_j}{\gamma_j - 1}} , \tag{14a}$$

with $j = 1, 2$ corresponding to the high- and low-speed streams, γ_j the ratios of specific heats, and a_j the speeds of sound for the high-speed and low-speed stream fluids, respectively. The superscript '(i)' over the convective Mach number denotes that the estimate is based on isentropic pressure recovery assumptions.

Approximately equal pressure recovery from each freestream at the large structure interstitial stagnation points then yields (again, for $p_1 \approx p_2$)

$$\frac{p_{t1}}{p_1} \approx \frac{p_{t2}}{p_2} . \tag{14b}$$

This is the same result as the one arrived at by Bogdanoff[37] using different arguments (see also discussion in Papamoschou and Roshko[38]). It also agrees with the linear stability estimates of this quantity,[41] at least for subsonic convective Mach numbers. Note that, for equal ratios of specific heats ($\gamma_1 = \gamma_2$), e.g., if both freestreams are composed of monatomic gases, we have

$$M_{c1}^{(i)} \approx M_{c2}^{(i)} , \tag{15a}$$

or

$$r_c^{(i)} = \frac{U_1 - U_c^{(i)}}{U_c^{(i)} - U_2} \approx \frac{a_1}{a_2} , \tag{15b}$$

where, as above, the superscript '(i)' denotes an estimate based on isentropic pressure recovery assumptions. As can be seen, for static freestream temperatures that are also matched, the compressible, isentropic pressure recovery relation reverts to the incompressible, Bernoulli equation result (cf. Eq. 13).

It should be noted that these results are actually more robust than the assumption that the pressure recovered at the large-scale structure interstitial stagnation points can be computed assuming isentropic pressure recovery from each freestream. If only a *fraction* of the total pressure is recovered at these points, it suffices to assume that the losses from each side are roughly the same, in the mean. Considering the symmetries of the flow, even when viewed as an unsteady process, such an assumption may well be justified, at least for subsonic convective Mach numbers, and yields results in accord with the experimental data cited, as well as the computational evidence (e.g., Lele[45]) to date. Secondly, we should note that the pressure matching condition (Eqs. 12, 14) is not a *force* balance condition; the large-scale structure is not some intervening impermeable body between the two streams. If that were the case, a small "imbalance" that would momentarily decrease the velocity difference with respect to one stream would provide a *positive* feedback and drive the velocity difference with respect to the same stream to zero. This relation should rather be viewed as a nonlinear, quasistationary phase condition for the large-scale structures: any nonabiding flow substructure is subject to accelerations in its own frame, that will ultimately convect it with one or the other stream by the force of the positive feedback mechanism argument just cited. Perhaps the robustness of the large-scale flow structures in shear layers can be understood in this light. The positive feedback convection velocity mechanism strips away all other structures! The success of linear stability analyses of these phenomena, at least for subsonic convective Mach numbers, can perhaps also be understood in the same light, since the dominant surviving mode must abide by the same considerations.

Papamoschou and Roshko[38] find that the compressible shear-layer growth rate, when normalized by the corresponding incompressible flow growth rate estimated at the same velocity and density ratio, is well represented as a function of the isentropically estimated convective Mach number only, say, $M_{c1}^{(i)}$, i.e.,

$$\frac{\frac{\delta}{x}\left[r, s; M_{c1}^{(i)}\right]}{\frac{\delta}{x}\left[r, s; M_{c1}^{(i)} = 0\right]} \approx \mathrm{fn}\left[M_{c1}^{(i)}\right] . \tag{16}$$

Figure 4 includes the data of Papamoschou and Roshko[38], shear-layer growth-rate estimates computed from the earlier data of Chinzei *et al.*[46] that were processed to estimate $M_{c1}^{(i)}$ for each of their runs and normalized to the value of $\delta[M_{c1}^{(i)}]\,/\,\delta[0]$ at one point (filled circle), the data of Clemens and Mungal[47], and of Hall *et al.*[48] (see also Ref. 37 for a compilation of earlier data). The smooth curve in Fig. 4 is a plot of the function

$$f(M_{c1}) \;=\; (1 - f_\infty)\, e^{-3M_{c1}^2} \;+\; f_\infty \;. \tag{17}$$

It is drawn as a rough estimate of the effect, with fitted values for the asymptotic value of $f_\infty = 0.2$ and the factor of 3 multiplying the convective Mach number in the exponent.

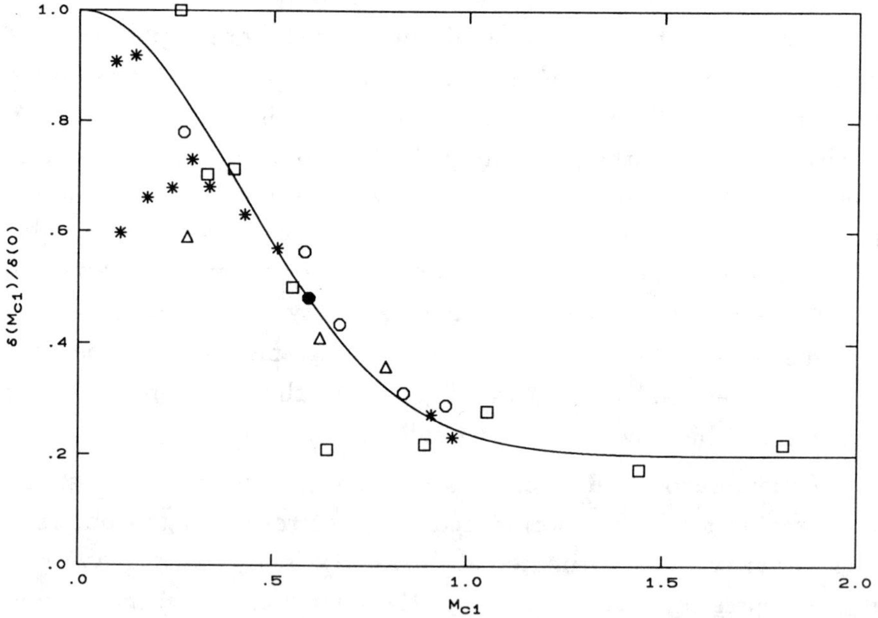

Fig. 4 Compressible shear-layer growth data, for a range of free-stream velocity and density ratios, as a function of $M_{c1}^{(i)}$ (superscript '(i)' omitted from figure legend). Squares: Ref. 38 data. Circles: estimated from Ref. 46 data (normalized to the filled point). Triangles: Ref. 47 data. Stars ($M_1 > 1$) and crosses ($M_1 < 1$): Ref. 48 data. Smooth curve is a plot of Eq. 17.

The data in Fig. 4 suggest that the convective Mach number need not be very large for compressibility effects to be significant. Secondly, for $M_{c1}^{(i)} > 0.8$, the growth rate appears to reach an asymptotic value roughly 0.2 of its incompressible counterpart. This is at variance with the results of two-dimensional, linear, shear-layer stability analyses (e.g., Refs. 49, 40, and 41), which find that the growth rate tends to very small values, as $M_{c1}^{(i)} \to \infty$. If the stability analysis results are accepted at face value, the applicability of such an analysis aside for the moment, the discrepancy could be attributable to other factors. In particular, it appears likely that three-dimensional modes are more unstable in the limit of large Mach numbers, as was suggested by Sandham and Reynolds.[50] Alternatively, the Papamoschou and Roshko[38] experiments were conducted in an enclosed test section, as opposed to the stability analyses which were carried out for unbounded flow. For supersonic convective Mach numbers, a closed test section can act as a wave guide, providing a feedback mechanism between the growing shear layer structures and the compression/expansion wave system whose energy would otherwise be radiated and lost to the far field.[51–53] Finally, recalling our aside, we should recognize that, for supersonic (or even transonic) convective Mach numbers, we expect shocks to form in the flow, a feature that cannot adequately be captured by *linear* stability analysis.

It should be noted that it is not clear at this writing whether the observed limiting value of the ratio $\delta[\,M_{c1}^{(i)}\,]\,/\,\delta[0]$, for $M_{c1}^{(i)} \gg 1$, is intrinsic to the behavior of the fully developed compressible shear layer, or depends on the details of the flow geometry, e.g., the distance of the upper and lower flow guide walls from the layer, whether only one or both streams are supersonic, etc. In the context of the potential for hypersonic air-breathing propulsion and flight, for example, whether the growth rate tends to zero with increasing Mach number is an important issue; an ever decreasing shear layer growth with increasing Mach number hardly bodes well for efficient supersonic mixing and combustion!

As can be seen in the data in Fig. 4, there are some "rogue" points at low convective Mach numbers, from the data of Hall *et al.*[48] It is significant that all the supersonic shear layer flow data of Hall *et al.* were taken at the same high-speed freestream Mach number of $M_1 \simeq 1.5$. For these shear layers, the various values of the convective Mach number were realized using different compositions for the freestream gases. The low growth rates of the "rogue" points corresponded to shear layers with low freestream density ratios, i.e., for $\rho_2/\rho_1 \ll 1$. On the basis of these data, we may infer that the utility of the convective Mach number as the scaling parameter of compressibility effects on shear-layer growth rates, which is based on a temporal growth picture in the Galilean frame moving with the turbulent structures, cannot describe all the factors that influence this behavior. Additional evidence of this shortcoming will be discussed later.

Subsequent to his initial investigations, Papamoschou conducted a series of experiments in which he investigated the convection velocity of the large-scale structures, for a range of freestream Mach numbers and various gases.[54] In those experiments, he found that, at high convective Mach numbers, the large-scale structures seemed to be "dragged" by one stream or another, at variance with the matched, isentropic pressure recovery model of Eq. 14. See Fig. 5a for a plot of the Papamoschou[54] and some additional, more recent, data.[♯] As can be seen, the experimental data are found to be close to the isentropically estimated values only for convective Mach numbers less than 0.5, or so. Papamoschou[56,54] offered a qualitative description of how shocks could be responsible for this behavior, crediting D. Coles for the suggestion, which was made before the experiments were conducted, that the effects of shocks needed to be incorporated in the analysis.

♯ The point (M_{c1}, M_{c2}) derived from the Fourguette *et al.* data[55] was computed using the quoted (directly measured) value for the convection velocity of $U_c = 352\,\mathrm{m/s}$.

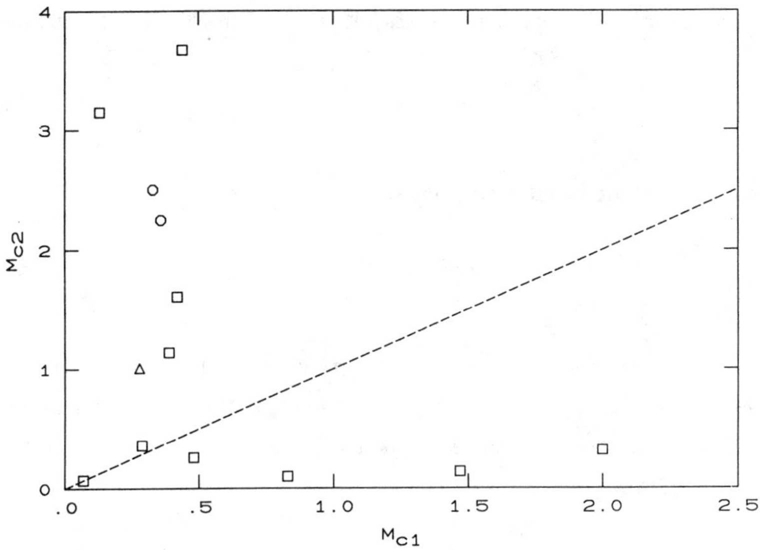

Fig. 5a Supersonic shear layer ($M_1 > 1$) convective Mach number experimental data of Papamoschou[54] (squares), Fourguette *et al.*[55] (triangle), and Hall *et al.*[48] (circles). Dashed line computed for $\gamma_1 = \gamma_2$, assuming $U_c = U_c^{(i)}$ (Eq. 15a).

We can appreciate that for supersonic, or transonic, convective Mach numbers the freestream flow over the turbulent large-scale structures can support a system of expansions and shocks. As a consequence, the assumption of isentropic, approximately matched, pressure recovery from each stream (Eq. 14) can be expected to be inadequate at supersonic convective Mach numbers. In that case, streamlines that end up on interstitial stagnation points from each freestream will have to traverse a shock, or, more likely, a system of shocks for turbulent flow, to connect to the freestream static conditions and will have suffered a loss in total pressure that may be reasonably well approximated by that of a normal shock.

This leads to the following possibilities, depending on which freestream has, or can support, shocks. In particular, we can have shocks in the high-speed stream, with a shock-free low-speed stream, i.e.,

$$\frac{p_{s1}}{p_1} \approx \frac{p_{t2}}{p_2} \; ; \tag{18a}$$

low-speed stream shocks, with a shock-free high-speed stream, i.e.,

$$\frac{p_{t1}}{p_1} \approx \frac{p_{s2}}{p_2} \tag{18b}$$

while for shocks in both streams, we must have

$$\frac{p_{s1}}{p_1} \approx \frac{p_{s2}}{p_2} \,. \tag{18c}$$

In these expressions, p_t/p is the isentropic total-to-static recovery pressure ratio (Eq. 14) and p_s/p is the ratio of the post-normal shock total pressure to the free stream static pressure, given by (Rayleigh pitot tube formula),

$$\frac{p_s}{p} = \left(\frac{\gamma+1}{2} M_s^2\right)^{\gamma/(\gamma-1)} \left(\frac{2\gamma}{\gamma+1} M_s^2 - \frac{\gamma-1}{\gamma+1}\right)^{-1/(\gamma-1)} \tag{19}$$

for $M_s > 1$, where M_s is the shock Mach number (e.g., Ref. 57, p. 149).

As was noted by Papamoschou,[54] the estimation of the convective velocity U_c of the turbulent structures using these relations requires the shock Mach number M_s to be specified, which is not known a priori. This issue was addressed in a recent proposal,[58] which is briefly outlined below.

Returning to Fig. 5a, for convective Mach numbers that are not small, we see that the data appear to fall in two groups: decidedly above, or below, the $M_{c1} = M_{c2}$ dashed line (cf. Eq. 15 and related discussion). These two groups correspond to supersonic shear layers with subsonic and supersonic low-speed streams, respectively. In the context of the previous discussion, we can understand this by assuming that, for $M_1 > 1$ and $M_2 < 1$, the shocks are borne by the low-speed freestream, whereas, for purely supersonic flow, i.e., $M_1, M_2 > 1$, the shocks are borne by the high speed stream. We will accept this as an empirical *stream selection rule*.

For subsonic, but near sonic, convective Mach numbers, evidence for the formation of shocks can be found in the calculations of Lele,[45] and Vandromme and Haminh (Ref. 59, cf. Fig. 2), for example, where one expects weak shocks (dubbed "shocklets") confined to the vicinity of the shear zone. See cartoon in Fig. 6a. No experimental evidence for these transonic shocklets is available at this writing.

For supersonic convective Mach numbers, experimental evidence has been available for turbulent-structure-generated shocks from the core region of supersonic jets, e.g., Lowson and Ollerhead,[60] Tam,[61] and Oertel.[62] More recently, such evidence has been documented in

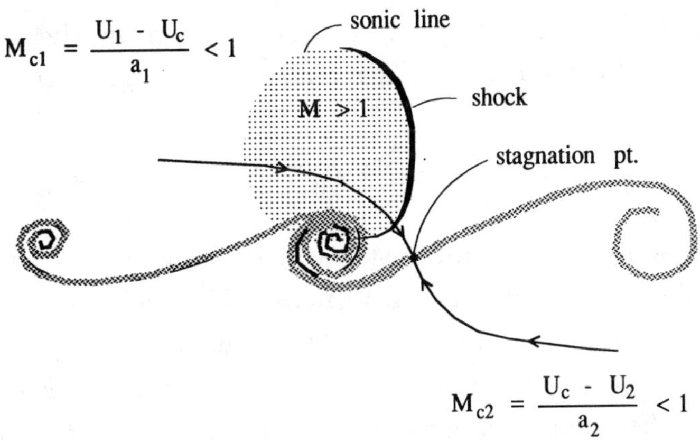

Fig. 6a Proposed vortex/shock configuration cartoon, sketched for a shock borne by the high speed stream and a transonic convective Mach number, i.e., $M_{c1} < 1$.

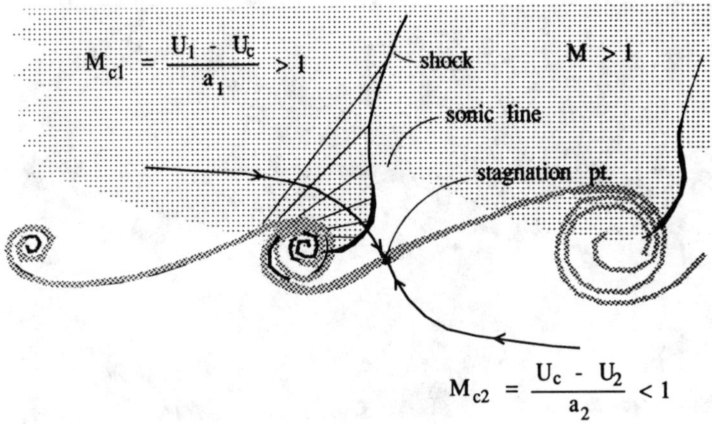

Fig. 6b Proposed supersonic vortex/shock configuration cartoon for a supersonic convective Mach number ($M_{c1} > 1$).

our laboratory for a two-dimensional, supersonic shear layer.[48,36] In the Hall *et al.* experiments, a shock/expansion wave system was found, extending into one of the free streams,[48] as sketched in Fig. 6b.

An example of such a wave system, for a $M_1 = 1.5$ He high-speed stream, over $M_2 = 0.35$ Ar low-speed stream shear layer, is reproduced in Fig. 7 (from Ref. 48, Fig. 5). See also Hall (Ref. 36, Fig. 4.11) for similar data from a $M_1 = 1.5$ He high-speed stream, over $M_2 = 0.3\,N_2$ low-speed stream shear layer.

Given the freestream j that carries the shocks and the shock Mach number, M_{sj}, or, equivalently, a given shock strength parameter

$$X_j \equiv \frac{M_{sj}}{M_{cj}} , \tag{20}$$

the convection velocity can be estimated by computing the total pressure loss through the shocks (cf. Eqs. 18). This yields a continuum of solutions for U_c vs. the shock strength parameter X_j.

With these assumptions, the strength of the shock can be estimated if the turning angle $\Delta\theta$ through which the flow has been expanded, prior to crossing the near-normal shock, is known. The

Fig. 7 Schlieren data for a $M_1 = 1.5$ He high-speed stream, over $M_2 = 0.35$ Ar low-speed stream shear layer. Note travelling oblique shock system in low-speed stream (Ref. 48, Fig. 5).

turning angle $\Delta\theta_j$ in the jth stream can be estimated, in turn, as the difference of the corresponding Prandtl-Meyer angles between the flow just ahead of the shock and the free stream (or sonic conditions), i.e.,

$$\Delta\theta_j = \theta_{\text{PM}}(M_{sj}) - \theta_{\text{PM}}(M_{cj}) , \qquad \text{for} \qquad M_{cj} \geq 1 , \tag{21a}$$

where $\theta_{\text{PM}}(M)$, defined for $M > 1$, is the Prandtl-Meyer angle function (e.g., Ref. 57, p. 99). If the convective Mach number M_{cj} in the jth stream is close to, but less than, unity (transonic M_{cj}), the turning angle $\Delta\theta_j$ will be computed using

$$\Delta\theta_j = \theta_{\text{PM}}(M_{sj}) , \qquad \text{for} \qquad M_{cj} < 1 . \tag{21b}$$

The latter is equivalent to starting the calculation at the location where the streamline crosses the sonic line to enter the supersonic bubble. See cartoon in Fig. 6a.

Depending on the flow parameters, the pressure-matching condition can lead to several solution branches. Given the free stream that carries the shock and the shock strength, several branches will typically exist, with a continuum of solutions for the convection velocity U_c as a function of the shock strength parameter X. The proposed ansatz is that *the convection velocity of the large scale structures is such as to render the flow stationary*. One can argue for this conjecture by noting that if the shock-generating flow structures are to represent a quasi-steady, convecting flow configuration, they must be able to persist in the presence of turbulent fluctuation disturbances.

We note here that a Galilean-invariant analysis, i.e., one based on the temporal behavior of the large scale structures in the convected frame, as in the cartoons in Figs. 6a and 6b, *cannot* capture the (laboratory frame) empirical stream selection rule for the stream that carries the shocks cited above. As a consequence, we will accept the value derived from the proposed stationary flow ansatz when it yields solutions in accord with the empirical selection rule.

When the flow is computed as a function of the shock strength parameter $X_j = M_{sj}/M_{cj}$, corresponding to a shock in the jth stream, one finds that the solution branches fall into two classes. In the first solution class, Type I flow, the turning angle $\Delta\theta$ can be computed by assuming that the flow chooses the stream j and the shock Mach number, i.e., the shock strength parameter $X_j = M_{sj}/M_{cj}$, so as to render the turning angle $\Delta\theta_j$ stationary (a maximum). This corresponds to a stable flow configuration wherein small changes in the shock Mach number M_{sj} result in quadratically small changes in $\Delta\theta_j$. Alternatively, in Type II solutions, it is the shock strength parameter X_j that is stationary with respect to small changes in the turning angle $\Delta\theta$, corresponding to the maximum admissible value for X_j that yields a solution for U_c.

It is found that both Type I and Type II solutions can be admissible (in the same flow). In the latter case, one can argue for a selection rule which favors the Type I branch, over the Type II solution branch, as being the more robust configuration of the two. If more than one solution branch of the same type is possible, the proposed selection rule is that the branch that yields the *lower* total pressure is chosen by the flow. In other words, the flow will try to satisfy the pressure matching condition at the lowest stagnation pressure possible, generating the shock with the requisite strength (see Ref. 58 for a more detailed discussion).

The results of calculations based on the proposed scheme are summarized in Fig. 5b, which compares the experimental data in Fig. 5a with the theoretical calculations. The estimates, based on the proposed scheme, for flows found to yield Type I solutions are denoted by asterisks, while those corresponding to Type II solutions are denoted by crosses. If the computed values are found to fall outside the extent of the experimental data point symbols, they are joined to the corresponding data points by straight lines. There is one case for which the stationary shock is borne by the stream that is not accord with the empirical stream selection rule. This corresponds to the $M_1 = 3.2\,\mathrm{Ar}$ and $M_2 = 0.2\,\mathrm{Ar}$ shear layer of Papamoschou. No computed (M_{c1}, M_{c2}) point is indicated for it.

It may be interesting to also ask for input from linear stability analyses of this flow, with the appreciation that finite amplitude wave

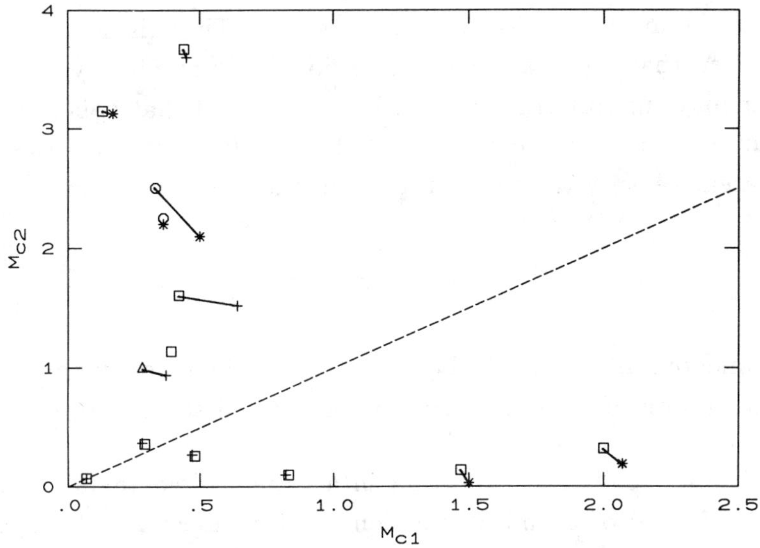

Fig. 5b Experimental data of (M_{c1}, M_{c2}) from Fig. 5a. Computed points are joined to corresponding flow data points by straight lines, asterisks corresponding to Type I and crosses corresponding to Type II flows.

effects, such as the loss in total pressure associated with entropy production in shock waves, cannot properly be captured by such analyses. Nevertheless, the very small entropy generation from weak oblique shocks, as would be expected under many flow conditions, might render linear stability analysis results useful for convective Mach numbers that are not too high. Such an investigation was specifically undertaken by Sandham and Reynolds.[63] The agreement for low convective Mach numbers is quite good. At higher convective Mach numbers, however, the linear stability analysis calculations underestimate the departure of the convection velocity from the isentropically computed values. As was noted, this is as one would anticipate. It is interesting that stability analysis appears to predict the correct shock-bearing stream for the case not computed in Fig. 5b, but predicts the wrong one for the flow with the highest M_{c2} (see Ref. 63, Fig. 2.25).

We may conclude that available data to date appear to be reasonably well accounted for by assuming the existence of a turbulent structure, convecting with a well-defined speed and generating a set of shocks in one of the two free streams. There is some evidence, however, that shear layers at higher flow Mach numbers yet the shear layer may support shocks in both streams.[62] If that is borne out by future experiments, we can expect that the large asymmetries in the apparent velocity ratio r_c in the turbulence convection frame, i.e.,

$$r_c \equiv \frac{U_1 - U_c}{U_c - U_2} , \tag{22}$$

documented in Fig. 5, will be mitigated by the more symmetric flow configuration of turbulent-structure-generated shocks borne by both freestreams.

These results are important in a variety of contexts. From the point of view of chemical reactions and combustion, the apparent velocity ratio r_c (Eq. 22) seen by the turbulent structures is an important factor in the shear layer entrainment ratio and the consequent stoichiometric composition of the molecularly mixed fluid in the layer, as we will discuss later. The stream selection rule and the proposed ansatz of shock strengths selected by the condition of stationarity suggest that one can expect jumps in r_c, potentially, as a result of small changes in the flow parameters.

From a theoretical vantage point, we should recall that the empirical stream selection rule is not Galilean-invariant and, therefore, it cannot be captured by the type of analysis outlined above. As was noted in the context of the discussion on shear layer growth rate, the fact that a supersonic/subsonic shear layer contains an elliptical region connecting the inflow and outflow boundary conditions needs to be incorporated in the analysis. Evidence of the need for a global, rather than local, description of these flows was also discussed earlier in the context of the use of the convective Mach number to account for the compressibility effects on shear-layer growth rates. At least for shear-layer flows which include an elliptical region, i.e., a subsonic low-speed freestream, it would appear that a more complicated description is required. The resolution of these and other issues must await the results of several investigations currently in progress.

Heat Release Effects

Some experimental investigations have focused on the effects of heat release on the growth rate of chemically reacting shear layers (e.g., Refs. 64, 65, as well as experiments with combusting shear layers at high levels of heat release (e.g., Refs. 66–69). Useful information has also been derived from computations (e.g., Refs. 70 and 71), which is in qualitative accord with the experimental findings, even though turbulent reacting flow computations, at least to date, have perforce been conducted at Reynolds numbers that do not meet the fully developed flow criterion of Eq. 1.

One might argue that dilatation owing to heat release in a chemically reacting shear layer, which is confined to the shear-layer wedge, might result in an *increase* in shear-layer growth. Although the basis of that intuition is well founded, the inference is not. One does observe a *displacement* velocity in the far field away from the shear layer, which increases with the amount of heat released. This can be measured experimentally as a displacement thickness δ^*/x by noting the angle of, say, the lower freestream flow guide wall required

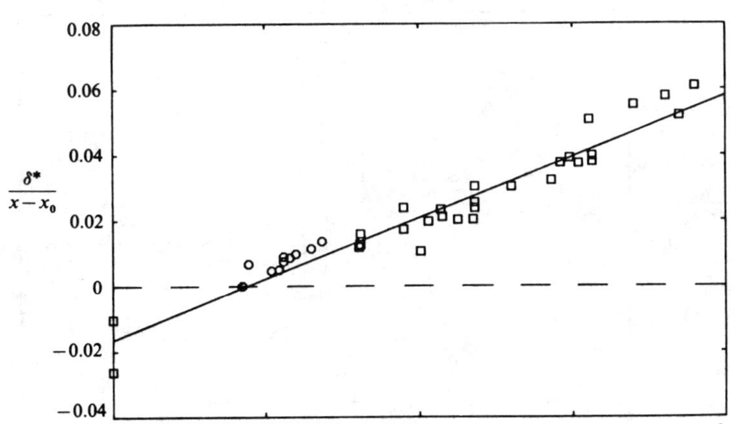

Fig. 8 Normalized shear-layer displacement thickness vs. heat release. Circles: Mungal data (unpublished). Squares: Ref. 65 data.

to maintain a nonaccelerating flow ($\mathrm{d}p/\mathrm{d}x = 0$), as a function of the amount of heat release. See Fig. 8.

At least for subsonic flow and equal freestream densities, other parameters held constant, one observes a *decrease* in the shear-layer growth rate with increasing heat release. This behavior is depicted in the data of the 1% thickness in Fig. 9 (taken from Ref. 65, Fig. 5). Note that, in these experiments, the pressure gradient was maintained close to zero by adjusting the lower stream guide wall as necessary. These data suggest that the decrease in the shear-layer growth rate with heat release is approximately given by ($q = \Delta\rho/\rho$, recall Eq. 7)

$$\frac{\frac{\delta}{x}(r = 0.4,\ s = 1;\ q)}{\frac{\delta}{x}(r = 0.4,\ s = 1;\ q = 0)} \approx 1 - C_q\, q\ , \tag{23a}$$

with

$$C_q \approx 0.05\ . \tag{23b}$$

While these experiments were conducted at a fixed velocity ratio $r = 0.4$ and matched freestream densities ($s = 1$), one can specu-

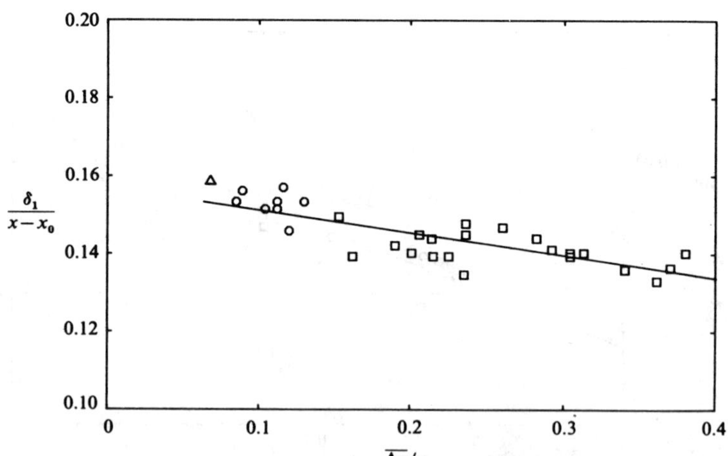

Fig. 9 Normalized 1% thickness vs. heat release (note displaced origin). Triangle: Ref. 64 data. Circles: Mungal data (unpublished). Squares: Ref. 65 data.

late that heat-release effects manifest themselves as a reduction in the growth-rate coefficient C_δ (Eqs. 9 and 10), with Eq. 23 expressing the dependence of C_δ on q. In any event, at least for subsonic shear layers, the effect of heat release on the growth rate is slight (see also Ref. 68).

The physical implication is that the outward displacement velocity owing to heat release impedes the entrainment process to an extent that more than offsets the effects of dilatation. It is interesting that this reduction in growth rate is also found to be consistent with the assumption that the heat release and dilatation effects leave the $\overline{u'v'}$ velocity correlation largely unaltered. The reduction in the growth rate can then be approximately accounted for by noting the reduction in the turbulent stress $\tau = \overline{\rho u'v'}$ in the layer; a result of the reduction in the density profile $\rho(y/x)$ owing to heat release. It should be noted that it would probably be difficult to argue for such an assumption a priori (see discussion in Ref. 65, Secs. 5.2 and 5.4).

At high Reynolds numbers, a substantial volume fraction in the turbulent shear layer is occupied by fluid that is *not* molecularly mixed (independently estimated to be roughly 1/2 at the conditions of the Hermanson and Dimotakis experiments) so that, even with fast chemical reactions characterized by large adiabatic flame temperatures, there will be a limiting value of the expected reduction in $\Delta\rho/\rho$ within the layer owing to heat release. This is a consequence of the large pockets of entrained, unmixed fluid whose density is essentially unaltered by the combustion process. This behavior is illustrated by the data in Fig. 10, where the estimated mean density reduction $\Delta\rho/\rho_0$ is plotted vs. the normalized adiabatic temperature rise $\Delta T_f/T_0$. These data were recorded at high Reynolds numbers ($Re \approx 6 \times 10^4$), for several values of the freestream stoichiometric mixture ratio ϕ, an important quantity that we will discuss later in the context of chemical reactions.

A useful test model of the behavior with increasing heat release is one in which vortical structures tend to a configuration of low-density (hot) cores and are surrounded by irrotational, recently entrained, unmixed (cold) fluid. Using this picture, the expected mean density

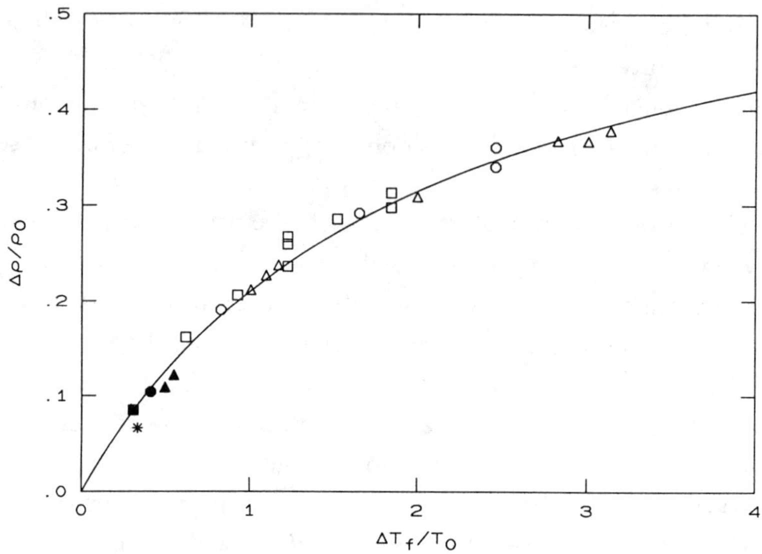

Fig. 10 Mean density reduction $\Delta\rho/\rho$ vs. adiabatic flame temperature rise $\Delta T_f/T_0$, where T_0 is the (common) freestream temperature. Squares/asterisk: $\phi = 1$. Circles: $\phi = 1/2$. Triangles: $\phi = 1/4$. Solid symbols: Ref. 7 data. Open symbols: Ref. 65 data. Asterisk: Ref. 64 data. Smooth line computed using Eq. 23.

reduction in the combustion zone can be estimated. In the notation of Eq. 4, we find

$$\frac{\Delta\rho}{\rho_0} \approx \frac{\left(\frac{\delta_{\mathrm{P}}}{\delta}\right)\frac{\Delta T_{\mathrm{f}}}{T_0}}{1 + \left(\frac{\delta_{\mathrm{P}}}{\delta_{\mathrm{m}}}\right)\frac{\Delta T_{\mathrm{f}}}{T_0}} \,. \tag{23a}$$

The smooth line in Fig. 10 was computed using this expression and constant, heat-release-independent values of

$$\frac{\delta_{\mathrm{m}}}{\delta} = 0.63 \,, \qquad \frac{\delta_{\mathrm{P}}}{\delta_{\mathrm{m}}} = 0.5 \,. \tag{23b}$$

It is interesting that the resulting curve does as well as it does, suggesting that the simple model may have merit even at moderately high values of the heat release. As we shall see later, the inferred value of 0.63 of the mixed fluid fraction $\delta_{\mathrm{m}}/\delta$ is a little high. It should be noted, however, that the mean density reduction values in Fig. 10 were estimated here using the reciprocal of the mean temperature measured in the combustion zone, as opposed to the mean of the reciprocal temperature, which would have provided better estimates. Additionally, of course, the mixed fluid fraction is not likely to be exactly constant, i.e., independent of $\Delta T_{\mathrm{f}}/T_0$.

Pressure Gradient Effects

The effects of pressure gradients on shear layer growth were discussed by Sabin[10] and have been investigated in non-reacting shear layers,[5] and in reacting shear layers[69,65] for incompressible flow conditions. In the case of a favorable pressure gradient, i.e., $\mathrm{d}p/\mathrm{d}x < 0$, and equal free stream densities it was found that the decrease in the growth could be accounted by interpreting Eq. 9 as a local relation.[65] The argument, which was suggested by M. Koochesfahani and is akin to some of the ideas put forth by Sabin[10], is reviewed and extended below.

A local rendition of the Abramowich-Sabin relation (Eq. 9) yields

$$\frac{\mathrm{d}\delta}{\mathrm{d}x} \approx C_\delta \frac{1-r(x)}{1+r(x)} , \tag{24a}$$

with

$$r(x) \equiv \frac{U_2(x)}{U_1(x)} \tag{24b}$$

the local velocity ratio. This can be computed using the Bernoulli equation in each of the freestreams, and yields

$$r(x) = r_0 \sqrt{\frac{1 - C_{p2}(x)}{1 - s\, r_0^2\, C_{p2}(x)}} , \tag{25a}$$

where

$$r_0 \equiv \frac{U_2(0)}{U_1(0)} \tag{25b}$$

is the freestream velocity ratio at $x = 0$,

$$C_{p2}(x) \equiv \frac{p(x) - p(0)}{\frac{1}{2}\rho_2 U_2^2(0)} \tag{25c}$$

is the local pressure coefficient normalized by the low-speed stream dynamic head at $x = 0$, and $s = \rho_2/\rho_1$ is the freestream density ratio. It can be seen that a favorable pressure gradient is expected to decrease, or increase, the shear-layer growth rate, depending on whether the product $s\,r_0^2$ is less, or greater than, unity, respectively. The converse is true for an unfavorable pressure gradient.

For a specified pressure gradient, Eq. 25 for the local velocity ratio can be used to integrate the local growth rate (Eq. 24) to yield the local shear layer width $\delta\,[C_{p2}(x)]$ at the station x. The results of this procedure are in accord with the observed effects, at least for the range of values of the pressure coefficient at the measuring station that were investigated (see Ref. 65, Sec. 8).

III. Mixing: δ_m/δ

As alluded to in the preceding discussions, one finds that, at sufficiently high Reynolds numbers, a substantial fraction of the fluid within the δ/x shear-layer wedge is not molecularly mixed. Additionally, at least for incompressible shear layers, the available evidence suggests that the mixed fluid in a turbulent shear layer exhibits the following characteristics:

1. The mixed fluid composition, averaged across the shear-layer width $\delta(x)$, is *not* generally centered around a 50:50 mixture but favors a composition that is a function of the freestream density and velocity ratio.

2. The amount of mixed fluid, beyond the downstream location where the shear layer has attained fully developed, three-dimensional flow status (e.g., Eq. 1), depends weakly on the local flow

Reynolds number. The evidence suggests that, at least for gas-phase shear layers, it decreases as the Reynolds number increases.

3. The mixed fluid fraction δ_m/δ is found to depend on the fluid Schmidt number $Sc \equiv \nu/\mathcal{D}$.

A large number of models are employed today in efforts to account for the observed turbulent shear-layer mixing phenomena and, whereas they all differ in the details of their implementation, they can be classified, in my opinion, into two main categories: models that ultimately rely on some form of Reynolds averaging and/or turbulent gradient transport, and models that do not. In the discussion that follows, models which cannot account for the characteristics just listed will not be considered in the attempt to account for shear-layer mixing phenomena. Although it could be argued that this need not be so, this criterion, to the best of my knowledge, presently eliminates models that rely on gradient transport ideas. For an opposing viewpoint, the reader is directed to the review article by Bilger[72] and references therein. Unfortunately, since molecular mixing takes place throughout the spectrum of spatial and temporal scales, direct computations at high Reynolds numbers are out of the question for now, at least, and we must continue resort to some kind of modeling for some time.

It may appear surprising that *Schmidt* number effects are given so much weight when it could be argued that most turbulent mixing/combustion phenomena are encountered in gas-phase flows. There are two issues here. First, although it may be that *most* turbulent combustion occurs in gas-phase flows, it is certainly not so that *all* of it does (cf. underwater, liquid metal, particulate combustion, etc.). Second, if we are to formulate models on which we must rely for predictions and design outside the range of experience that was used to validate them, and not just use them as interpolative french curves, we must at least require that they adequately account for the *known* turbulent mixing behavior. In the case of Schmidt number effects, the issue is particularly important, inasmuch as those effects are a direct manifestation of, and provide important clues to, the role of the small mixing scales of the problem, which must be correctly accounted for by turbulent mixing models.

The Mixing Transition

The flow in a two-dimensional shear layer issuing from a smooth splitter plate with a sharp trailing edge and low-turbulence-level free-streams originally develops as two-dimensional flow. This flow is characterized by large, two-dimensional, vortical structures (e.g., Refs. 73, 74), but is susceptible to a three-dimensional instability mode of counter-rotating streamwise vortices,[11,75–77,68,78–83] which spawn the transition to three-dimensional, fully developed turbulent flow, leading to substantial increases in the mixed fluid fraction.[11,84,35,18,8] This is illustrated in the liquid-phase shear-layer flow laser-induced fluorescence data,[8] reproduced in Fig. 11, recorded before ($Re \approx 2 \times 10^3$) and after ($Re \approx 2.3 \times 10^4$) the mixing transition, respectively. Note that, in the pre-mixing-transition data, the entrained fluids participate in the large-scale motion but remain essentially unmixed. Under these conditions, the surface-to-volume ratio of the two-dimensional interfacial area between the two entrained fluids is relatively small. In particular, when multiplied in water ($\mathcal{D} \approx \nu/10^3$) with the small local transverse diffusion thickness straddling this interface, i.e.,

$$\lambda_{\mathcal{D}} \propto \sqrt{\mathcal{D}/\sigma} \,, \tag{26}$$

where σ is the (local) strain rate (see Refs. 85; 86; and Ref. 87, Sec. 2.2), it yields a negligible mixed fluid volume fraction $\delta_{\mathrm{m}}/\delta$ within the shear layer width $\delta(x)$.

The large increase in interfacial area following the mixing transition changes this tally, resulting in a mixed fluid fraction, under these conditions, of

$$\delta_{\mathrm{m}}/\delta \approx \begin{cases} 0.26 \,, & \text{in water;} \\ 0.49 \,, & \text{in gas-phase flow.} \end{cases} \tag{27}$$

We will substantiate these values later.

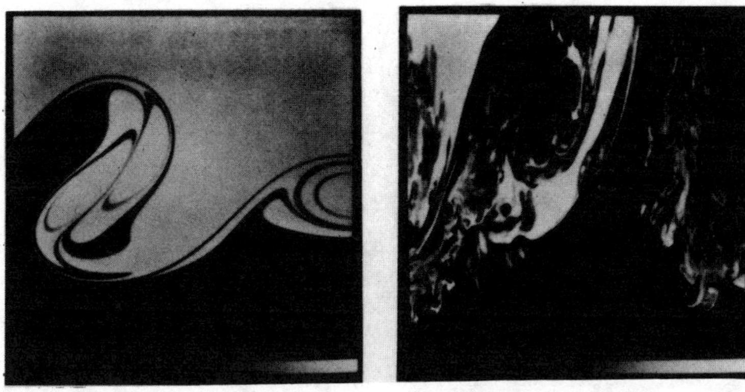

Fig. 11 Liquid-phase shear-layer mixing digital (y, t) image data[8] at a fixed streamwise location x. Left image: pre-mixing-transition ($Re \approx 2 \times 10^3$). Right image: post-mixing-transition ($Re \approx 2.3 \times 10^4$).

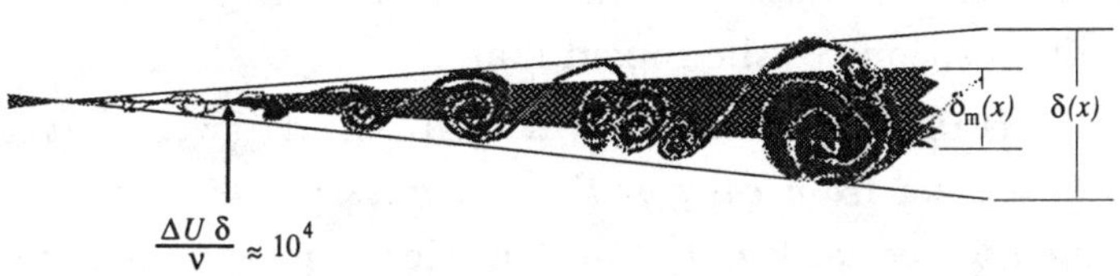

Fig. 12 Sketch of shear layer and gas-phase mixed fluid thickness growth through the mixing transition.

It is interesting that the growth rate of the shear layer does not appear to respond to this mixing transition, suggesting that it is dominated by the two-dimensional large-scale dynamics (see Fig. 3 and also discussion in Refs. 77 and 83). A sketch of the evolution of $\delta(x)$ and $\delta_{\mathrm{m}}(x)$ through the mixing transition appears in Fig. 12.

It is not clear, at this time, how this picture is altered by compressibility effects, or even whether the criterion of a minimum local Reynolds number of 10^4 will be good when the convective Mach numbers become large. The depressed growth rate of the two-dimensional Kelvin-Helmholtz disturbances, discussed in the previous section, may well alter the environment in which the three-dimensional motions develop, which are vital for the large interfacial area generation (see also Refs. 88–90). At higher convective Mach numbers, shocks can certainly be expected to play an important role in this process. Whether that role enhances a transition to three-dimensionality and improved mixing must also await future investigations.

Entrainment Ratio

We can think of the growth of the shear-layer region $\delta(x)/x$ as the increasing participation of freestream fluid in the turbulent process, i.e., the *entrainment*, as the downstream distance from the splitter plate increases. In this context, the preceding discussion on shear-layer growth addresses the *total* entrainment flux from each of the two freestreams, without regard as to the relative amounts from each freestream, i.e., the entrainment flux *ratio*. It is clear, however, that the entrainment flux ratio, that is supplied to the mixing processes of turbulence, must be taken into account in the tally of the resulting range of compositions of the mixed fluid.

An important conclusion drawn by Konrad[11] was that *a shear layer entrains fluid from each of the two freestreams in an asymmetric way*, even for equal freestream densities. In particular, for equal freestream densities ($s = 1$) and a freestream speed ratio of $r = 0.38$, Konrad estimated a volume flux entrainment ratio of $E_v \approx 1.3$. For a freestream density ratio of $s = 7$ (high speed He, low speed N_2) and the same velocity ratio, he estimated an entrainment ratio of $E_v \approx 3.4$.

Brown[12] proposed an estimate for the entrainment ratio based on the freestream velocity ratio, as seen from the frame of the large scale structures, i.e.,

$$E_v \approx r_c = \frac{U_1 - U_c}{U_c - U_2} \approx s^{1/2}$$

(cf. Eq. 13). Although the density ratio dependence of this proposal is in accord with the *ratio* of the two E_v experimental estimates of Konrad, i.e., $3.4/1.3 \approx 2.6 \approx \sqrt{7}$, it cannot account for the asymmetric entrainment ratio that was observed with equal freestream densities.

This behavior can be understood in terms of the upstream/downstream asymmetry that a given large-scale vortical structure sees in a spatially growing shear layer and the fact, also noted by Fiedler[91] in a different context, that a vortex entrains from each stream from its "lee side" (see Fig. 13). For incompressible flow, simple arguments based on the symmetry of the flow in the large-scale structure convection frame, suggest that, for a similarly growing shear layer, the volume flux entrainment ratio can be estimated by the expression

$$E_{\mathrm{v}} \approx \frac{U_1 - U_{\mathrm{c}}}{U_{\mathrm{c}} - U_2}\left(1 + \frac{\ell}{x}\right), \tag{28}$$

where ℓ/x is the large-structure spacing-to-position ratio.[13] In this expression, the quantity in parentheses is always greater than unity and the consequence of the spatial growth of the shear layer and the self-similarly increasing large-structure spacing with streamwise distance. It would be equal to unity for a temporally growing layer. Fitting the available data, one finds that the relation ($r = U_2/U_1$)

$$\frac{\ell}{x} \approx C_\ell \frac{1-r}{1+r}, \qquad C_\ell \approx 0.68, \tag{29}$$

is a good representation for ℓ/x, independently of the freestream density ratio.

We argued earlier that, for incompressible flow, $r_{\mathrm{c}} = (U_1 - U_{\mathrm{c}})/(U_{\mathrm{c}} - U_2) \approx s^{1/2}$. Consequently, for $s = 1$ and Konrad's freestream speed ratio of $r = 0.38$, we estimate (Eq. 28)

$$E_{\mathrm{v}}\,(r = 0.38,\ s = 1) \approx (1 + \ell/x) \approx 1.3,$$

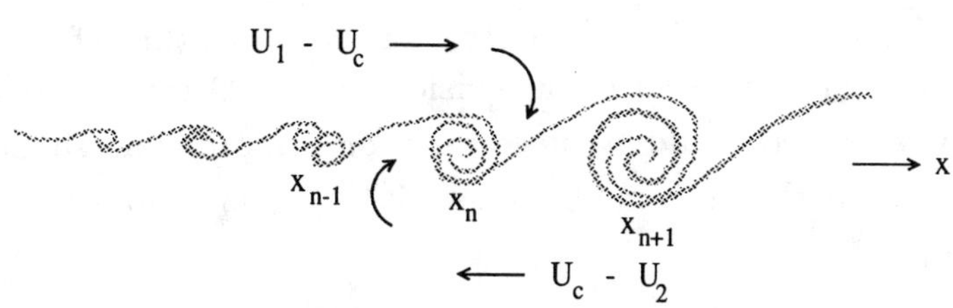

Fig. 13 Large-structure array and induction velocities in vortex convection frame.

while, for the He/N_2 shear-layer data at the same speed ratio,

$$E_v(r = 0.38,\ s = 7) \approx 7^{1/2} \times 1.3 \approx 3.4 ,$$

in rather good agreement with Konrad's experimental values.

For incompressible flow, the *mass* flux entrainment ratio E_m can also be similarly estimated, i.e.,

$$E_m = \frac{\rho_1}{\rho_2}\left(\frac{U_1 - U_c}{U_c - U_2}\right)\left(1 + \frac{\ell}{x}\right) , \tag{30a}$$

while for gas-phase flows, the *molar* entrainment ratio E_n would be given by,

$$E_n = \frac{\mathcal{M}_2}{\mathcal{M}_1} E_m , \tag{30b}$$

where $\mathcal{M}_j$ denotes the molecular mass of the gas comprising the jth freestream.

The arguments that led to the expression for the entrainment ratio for incompressible flow (Eq. 28) should also be useful for compressible flow, noting that, in this case, U_c must be computed accordingly (i.e., Eq. 14 or 18, as appropriate). Additionally, in the presence of shocks, one should not rely on the volumetric entrainment ratio E_v but rather on the mass entrainment ratio E_m; the product ρu is conserved across a (normal) shock. Nevertheless, we recognize that in the presence of shocks borne by one freestream, or the other, but not both, the underlying symmetry of the flow, in the large-scale structure convection frame, is lost (e.g., Figs. 6). As a consequence, the correct expression for E_m will also include an as yet undetermined multiplicative factor, of order unity, that captures this effect.

We can also expect that a revision of Eq. 29 for the spacing-to-position ratio ℓ/x would be necessary for compressible shear layers. A first guess is that ℓ/x might scale with δ/x, as it does for subsonic flow, i.e.,

$$\frac{\ell}{x}[r,\ s;\ M_{c1}^{(i)} \to 0] \propto \frac{\delta}{x}[r,\ s = 1;\ M_{c1}^{(i)} \to 0] ,$$

with a plausible extension of the form, as was assumed by Dimotakis and Hall,[92]

$$\frac{\ell}{x}[\,r,\ s;\ M_{c1}^{(i)}\,] \approx C_\ell\,\frac{1-r}{1+r}\,f[\,M_{c1}^{(i)}\,]\ , \tag{31}$$

where $f[\,M_{c1}^{(i)}\,]$ is an estimate of the Papamoschou and Roshko compressibility effect in the shear-layer growth, e.g., Eq. 17.

If one were to assess the relative importance of these different effects for compressible shear layers, one effect is likely to dominate. It is the asymmetry in the convective Mach numbers for flows that support shocks on one of the shear-layer freestreams (recall Fig. 5b) and the relative freestream speed ratio, r_c (Eq. 22), in the turbulent-structure convective frame. This can be illustrated by considering the $M_1 = 1.5\,\mathrm{He}$, $M_2 = 0.3\,\mathrm{N_2}$ supersonic shear layer, documented by Hall *et al.*[48] by way of example. For this shear layer, we might have predicted a relative velocity ratio, based on an isentropic estimate for U_c, of

$$\frac{U_1 - U_c^{(i)}}{U_c^{(i)} - U_2} \simeq 2.1\ , \quad \text{vs.} \quad \frac{U_1 - U_c}{U_c - U_2} \simeq 0.36\ , \tag{32}$$

using the convection velocity estimate of $U_c \approx 880\,\mathrm{m/s}$ that is suggested by the data and also derived using the stationary flow ansatz described earlier (see Ref. 58 for more details). The two estimates of this important factor span unity and differ by a factor of 6. Such a layer, rather than being high-speed fluid rich may be *low*-speed fluid (depending on the other factors that enter in the entrainment ratio estimate).

In the context of mixing and the resulting range of mixed fluid compositions within the shear layer, it is useful to define a *conserved scalar* ξ which denotes the mole fraction of high speed stream fluid in the molecularly mixed fluid (e.g., Ref. 93). Accordingly, $\xi = 0$ corresponds to pure low-speed-stream fluid, $\xi = 1$ represents pure high-speed-stream fluid, and $\xi = 1/2$ represents a 50:50 mixture. In this notation, the entrainment ratio E measures the flux of $\xi = 1$ fluid entering the turbulent mixing region, per unit flux of $\xi = 0$ fluid.

The asymmetric entrainment ratio suggests a zeroth order model for mixing in a two-dimensional shear layer, which entrains $\xi = 1$ and

$\xi = 0$ fluid from each of the freestreams at a ratio E, respectively, that was employed by Konrad[11] in his discussion of his concentration fluctuation data. The two entrained fluids are mixed by the efficient action of turbulence and can be expected to tend toward a mixed fluid composition of

$$\xi_E \equiv \frac{E_n}{E_n + 1} . \tag{33}$$

A useful cartoon is that of a bucket being filled by two faucets running with unequal flow rates, as a laboratory stirring device mixes the effluents (Fig. 14). We can also think of a hot/cold water faucet and the (average) temperature in the bucket; it is only a function of the *ratio* of the two flow rates. For all the complexity of the ensuing turbulent flow, we would expect to find a probability density function (PDF) of mixed fluid compositions in the bucket clustered around the value of the mixture fraction given by Eq. 33, where E corresponds to the ratio of the flux from each of the two faucets. Fluid homogenized at this composition is an important component in the mixing model by Broadwell and Breidenthal,[86] as we will discuss later.

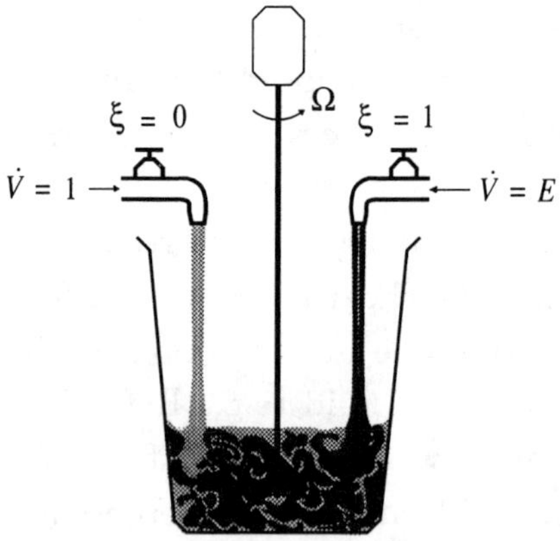

Fig. 14 Stirred bucket mixing. E corresponds to the right/left faucet flow-rate ratio. Note that as the stirring rate is increased, at fixed faucet flow rate, $p(\xi)\,d\xi \rightarrow \delta_D(\xi - \xi_E)\,d\xi$, where $\xi_E = E/(1+E)$.

One can appreciate that the range of compositions one should expect to encounter in the bucket depends on the relative rate of inflow to mixing. One can also appreciate that as we lower the combined faucet flow rate, keeping the ratio and the stirring fixed, we can expect the mixed fluid to be homogenized with a composition PDF tending to a Dirac delta function centered at ξ_E, i.e.,

$$p(\xi)\,\mathrm{d}\xi \;\rightarrow\; p_{\mathrm{H}}(\xi)\,\mathrm{d}\xi \;=\; \delta_{\mathrm{D}}(\xi - \xi_E)\,\mathrm{d}\xi \;, \tag{34}$$

in the limit. Similarly for a fixed faucet flow rate as the stirring rate is increased (see discussion, for example, in Ref. 94).

The effects of the asymmetric entrainment ratio can be seen in the PDF measurements of Konrad[11] made in a gas-phase, matched freestream density [$^1\!/_3$ He : $^2\!/_3$ Ar] / N_2 shear-layer, using an aspirating probe.[95] See Fig. 15. Note that the most probable value of the high-speed fluid fraction ξ, denoted as $C(N_2)$ in the figure, is very close to $\xi_E \approx E/(1+E) = 0.57$, corresponding to the estimated entrainment ratio of $E \approx 1.3$ for this case (see Ref. 11 for details).

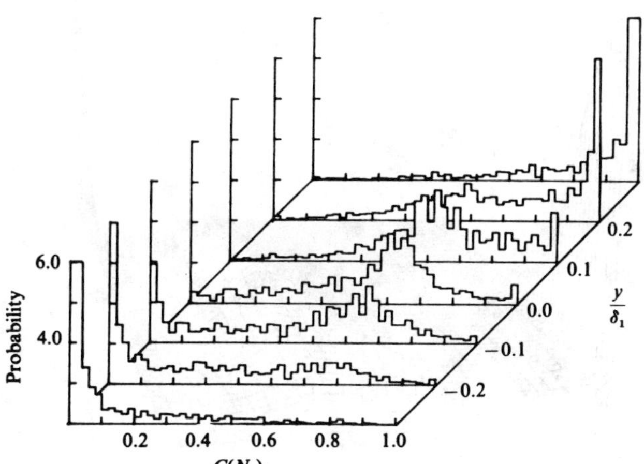

Fig. 15 Gas-phase PDF measurements in a matched density, $r = U_2/U_1 \approx 0.4$ shear layer.[11] High-speed fluid mixture fraction ξ is denoted by $C(N_2)$.

Similar measurements were obtained in a shear layer in water, at the same velocity ratio, in which the PDF was measured using laser-induced fluorescence techniques.[8] The most probable value of the composition is again very close to the value $\xi_E \approx 0.57$ (Fig. 16).

Notable in both sets of measurements is that this most probable value is observed across the whole transverse extent of the shear layer. This can be understood in terms of the circumferential velocities, in the frame of the large-scale vortical structures, which can transport a fluid element across the shear layer, before it has much chance to alter its own internal composition. A comparison of the gas-phase and liquid-phase data suggests that, as expected at the higher Schmidt numbers in the latter case, this is more the case in the liquid than in the gas phase. It should be noted, however, that lower resolution in the gas-phase measurements could account for some of the observed trends.

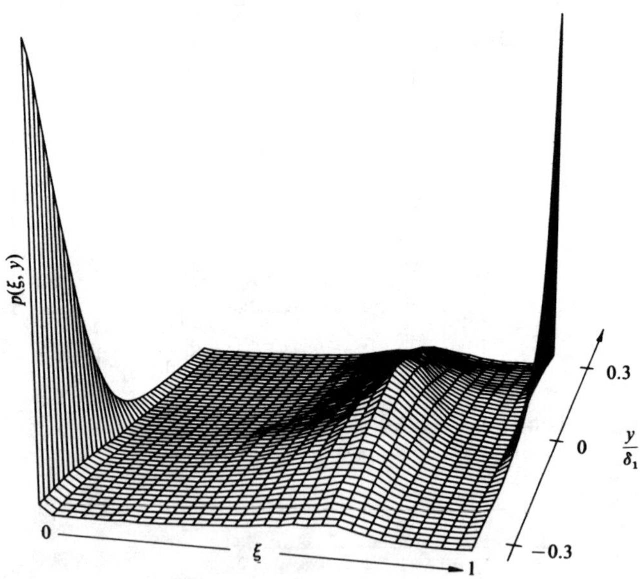

Fig. 16 High-speed fluid mixture fraction PDF measurements[8] in a liquid-phase (matched density), $r = U_2/U_1 \approx 0.38$ shear layer at $Re = 2.3 \times 10^4$. PDF computed from data that yielded the post-mixing-transition image in Fig. 11.

These observations are at variance with the results of gradient-transport-based PDF modeling efforts (e.g., Refs. 96 and 97). Those models yield a most probable value of ξ *for the mixed fluid* that is close to the local value of the mean mixture fraction profile (mixed and unmixed), i.e., $\bar{\xi}(y)$.

Direct consequences of the asymmetric entrainment ratio, as reflected in the mixed fluid composition, are to be found in chemically reacting shear layers. In particular, in the case in which the reactant concentrations are not carried by the freestreams at the stoichiometric ratio, which side carries the *lean* reactant can make an easily discernible difference in the amount of chemical product formed in the layer. This was illustrated in liquid-phase "flip" experiments[98] in which mixed fluid in the range of compositions $0 \leq \xi < 0.36$ was compared in a complementary run to mixed fluid in a range of compositions $0.64 < \xi \leq 1$, using a pH-sensitive, laser-induced fluorescence technique (see Fig. 17). The "chemical product" is found to be many times larger in the second case, which marks high values of ξ in the local composition. Note also that there is no systematic gradient in the labeled mixed fluid concentration across the shear-layer normalized width δ/x. It should be emphasized that these experiments, which were designed to illustrate the potential of this effect, were conducted in the mixing transition region, where the remnants of the much larger asymmetries in the initial roll-up have yet to be amor-

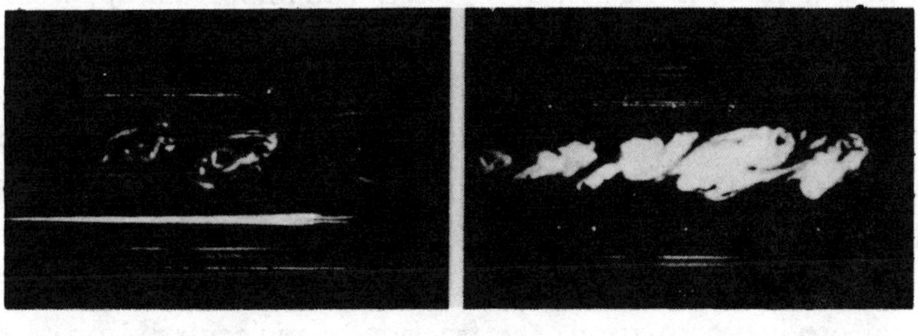

Fig. 17 Mixing transition laser-induced fluorescence "flip" experiment. Left: Fluorescence from mixed fluid compositions in the range $0 \leq \xi < 0.36$. Right: fluorescence from the range $0.64 < \xi \leq 1$. From Ref. 98.

tized with entrainment at the asymptotic values of the entrainment ratio (e.g., Eq. 28). See pre-mixing-transition image in Fig. 11 data, as well as Fig. 12 in Ref. 8 and related discussion. Similar conclusions were drawn from gas-phase measurements in the mixing transition region,[99] and from computational modeling of the shear layer at low Reynolds number.[34]

A sufficient distance beyond the mixing transition, the observed asymmetries are consistent with the asymptotic values of E and the associated "tilt" in the mixed fluid composition PDF in Fig. 16. See chemically reacting data at higher Reynolds numbers in Ref. 8, Figs. 16 and 17. These and other mixing issues in subsonic turbulent shear layers will be discussed in the context of chemically reacting shear layers, which must be relied on for data at the high Reynolds numbers of interest here.

IV. Chemical Reactions: $\delta_{\mathrm{P}}/\delta$

In the case in which the entrained fluids are not premixed and can react, the associated chemical product formation can obviously proceed no faster than the rate at which the reactants are mixed on a molecular scale by the turbulent flow. Considering a vertical slice of the turbulent region of streamwise extent $\mathrm{d}x$, located at some downstream location x, the (expected) mixed fluid fraction $\delta_{\mathrm{m}}/\delta$ within the transverse extent $\delta(x)$ of the turbulent region occupied by molecularly mixed fluid (in the mean) represents an important upper bound for the expected chemical product fraction $\delta_{\mathrm{P}}/\delta$ within the layer at that location. In the case of combustion of nonpremixed reactants, it also bounds the heat release corresponding to the amount of chemical product formed.

In the limit of fast chemistry, i.e., at a chemical kinetic rate sufficiently large so as not to serve as the limiting process in the rate of chemical product formation, the fraction of molecularly mixed fluid that is converted to chemical product, i.e., $\delta_{\mathrm{P}}/\delta_{\mathrm{m}}$, depends on the resulting PDF, i.e., $p(\xi)\,\mathrm{d}\xi$, of molecular mixture compositions within the turbulent region. In particular, it depends on the distribution of mixture fractions ξ of high-speed fluid to low-speed fluid in the molecularly mixed fluid, relative to the stoichiometric mixture fraction ξ_ϕ required for complete consumption of the available (entrained) reactants, as we will discuss later.

If the chemical kinetics are not sufficiently fast by the previous measure, the chemical product formation will lag behind the rate at which the reactants are mixed on a molecular scale by the turbulence. As a consequence, $\delta_{\mathrm{P}}/\delta_{\mathrm{m}}$ will be smaller, depending also on the ratio (Damköhler number)

$$Da \equiv \frac{\tau_{\mathrm{m}}}{\tau_{\mathrm{ch}}}, \tag{35}$$

of the expected time τ_{m} required for mixing, to the time τ_{ch} required to complete the ensuing chemical reactions. What is also important from a diagnostics vantage point is the recognition that, for chemical/flow systems that *can* be regarded as kinetically fast, i.e., in the limit of $Da \to \infty$, measurements of the chemical product volume fraction $\delta_{\mathrm{P}}/\delta$ can be combined to provide us with reliable estimates of molecular mixing and the mixed fluid fraction $\delta_{\mathrm{m}}/\delta$, as well as the distribution of compositions of the molecularly mixed fluid, as we will also discuss later. This often obviates the need for direct measurements of these quantities, which would, for the most part, have been anyway infeasible at the high Reynolds numbers of interest here. Direct computations fare no better, as the behavior of fast chemical systems can result in reaction zones that are even thinner than the expected diffusion scales (e.g., Eq. 26), under these conditions, and an intractably stiff problem numerically.

In the context of mixing, we will restrict the discussion that follows to the behavior in the limit of fast chemical kinetics ($Da \to \infty$). Chemical product formation for finite Damköhler numbers, however, is important theoretically inasmuch as it depends not only on the state of the flow at the measurement location but also on the flow history, which, in turn, prescribes the local molecular mixing (scalar dissipation) rate. See Bilger (Ref. 100, Sec. 2.5) and Williams (Ref. 101, Sec. 10.2.4) for a general discussion. It is also important from an applications vantage point, as the impetus for ever-increasing flight speeds is forcing us to consider chemical product formation at higher flow velocities and Mach numbers. In that regime, chemical product formation may, perforce, ultimately be limited by the fixed available chemical kinetic rates.

Data on the Damköhler number dependence of the product volume fraction in a gas-phase, subsonic shear layer were documented by Mungal and Frieler.[102] An analysis of these data, in terms of the Broadwell-Breidenthal-Mungal model we will discuss later, can be found in Ref. 103. An attempt to incorporate a more realistic account of the complex combustion chemistry was made by Dimotakis and Hall,[92] using the bucket zeroth-order mixing model described earlier (Fig. 14). The reader is directed to those references for an account.

Dependence on the Stoichiometric Mixture Ratio

Consider the idealized case of the high-speed stream carrying a reactant at a concentration (mole fraction) X_{01}, and the low-speed stream carrying a reactant at a concentration X_{02}, which can react infinitely fast to form a chemical product, associated with an enthalpy release $\Delta\mathcal{H}$. An important quantity, in this context, is the *stoichiometric mixture ratio* ϕ, defined as the volume (number of moles) of high speed fluid that carries sufficient reactants to consume a unit volume (mole) of low speed fluid, i.e.,

$$\phi \equiv \frac{X_{02}/X_{01}}{(X_{02}/X_{01})_{\text{st}}} , \tag{36}$$

where the subscript "st" in the denominator denotes a stoichiometric mixture. For example, a (free-stream) stoichiometric mixture ratio of $\phi = 4$ implies that a mixture of four parts of high-speed fluid per part of low-speed fluid is required for complete consumption of all reactants. Accordingly, complete consumption of all reactants will occur at a stoichiometric (high speed fluid) mixture *mole fraction*

$$\xi_\phi = \frac{\phi}{\phi + 1} . \tag{37}$$

We can see that a mixture fraction of $\xi < \xi_\phi$, for example, will be lean in high-speed stream reactants and result in unreacted low-speed stream reactants. Similar definitions can also be employed on a mass basis (e.g., Ref. 104, Sec. 1.8).

Consider, for example, the chemical reaction between hydrogen and fluorine which, in the limit of fast chemistry, we can simplify as a one-step reaction (see Ref. 7 for details)

$$H_2 + F_2 \longrightarrow 2\,HF \ , \qquad \Delta\mathcal{H} = -\,130\,\text{kcal/mole} \ , \qquad (38)$$

and which was used in many of the experiments that will be cited below. A mixture of 1% H_2 in 99% N_2 and an equal volume of 1% F_2 in 99% N_2 is stoichiometric and will result in an adiabatic (flame) temperature rise owing to the heat released of $\Delta T_f = 93$ K. A shear layer with a high-speed stream fluid composed of 4% H_2 + 96% N_2, and a low-speed stream fluid of 1% F_2 + 99% N_2 would be characterized by $\phi = 1/4$, i.e., $1/4$ parts of high-speed fluid must be mixed per part of low-speed fluid for complete reaction.

For equal heat and species diffusivities, i.e., for Lewis numbers $Le \equiv \kappa/\mathcal{D} = Sc/Pr = 1$, the adiabatic flame temperature rise $\Delta T_f(\phi)$ is the highest temperature rise that can be observed in the flow and serves as a convenient normalization of the observed mean temperature rise $\Delta T(y,\phi)$ in the reaction zone. Note that in a mixture in which X_{01} is kept constant and the stoichiometric mixture ratio ϕ is changed by varying X_{02} (Eq. 36), keeping the heat capacities constant in the process, the dependence of the adiabatic flame temperature rise on ϕ is given by

$$\Delta T_f(\phi) = \frac{2\phi}{\phi+1}\,\Delta T_f(1) = 2\,\xi_\phi\,\Delta T_f(1) \qquad (39)$$

(recall Eq. 37).

Experimental data for the normalized mean temperature rise, for matched freestream density ($s = 1$), a freestream velocity ratio of $r \approx 0.4$, gas-phase reacting shear layers at low heat release, are plotted in Fig. 18 for $\phi = 1/8$, 1, and 8. The plotted quantity reflects the local mean fraction of the total chemical product (heat release) possible under the circumstances. There is a shift towards the lean side of the location of the peak mean temperature rise. There is also a marked asymmetry in the total amount of product (heat release) between the low-ϕ and the high-ϕ runs, which, in view of the relation between the entrainment ratio E and the (required) stoichiometric mixture ratio ϕ, is clear in this context. In particular, for $\phi < E_n$, for example, fluid homogenized at the entrainment ratio will be low-speed stream reactant lean ($\xi_\phi < \xi_E$) and result in unconsumed high-speed stream reactants. The maximum amount of product is expected at $\phi \approx E_n$, with more product for $\phi \gg 1$ than for $\phi \ll 1$, for $E_n > 1$ (recall that $E_n \approx 1.3$ under these conditions). Note also that, consistent with our observation that a substantial fraction of fluid is unmixed within the shear layer, the mean temperature rise is everywhere less than $0.65\,\Delta T_{\mathrm{f}}$.

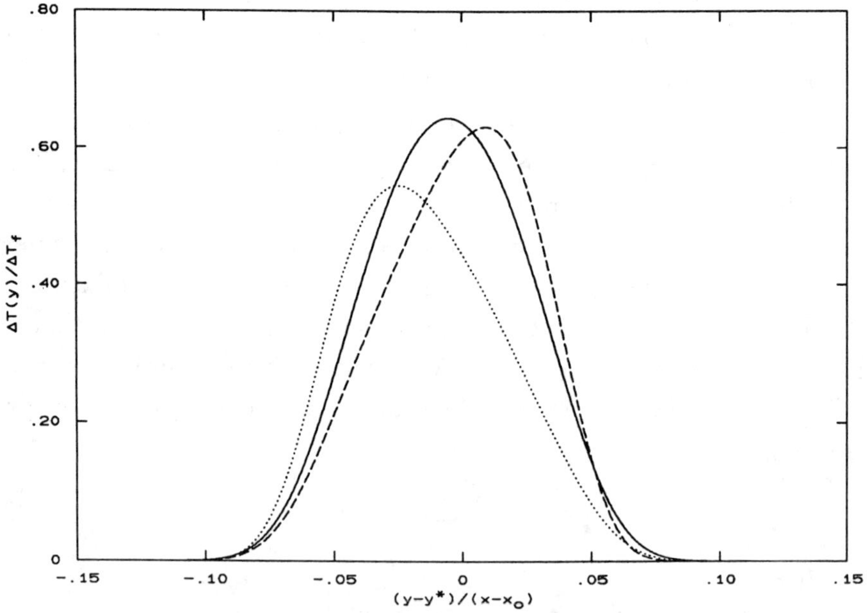

Fig. 18 Gas-phase shear layer: Normalized chemical product at low heat release. Solid line: $\phi = 1$; dashed line: $\phi = 8$; dotted line: $\phi = {}^1/_8$ (Ref. 7; $s = 1$, $r \approx 0.4$). Note the peak temperature rise tilt towards the lean side for $\phi \neq 1$ and the larger total chemical product for $\phi \gg 1$ relative to $\phi \ll 1$, corresponding to an entrainment ratio of $E_n > 1$.

It is interesting to compare these results to the corresponding data from a liquid-phase ($s = 1$), chemically reacting shear layer, at the same freestream speed ratio ($r = 0.4$). These are depicted in Fig. 19, for $\phi = {}^1/_{10}$ and $\phi = 10$. Note the reduced amount of product relative to the gas-phase results, the asymmetry between the high-ϕ and low-ϕ runs ($E_n \approx 1.3$ here also), but note that the tilt towards the lean side is no longer there. We can trace these differences to Schmidt number effects on the basis of our preceding discussions. In particular, the reduction in the total chemical product is attributable to the reduction in the amount of molecularly mixed fluid at the higher Schmidt number (lower species diffusivity). The absence of a tilt of the peak mean temperature toward the lean side is the result of the delayed (slower) molecular mixing, which allows a longer Langrangian time for homogenization to occur at the larger-than-diffusion scales across the whole shear-layer transverse extent, owing to the large-structure motion (recall discussion of data in Fig. 16). See also discussion in Ref. 103.

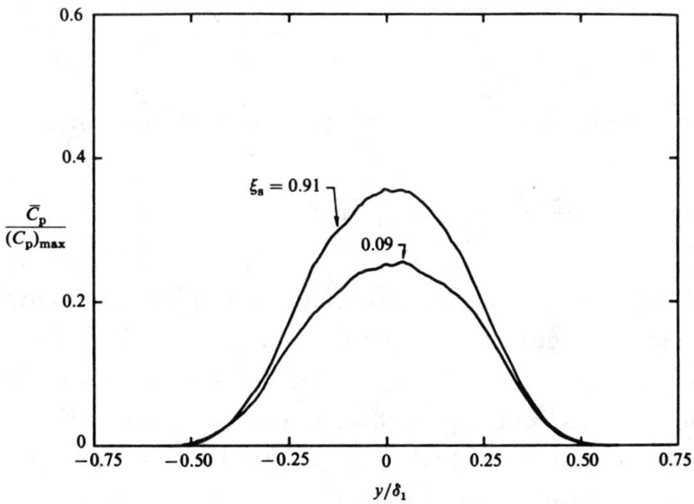

Fig. 19 Liquid-phase shear layer: Normalized chemical product at low heat release. for $\phi = {}^1/_{10}, 10$ (Ref. 8; $s = 1$, $r \approx 0.4$). Note symmetric chemical product distribution for both $\phi \gg 1$ and $\phi \ll 1$. ξ_s in the figure denotes ξ_ϕ (Eq. 37).

Relation to the PDF and Schmidt Number Effects

These results would all be derivable from the local PDF $p(\xi, y)$ of the mixture fraction at the measuring station at x, if that were available. In particular, the product (or heat release) that can be produced corresponding to a particular value of the mixture fraction ξ is easily computed by assuming complete consumption of the lean reactant. This yields two straight lines in ξ, joined at ξ_ϕ, where the normalized product is equal to unity, i.e.,

$$\theta(\xi;\xi_o) = \begin{cases} \dfrac{\xi}{\xi_\phi}\,, & \text{for } 0 \le \xi \le \xi_\phi\ ; \\[2ex] \dfrac{1-\xi}{1-\xi_\phi}\,, & \text{for } \xi_\phi \le \xi \le 1 \end{cases} \tag{40}$$

(see, for example, Ref. 104, Sec. 1.9). This dependence is depicted in Fig. 20 for $\phi = {}^1/_8$, 1, and 8.

The average chemical product volume (mole) fraction δ_P/δ can be computed as the integral of the normalized product profile in the interior of the shear layer,

$$\frac{\delta_P(\xi_\phi)}{\delta} = \frac{1}{\delta}\int_{-\infty}^{\infty} \frac{\Delta T(y,\phi)}{\Delta T_f(\phi)}\,\mathrm{d}y\ , \tag{41a}$$

or as an integral over the PDF of mixture fractions, since

$$\frac{\Delta T(y,\phi)}{\Delta T_f(\phi)} = \int_0^1 \theta(\xi;\xi_\phi)\,p(\xi,y)\,\mathrm{d}\xi\ , \tag{41b}$$

where $\theta(\xi;\xi_\phi)$ is the triangular normalized product function (Eq. 40). Experimental values of this quantity are included in Fig. 21 for a gas-phase reacting shear layer at $Re = 6.4 \times 10^4$, as a function of the stoichiometric mixture fraction ξ_ϕ (Ref. 7). Also included in that figure is a point at $\phi = 10$ (Ref. 8, $\xi_\phi = 0.91$) for a liquid-phase shear layer at a comparable Reynolds number ($Re = 7.8 \times 10^4$).

The triangular normalized product function $\theta(\xi;\xi_\phi)$ suggests the use of chemically reacting experiments to estimate some statistics of the mixed fluid PDF. In particular, for $\xi_\phi \to 0$ and $\xi_\phi \to 1$, the

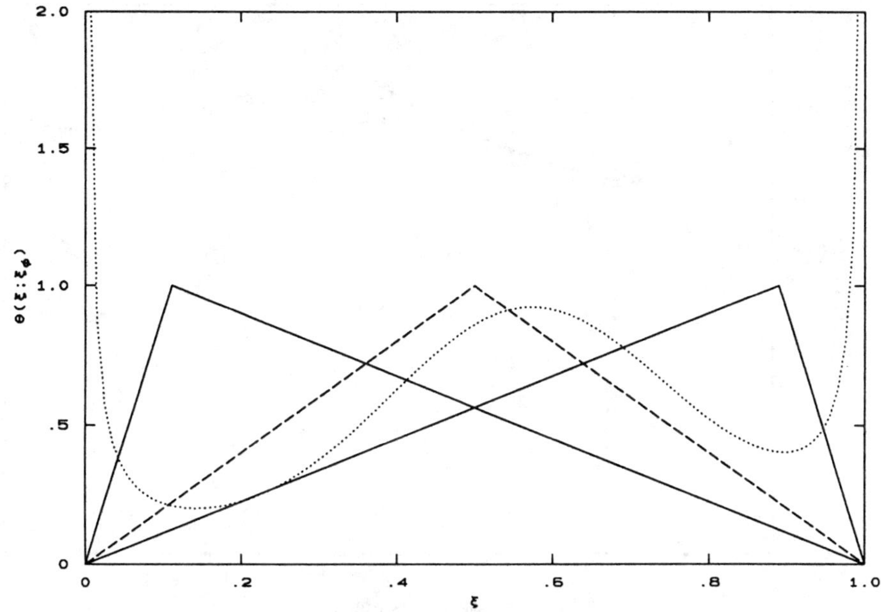

Fig. 20 Normalized chemical product function $\theta(\xi;\xi_\phi)$ for $\phi = 1/8$, and $\phi = 8$ (solid lines), and $\phi = 1$ (dashed line). PDF (dotted line) is sketched for reference, corresponding to $E_n \approx 1.3$ (cf. Fig. 18).

ratio of the corresponding product volume fractions can be used to estimate the average composition $\overline{\xi}_{\text{m}}$ *in the mixed fluid*. For a small $\xi = \xi_0 \to 0$, we find

$$\overline{\xi}_{\text{m}} \approx \frac{\delta_{\text{P}}(1-\xi_0)}{\delta_{\text{P}}(\xi_0) + \delta_{\text{P}}(1-\xi_0)} . \tag{42}$$

Using the experimentally determined liquid-phase values of

$$\frac{\delta_{\text{P}}(\xi_\phi)}{\delta} = \begin{cases} 0.125\,, & \text{at } \xi_\phi = \xi_0 = 0.09\,; \\ 0.165\,, & \text{at } \xi_\phi = 1 - \xi_0 = 0.91\,, \end{cases} \tag{43}$$

we then estimate a value of $\overline{\xi}_{\text{m}} \approx 0.57$.[8] This agrees with the value of $\xi_E = E/(E+1)$, calculated using the independently estimated value of the volume flux entrainment ratio $E \simeq s^{1/2}(1 + \ell/x) \approx 1.3$ (cf. Eqs. 28, 29, 13).

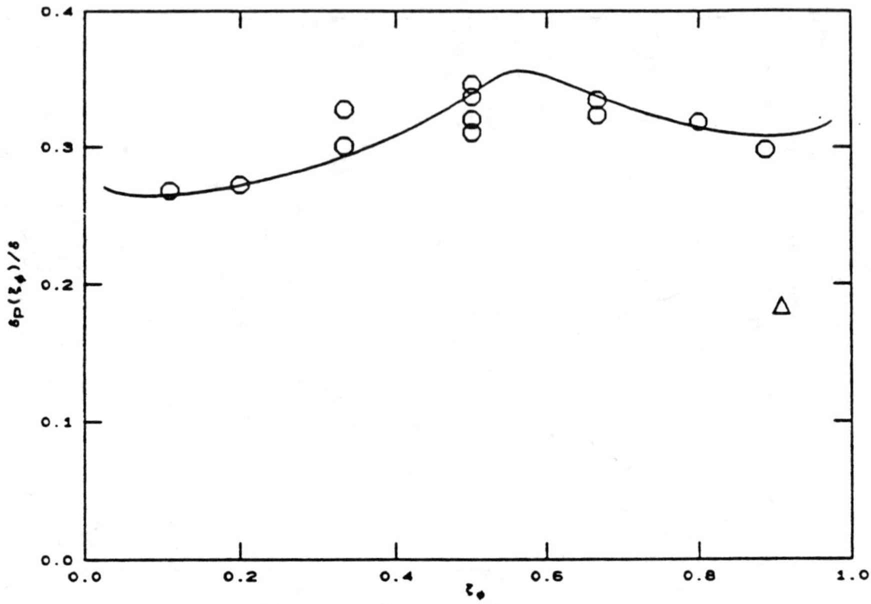

Fig. 21 Chemical product volume fraction δ_P/δ vs. ξ_ϕ for $r \approx 0.4$ and matched freestream densities. Circles: gas-phase data,[7] $Re = 6.4 \times 10^4$. Triangle: liquid-phase data,[8] $Re = 7.8 \times 10^4$. Smooth curve drawn to aid the eye.

This idea was also used to estimate the dependence of the mean mixed fluid composition in a recent set of experiments[105] in subsonic, low-heat-release, gas-phase shear layers with unequal freestream densities ($s \neq 1$), for which the expected asymmetries in the entrainment ratio can be large (Eqs. 13 and 28). The resulting data are plotted in Fig. 22 for freestream density ratios in the range of $0.1 < s < 4$, and compared to the estimated value of ξ_E, using the subsonic expression for the volume flux entrainment ratio discussed in the previous paragraph.

Comparing the triangular product functions for small and large ξ_ϕ, we also note that, except for omitting the endpoints, they are essentially complements of each other. Consequently, for the case of negligible heat release, we find

$$\frac{\delta_m}{\delta} \approx (1 - \xi_0) \left[\frac{\delta_P(\xi_0)}{\delta} + \frac{\delta_P(1 - \xi_0)}{\delta} \right] .$$

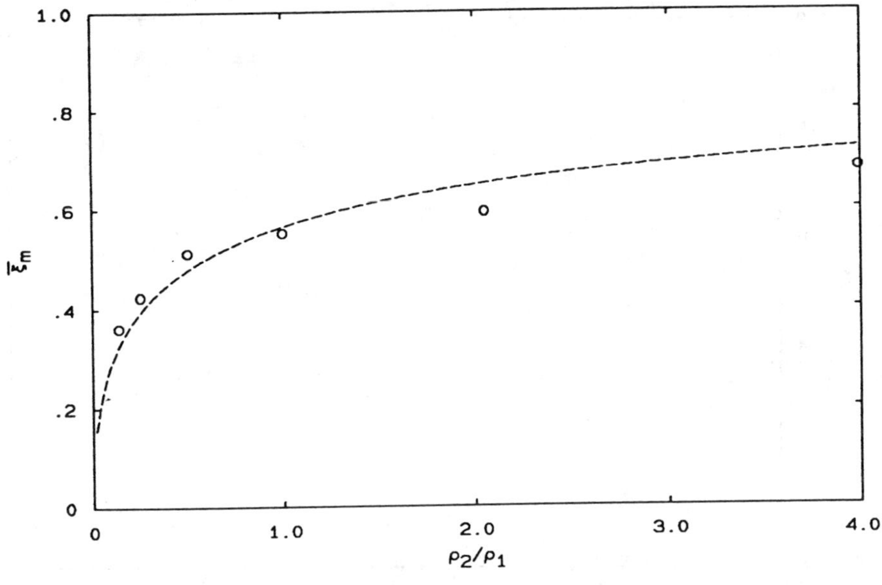

Fig. 22 Experimentally estimated mixed-fluid mixture fraction $\overline{\xi_m}$ as a function of the density ratio. Dashed line depicts the estimated dependence of ξ_E on the density ratio.[105]

This represents the mixed fluid fraction, if the edge contributions from the regions $0 \leq \xi < \epsilon$ and $1 - \epsilon < \xi \leq 1$ are excluded from the mixed fluid tally. In this approximation, $\epsilon \approx \xi_0/2$, corresponding to the gas-phase chemical reaction product function, and $\epsilon \approx \xi_0$ for the liquid-phase data. Using the values for the liquid-phase chemically reacting layer (Eq. 43), with $\epsilon \approx \xi_0 = 0.09$, we then estimate ($s = 1$)

$$\left(\frac{\delta_m}{\delta}\right)_{\text{liq}} \approx 0.26 \ . \tag{44a}$$

A similar calculation was also made using the results of the low-heat-release gas-phase data vs. freestream density ratio of Frieler and Dimotakis. Small dilatation corrections were applied to those data, which are of first order for this quantity. The results are plotted in Fig. 23 as a function of the freestream density ratio. It is significant that the mixed fluid fraction is found to be essentially independent of the density ratio, even as the mixed fluid composition depends rather strongly

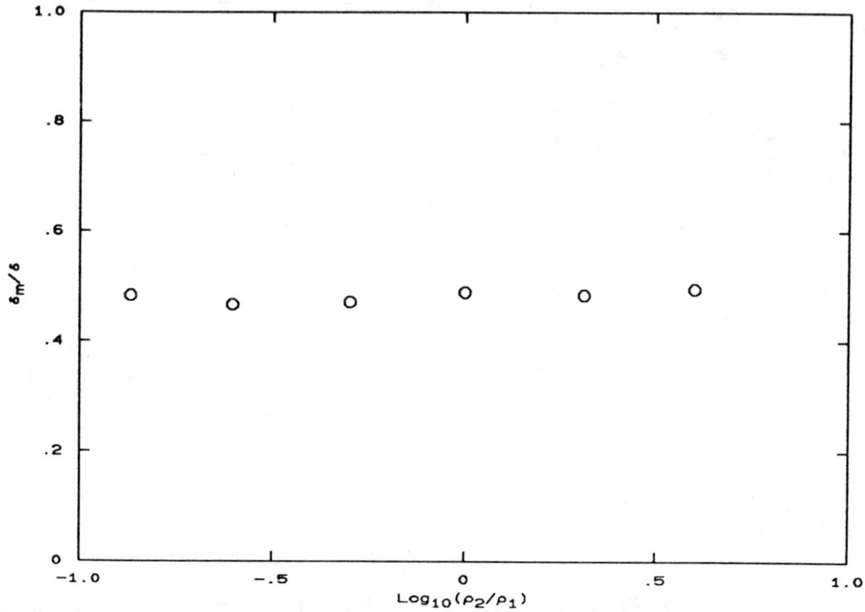

Fig. 23 **Mixed fluid fraction δ_m/δ as a function of the freestream density ratio $s = \rho_2/\rho_1$. Gas-phase data.[105]**

on it. The mixed fluid fraction derived from these data for matched freestream densities is then found to be (note that $\epsilon \approx \xi_0/2 \approx 0.05$)

$$\left(\frac{\delta_m}{\delta}\right)_{gas} \approx 0.49 \ . \tag{44b}$$

The estimates in Eqs. 44a and 44b were the values quoted in Eq. 27 for this quantity.

Reynolds Number Effects

The existing experimental evidence suggests that chemical product formation and the resulting chemical product mole fraction δ_P/δ observed at a station x is a (weak) function of the local Reynolds number, at least for gas-phase flows. Available gas-phase and liquid-phase data, for a range of stoichiometric mixture ratio ϕ, heat release (indicated by the adiabatic flame temperature rise ΔT_{f}), and Mach number are plotted in Fig. 24.

Although these data span a range of flow and chemical reaction parameters, one can discern that the product fraction $\delta_{\mathrm{P}}/\delta$ is found to decrease slowly with increasing Reynolds number for gas-phase shear layers. Liquid-phase results exhibit an even weaker Reynolds number dependence, if any, even though the lower Reynolds number value of $Re \approx 2.3 \times 10^4$ in those data may not be sufficiently above the mixing transition to be used as a reference for the comparison.

It should be noted that the shear layer thickness δ, at the measuring station x, was also changing with Reynolds number in these experiments. Although this variation was normalized out by taking the *ratio* of the product thickness δ_{P} and the shear layer (1%) thickness $\delta_{1\%} \approx \delta_{\mathrm{vis}}$ to compute the chemical product mole fraction, we

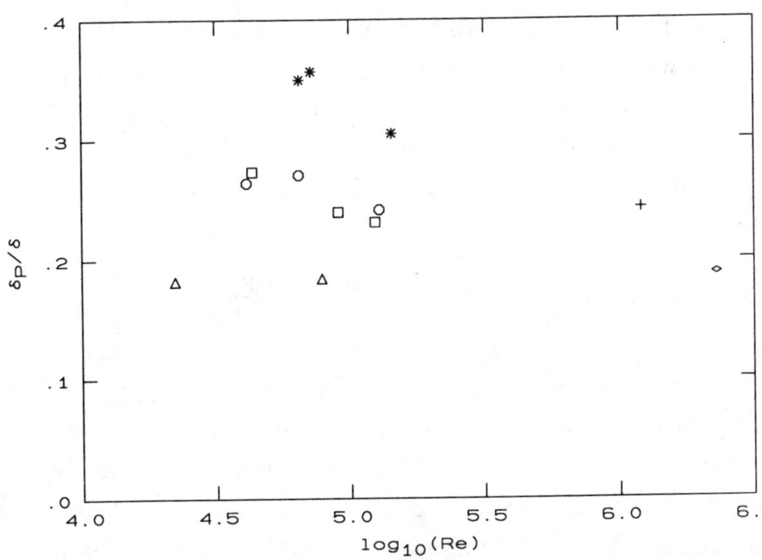

Fig. 24 Chemical product fraction vs. Reynolds number. Subsonic, gas-phase: circles (laminar boundary layer) and squares (turbulent boundary layer), $\phi = 1/8$, $\Delta T_f \approx 190\,\mathrm{K}$.[106] Stars: higher heat release, $\phi = 1$, $\Delta T_f \approx 370\,K$.[65] Supersonic: cross for $M_1 = 1.5\,N_2$, $M_2 = 0.3\,N_2$, $\phi = 1/4$, $\Delta T_{\mathrm{f}} \approx 300\,\mathrm{K}$; diamond for $M_1 = 1.5\,\mathrm{He}$, $M_2 = 0.3\,\mathrm{Ar}$, $\phi = 1/3$, $\Delta T_{\mathrm{f}} \approx 580\,\mathrm{K}$.[107] Liquid-phase: triangles, $\phi = 10$.[8]

should appreciate that the change in the shear-layer thickness was sometimes larger than the change in the estimated product volume fraction documented in Fig. 24 (recall Figs. 2 and 3 and related discussion).

The highest Reynolds number data included in Fig. 24 are derived from the measurements by Hall *et al.* in supersonic shear layers.[107] While it would appear that those values are consistent with the general trend, it is difficult to say, at this time, if the reason for the lower values of $\delta_{\mathrm{P}}/\delta$ is attributable to the higher Reynolds numbers, or to compressibility. Further experiments, directed at separating these two important effects, are required.

A proposal for an explanation of Reynolds number and Schmidt number effects was first made by Broadwell and Breidenthal.[86] The suggestion in that model was that the mixed fluid PDF can be modeled as a superposition of the $p_{\mathrm{H}}(\xi)$ PDF corresponding to the homogeneously mixed fluid in the bucket cartoon and the contribution from thin interfacial diffusion layers interspersed in the shear layer, and which separate pure $\xi = 0$ and $\xi = 1$ fluid. Some revisions and clarifications were made in the more recent discussion of this model by Broadwell and Mungal.[103] The Broadwell-Breidenthal-Mungal (BBM) model then yields for the mixed fluid PDF, i.e., for $\xi \neq 0$ and $\xi \neq 1$,

$$p(\xi)\,\mathrm{d}\xi \;\approx\; \left[C_{\mathrm{H}}\,\delta_{\mathrm{D}}(\xi - \xi_E) \,+\, \frac{C_{\mathrm{F}}}{\sqrt{Sc\,Re}}\,p_{\mathrm{F}}(\xi) \right] \mathrm{d}\xi \,, \tag{45}$$

where $\delta_{\mathrm{D}}(\xi)$ denotes the Dirac delta function and $p_{\mathrm{F}}(\xi)$ is the PDF of composition as would arise in a laminar strained interface ("flame sheet") between $\xi = 0$ and $\xi = 1$ interdiffusing fluids. The dependence on the Reynolds number and Schmidt number in this superposition arises from modeling the amount of mixed fluid taken as residing in the diffusive interfaces. The coefficients C_{H} and C_{F} are assumed to be constants of the flow and, in particular, independent of the Schmidt and Reynolds numbers.

The proposed Schmidt and Reynolds number dependence in the BBM model is equivalent to the assumption that interfacial diffusion-layer thicknesses can be accounted for by modeling the relevant strain rate σ as Reynolds-number-independent using the local outer flow variables, i.e., $\sigma \propto \Delta U/\delta$ (see Eq. 26), and that the associated expected diffusion interface surface-to-volume ratio is also independent of the Reynolds number. These authors suggest that the model should apply for $Re \gg 1$, $\sqrt{Sc\,Re} \gg 1$, and $\sqrt{Re} \gg \ln Sc$, even though the latter inequality would automatically be satisfied at the Reynolds numbers of interest here.[103]

Integrating the proposed model PDF over ξ, excluding the contributions of the unmixed fluid at $\xi = 0$ and $\xi = 1$, we then obtain, for the BBM model estimate of the mixed fluid fraction,

$$\frac{\delta_{\mathrm{m}}}{\delta} \approx C_{\mathrm{H}} + \frac{C_{\mathrm{F}}}{\sqrt{Sc\,Re}} \; . \tag{46}$$

A similar result is obtained for the product fraction $\delta_{\mathrm{P}}(\xi_\phi)/\delta$, in which the dependence of the homogeneous mixture contribution on ξ_ϕ is given by (see Eq. 40) $\theta(\xi_E;\xi_\phi)$, and the dependence of the flame sheet contribution is given by[103]

$$F(\xi_\phi) = \frac{e^{-z_\phi^2}}{\sqrt{\pi}\,\xi_\phi\,(1-\xi_\phi)} \; , \tag{47a}$$

with z_ϕ implicitly defined by

$$\operatorname{erf} z_\phi = 2(\,\xi_\phi - 1/2\,) \; . \tag{47b}$$

The constants C_{H} and C_{F} in the BBM model are to be determined by fitting the data, e.g., the Schmidt number dependence of $\delta_{\mathrm{m}}/\delta$ (Eq. 46). The proposed model PDF is depicted in Fig. 25, with the Dirac delta function contribution represented by a narrow Gaussian of the appropriate area, centered on a $\xi = \xi_E$ corresponding to an entrainment ratio of $E = 1.3$.

In the BBM estimates for $\delta_{\mathrm{m}}/\delta$ and $\delta_{\mathrm{P}}/\delta$, the "flame sheet" contribution vanishes at large Schmidt numbers (cf. Eq. 46). The

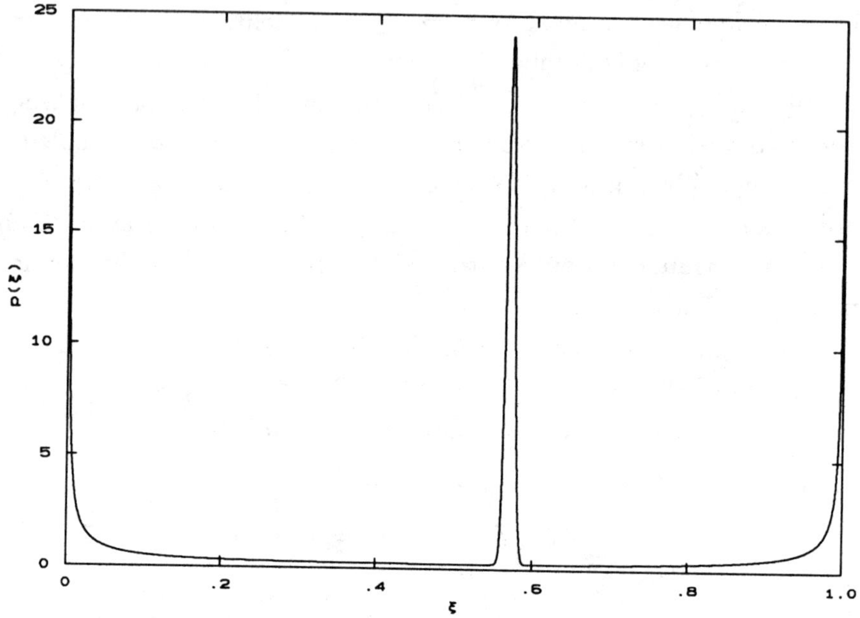

Fig. 25 Broadwell-Breidenthal-Mungal model PDF.[86]

gas/liquid difference is then accounted for by noting that the mixed fluid, in that case, is solely composed of the homogeneously mixed fluid at the composition $\xi \approx \xi_E$. Similarly, the gas-phase expressions asymptote to the liquid value at high Reynolds numbers. Conversely, the BBM model predicts that there should be no Schmidt number dependence at high Reynolds numbers. For gas phase flow, the predicted BBM dependence on Reynolds number is stronger ($Re^{-1/2}$) than the logarithmic dependence suggested by the data (Fig. 24) but nicely simulates the much weaker Reynolds number dependence of the liquid phase data. Finally, we should note that, according to the BBM model, the mixed fluid and chemical product volume fraction does not depend on the fluid kinematic viscosity, being a function of the Peclet number,

$$Pe \equiv \frac{\Delta U \, \delta}{\mathcal{D}} = Sc \times Re \ ,$$

in which only the species diffusivity $\mathcal{D}$ enters. See Ref. 103 for more details.

It is interesting that similar conclusions have also been arrived at by Kerstein,[108,109] using a phenomenological Monte Carlo model to represent the mechanics of turbulent transport and mixing. In his numerical simulations, Kerstein arrives at results for the mixed fluid and chemical product formed in a shear layer that are in accord with the BBM result expressed in the form of Eq. 46.

It could be argued that the strain rate at the interfacial surface formed by the turbulent flow between the entrained pure fluids from each of the freestreams should be estimated as a function of the distribution of spatial scales associated with that interface. In particular, one could argue that the predominant fraction of the surface-to-volume ratio $\mathcal{S}$ would be associated with the smallest scales in the flow which, for $Sc \approx 1$, would be in the vicinity of the Kolmogorov scale[110]

$$\lambda_{\mathrm{K}} \equiv \left(\frac{\varepsilon^3}{\nu}\right)^{1/4} , \tag{48a}$$

where $\varepsilon \propto (\Delta U)^3/\delta$ is the expected local kinetic energy dissipation rate per unit mass. This yields

$$\mathcal{S} \sim \frac{1}{\lambda_K} \sim \frac{1}{\delta}\, Re^{3/4} , \tag{49}$$

in the limit of large Reynolds numbers. The expected strain rate at those spatial scales would be proportional to the reciprocal of the Kolmogorov time $t_{\mathrm{K}} \equiv \sqrt{\nu/\varepsilon}$ or, in terms of the outer variables of the flow,

$$\sigma_{\mathrm{K}} \propto \frac{\Delta U}{\delta}\, Re^{1/2} . \tag{48b}$$

For $Sc \approx 1$, an estimate of the mixed fluid fraction scaling might be obtained as the product of the expected diffusion thickness $\lambda_{\mathcal{D}} \sim \sqrt{\mathcal{D}/\sigma}$ (Eq. 26), at the small scales, and the surface-to-volume ratio $\mathcal{S}$ of Eq. 49. It is interesting that this simple tally yields a *Reynolds-number-independent estimate*, to leading behavior, *for the mixed fluid fraction* (see also discussion in Ref. 87, Sec. 3.3).

As a rebuttal to this argument, we recognize that the interfacial surface will be characterized by the full spectrum of turbulent scales and the associated distribution of strain rates. Accordingly, one might attempt a tally in which this distribution is accounted for, with closure requiring the assignment of a statistical weight to each scale λ within the bounds of the turbulent flow. Such a model has been attempted,[87] where it was argued that the statistical weight $w(\lambda)\,\mathrm{d}\lambda$ of a scale λ in the self-similar inertial range must be given by

$$w(\lambda)\,\mathrm{d}\lambda \;\propto\; \frac{\mathrm{d}\lambda}{\lambda}\,, \tag{50}$$

as the only scale-invariant group that can be formed, and that the flow behavior below the Kolmogorov scale cannot alter this distribution in the range $\lambda_{\mathrm{B}} < \lambda < \lambda_{\mathrm{K}}$, where $\lambda_{\mathrm{B}} = \lambda_{\mathrm{K}}/\sqrt{Sc}$ is the scalar species Batchelor diffusion scale.[111] This is equivalent to assuming that all scales are equally probable and that the statistical weight of a scale λ is therefore given by the surface-to-volume ratio of that scale, i.e., $\mathcal{S}(\lambda) \propto 1/\lambda$, with the constant of proportionality determined by normalization This leads to an estimate for the mixed fluid fraction of

$$\frac{\delta_{\mathrm{m}}}{\delta} \approx \frac{B_1(Sc)}{B_0(Sc) \,+\, \frac{3}{4}\left(1-\frac{\mu}{8}\right)\ln\left(Re/Re_{\mathrm{cr}}\right)}\,, \tag{51}$$

where the functions $B_0(Sc)$ and $B_1(Sc)$ are determined by the calculation, $\mu \approx 0.3$ is the dissipation rate fluctuation coefficient,[112,113] also known as the intermittency exponent (e.g., Ref. 114) and $Re_{\mathrm{cr}} \approx 26$ (see Ref. 87 for details).

The model predictions are in accord with the observed dependence of the chemical product on the stoichiometry of the free streams, as well as the Schmidt number and Reynolds number dependence of the chemical product and mixed fluid fractions. It also predicts an ever-decreasing chemical product and mixed fluid fraction with increasing Schmidt number (decreasing species diffusivity). As can be seen in the resulting expression for $\delta_{\mathrm{m}}/\delta$, however, it predicts that the mixed fluid volume fraction is also an ever-decreasing (albeit slowly) function of the Reynolds number. This is a rather robust consequence of the $w(\lambda)\,\mathrm{d}\lambda$ statistical weight distribution that was assumed (Eq. 50). On the other hand, even a small departure from this distribution would alter this behavior in the limit, with no discernible differences within the range of Reynolds numbers that have been investigated and are typically achievable in the laboratory.

It need not be emphasized that the dependence and limiting behavior of turbulent mixing processes on Reynolds number is of considerable significance not only theoretically but also from an applications vantage point.

V. Discussion and Conclusions

For all the progress that has been made in addressing problems of mixing and combustion in turbulent shear flows, it is clear that important issues remain to be resolved. Many of these arise from the complexity and constraints imposed on these flows for a diverse set of reasons, which fundamental research often has the luxury of ignoring. Just as significant, however, are the problems that can be considered important and fundamental from any perspective, whose resolution would not only advance our understanding of turbulence, mixing, and combustion but would also have a direct impact on technology and applications. As Boltzmann used to say: "There is nothing more practical than a good theory."

Of the many problems that emerge from the preceding discussion, there are three, in my opinion, that merit close future scrutiny. These are: the apparent dependence of the far field behavior of high Reynolds number flows on initial conditions, the limiting behavior of high Reynolds number turbulence as the Reynolds number is increased to very large values, and the nature of turbulence under compressible flow conditions. The discussion that follows on these is necessarily more in the nature of speculation. In deference to Sir Arthur's admonition: "It is dangerous to theorize without data."

To paraphrase one of the conclusions of the discussion on shear-layer growth, it is surprising that the initial conditions seem to determine the far-field behavior of the turbulent shear layer. Should that conclusion survive future scrutiny, it will be a remarkable manifestation of what the equations of motion are capable of admitting in principle. Nevertheless, it flies in the face of traditional assumptions about the behavior of turbulence in the limit of high Reynolds numbers. Additionally, unless an explanation can be formulated for this behavior, it also complicates the analysis, simulation, and modeling of these flows in that this behavior must be characterized as a nonunique response to seemingly similar flow conditions. To the extent that we cannot mix any faster than the shear layer grows, the stakes, both theoretically and from an applications standpoint, are not small, as measured by the range of the empirical shear layer growth coefficient C_δ of almost a factor of 2 (recall Eqs. 10, 11).

Turning the coin over, we can see the potential for substantial benefits from flow control: if we can get, or ...lose, a factor of 2 by doing hardly anything, think what we can do if we try! Some recent results in our laboratory in what has previously been regarded as canonical turbulent jet mixing also substantiate this conclusion (see Ref. 115, Ch. 6) and suggest that this behavior is not peculiar to shear layers. There is a growing body of evidence that the possibilities that arise with active flow control are very significant, as can be seen from the work cited on the response of shear layers to external forcing (footnote ‡, Sec. II), the flow control and resulting mixing control in jets,[116] the wake of a circular cylinder,[117] and many others that have not been included here.

Returning to shear layers, it is important to understand the mechanism by which the initial conditions are felt by the flow thousands of momentum thicknesses downstream of the splitter plate trailing edge and how this observation is reconcilable with classical theories and descriptions of turbulence. It will be interesting to examine this issue and the clues that may be offered by the various research efforts in supersonic shear layers in progress. As the flow changes from elliptic to hyperbolic, the communication channels between different portions of the flow become a function of the Mach number of the two streams, as well as the respective convective Mach numbers that result.

In considering Reynolds number effects, it may be useful to think about a *gedanken* experiment in which the Reynolds number is varied by controlling the test section or combustor *pressure* at fixed geometry and freestream speeds. This would control the Reynolds number (at fixed Schmidt number) through a change in the molecular diffusivity coefficients, leaving most other flow parameters unaltered. Such a scheme would still change the Reynolds number of the initial conditions (recall Figs. 2 and 3, related caveats, and previous discussion), as well as the expected number of large-scale structures between the splitter-plate trailing edge and a fixed measuring station.[28] Nevertheless, it would leave the scaling with respect to the local *outer* flow variables unaltered and make it easier to argue for (or even discern, should such experiments be undertaken in the future) the subtle dependence of turbulence and mixing on the fluid Reynolds number, at fixed Schmidt number. If the dynamic range of Reynolds numbers in such experiments is large enough, one might obtain important clues about the behavior of turbulence as the Reynolds number is increased without limit. This is all the more important because the experimental evidence suggests that the mixed fluid fraction in shear layers is decreasing with increasing Reynolds number, in a Reynolds number regime untouchable by the foreseeable computing community, most models silent on the issue, and disagreement between two models that have stuck their necks out at this writing!

To complicate matters further, we should mention that recent experimental investigations of turbulent mixing in turbulent jets in our laboratory, in both gas-phase and liquid-phase flows, suggest that the dependence of turbulent *jet* mixing on Reynolds number *has the opposite sign*. Specifically, we have found that, for gas-phase flows, turbulent jet diffusion flame length *decreases* with Reynolds number, up to a jet Reynolds number of $Re \approx 2.5 \times 10^4$, indicating better mixing in the far field with increasing Reynolds number for turbulent jets. The near field behavior is more complicated, as evidenced by the Reynolds number dependence of the virtual origin of the jet flame length with respect to stoichiometric mixture ratio.[118] Measurements in liquid-phase jets have shown similar trends, as manifested by a *decreasing* variance of the jet fluid concentration fluctuations in the far field of the jet with increasing Reynolds number, with no Reynolds-number-independent mixing regime, beyond some minimum Reynolds number, at least within the range of Reynolds numbers attained in the experiments.[119,115] These results are very significant, in light of the opposite behavior found for shear layers, because they suggest that *the dependence of turbulent mixing processes on Reynolds number is not universal.* This behavior can probably be traced to differences in the behavior at the largest flow scales and the interplay between the large-scale organized structures, that are manifestly peculiar to each flow geometry, and the behavior at the smallest small scales, where the actual molecular mixing largely takes place. It goes without saying that this behavior must also be captured by turbulent mixing models, if they are to account for the observed phenomena.

Even in the unlikely event that compressible turbulence proves to be an even better mixer than its incompressible counterpart, it seems clear that we should expect to find a reduced overall mixing rate at high Mach numbers relative to incompressible flow. If the growth rate, absent external disturbances and mixing devices, is diminished by a factor of five, or so, there is little the turbulent interior can do to offset this reduction. In fact, the scant experimental evidence presently available from chemically reacting experiments in supersonic shear layers (see Fig. 24) would indicate that one should probably not expect any spectacular surprises on this score from canonical shear-layer flow configurations.

A second potentially important difference in behavior can be gleaned from the discussion of the entrainment ratio and the behavior of the large scale structure convection velocity under supersonic conditions. In particular, in a flow regime where the evidence and simple arguments suggest that shocks may be borne by one free stream or another, but not both, we expect much larger potential asymmetries in the entrainment ratio than in incompressible flow. To make matters worse, it is possible that it may prove difficult to predict or control in which direction the asymmetry may be realized. This is of considerable significance in the context of the expected composition of the mixed fluid, to the extent that one may not be able to predict and design for even the stoichiometry at which the combustion may have to be asked to take place within a shear layer zone. Moreover, it is a behavior that may well exhibit large changes in response with small changes in Mach number.

As for the mixing process itself, we can only speculate about it, at present. Our views of compressible turbulence are limited and not well substantiated. Some of the mainstays of incompressible turbulence, like the Kolmogorov similarity theories, may have to be revised if not abandoned as supersonic convective Mach numbers admit shocks running through the flow. There is evidence that in cases in which the driving field can generate eddies at intermediate scales, these similarity ideas cease to apply. In the case of shocks running through a turbulent flow characterized by density inhomogeneities, baroclinic generation of vorticity will form such eddies (e.g., Refs. 120–123), which will both influence and be influenced by subsequent shocks that visit (e.g., Refs. 124 and 125). Secondly, in an environment that is characterized by limits in the *speed*, on the one hand, and couples density fluctuations particularly efficiently to acoustically radiated power, as the Mach number increases, fluctuations and mixing may become dear commodities. On the other hand, this behavior could depend rather strongly on whether this is confined or open flow, as noted in our discussion on stability analysis,[51,52] and by H. Hornung in private discussions, with the *sign* of the outcome possibly dependent on the details!

It may be worth concluding by stating what is perhaps obvious, namely, that, from an applications point of view, *enhancement* of shear-layer growth and mixing is not always the objective. Although it may be, if one is interested in combustion efficiency and propulsion, it certainly is not the objective in the case of film cooling of hypersonic propulsion devices, aerodynamic windows for high-power chemical lasers, etc. What is at a premium here is the mastering of the fundamental physics of these phenomena, which will permit the optimization and control of their behavior, in each case, for the specific, complex, and sometimes purposes unanticipated at the outset.

Acknowledgments

I would like to acknowledge the continuous discussions within the GALCIT community, which directly or indirectly have contributed to the work reviewed and to this paper. I would, however, specifically like to recognize the many, often heated, discussions with Dr. J. Broadwell over the years, which have contributed to this work and the evolution of the ideas described in this discussion. I would also like to thank Mr. C. Frieler for his help with the preparation of some of the figures. Finally, I would like to acknowledge the many contributions and expertise of Dr. Dan Lang, who has been responsible for many of the electronics and computer-data-acquisition developments throughout the experimental effort at GALCIT summarized in this paper, without which many of the experiments would have had to wait for the corresponding capabilities to have become available commercially.

This work is part of a larger effort to investigate mixing and combustion in turbulent shear flows, sponsored by the Air Force Office of Scientific Research Contract No. F49620–79–C–0159 and Grants AFOSR–83–0213 and AFOSR–88–0155, whose support is gratefully acknowledged.

References

[1] Brown, G. L., and Roshko, A., 1974, "On Density Effects and Large Structure in Turbulent Mixing Layers," *J. Fluid Mech.* **64**(4), 775–816.

[2] Von Neumann, J., 1949, "Recent Theories of Turbulence," *Collected Works* **VI** (McMillan, New York, 1963), 437–472.

[3] Leonard, A., 1983, "Numerical simulation of Turbulent Fluid Flows," *Proceedings of the* 3rd *International Symposium on Numerical Methods in Engineering* (Pluralis, Paris), 45–63.

[4] Rogallo, R. S., and Moin, P., 1984, "Numerical Simulation of Turbulent Flows," *Ann. Rev. Fluid Mech.* **16**, 99-137.

[5] Rebollo, M. R., 1973, *Analytical and Experimental Investigation of a Turbulent Mixing Layer of Different Gases in a Pressure Gradient*, Ph.D. thesis, California Institute of Technology.

[6] Brown, G. L., and Roshko, A., 1971, "The Effect of Density Difference on the Turbulent Mixing Layer," *Turbulent Shear Flows*, AGARD–CP–93, 23.1–12.

[7] Mungal, M. G., and Dimotakis, P. E., 1984, "Mixing and combustion with low heat release in a turbulent mixing layer," *J. Fluid Mech.* **148**, 349–382.

[8] Koochesfahani, M. M., and Dimotakis, P. E., 1986, "Mixing and chemical reactions in a turbulent liquid mixing layer," *J. Fluid Mech.* **170**, 83–112.

[9] Abramowich, G. N., 1963, *The Theory of Turbulent Jets* (MIT Press, Cambridge, MA).

[10] Sabin, C. M., 1965, "An analytical and experimental investigation of the plane, incompressible, turbulent free-shear layer with arbitrary velocity ratio and pressure gradient," *Transactions of the ASME.* D **87**, 421–428.

[11] Konrad, J. H., 1976, *An Experimental Investigation of Mixing in Two-Dimensional Turbulent Shear Flows with Applications to Diffusion-Limited Chemical Reactions*, Ph.D. thesis, California Institute of Technology.

[12] Brown, G. L., 1974, "The Entrainment and Large Structure in Turbulent Mixing Layers," 5th *Australasian Conf. on Hydraulics and Fluid Mechanics*, 352–359.

[13] Dimotakis, P. E., 1984, "Two-dimensional shear-layer entrainment," *AIAA* 22nd *Aerospace Sciences Meeting*, *AIAA J.* **24**(11), 1791–1796 (1986).

[14] Oster, D., and Wygnanski, I., 1982, "The forced mixing layer between parallel streams," *J. Fluid Mech.* **123**, 91–130.

[15] Husain, Z. D., and Hussain, A. K. M. F., 1983, "Natural Instability of Free Shear Layers," *AIAA J.* **21**(11), 1512–1517.

[16] Ho, C.-M., and Huerre, P., 1984, "Perturbed Free Shear Layers," *Ann. Rev. Fluid Mech.* **16**, 365–424.

[17] Roberts, F. A., and Roshko, A., 1985, "Effects of Periodic Forcing on Mixing in Turbulent Shear Layers and Wakes," *AIAA Shear Flow Control Conference*, Paper 85–0570.

[18] Roberts, F. A., 1985, *Effects of a Periodic Disturbance on Structure and Mixing in Turbulent Shear Layers and Wakes*, Ph.D. thesis, California Institute of Technology.

[19] Wygnanski, I. J., and Petersen, R. A., 1987, "Coherent Motion in Excited Free Shear Flows," *AIAA J.* **25**(2), 201–213.

[20] Koochesfahani, M. M., and Dimotakis, P. E., 1988, "A Cancellation Experiment in a Forced Turbulent Shear Layer," Proceedings, *First National Fluid Dynamics Congress*, 1204–1208.

[21] Batt, R. G., 1975, "Some Measurements on the Effect of Tripping the Two-Dimensional Shear Layer," *AIAA J.* **13**(2), 245–247.

[22] Hussain, A. K. M. F., 1977, "Initial Condition Effect on Free Turbulent Shear Flows," Proceedings, *Structure and Mechanisms of Turbulence* **I** (Springer-Verlag, 1978), 103–107.

[23] Browand, F. K., and Latigo, B. O., 1979, "Growth of the two-dimensional mixing layer from a turbulent and non-turbulent boundary layer," *Phys. Fluids* **22**(6), 1011–1019.

[24] Weisbrot, I., Einav. S., and Wygnanski, I., 1982, "The non unique rate of spread of the two-dimensional mixing layer," *Phys. Fluids* **25**(10), 1691–1693.

[25] Dziomba, B., and Fiedler, H. E., 1985, "Effect of initial conditions on two-dimensional free shear layers," *J. Fluid Mech.* **152**, 419–442.

[26] Lang, D. B., 1985, *Laser Doppler Velocity and Vorticity Measurements in Turbulent Shear Layers*, Ph.D. thesis, California Institute of Technology.

[27] Bradshaw, P., 1966, "The effect of initial conditions on the development of a free shear layer," *J. Fluid Mech.* **26**(2), 225–236.

[28] Dimotakis, P. E., and Brown, G. L., 1976, "The mixing layer at high Reynolds number: Large-structure dynamics and entrainment," *J. Fluid Mech.* **78**(3), 535–560 + 2 plates.

[29] Briggs, R. J., 1964, "Electron-stream interaction in plasmas," Research Monograph No. 29 (M.I.T. Press, Cambridge, MA).

[30] Bers, A., 1975, "Linear waves and instabilities," *Physique des Plasmas* (Gordon and Breach, New York), 117–213.

[31] Huerre, P., and Monkewitz, P. A., 1985, "Absolute and convective instabilities in free shear layers," *J. Fluid Mech.* **159**, 151-168.

[32] Koch, W., 1985, "Local instability characteristics and frequency determination of self-excited wake flows," *J. Sound and Vibration* **99**(1), 53–83.

[33] Koochesfahani, M. M., and Frieler, C. E., 1987, "Inviscid Instability Characteristics of Free Shear Layers with non-Uniform Density," *AIAA* 25th *Aerospace Sciences Meeting*, Paper 87-0047.

[34] Sandham, N. D., and Reynolds, W. C., 1987, "Some Inlet-Plane Effects on the Numerically Simulated Spatially-Developing Mixing Layer," 6th *International Symp. on Turb. Shear Flows* (Springer Verlag, 1989), 441–454.

[35] Breidenthal, R. E., 1981, "Structure in Turbulent Mixing Layers and Wakes Using a Chemical Reaction," *J. Fluid Mech.* **109**, 1–24.

[36] Hall, J. L., 1991, *An Experimental Investigation of Structure, Mixing and Combustion in Compressible Turbulent Shear Layers*, Ph.D. thesis, California Institute of Technology.

[37] Bogdanoff, D. W., 1983, "Compressibility Effects in Turbulent Shear Layers," *AIAA J.* **21**(6), 926–927.

[38] Papamoschou, D. , and Roshko, A., 1988, "The Compressible Turbulent Shear Layer: An Experimental Study," *J. Fluid Mech.* **197**, 453–477.

[39] Mack, L. M., 1975, "Linear Stability and the Problem of Supersonic Boundary-layer Transition," *AIAA J.* **13**, 278–289.

[40] Ragab, S. A., and Wu, J. L., 1988, "Instabilities in the Free Shear Layer Formed by Two Supersonic Streams," *AIAA* 26th *Aerospace Sciences Meeting*, Paper 88–0038.

[41] Zhuang, M., Kubota, T., and Dimotakis, P. E., 1988, "On the Stability of Inviscid, Compressible Free Shear Layers," Proceedings, *First National Fluid Dynamics Congress*, 768–773.

[42] Coles, D., 1981, "Prospects for Useful Research on Coherent Structure in Turbulent Shear Flow," *Proc. Indian Acad. Sci. (Eng. Sci.)* **4**(2), 111–127.

[43] Coles, D., 1985, "Dryden Lecture: The Uses of Coherent Structure," *AIAA* 23rd *Aerospace Sciences Meeting*, Paper 85–0506.

[44] Wang, C., 1984, *The effects of curvature on turbulent mixing layers*, Ph.D. thesis, California Institute of Technology.

[45] Lele, S. K., 1989, "Direct Numerical Simulation of Compressible Free Shear Flows," *AIAA* 27th *Aerospace Sciences Meeting*, Paper 89–0374.

[46] Chinzei, N., Masua, G., Komuro, T. Murakami, A., and Kudou, K., 1986, "Spreading of two-stream supersonic turbulent mixing layers," *Phys. Fluids* **29**(5), 1345-1347.

[47] Clemens, N. T., and Mungal, M. G., 1990, "Two- and Three-Dimensional Effects in the Supersonic Mixing Layer," Paper 90–1978.

[48] Hall, J. L., Dimotakis, P. E., and Rosemann, H., 1991, "Experiments in non-reacting compressible shear layers," *AIAA* 29th *Aerospace Sciences Meeting*, Paper 91–0629.

[49] Gropengiesser, H., 1970, "Study of the Stability of Boundary Layers in Compressible Fluids," NASA–TT–F–12, 786.

[50] Sandham, N., and Reynolds, W. C., 1989, "The Compressible Mixing Layer: Linear Theory and Direct Simulation," *AIAA* 27th *Aerospace Sciences Meeting*, Paper 89–0371.

[51] Tam, C. K. W., and Morris, P. J., 1980, "The radiation of sound by the instability waves of a compressible plane turbulent shear layer," *J. Fluid Mech.* **98**, 349-381.

[52] Tam, C. K. W., and Hu, F. Q., 1988, "Instabilities of supersonic mixing layers inside a rectangular channel," Proceedings, *First National Fluid Dynamics Congress*, 1073–1086.

[53] Zhuang, M., Dimotakis, P. E., and Kubota, T., 1990, "The Effect of Walls on a Spatially Growing Supersonic Shear Layer," *Phys. Fluids* A **2**(4), 599–604.

[54] Papamoschou, D., 1989, "Structure of the compressible turbulent shear layer," *AIAA* 27th *Aerospace Sciences Meeting*, *AIAA J.* **29**(5), 680–681 (1991).

[55] Fourguette, D., Mungal, M. G., and Dibble, R., 1990, "Time Evolution of the Shear Layer of an Axisymmetric Supersonic Jet at Matched Conditions," *AIAA* 28th *Aerospace Sciences Meeting*, Paper 90–0508.

[56] Papamoschou, D., 1988, "Outstanding issues in the area of compressible mixing," *International Workshop on the Physics of Compressible Mixing*(Proceedings to be published by Springer-Verlag).

[57] Liepmann, H. W., and Roshko, A., 1957, *Elements of Gasdynamics* (John Wiley, New York).

[58] Dimotakis, P. E., 1991, "On the convection velocity of turbulent structures in supersonic shear layers," *AIAA* 22nd *Fluid Dynamics, Plasma Dynamics and Lasers Conference*, Paper 91–1724.

[59] Vandromme, D., and Haminh, H., 1989, "The compressible mixing layer," *Turbulence and Coherent Structures* (Kluwer AP, Dordrecht, 1991), 507–523.

[60] Lowson, M. V., and Ollerhead, J. B., 1968, "Visualization of noise from cold supersonic jets," *J. Acoust. Soc. Am.* **44**, 624.

[61] Tam, C. K. W., 1971, "Directional acoustic radiation from a supersonic jet," *J. Fluid Mech.* **46**(4), 757–768.

[62] Oertel, H., 1979, "Mach wave radiation of hot supersonic jets investigated by means of the shock tube and new optical techniques," 12th *Int. Symp. on Shock Tubes and Waves*, 266–275.

[63] Sandham, N. D., and Reynolds, W. C., 1989, "A Numerical Investigation of the Compressible Mixing Layer," Stanford Report TF–45.

[64] Wallace, A. K., 1981, *Experimental Investigation on the Effects of Chemical Heat Release in the Reacting Turbulent Plane Shear Layer*, Ph.D. thesis, U. Adelaide.

[65] Hermanson, J. C., and Dimotakis, P. E., 1989, "Effects of heat release in a turbulent reacting shear layer," *J. Fluid Mech.* **199**, 333–375.

[66] Ganji, A. R., and Sawyer, R. F., 1980, "Experimental Study of a Two-Dimensional Premixed Turbulent Flame," *AIAA J.* **18**(7), 817–824.

[67] Pitz, R. W., and Daily, J. W., 1983, "Combustion in a Turbulent Mixing Layer at a Rearward-Facing Step," *AIAA J.* **21**(11), 1565–1570.

[68] Daily, J. W., and Lundquist, W. J., 1984, "Three dimensional structure in a turbulent combusting mixing layer," 20th *Symposium (International) on Combustion* (The Combustion Institute, Pittsburgh), 487–494.

[69] Keller, J. O., and Daily, J. W., 1985, "The Effect of Highly Exothermic Chemical Reaction on a Two-Dimensional Mixing Layer," *AIAA J.* **23**(12), 1937–1945.

[70] McMurtry, P. A., Jou, W. H., Riley, J. W., and Metcalfe, R. W., 1986, "Direct Numerical Simulations of a Reacting Mixing Layer with Chemical Heat Release," *AIAA J.* **24**(6), 962–970.

[71] McMurtry, P. A., and Riley, J. J., 1987, "Mechanisms by which heat release affects the flow field in a chemically reacting turbulent mixing layer," *AIAA* 25th *Aerospace Sciences Meeting*, Paper 87–0131.

[72] Bilger, R. W., 1989, "Turbulent Diffusion Flames," *Ann. Rev. Fluid Mech.* **21**, 101–135.

[73] Winant, C. D., and Browand, F. K., 1974, "Vortex pairing: the mechanism of turbulent mixing layer growth at moderate Reynolds number," *J. Fluid Mech.* **63**(2), 237–255.

[74] Corcos, G. M., and Sherman, F. S., 1984, "The mixing layer: deterministic models of the turbulent flow. Part 1. Introduction and the two-dimensional flow," *J. Fluid Mech.* **139**, 29–65.

[75] Bernal, L. P., 1981, *The Coherent Structure of Turbulent Mixing Layers. I. Similarity of the Primary Vortex Structure, II. Secondary Streamwise Vortex Structure*, Ph.D. thesis, California Institute of Technology.

[76] Alvarez, C., and Martinez-val, R., 1984, "Visual measurement of streamwise vorticity in the mixing layer," *Phys. Fluids* **27**(9), 2367–2368.

[77] Corcos, G. M., and Lin, S. J., 1984, "The mixing layer: deterministic models of the turbulent flow. Part 2. The origin of the three-dimensional motion.," *J. Fluid Mech.* **139**, 67–95.

[78] Browand, F. K., 1986, "The Structure of the Turbulent Mixing Layer," *Physica* D **18**, 135–148.

[79] Bernal, L. P., and Roshko, A., 1986, "Streamwise vortex structure in plane mixing layers," *J. Fluid Mech.* **170**, 499–525.

[80] Metcalfe, R. W., Orzag, S. A., Brachet, M. E., Menon, S., and Riley, J. J., 1987, "Secondary instability of a temporally growing mixing layer," *J. Fluid Mech.* **184**, 207–244.

[81] Knio, O. M., and Ghoniem, A. F., 1988, "On the formation of streamwise vorticity in turbulent shear flows," *AIAA* 26th *Aerospace Sciences Meeting*, Paper 88–0728.

[82] Lasheras, J. C., and Choi, H., 1988, "Three-dimensional instability of a plane shear layer: an experimental study of the formation and evolution of streamwise vortices," *J. Fluid Mech.* **189**, 53–86.

[83] Rogers, M. M., and Moser, R. D., 1991, "The three-dimensional evolution of a plane mixing layer. Part 1: The Kelvin-Helmholtz rollup," NASA TM–103856.

[84] Bernal, L. P., Breidenthal, R. E., Brown, G. L., Konrad, J. H., and Roshko, A., 1979, "On the Development of Three-Dimensional Small Scales in Turbulent Mixing Layers," 2nd *Int. Symposium on Turb. Shear Flows*(Springer-Verlag, New York, 1980), 305–313.

[85] Marble, F. E., and Broadwell, J. E., 1977, "The Coherent Flame Model for Turbulent Chemical Reactions," Project SQUID TRW–9–PU.

[86] Broadwell, J. E., and Breidenthal, R. E., 1982, "A simple model of mixing and chemical reaction in a turbulent shear layer," *J. Fluid Mech.* **125**, 397–410.

[87] Dimotakis, P. E., 1987, "Turbulent shear layer mixing with fast chemical reactions," US-France Workshop on *Turbulent Reactive Flows* (Springer-Verlag, New York, 1989), 417–485.

[88] Demetriades, A., 1980, "Necessary conditions for transition in a free shear layer," AFOSR–TR–80–0442.

[89] Demetriades, A., Ortwerth, P. J., and Moeny, W. M., 1981, "Laminar-Turbulent Transition in Free Shear Layers," *AIAA J.* **19**(9), 1091–1092.

[90] Demetriades, A., and Brower, T. L., 1982, "Experimental Study of Transition in a Compressible Shear Layer," Montana State U. Annual Report, AFOSR–TR–83–0144.

[91] Fiedler, H. E., 1975, "On Turbulence Structure and Mixing Mechanism in Free Turbulent Shear Flows," *Turbulent Mixing in Non-Reactive and Reactive Flows* (Plenum Press, New York), 381–409.

[92] Dimotakis, P. E., and Hall, J. L., 1987, "A simple model for finite chemical kinetics analysis of supersonic turbulent shear layer com-

bustion," AIAA/SAE/ASME/ASEE 23rd Joint Propulsion Meeting, Paper 87–1879.

93 Bilger, R. W., 1980, "Turbulent Flows with Nonpremixed Reactants," *Turbulent Reacting Flows* (Springer-Verlag, Topics in Applied Physics **44**, Eds. P. A. Libby, F. A. Williams, New York), 65–113.

94 Levenspiel, O., 1962, *Chemical Reaction Engineering. An Introduction to the Design of Chemical Reactors.* (John Wiley, New York).

95 Brown, G. L., and Rebollo, M. R., 1972, "A Small, Fast-Response Probe to Measure Composition of a Binary Gas Mixture," *AIAA J.* **10**(5), 649–652.

96 Pope, S. B., 1981, "A Monte Carlo method for the PDF equations of turbulent reactive flow," *Combust. Sci. and Technol.* **25**, 159-174.

97 Kollmann, W., and Janicka, J., 1982, "The Probability Density Function of a Passive Scalar in Turbulent Shear Flows," *Phys. Fluids* **25**, 1755–1769.

98 Koochesfahani, M. M., Dimotakis, P. E., and Broadwell, J. E., 1985, "A 'Flip' Experiment in a Chemically Reacting Turbulent Mixing Layer," *AIAA J.* **23**(8), 1191–1194.

99 Masutani, S. M., and Bowman, C. T., 1986, "The structure of a chemically reacting plane mixing layer," *J. Fluid Mech.* **172**, 93–126.

100 Bilger, R. W., 1979, "Turbulent Jet Diffusion Flames," *En. and Comb. Science* (Student Ed. **1**, N. Chigier, ed.), 109–131.

101 Williams, F. A., 1988, *Combustion Theory. The Fundamental Theory of Chemically Reacting Flow Systems* (2nd edition, Addison-Wesley, Menlo Park, CA).

102 Mungal, M. G., and Frieler, C. E., 1988, "The Effects of Damköhler Number in a Turbulent Shear Layer," *Comb. and Flame* **71**, 23–34.

103 Broadwell, J. E., and Mungal, M. G., 1988, "Molecular Mixing and Chemical Reactions in Turbulent Shear Layers," 22nd *Symposium (International) on Combustion* (The Combustion Institute, Pittsburgh), 579–587.

104 Kuo, K. K., 1986, *Principles of Combustion* (John Wiley, New York).

[105] Frieler, C. E., and Dimotakis, P. E., 1988, "Mixing and Reaction at Low Heat Release in the Non-Homogeneous Shear Layer," *First National Fluid Dynamics Congress*, Paper 88–3626.

[106] Mungal, M. G, Hermanson, J. C., and Dimotakis, P. E., 1985, "Reynolds Number Effects on Mixing and Combustion in a Reacting Shear Layer," *AIAA J.* **23**(9), 1418–1423.

[107] Hall, J. L., Dimotakis, P. E., and Rosemann, H., 1991, "Some measurements of molecular mixing in compressible turbulent mixing layers," *AIAA 22nd Fluid Dynamics, Plasma Dynamics and Lasers Conference*, Paper 91–1719.

[108] Kerstein, A., 1988, "A linear-eddy model of turbulent scalar transport and mixing," *Combust. Sci. and Technol.* **60**, 391.

[109] Kerstein, A., 1989, "Linear-Eddy Modeling of Turbulent Transport II: Application to Shear Layer Mixing," *Comb. and Flame* **75**, 397–413.

[110] Kolmogorov, A. N., 1941, "Local Structure of Turbulence in an Incompressible Viscous Fluid at Very High Reynolds Numbers," *Dokl. Akad. Nauk SSSR* **30**, 299.

[111] Batchelor, G. K., 1959, "Small-scale variation of convected quantities like temperature in turbulent fluid. Part 1. General discussion and the case of small conductivity," *J. Fluid Mech.* **5**, 113–133.

[112] Kolmogorov, A. N., 1962, "A refinement of previous hypotheses concerning the local structure of turbulence in a viscous incompressible fluid at high Reynolds number," *J. Fluid Mech.* **13**, 82–85.

[113] Oboukhov, A. M., 1962, "Some specific features of atmospheric turbulence," *J. Fluid Mech.* **13**, 77–81.

[114] Monin, A. S., and Yaglom, A. M., 1975, *Statistical Fluid Mechanics: Mechanics of Turbulence* **II**, Ed. J. Lumley (MIT Press, Cambridge, MA).

[115] Miller, P. L., 1991, *Mixing in High Schmidt Number Turbulent Jets*, Ph.D. thesis, California Institute of Technology.

[116] Parekh, D. E., and Reynolds, W. C., 1989, "Forced Instability Modes in a Round Jet at High Reynolds Numbers," *Phys. Fluids* A **1**(9), 1447.

[117] Tokumaru, P. T., and Dimotakis, P. E., 1991, "Rotary Oscillation Control of a Cylinder Wake," *J. Fluid Mech.* **224**, 77–90.

[118] Gilbrech, R. J., 1991, *An Experimental Investigation of Chemically-Reacting, Gas-Phase Turbulent Jets*, Ph.D. thesis, California Institute of Technology.

[119] Miller, P. L., and Dimotakis, P. E., 1991, "Reynolds number dependence of scalar fluctuations in a high Schmidt number turbulent jet," *Phys. Fluids* A **3**(5), Pt. 2, 1156–1163.

[120] Haas, J. F., and Sturtevant, B., 1987, "Interaction of weak shock waves with cylindrical and spherical gas inhomogeneities," *J. Fluid Mech.* **181**, 41–76.

[121] Brouillette, M., 1989, *On the Interaction of Shock Waves with Contact Surfaces Between Gases of Different Densities*, Ph.D. thesis, California Institute of Technology.

[122] Waitz, I., 1991, *An Investigation of Contoured Wall Injectors for Hypervelocity Mixing Augmentation*, Ph.D. thesis, California Institute of Technology.

[123] Yang, J., 1991, *An Analytical and Computational Investigation of Shock-Induced Vortical Flows with Applications to Supersonic Combustion*, Ph.D. thesis, California Institute of Technology.

[124] Hesselink, L., and Sturtevant, B., 1988, "Propagation of weak shocks through a random medium," *J. Fluid Mech.* **196**, 513–553.

[125] Rotman, D., 1991, "Shock wave effects on a turbulent flow," *Phys. Fluids A* **3**(7), 1792–1806.

Chapter 6

Turbulent Mixing in Supersonic Combustion Systems

J. Swithenbank,* I. W. Eames,† S. B. Chin,† B. C. R. Ewan,† Z. Yang,§ J. Cao,‡ and X. Zhao§

University of Sheffield, England, United Kingdom

Abstract

Optimization of scramjet combustor design must take into account the requirement that the fuel be well mixed with the air within a few tens of microseconds. The supersonic combustion process is controlled by both chemical kinetics and mixing. To minimize the chemical kinetic limitations, the designer must insure that the local static temperatures are sufficiently high that the ignition delay time plus the time to complete the reaction are less than the time of flow through the combustor. However, the combustion efficiency is also mixing limited so that, for example, 80% mixedness would correspond to 80% complete combustion.

In scramjet combustors there is a trade-off between engine thrust loss due to low combustion efficiency and thrust loss due to mixing energy, resulting in an optimum combustion efficiency that is usually much less than 100%.

In order to investigate the fundamentals of scramjet combustor design, a test facility is needed in which the flow energy is about 50,000 MW/m^2 flow area. The combustion-driven tailored interface hypersonic shock tunnel can provide, at modest cost, the severe test conditions required. The shock tunnel at Sheffield University has been specially designed and instrumented for scramjet combustor testing. The results obtained to date confirm the design principles based on our

*Professor.

†Academic Staff.

‡ Research Staff.

§ Research Student.

mixing model and demonstrate the advantage of carefully controlling the location of turbulence generation. To date, our best fuel-injector/turbulence generator design used dogtooth trailing edges, swept back behind the local Mach angle, to produce the required multivortex flowfield.

INTRODUCTION

Scramjets are attractive engines for both cruise and boost missions in the hypersonic portion of the flight corridor (e.g., Refs. 1 and 2). At Sheffield University, our present research program is designed to establish high-Mach-number design criteria that minimize overall combustor loss mechanisms and maximize scramjet performance.

The basic components of a scramjet (Fig. 1) consist of an air intake, a combustion chamber and a propelling nozzle. The intake compresses the air, raising its static pressure by one to two orders of magnitude and reducing its velocity (relative to the engine) by a few percent only. Fuel is injected, mixed and burned with the air in the combustion chamber, and the majority of the heat release occurs in supersonic flow. (Local regions of subsonic combustion may exist near the fuel injectors and in the boundary layers.) The combustion products are accelerated in the nozzle to a pressure that is usually just above freestream static pressure, and the velocity increases to a few percent above the flight velocity. The net thrust of the engine results from the increase in gas velocity, and is transmitted as pressure on the engine surfaces. The flow through the engine exhibits the characteristic property of hypersonic flowfields, in that the velocity varies

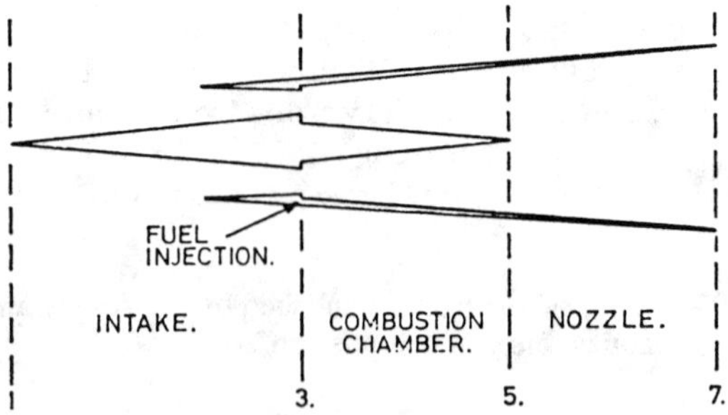

Fig. 1 The basic components of a scramjet.

only by a few percent from its freestream value. The corresponding Mach number of the flow through the engine may vary by up to an order of magnitude, mainly because the static temperature (and, hence sound speed) vary widely. Two conclusions can be drawn from these considerations: first, since the combustion chamber velocity will be almost equal to flight velocity, the time available for mixing and combustion can be approximated by combustion chamber length/flight velocity; second, the existence of net thrust can be appreciated readily since the exit momentum is approximately equal to the inlet momentum plus an increment derived from the fuel enthalpy. Thus, the intake and nozzle efficiencies of a scramjet are much greater than in a conventional ramjet, in which large variations in velocity are encountered. (There is, however, a greater fundamental stagnation pressure loss in the combustion chamber due to heat addition to a supersonic flow.)

All power plants that depend on ram air for operation give zero thrust at zero speed. Therefore, some auxiliary engine is required to achieve the takeover speed. A mixed propulsion system is an inevitable consequence and, for the first-stage engine several alternatives are being investigated:

(1) Solid or liquid fuel rocket.
(2) Turborocket.
(3) Turbojet or turboramjet.
(4) Precooled turbojet.
(5) Liquid-air cycle engine.
(6) Liquid-air collection for use later in flight.

At present, the development of a scramjet capable of good performance from Mach 5 to satellite velocity appears to be more promising than that of other hypersonic engine systems. It is therefore receiving considerable attention.

The upper flight velocity likely to be of interest can be determined by considering the probable applications, which divide broadly into boost and cruise missions. The upper velocity limit for space booster missions will be satellite velocity (with higher velocities of potential interest for deep space missions). The upper limit for cruise vehicles is that speed that would allow ranges of about half the Earth's circumference to be achieved. For such a range, with a mean acceleration phase at around 0.2g, and a glide phase at about 0.15g deceleration (i.e., L/D ~

6), the maximum speed attained would be about Mach 18, with no cruise phase. With a reasonable length of cruise phase, the maximum speed would probably be below Mach 15.

At the present time, most groups studying scramjets have discontinued work on both external combustion and standing detonations in favor of internal supersonic combustion because the latter offers better performance for aircraft boost and cruise missions.

Various early studies on the relative efficiency of the ramjet and scramjet gave estimates of the 'break-even' Mach number varying between 5 and 10, depending on the component efficiencies assumed. Fortunately, this shows that over a wide range of Mach numbers, the choice of ramjet or scramjet can be dictated by considerations other than thermodynamic efficiency.

Although there is a thermodynamic penalty due to heat additions at higher supersonic Mach numbers, it can be more than offset by improvement in the intake and nozzle efficiencies and decreased chemical dissociation due to lower static temperatures. Furthermore, the lower static pressures that result serve to improve the feasibility of engine construction because of lower pressure loads and lower heat fluxes.

Since the early 1960s, scramjet components and even complete engine models have been tested by many laboratories and companies with mixed success. However, a considerable amount of both research and development are still required before scramjets can be designed reliably and operated on routine missions.

Theoretical Cycle Analysis

To date, the research effort and expenditure on scramjets has been justified by their high potential performance when compared with rockets. A brief description of the methods used to compute the theoretical performance is presented here, followed by a discussion of the limitations of the analysis.

Two different methods of calculation can be used:

1) Approximate methods, which indicate the flight speed range within which the scramjet is superior to the conventional ramjet.

2) Exact analysis, in which conditions throughout the engine are evaluated. These studies form an essential

background to detailed investigation of the performance of individual components. When these investigations yield better values of component efficiency, the exact analysis can be improved until an optimum engine design can be formulated. The fundamental requirement governing the engine design is mission achievement at minimum overall cost. The interaction between the engine and the rest of the system is particularly important at hypersonic velocities; for example, the engine must be completely integrated with the airframe.

In this paper, the approximate method will be used since this demonstrates clearly the fundamental factors controlling performance. Full details of the exact method may be found in Ref. 1.

Approximate Thermodynamic Analysis

An upper limit to the performance available from chemical propulsion systems can be obtained from a simple energy balance. For example, the potential and kinetic energy per pound of a body in Earth orbit is one-fifth of the chemical energy contained in one pound of hydrogen when burned with air. Thus an absolute upper limit to the payload in orbit using hydrogen is 80%. If the oxidant must be carried as in a rocket, the maximum fraction in orbit would be about 30% of takeoff weight.

It is well known that the optimum can be achieved only if the propulsive jet is ejected exactly at flight velocity, thus giving 100% propulsive efficiency. In the case of a rocket, the characteristic exhaust velocity lies between 3000 and 5500 m/s for current propellant combinations and, therefore, high propulsive efficiency is achieved only at the upper end of the boost velocity spectrum (note that orbital speed is 8000 m/s). At takeoff speeds, the propulsive efficiency can be improved if the rocket interacts with the air as in a turborocket. It would also be logical to use the oxidizing properties of the air to burn fuel; thus, an air-breathing engine is a logical development from the rocket. For hydrogen fuel, the characteristic velocity when burned with atmospheric air is 15,600 m/s.

To derive the maximum, or ideal, performance of an air-breathing system as a function of flight velocity, it is necessary to include the kinetic energy the fuel possesses in the moving vehicle in addition to its chemical energy. Thus, the work rate from the chemical energy of the fuel is

$$F_c V_1 = W_f h_f$$

and the work rate from the kinetic energy of the fuel is

$$F_f V_1 = W_f V_1{}^2/2$$

To these can be added a contribution due to heat recovered from the Atmosphere. At subsonic speeds, this contribution to a simple heat-addition cycle is negligible but, as the flight speed increases, this contribution can become very significant. In the present instance, we will assume that the heat is transferred to the fuel before it is burned. The absolute maximum level of fuel heating is limited by the stagnation temperature of the external stream; then the work rate from this source of recycled heat is

$$F_r V_1 = W_f h_r$$

where

$$\begin{aligned} h_r &= C_p \ \Delta t \\ &= C_p (T - t_f) \\ &= C_p [(t - t_f) + t((\gamma-1)/2)M^2)] \end{aligned}$$

The term $(t - t_f)$ represents heat recovered from the atmosphere because cryogenic fuel is heated up to ambient static temperature. Bearing in mind that h_f in the previous equation could have been based on liquid hydrogen fuel at cryogenic temperatures, we note that it is more convenient to base calculations on the enthalpy of gaseous hydrogen fuel at ambient static temperature, thus canceling the term $(t - t_f)$ in this equation. However, it should be noted that we could recover thrust from the liquefaction energy of the fuel if we were to use one of the engine cycles such as the reverse turbojet. In such a system we expand the airflow through a turbine, then remove heat to the cryogenic fuel as a heat sink, followed by compression resulting in excess jet energy in the exhaust and a stream in which we can burn the vaporized fuel. Due to the high stagnation temperature such a cycle is irrelevant at hypersonic speed, and this contribution will be neglected here. On the other hand, the heat recovery from the atmosphere must take place through the boundary layer, and we should use the recovery temperature rather than the total temperature. This introduces a fundamental recovery efficiency factor of 0.89.

Therefore, the ideal thrust F_{max} is given by

$$F_{max} = F_c + F_k + F_h$$

$$= W_f \left[\frac{h_f}{V_1} + \frac{V_1}{2} + 0.89\ C_p [\frac{t(\gamma-1)}{2\ V_1} \quad M^2]\right]$$

and the ideal specific impulse (I_{max})

$$I_{max} = \frac{F_{max}}{W_f} = \left[\frac{h_f}{V_1} + \frac{V_1}{2} + 0.89\ C_p\ t\ \frac{(\gamma-1)}{2\ c^2}\ V_1\right]$$

This relation is plotted as a function of flight velocity in Fig. 2a for hydrogen fuel, and it can be seen that while the kinetic energy (KE) contribution becomes significant at hypersonic velocities, the potential contribution from recycled heat energy could be very dramatic. However, from the practical point of view, the recycled heat must be transferred to the fuel, and material limitations impose a limit on the hydrogen temperature at the injector. The curve labeled "T limit" corresponds to a hydrogen injection temperature of 1800 K. The same data are plotted as a function of flight Mach number in Fig. 2b. The recycled heat contribution is a very important bonus since it is well known that small increments in engine performance result in disproportionately large benefits in aircraft performance parameters such as range.

If the recycle heat contribution is neglected, the ideal specific impulse decreases gradually to the characteristic velocity, after which it increases again as the fuel kinetic energy contribution starts to dominate. The increase occurs well above satellite velocity for a scramjet cycle but, in the case of an oxygen-addition ramjet, the characteristic velocity is reduced to below satellite speed, and a large fraction of the impulse is due to the kinetic energy of the fuel. In the hypersonic speed range of interest (1600-9000 m/s) the potential specific impulse varies from 100,000 to 25,000 m/s and is therefore very much greater than the value of 4000 m/s available from a chemical rocket. (Since a rocket combustion chamber does not interact with the atmosphere, a rocket gives about the same impulse at all speeds.)

The specific impulse contribution due to the kinetic energy of the fuel is the same for all fuels including, for example, water. Thus, water or additional hydrogen, when carried as a coolant, could be used in

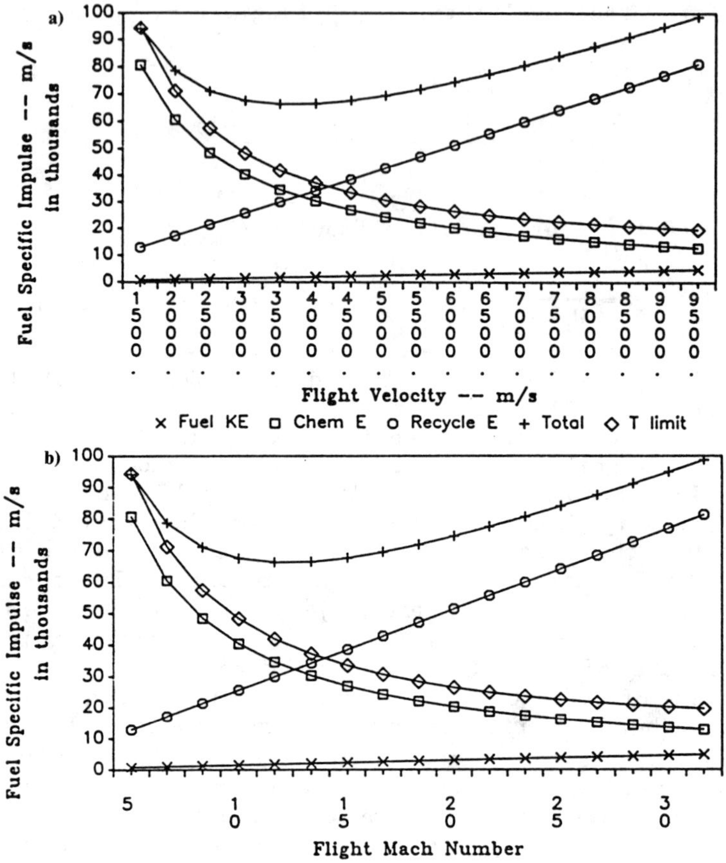

Fig. 2 Maximum specific impulse of hydrogen vs a) flight velocity and b) flight Mach number.

the engine to contribute to the thrust. Fuel-rich operation is particularly attractive when hydrogen is the fuel because, in addition to reducing dissociation losses, it also reduces the molecular weight of the exhaust gas significantly and:

$$I_f \sim (M_{exhaust})^{-1/2}$$

This mass-addition effect becomes more attractive as speed is increased.

Figures 2a and 2b tend to give the impression that the performance of a scramjet falls at higher speeds since the curve for overall specific impulse falls at Mach numbers approaching orbital speed. Specific impulse can be regarded as the thrust force (N) per unit fuel flow (kg/s). A more relevant parameter is the work rate of the engine, which is the product of impulse and velocity. This parameter is plotted in Fig. 3, which illustrates that performance is actually maintained at high speeds.

As the chemical energy of the fuel is decreased (for example, by using kerosene), the maximum possible specific impulse is decreased as shown in Fig. 4, and

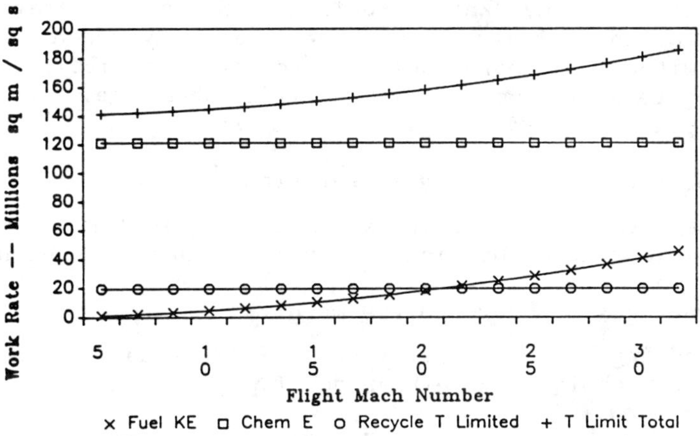

Fig. 3 Maximum work rate for hydrogen fuel vs flight Mach number for chemical, KE, and T limit recycle energy.

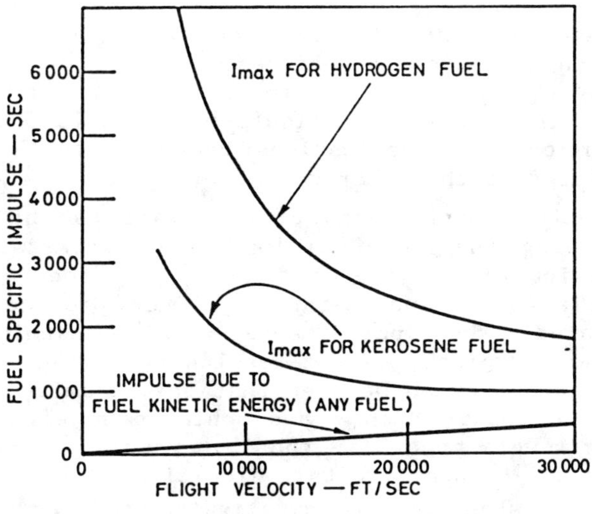

Fig. 4 Maximum specific impulse of kerosene vs flight velocity.

the contribution from the kinetic energy of the fuel is a larger fraction of the whole. The absolute contribution to specific impulse from recycled heat is reduced because of the lower specific heat of kerosene, although the proportional contribution increases.

The use of hydrogen fuel for hypersonic propulsion is a logical choice, especially when the following additional advantages are considered:

1) The cooling capacity of liquid hydrogen is the greatest for any known fuel. Since the favorable mass-addition effects may allow excess hydrogen to be used without appreciable performance penalty, this cooling capacity can be used to full advantage. No other fuel approaches hydrogen in this regard.

2) The thermal stability is satisfactory.

3) The reaction kinetics are particularly rapid, thus minimizing combustion length in a supersonic burner.

Disadvantages of hydrogen are mainly:

1) Handling and storage difficulties, although suitable techniques have been developed.

2) The low density of hydrogen means that about four times more fuel tank volume is required than for kerosene for the same heat content. Fortunately, this volume can be provided in large vehicles.

3) The scaling laws dictate that denser fuels must be used for small hypersonic vehicles.

4) The high cost of liquid hydrogen is a secondary factor for booster applications because fuel cost is small compared to the other costs.

The actual performance of an engine can be obtained by multiplying the maximum impulse or work-rate values given previously by the efficiency of the particular engine cycle under consideration. Many previous studies have compared the performance of air-breathing engine cycles, and a typical set of results are given in Fig. 5. In this figure, it can be seen that the scramjet overtakes the ramjet at about Mach 7 and achieves an efficiency of about 50% over much of the hypersonic speed range, extending up to near-orbital velocities.

The next step is to investigate the source of the losses, and here the analysis given in Ref. 3 is

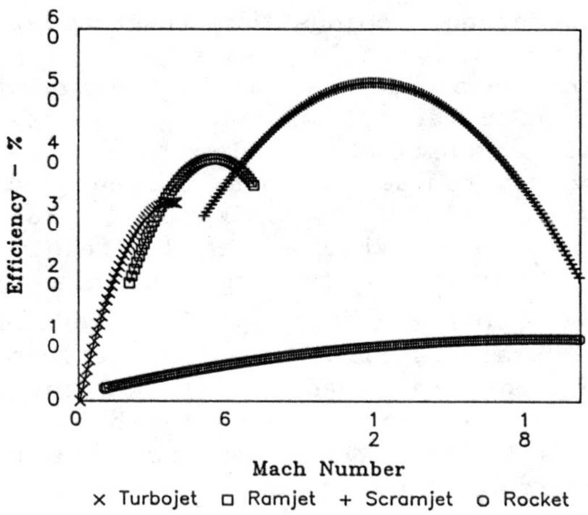

Fig. 5 Engine efficiency vs Mach number.

invaluable. In this treatment, the thrust expressions are expanded, and small terms deleted, subject to hypersonic values of certain variables. The resulting relation, applicable to both ramjets and scramjets, is extremely simple:

$$I_f = I_{max} - \frac{V_1}{\alpha}\left\{ 2 - (\eta_D + \eta_N) + \frac{1}{4}\left[\frac{2\alpha h_f}{V_1^2}\right]^2 + \frac{(V_3/V_1)^2}{1 - (V_3/V_1)^2}\,\frac{2\alpha h_f}{V_1^2}\left[1 - \eta_D + \frac{2\,h_1}{V_1^2}\right]\right\}$$

Inspection of this relation shows that the terms in braces represent the losses that reduce the performance below ideal. When $V_3 \ll V_1$, as with subsonic combustion, the fourth term within the braces vanishes, and this term represents losses associated with supersonic combustion. The other losses are common to both cycles, and the superiority of the scramjet at high speed arises from the improved intake and nozzle efficiencies. From the second term, it is apparent that intake and nozzle efficiencies are equally significant and that all engines with the same sum $(\eta_D + \eta_N)$ will have the same performance.

The relative magnitude of these losses is shown graphically on Figs. 6a and 6b for values typical of subsonic and supersonic combustion, respectively. The way in which intake and nozzle losses ($\eta_D + \eta_N$) may absorb all the available impulse at high speed is apparent from Fig. 6a. Similarly, the large losses associated with supersonic combustion at low speed are shown on Fig. 6b. The compromise between intake and combustion losses sets a difficult optimization problem in the design of scramjets, which can be resolved only by experimental determination of intake efficiencies.

It must be emphasized that the above-mentioned methods of approximate analysis illustrate only performance trends and should not be used for accurate performance estimates. In particular, nonequilibrium flow in the nozzle can result in very serious losses in ramjet impulse at hypersonic speeds.

Friction and fuel/air mixing losses have been omitted in the preceding discussion to avoid complicating the issues. Considering these in turn; the approximate effect of skin friction on specific impulse can be demonstrated very simply as follows. The friction drag D reduces the thrust F to (F - D), and the modified specific impulse I'_f then becomes

$$I'_f = \frac{F - D}{W_f}$$

$$= I_f - \frac{D}{\alpha \, W_a}$$

$$= I_f - C_f \frac{(\rho V_1^2 A_{wetted}/2)}{\alpha \, \rho \, A_1 \, V_1}$$

If we approximate the scramjet drag surface by a tube of radius r and length l, then,

$$I'_f = I_f - C_f \frac{(2\pi r l V_1/2)}{\alpha \, \pi \, r^2}$$

$$= I_f - \frac{V_1 \, l}{\alpha \, r} C_f$$

To this order of approximation, we can assume that C_f is constant. The specific impulse including friction is therefore I_f reduced by an amount that increases

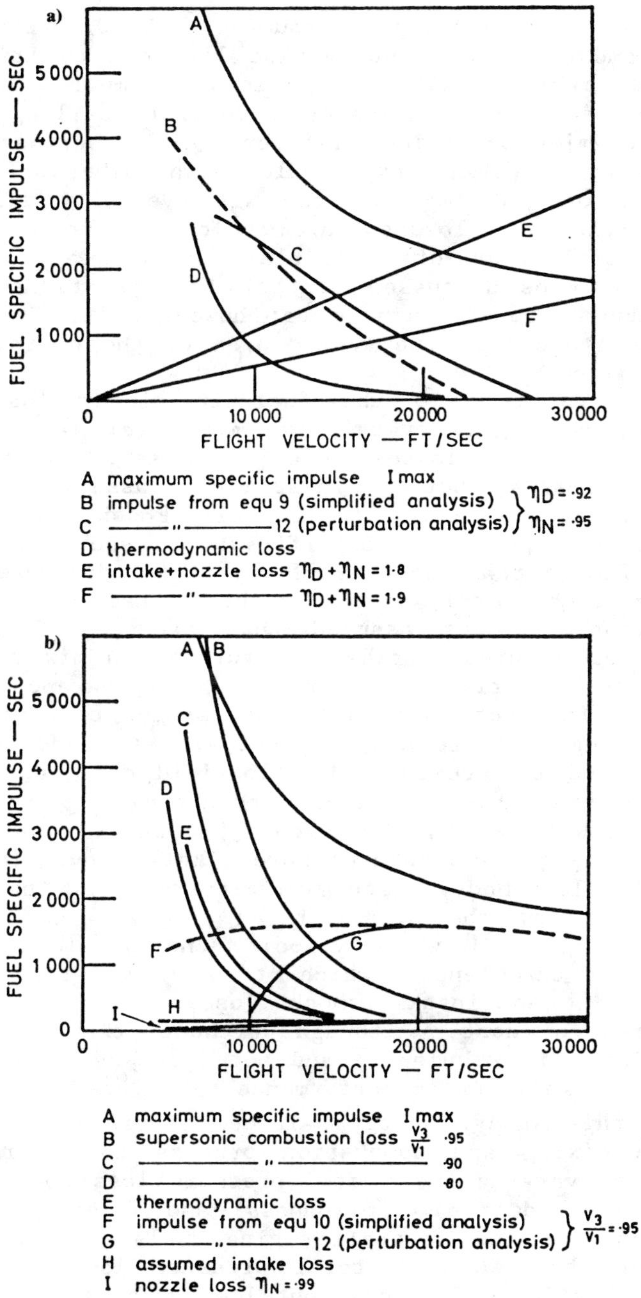

Fig. 6 Combustion: a) subsonic, b) supersonic. Losses to be subtracted from I_{max} to obtain specific impulse.

linearly with $\bar{v}$elocity. Assuming that $C_f = 0.0025$, $l/r = 15$, and $\alpha = 0.03$, the impulse loss due to friction is plotted on Fig. 7. The large loss of impulse at orbital velocity and above indicates a possible ceiling velocity for scramjet operation, and certainly a serious loss of impulse at the higher hypersonic Mach numbers. The linear behavior of the impulse loss with velocity illustrates that friction losses are analogous to the kinetic energy efficiency of the inlet and nozzle. Friction losses will be discussed later in the sections covering individual scramjet components; however, it is apparent at this stage that the wetted area of the engines should be minimized.

We turn next to the fuel/air mixing loss; early analysis of the overall scramjet cycle allowed for intake losses, losses due to heat addition, and nozzle losses, but invariably assumed negligible mixing loss. Although it is well known that subsonic combustors require a significant pressure drop across the baffle (e.g., about 5% of the total pressure in gas turbine combustors), in the case of supersonic combustion, it was assumed that hydrogen fuel from downstream or cross-stream injectors would mix readily in the jets. Early photographs of the appearance of flame at the fuel/hot air interface seemed to confirm the zero mixing pressure drop concept, and the eddy-viscosity models used to compute the flowfield did not highlight the pressure losses actually occurring. In fact, as shown in Ref. 4, the hydrogen jet carries very little "mixing energy" and behaves more like a conventional wake behind a bluff body. The mixing energy is thus largely extracted from the airstream and appears as a total pressure loss in which velocity energy is converted first to turbulence, which then decays to heat. By comparison, the intake shock losses also convert total pressure to heat, although without the intermediate formation of turbulence, and both processes result in a loss of overall engine performance.

In this study, energy balance principles are applied to the mixing and combustion process to determine the effect of varying the total pressure loss due to mixing (i.e., the degree of mixedness) and, hence, combustion efficiency on the overall engine cycle performance. The trade-off between high total pressure loss, leading to high combustion efficiency, but low cycle performance, and low pressure loss causing low combustion efficiency again leading to low cycle performance, is investigated and optimized. The practical realization of an optimum combustor requires not only the correct amount of mixing

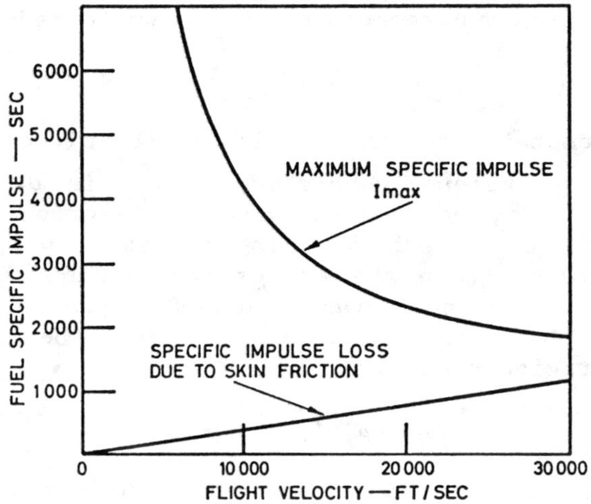

Fig. 7 Specific impulse loss due to skin friction (approximate method).

but also the attainment of this mixing in a short length since wall friction losses in ducted supersonic flow are extremely large. Turbulence theory [7] can again be used to determine the length of the mixing region and, hence, the required scale of turbulence. Experimental evidence of the validity of this concept obtained at simulated Mach 10 flight conditions is reported later.

In order to incorporate combustion-chamber performance into an overall engine cycle calculation, it is necessary to define further efficiency parameters that describe the mixing and combustion processes. If the combustion chamber is considered to consist of a duct into which a mixing device such as a center-body fuel injector has been placed, then a kinetic energy efficiency η_m, due to the turbulent mixing processes, may be defined

$$\eta_m = \frac{h_t - h_{ms}}{h_t - h}$$

where h_t is the total and h the static enthalpy respectively and h_{ms} is the static enthalpy in the mixed region when expanded isentropically to the initial pressure.

This definition is then compatible with the intake kinetic energy efficiency η_d defined previously. If now a constant pressure mixing system is assumed, then the mixing efficiency parameter can be shown to reduce to

$$\eta_m = (U_m/U)^2$$

where U_m denotes the mean velocity in the mixed region.

If the turbulence-generating device is designed with an aerodynamically shaped upstream surface and the momentum of the fluid lost in injector wall friction is negligible, then the energy loss occurs in the wake (because of base drag) and appears first as turbulence. An energy balance across the turbulence-producing region then gives:

$$3/2\ (\rho\ u'_{max}{}^2) = \Delta P$$

where u'_{max} is a reference rms turbulence velocity, assuming that the turbulence generated is isotropic and all the energy becomes turbulence, i.e. there is no dissipation. At hypersonic flow conditions, this pressure loss can also be equated with the overall loss in mean kinetic energy:

$$3/2\ (\rho u'_{max}{}^2) = \Delta P = 1/2\ \rho U^2 - 1/2\ \rho U_m{}^2$$

hence

$$U_m/U = (1 - 3(u'_{max}/U)^2)^{1/2}$$

The mixing efficiency parameter then becomes

$$\eta_m = 1 - 3(u'/U)^2{}_{max}$$

which can then be incorporated into engine cycle calculations in the usual manner.

To show what this means in terms of loss of specific impulse, we can write, approximately,

$$\Delta I_f = (W\ \Delta U)/(g\ W_f)$$

$$\sim (3\ U\ /\ 2\ \alpha g)(u'\ /\ U_{max})^2$$

This function is plotted in Fig. 8, where a loss in specific impulse amounting to a few thousand m/s is apparent. It is interesting to note that this loss of impulse increases linearly with flight velocity in the same way that the loss of impulse due to skin friction increases linearly with flight velocity as shown earlier. Thus, turbulence-generating devices must

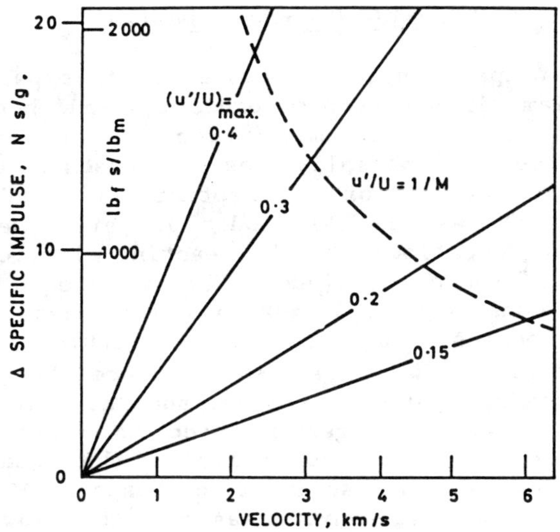

Fig. 8 Specific impulse loss due to mixing.

be designed to produce relatively low turbulence intensities to avoid large thrust penalties, and they must also be designed to utilize the turbulence efficiently.

The principal results of the approximate analysis can be summarized as follows:

1) Scramjet performance is poor in the lower supersonic speed range ($M < 4$) because of the high losses associated with adding heat to a supersonic stream.

2) Between Mach 4 and 8, either the ramjet or scramjet cycles may be used; the choice will depend on the particular mission requirements. However, the scramjet performance always overtakes the ramjet performance at higher Mach numbers.

3) Between Mach 8 and 18, good performance can be anticipated for the scramjet.

4) At higher Mach numbers, fuel/air mixing and friction losses become increasingly serious, and the attainment of significant impulse depends on the optimization of many factors.

The preceding discussion highlights the role of energy flows in hypersonic propulsion systems.

Scramjet Combustor Design

The principal topic discussed in this paper is the mixing problem in supersonic combustion and, hence, major emphasis is given to scramjet combustor design. At the present time, we normally design subsonic combustion systems with the aid of a computational fluid dynamics (CFD) code such as FLUENT. Codes of this type evaluate the governing equations of the reacting flowfield using a finite-difference technique. The major equations used are the conservation equations for mass, momentum, energy, and species, together with a turbulence model such as the well-known k,ϵ model. Since the combustion is usually mixing-limited, a model such as that proposed by Magnussen and Hertager is used to represent the mixing process. In supersonic flow, the equations are parabolic; however, in scramjet combustors, there will usually be some subsonic areas of the flow in which recirculation is present, necessitating a mixed flow analysis. The use of CFD codes insure that factors such as energy balance are satisfied; however, they can conceal some of the fundamental features of the reacting flowfield. In this paper, simple approximations are used in order to draw attention to these fundamentals, although the need for a general computational fluid dynamic code for supersonic combustor design will be emphasized.

An important parameter in the design of scramjets is the performance of the intake, where some compromise is usually necessary between the amount of intake diffusion and the efficiency with which the air is diffused. The amount of diffusion is normally specified by the diffusion factor (M_1/M_3). This parameter has the advantage that its optimum value remains almost constant throughout the hypersonic speed range. The intake efficiency may be specified by the process parameter K_d, which relates efficiency to diffusion and is defined as

$$K_d = \frac{h_3 - h_1'}{h_3 - h_1}$$

$$= \frac{\eta_d - (V_3/V_1)^2}{1 - (V_3/V_1)^2}$$

Some typical performance curves are shown in Fig. 9 for four values of η_d spanning the range of likely intake efficiencies. These performance curves were calculated by the exact analysis outlined in Ref. 1 and

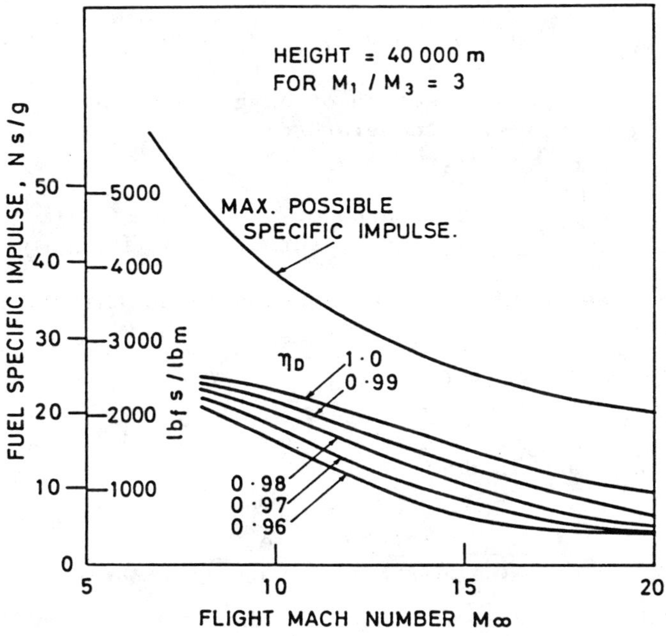

Fig. 9 Scramjet performance.

assume an optimum diffusion factor of $(M_1/M_3) = 3$. A simple two-shock inlet was also assumed with stoichiometric hydrogen-fueled combustion at 100% efficiency and an isentropic nozzle expanding to $p_7/p_1 = 2$. These results were calculated for a height of 40,000 m but are not very sensitive to altitude. Data such as those given in Fig. 9 provide a starting point from which design studies can be evolved since the results can be recalculated to incorporate component performance details as they becomes available.

In scramjet systems currently envisaged for flight at $M > 8$, the intake processes will generally result in static temperatures at the combustion-chamber entry in excess of 1000 K such that spontaneous ignition of hot hydrogen (or other) fuel is possible. Thus, the chemical kinetic requirements of combustion can usually be met, providing the fuel and air can be mixed satisfactorily. In such a mixing-limited system, the maximum achievable combustion efficiency must be related to the turbulence dissipation mechanism, as discussed earlier. A study of stirred reactor theory [5] allows the completeness of combustion to be expressed in terms of the dimensionless time ratio τ_{sd} (Ref. 6):

$$1 - \psi = 1 / [1 + 1/\tau_{sd} + \exp(1/\theta)/\tau_{sk}]$$

where

ψ = fraction of oxygen untreated
θ = f(temperature)
τ_{sd} = τ_s/τ_d
τ_{sk} = τ_s/τ_k
τ_s = residence or stay time
τ_d = mixing or turbulent diffusion time

If the kinetic time is short compared to the mixing time, then this becomes

$$\eta_c = 1 - \psi = 1 / (1 + 1/\tau_{sd})$$

The stay time τ_s is determined by the reactant throughput and is defined as

$$\tau_s = \frac{\text{Reactor volume}}{\text{Volumetric flow rate of reactant}}$$

$$\sim \frac{X}{U}$$

where the reactor volume is confined to the region in which turbulence is being dissipated. In the present simplified model, mixing rate is assumed to be directly proportional to the dissipation rate of the turbulent fluctuations by Reynolds analogy. It can be shown [6] that the diffusion time τ_d is given by

$$\tau_d \sim l_e / u'_{max}$$

where l_e is the average size of the energy containing eddies and is directly proportional to some characteristic dimension (λ) of the turbulence-producing device. Thus,

$$\tau_{sd} = (X/U)(u'_{max} / l_e)$$

In order to ascribe values to τ_{sd}, it is reasonable to assume at this stage that, for a simple shear-layer turbulence-generation system, $l_e \sim 0.2\lambda$ and $X \sim 10\lambda$ [6] so that

$$\tau_{sd} = 50(u'/U)_{max}$$

and is independent of λ!. The mixing-limited combustion efficiency is then

$$\eta_c = \frac{1}{1 + 1/50(u'/U)_{max}}$$

The two efficiency parameters η_m and η_c just defined may now be included in the engine-cycle analysis. Both parameters relate the efficiency to the turbulence-producing region through the turbulence intensity $(u'/U)_{max}$, which is a function of the energy loss across the turbulence-generating device and, for simple systems, may be related directly to the size of the fuel-injection step or baffle.[6] The effects of including η_m and η_c into the engine-cycle calculation are shown in Figs. 10 - 13. The results are for a specific engine geometry at a specific flight condition and illustrate the effect on performance of varying the intensity of turbulence produced in the combustion chamber.

In Fig. 10, engine specific impulse is plotted as a function of turbulence intensity $(u'/U)_{nax}$ for a flight Mach number of 10, assuming an intake diffusion factor of 3. The results are plotted for four values of η_d covering the likely range of intake efficiencies. The curves show that an optimum value of $(u'/U)_{max}$ exists and that this value decreases as the intake efficiency is reduced. For turbulence intensities above the optimum value, so much mixing energy is taken from the airstream that the losses become

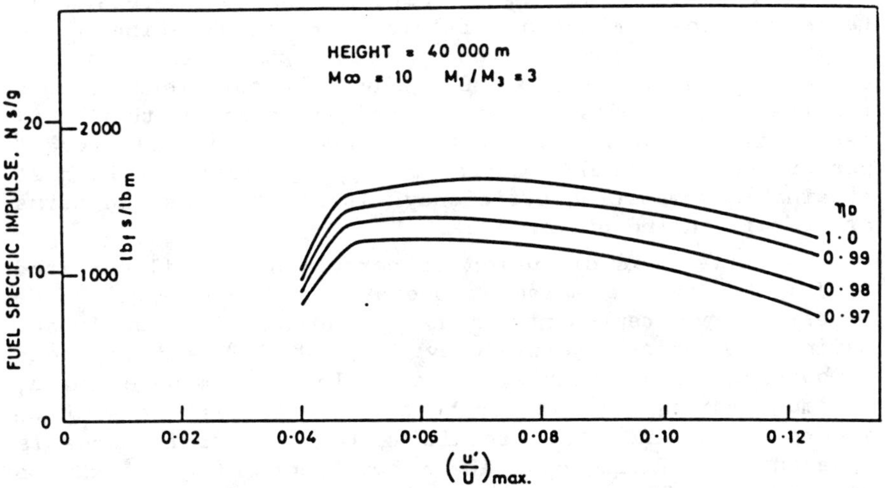

Fig. 10 Variation of engine performance with turbulence intensity.

prohibitive and the engine performance falls off. For values of $(u'/U)_{max}$ below the optimum, there is insufficient mixing within the combustor and the combustion efficiency itself suffers, again reducing the performance. The optimum value of $(u'/U)_{max}$ thus represents a condition at which the mixing and combustion losses together have been minimized and engine performance maximized.

At this stage, it should also be noted that not only is the intensity of turbulence important to the combustion process but also that the scale of the turbulent fluctuations is important. Since the rate of mixing is here assumed to be directly proportional to the rate of turbulence dissipation, [6] then the position of the dissipation zone will depend upon the nature of the initial eddies. If the initial eddies are very large, then they will travel a considerable distance before they have dissipated their turbulent kinetic energy, and the combustor will be long. The combustor must therefore be designed to generate the optimum turbulence intensity and to provide a scale of turbulence sufficiently low that the combustor can be kept as short as possible. On the other hand, the turbulence scale must initially be as large as the inhomogeneities that must be mixed. Thus, reducing the scale leads to the requirement for many small fuel jets (or a narrow two-dimensional slot).

In Fig. 10, the engine performance is expressed in terms of the combustion efficiency η_c by relating η_c to $(u'/U)_{max}$. The curves suggest that the optimum efficiency is less than 80% since the turbulence levels required to obtain efficiencies greater than this result in greater pressure losses and reduced overall performance. It would therefore appear that the highest attainable combustion efficiency is not the best in terms of overall engine performance.

The effect of diffusion on performance is illustrated in Fig. 13 for a scramjet operating at 4900 m/s. The upper curve represents the performance of an intake having a process efficiency K_d of 0.9 and neglects combustion and nozzle losses. In the middle curve, optimum mixing and combustion losses have been incorporated, and the resulting loss in performance is apparent. Including the combustor losses has no significant effect on the optimum value of (V_3/V_1) since the effect on the combustor of changing V_3 is included in the upper curve. In order to present a

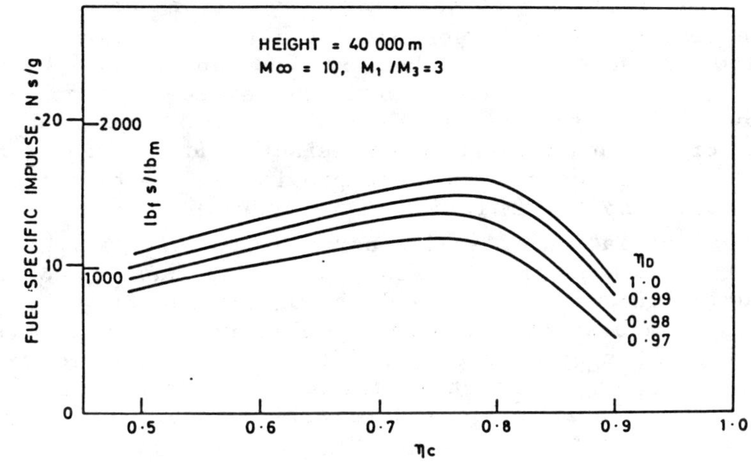

Fig. 11 Variation of engine performance with combustion efficiency.

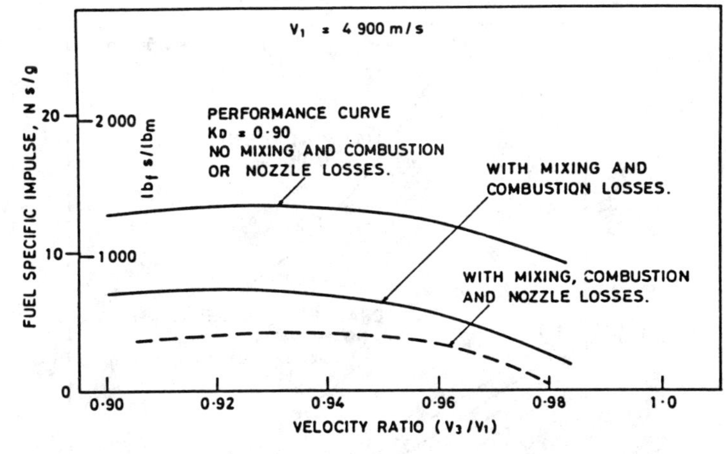

Fig. 12 Variation of engine performance with inake diffusion illustrating the effect of mixing and combustion and nozzle losses.

more complete picture, the effect of including friction and nonequilibrium losses in the nozzle, according to the scheme of Ref. 1, is illustrated in the lower curve. Since the nozzle losses increase with decreasing velocity ratio, the optimum (V_3/V_1) has increased above the value previously indicated. The important feature of Fig. 13, however, is the large loss in performance incurred by incorporating the mixing and combustion losses. At the flight condition of Fig. 13 these losses are so large that the final performance value is low enough to be matched by conventional chemical rockets and to make scramjet systems apparently uneconomic at this speed. This observation is based on a very crude analysis, and more exact study is required to validate the conclusion and provide a firm estimate for the upper speed limitation of the scramjet.

The variation of engine performance with flight Mach number is shown in Fig. 13 which assumes a constant diffusion factor of $M_1/M_3 = 3$ throughout the flight range. The optimum overall combustor efficiency for each flight condition has been chosen, and the performance, expressed as $I_{sp(optimum)}$, is the maximum attainable at that condition. Nozzle losses are neglected in this figure. As can be seen from Fig. 13, the engine performance decreases approximately linearly with increasing flight Mach number > 5, finally reducing to zero between Mach 19 and 22, depending on the intake performance. If nozzle

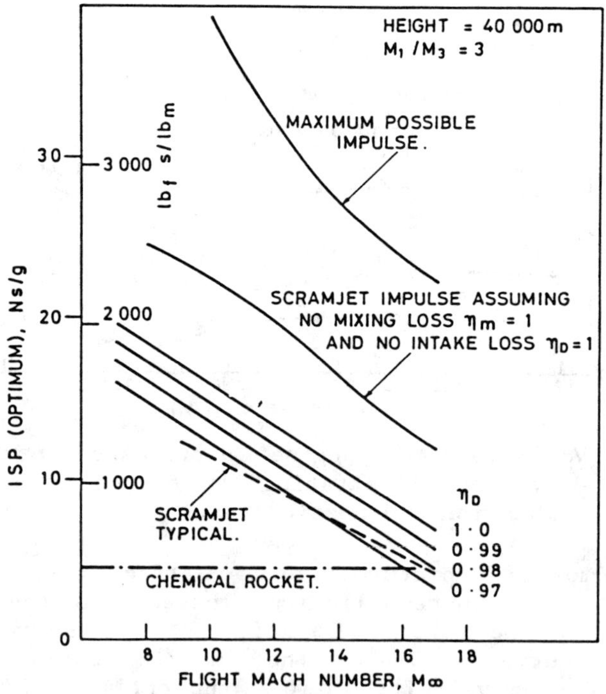

Fig. 13 Variation of engine specific impulse with flight Mach number, neglecting nozzle losses.

losses are taken into account and reasonable values of intake performance assumed, then, if we neglect the contribution from recycled heat, scramjet performance approaches that of the chemical rocket at a Mach number of about 17. When the recycled heat contribution is included, then scramjet performance exceeds that of a chemical rocket up to speeds approaching satellite velocity.

The data that have been presented thus show the general trends in engine performance with varying combustor performance and illustrate the significant impulse losses for scramjets, which inevitably operate with mixing-limited combustion systems. The significant features of the data follow:

1) An optimum level of turbulence exists at which mixing and combustion losses, taken together, are minimized.
2) The best combustion efficiency in terms of engine performance is considerably less than 100%.
3) When combustor losses are included in the theoretical scramjet cycle, engine specific impulse varies approximately linearly with flight Mach number.
4) The crude theory used directly relates the combustor performance to the geometrical arrangement and pressure loss and may be extended to include droplet evaporation, chemical kinetics, and flow features such as recirculation, so that it can be applied to many other combustor configurations.

Experimental Approach

An investigation into the effects of initial turbulence generation on the position and extent of the mixing and combustion zone was carried out using our shock-tunnel test facility. For the tests, the tunnel was fitted with a two-dimensional Mach 3.5 test section, and supplied air at stagnation conditions of approximately 6000 K at 200 bars. This permits simulation of scramjet combustor inlet conditions for flight at Mach 10 at an altitude of 25-30 km. These conditions are obtained using a combustion-driven hypersonic shock tunnel operated at the tailored interface condition ($10.2 < M_s < 10.6$) to give a contamination-free test time of 2 ms.

In general, the time required for the heat-release process to go to completion and, hence the length of the combustor, depends on the time needed to mix the fuel and air, and the time needed for it to react. In practice, these two processes are occurring simultaneously; however, with the test section conditions used in this study (T_{static} = 2000 K, p_{static} = 1 atm, M = 3.5), the chemical kinetic time for the hydrogen/air reaction is of the order of 10^{-6} s. Consequently, once reaction is initiated, complete combustion can occur within a very short distance (approx. 3 mm). However, in order to obtain complete combustion, all the fuel and air must be mixed in the correct proportions down to the molecular level. The size of the mixing region can be estimated as follows:

As stated previously, the mixing rate is assumed to be directly proportional to the turbulence dissipation rate. Structural turbulence theory [7] shows:

$$dD/dt = \rho \, A \, u'^3 \; / \; le$$

Consequently, the dissipation rate and, hence, the mixing rate at any axial position for a given system is dependent on the turbulence level (u') and a constant (l_e or λ). Integration of the preceding equation gives (Ref. 5)

$$\text{Mixing rate} \sim (\Delta D/q)/(\Delta x/\lambda)$$

$$\sim B \, (3(u'/U)^2/4)^{3/2}$$

The mixing rate and, hence, combustion rate can be related to the geometry by carrying out an aerodynamic energy balance. This reasoning shows that for a given amount of freestream energy available for turbulence production and a given throughput of fuel, the shortest combustor will be obtained when the turbulence dissipation rate is maximized over as short a distance as possible.

The experimental study investigated the effects of initial turbulence generation on the position and extent of the turbulence dissipation rate for a nonreacting hydrogen/nitrogen system and compared the results with the observed positions and extent of the combustion zones obtained using hydrogen/air. Two separate midstream injector configurations were tested, having equal geometric blockage ratios (and similar drag) in order that the total pressure energy theoretically available for turbulence generation be the same in each case. One injector consisted of a plain 12-deg wedge with fuel injected from the base. The other injector was specifically designed to be an efficient turbulence generator and consisted of a plain 12 degree wedge with backward-facing deltas attached to the trailing edges. These deltas sweep back behind the

Mach angle and produce swiss-roll-type vortices that interact in the near wake. The hydrogen was injected from the base into this highly turbulent region.

The nature and duration of the test flow precludes conventional velocity, turbulence, and concentration traverses in the wakes. However, midstream pitot pressures were obtained using a probe designed specifically to detect changes in test gas molecular weight. The aerodynamic energy balance across the injector is

$$p_1 + KE_1 = p_2 + KE_{wake} + KE_{turb} + D$$

Non dimensionalizing throughout by $q = \rho U^2/2$ gives

$$\frac{\Delta P}{q} + \frac{KE_{def}}{q} = 3\left(\frac{u'}{U}\right)^2 + \frac{D}{q} \qquad *$$

This equation permits calculation of the turbulence generation and dissipation and, hence, the mixing zones in the wakes of the injectors if the total pressure drop ($\Delta P/q$) across the injector is known, together with the axial distribution of the wake KE deficit (KE_{def}/q).

The pressure drop across the injector appears first as base drag ΔP = drag / A_{ref}, where A_{ref} is a reference area, which ideally should be taken as the initial area of the streamtube within which all the subsequent turbulence generation and mixing occur.

The base drag is given by drag = $C_D\ q\ A_B$, where C_D is an experimentally determined drag coefficient and A_B is the base area then,

$$\frac{\Delta P}{q} = C_D \frac{A_B}{A_{ref}}$$

The photographic results of the combustion zones indicate that the reacting wake does not spread appreciably; hence, A_{ref} is assumed equal to A_B, and then,

$$\frac{\Delta P}{q} = C_D$$

For a 12 degree wedge, the experimental value of $C_D \sim 0.5$ (Ref 8). KE_{def} / q is the nondimensionalised total

KE of the wake velocity deficit at any given axial position. For a two-dimensional wake, the axial decay of the velocity deficit is proportional to $1/x^{1/2}$. [9] Hence, $KE_{def} / q \sim 1/x$

The constant of proportionality in the preceding equation differs for different wakes and, in general, needs to be determined experimentally. Values of this constant were determined for the two injector configurations using measured wake pitot pressures.

Knowing both $\Delta P / q$ and the axial dependence of KE_{ref} / q, it is possible to determine the sum of the turbulence kinetic energy $[3(u'/U)^2]$ and the dissipation energy (D/q) from the energy balance equation. Absolute values of these two terms are then found from the relationship between these two terms from Eq. * where an initial guess at the value of (u'/U) results in a corresponding value of (D/q). Hence, for a given $(\Delta x/\lambda)$ and assuming $l_e = 0.2\lambda$, an iterative routine is followed until the overall energy balance equation is satisfied numerically.

The resultant curves of (D/q), (KE_{ref}/q) and $(3(u'/U)^2)$ are shown in Figs. 14 and 15 for the wakes of both the vortex-generating injector and the plain wedge injector. In each case, it can be seen that the turbulent KE is only a very small proportion of the total available energy, and the dissipation energy rises rapidly, asymptotically approaching the total energy in the far wake. The turbulence dissipation rates and, hence, the mixing rates for the two wakes are shown in Fig. 16, where it can be seen that the maximum mixing rate for the vortex generator is over twice that for the plain wedge and occurs at $x/\lambda \sim 2.5$ compared to $x/\lambda \sim 5$ for the plain wedge. It is clear that, in both cases, the mixing rate decreases rapidly after reaching a maximum value. It is the shape of these mixing curves that shows that the near-wake region closely approximates a well-stirred reactor containing the mixing rate maximum, followed by a plug flow region in the far wake where the mixing rate has fallen to negligible proportions.

An estimate of the amount of mixing occurring in the well-stirred region can be obtained from the area under the dissipation-rate curve. Assuming a minimum dissipation rate beyond which mixing is negligible (taken to be < 0.015), the stirred reactor zone for the wake of the vortex generator is seen to extend from $x/\lambda \sim 0$ to 8, with a mean dissipation rate of about 0.035.

Similarly, for the wedge wake, the mixing region extends from $x/\lambda \sim 3$ to 11, with a mean rate of about 0.025. This shows that the mixing in the well-stirred region is 30% more efficient for the vortex generator. These estimated positions of the well-stirred reactor zones agree well with the observed positions of the combustion zones, which show that, for the wedge, the reaction zone starts at $x/\lambda \sim 8$ and is well established by $x/\lambda = 11$ whereas, for the vortex generator, it begins at $x/\lambda = 5$ and is well established by $x/\lambda \sim 8$.

The combustion efficiency of the system can be estimated from the equation

$$\eta_c = 1 - \psi = 1/(1 + 1/\tau_{sd})$$

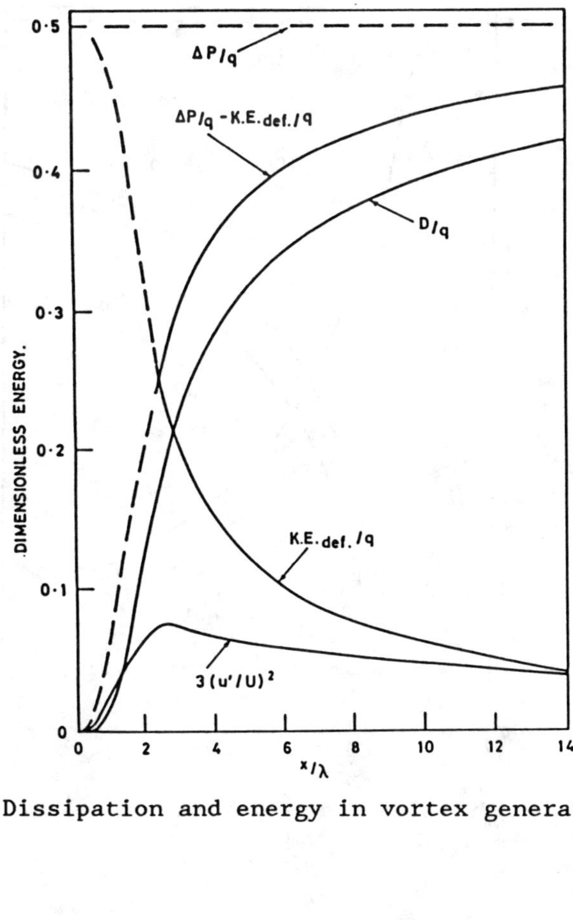

Fig. 14 Dissipation and energy in vortex generator wake.

hence, η_c increases with increasing τ_{sd} which, for a mixing-limited system, is given by

$$\tau_{sd} = (x/l_e)(u'/U)$$

Taking values of X and u'/U from Figs. 14 and 15 and assuming that le = 0.2λ, η_c is found to be about 86% for the vortex generator and about 84% for the plain wedge. This close agreement is not surprising when it is realized that the total pressure drop available for both configurations is equal, and, as the total dissipation curve becomes asymptotic to ΔP/q, the total mixing will be the same (Figs. 17-19). However, the important difference between the two from the propulsive efficiency viewpoint is the extent of the mixing zone X and the turbulence intensity u'/U. It is clear that large X and small u'/U will give good combustion efficiency but large friction, heat

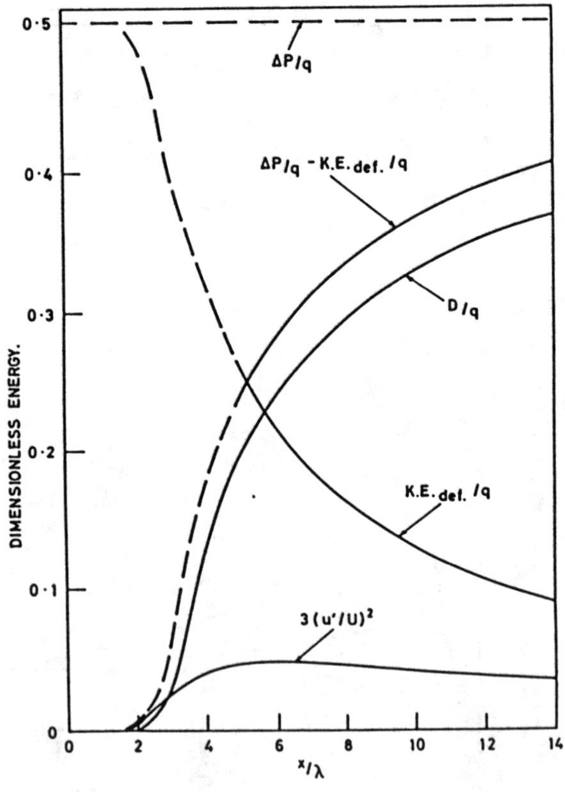

Fig. 15 Dissipation and energy in the wake of a wedge.

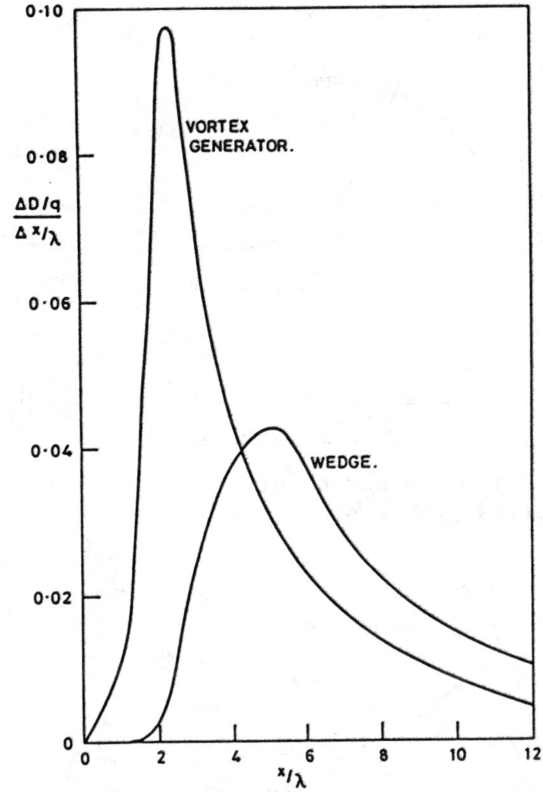

Fig. 16 Comparison of mixing rates in the wakes of the two injectors.

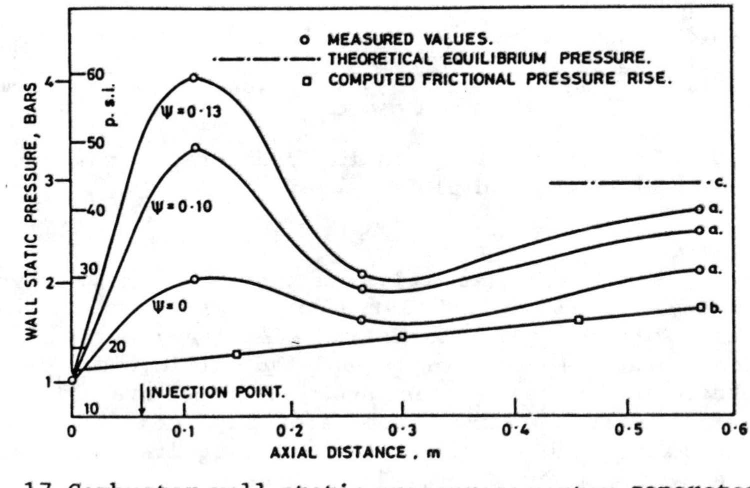

Fig. 17 Combustor wall static pressures; vortex generator.

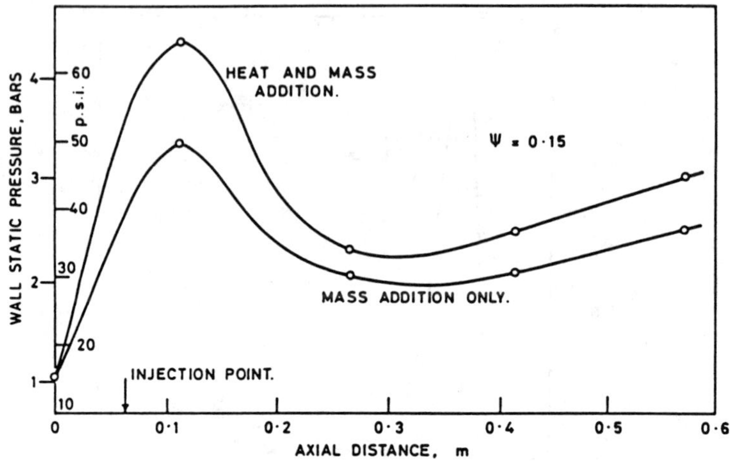

Fig. 18 Static pressures with and without combustion vortex generator.

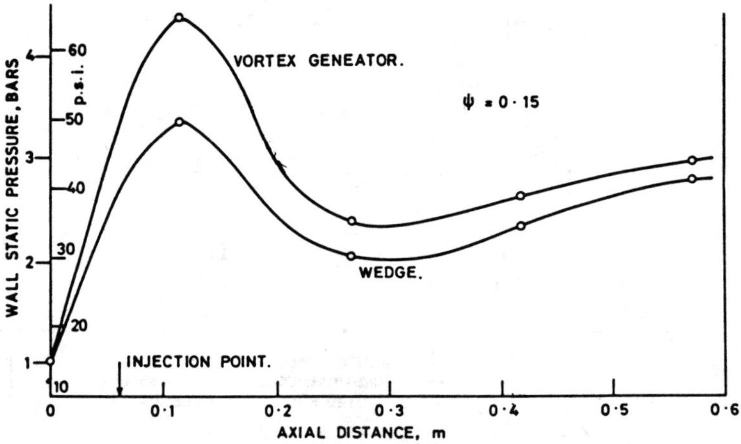

Fig. 19 Comparison of static pressures from vortex generator and plain wedge.

transfer, and weight penalties on the overall system efficiency. Small X and large u'/U will offset these penalties for the same combustion efficiency but, if u'/U is too large, losses in propulsive efficiency can be expected, as has been shown in the overall cycle analysis. The optimum combustor must combine small X and optimize u'/U to give the required combustion efficiency for maximum specific impulse.

An interesting new development in supersonic mixing technology is based on the instability of missmatched supersonic nozzle flow in the presence of an acoustic reflector. A typical arrangement is shown in Fig. 20 where the acoustic reflector is located around the nozzle. Alternative configurations are reflectors placed concentric with the transonic jet, or even upstream of the sonic throat. An example of this phenomenon was reported in Ref. 10. The effect of the reflector is to cause the jet to gyrate rapidly, probably as a result of the pressure field deflecting the jet, which then changes the flow to the first shock "diamond", leading in turn to the emission of a pressure pulse toward the reflector. The result of the jet instability can be easily detected in the rapid decay of the pressure on the jet axis when the reflector position is optimized (Fig. 21). A rapid decay in species concentration on the jet axis also accompanies the instability (Fig. 22), and optical diagnostics confirms the rapid spread of the precessing jet. We are presently investigating the application of this novel mixing technique to scramjet combustor design.

Experimental Results

Qualitative interpretation of wall static pressure measurements and direct glow photographs supports the assumption that mixing rate is approximately proportional to dissipation rate and shows the importance of the initial turbulence generation on the subsequent mixing and reaction zones. Figure 17 shows;
1) Measured wall static pressure distributions for the vortex generator.
2) The computed pressure increase due to wall friction alone.

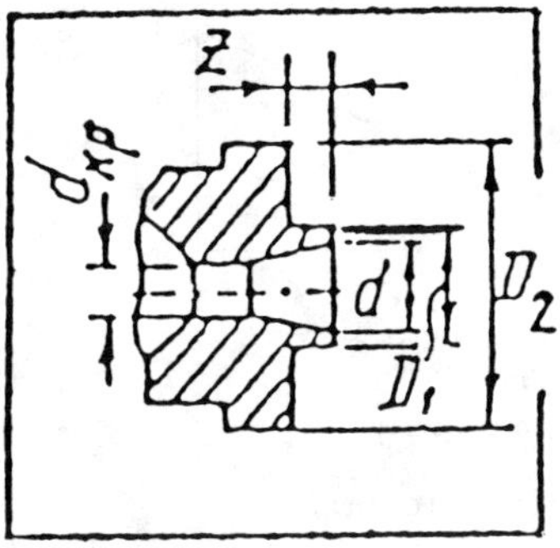

Fig. 20 Multichannel pulse LDA.

3) The final equilibrium combustion pressure as calculated from one dimensional Mollier chart calculations. It can be seen that the measured wall static pressure closely approaches the final pressure calculated for complete combustion. It is worth noting that at these conditions, the wall friction effects produce pressure

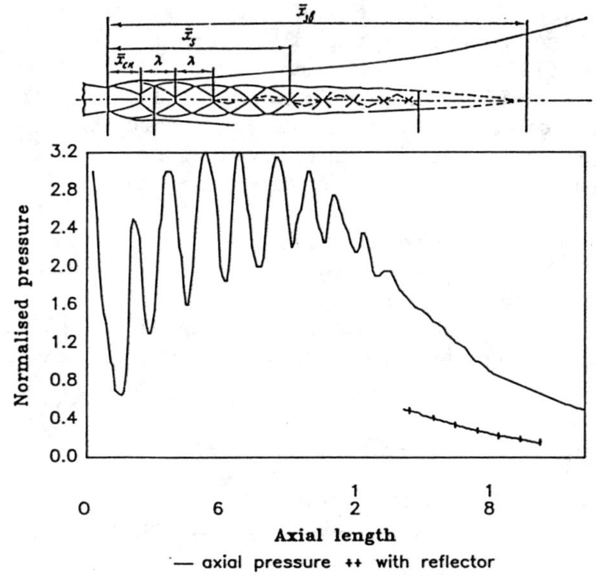

Fig. 21 Jet axial pressure.

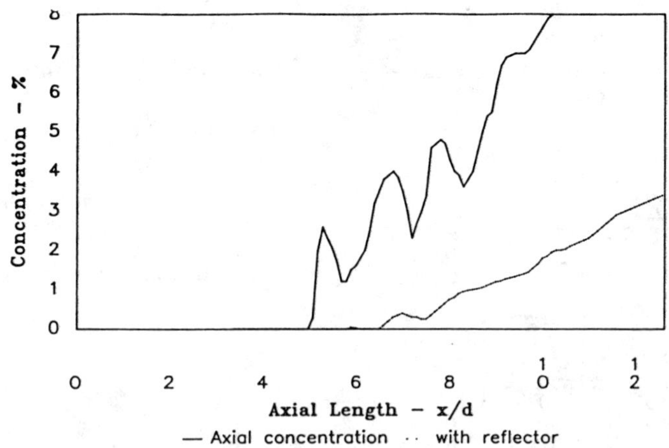

Fig. 22 Jet axial concentration.

rises comparable to the theoretical combustion pressure rise.

The steep compression zone following injection is due to the combined effects of heat and mass addition and the presence of the fuel injector. The theoretical equivalence between heat and mass addition [11] shows that, if dQ/dx is large, steep compression of the external flowfield will result. In order to investigate the effects of heat and mass addition, nitrogen was used in place of air as the test gas. [12] A comparison of the pressures obtained with equivalent hydrogen flow rates using air and nitrogen is shown in Fig. 18. These results show that there is a considerable pressure rise with mass addition alone; however, the increased pressure obtained with combustion indicates that mixing and combustion occur within a very short distance of injection.

The general shape of the wall static pressure distribution can be explained qualitatively in terms of the stirred reactor and turbulence concepts already presented. The most important factor affecting the external pressure field is the rate of heat release per unit length (dQ/dx), the absolute magnitude of which is determined by the mean flow velocity and the "reaction rate." The reaction rate is mixing-limited and, hence, depends on the turbulence dissipation rate. In the near wake of either injector, the mean axial velocity will be less than the freestream value. This relatively low-velocity, highly turbulent region corresponds to the well-stirred reactor zone. Consequently, if combustion occurs in this region, dQ/dx will be large, and the resultant pressure field will give rise to a high wall static pressure. Further downstream, the turbulence has all been dissipated, and the mean axial velocity increases. These conditions approximate those of a "plug flow" region, in which dQ/dx will be relatively small, resulting in lower wall static pressure. Comparison of wall static pressures obtained from both injectors for equivalent stoichiometry is shown in Fig. 19, where it can be seen that the pressure induced by the well-stirred region is lower at station 2 for the plain wedge, indicating that dQ/dx is lower in this region. Pitot pressure measurements showed that the axial velocity in the near wake of the wedge was less than that for the vortex generator and, hence, the residence time was higher in this region. Consequently, the inferred reduction in dQ/dx must be due to a decreased reaction rate brought about by lower mixing rates in this near-wake region. [11]

Multichannel Pulse LDA

The possibility of carrying out laser Doppler anemometry over an extended measuring volume using simultaneous data collection from several points has been investigated. The novel technique we have developed is based on the use of a diffraction grating located in the split beams of a conventional LDA configuration. This grating is specially blazed to distribute the light energy approximately equally into the first five orders. This results in eleven beam intersections in the focal plane of the main lens; i.e., five positive orders, five negative orders, and the undiffracted beam (Fig. 23). The introduction of a second grating perpendicular to the first splits each of these beams into eleven, giving a total of 121 measurement volumes (Fig. 24), all of which are located at precisely uniform separations on a rectangular grid. The Doppler signals resulting from the passage of particles through the fringes formed by the beam intersections may be transmitted to a group of detectors by means of fiber optics. The laser power at each intersection is only about 1% of the total laser power; however, the use of a pulsed ruby laser means that there is abundant laser power available over the pump flash duration of 1 ms. Since we need to use this system with our shock tunnel, in which each test lasts about 1 ms, there is good matching between the capabilities of the two systems. In fact, a non-Q-switched pulsed ruby laser does not emit a continuous beam for 1 ms but actually emits a series of about 50 separate spikes, each of which has a duration

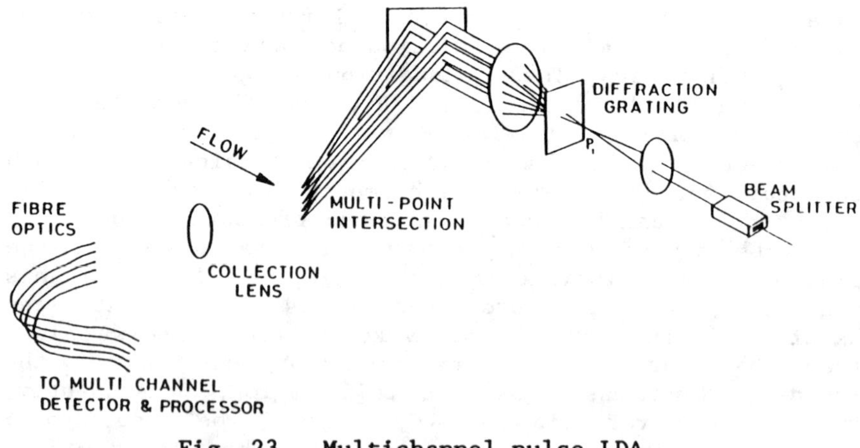

Fig. 23 Multichannel pulse LDA.

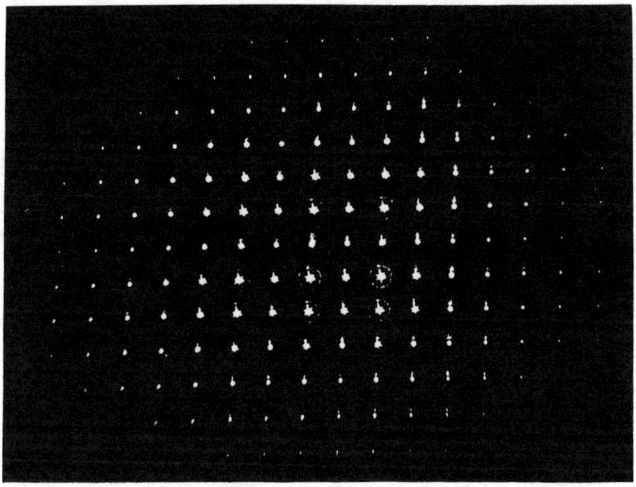

Fig. 24 Finite difference modelling.

of about 2 μs. Since the Doppler frequency we wish to measure is much greater than 1 MHz, there is adequate resolution in each burst. We have already demonstrated that it is possible to obtain samples of Doppler bursts from flow seeded with 1-2-μm particles by means of this system (Fig. 25). At a future stage in our scramjet combustor investigations, we propose to use this system to study the details of the steady and unsteady flowfields in the mixing region. This information will help us understand the anisotropic features of the turbulence in the strong shear layers where the mixing takes place.

Finite-Difference Modelling

Space precludes a detailed presentation of the application of computational fluid dynamic (CFD) modeling to scramjet systems, and the main purpose of this paper is to draw attention to the governing principles underlying scramjet combustor performance.

Our development of a CFD code for quantitative scramjet combustor design work is based on an existing finite-difference code, which has been modified to handle steady compressible flows in which both subsonic and supersonic regions may be present. This occurs, for example, in the swept rearward-facing wedge type of scramjet fuel injector, where the flow in the

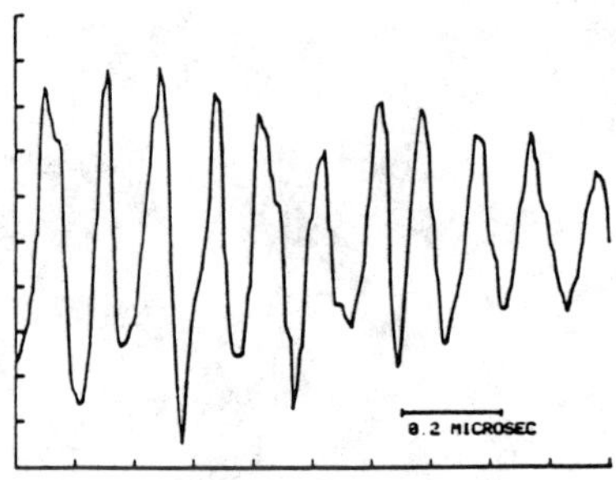

Fig. 25 Sample of Doppler bursts using 1-2 μ titanium dioxide particles and a line of diffracted crossing points. 3rd diffracted order used. Fringe spacing is 10 μ.

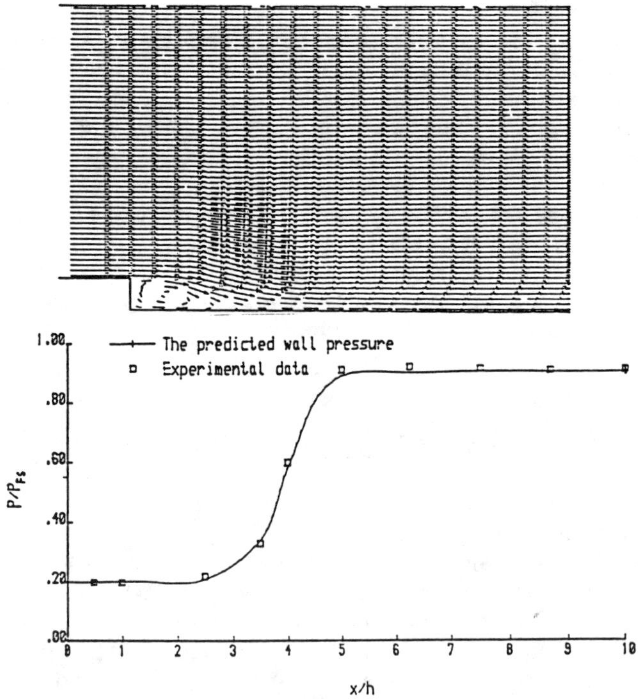

Fig. 26 Comparison between the predicted wall pressure distribution and experimental data for the supersonic laminar flow over a rearward-facing step.

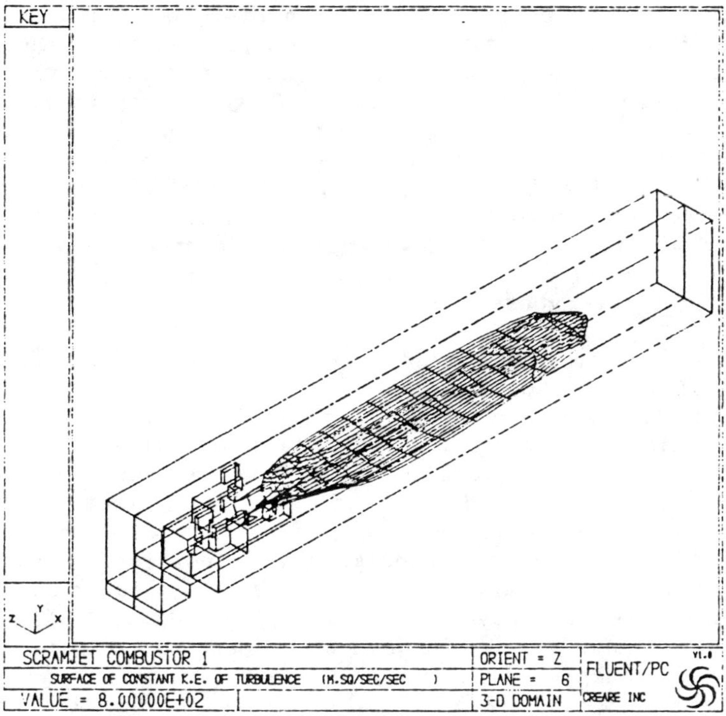

Fig. 27 Scramjet combustor - surface of constant kinetic energy of turbulence.

base region of the wedge deliberately includes a zone of subsonic flow to assist ignition at low flight Mach numbers. An example of the mixed flow over a rearward-facing step is shown in Figs. 26 and 27.

Inspection of the graphics presentation from some of our injector calculations (Fig. 27) shows that the turbulence kinetic energy, turbulence dissipation and, hence, mixing rates are all concentrated in the near wake of the turbulence-generating system. Thus, the same conclusions could be drawn from the application of these modern codes, and they should, of course, be used to quantify actual designs.

Conclusions

Fundamental principles of turbulence generation, dissipation, and mixing-limited combustion have been applied to the scramjet, with particular emphasis on the combustor. It has been shown that severe penalties may be incurred in overall engine performance in order to achieve adequate mixing and, hence, good combustion efficiency in an acceptable length of combustor. Cycle optimization studies show that the highest performance, in terms of fuel specific impulse, is obtained at a combustion efficiency of about 80% at M = 10. Optimum intake diffusion factors are slightly smaller than those estimated in previous studies, in which mixing losses were neglected. The average performance loss due to mixing amounts to about 6 Ns/g (600 lbf sec/lbm) throughout the hypersonic speed range.

In experimental studies, it has been shown that it is possible to deduce some of the mean turbulence characteristics and mixing rates in the wake region of fuel injectors using only mean flow measurements. The predicted axial positions of the mixing zones agree well with the observed combustion zones. These results support the basic assumption of the mixing model that the mixing rate is approximately proportional to the turbulence dissipation rate. They also show that propulsive efficiency will be optimized when the required amount of turbulence needed to produce the optimum combustion efficiency is dissipated in as short a distance as possible. The basic turbulence theory indicates that this can be achieved only by producing initial turbulent eddies that are as small as possible (consistent with the number of injection points). Consequently, the injection system must be designed with this in view, as well as the other problems of penetration, interaction, and low-temperature ignition.

References

[1] Swithenbank, J., "Hypersonic Air Breathing Propulsion," Progress in Aeronautical Science, Vol. 8, edited by D. Kuchemann, Pergamon, New York 1967.

[2] Curran, E.T., and Swithenbank, J., "Really High Speed Propulsion by Scramjets," Aircraft Engineering, Jan. 1966.

[3] Lindley, C.A., Elements of Space Propulsion edited by Simkin and Szego, Wiley, New York.

[4] Swithenbank, J., and Chigier, N.A., "Vortex Mixing for Supersonic Combustion," 12th International Symposium on Combustion, 1969 pp. 1153-1162.

[5] Swithenbank, J., "Combustion Fundamentals," Dept. of Chemical Engineering, Sheffield Univ., U.K., Rept. H.I.C. 150, 1970.

[6] Swithenbank, J., "Flame Stabilization in High Velocity Flow," Combustion Technology, edited by H. Palmer and J.M. Beer, Academic, New York, 1974, pp. 91-127.

[7] Hinze, J.O., Turbulence, McGraw-Hill, New York, 1959.

[8] Hoerner, S.F., Fluid Dynamic Drag. Published by author, 1958.

[9] Schlichting, H., Boundary Layer Theory, McGraw-Hill, New York, 1968.

[10] Mamin, C, and Rimsky-Dorsakov, "Some Experimental Studies of Whistling in Supersonic Air Flows," Gas Flow and Gas Flow with Solid Particles, 1968 (In Russian).

[11] Rues, D., "Concerning the Equivalence between Heat, Force and Mass Sources," R.A.E. Library Translation 1119, July 1965.

[12] Swithenbank, J., Jaques, M.T., and Payne, R. "The Role of Turbulence in Scramjet Combustor Design," Proceedings of the 1st International Symposium on Air Breathing Engines, Marseilles, June 1972. Published by Sheffield University, 1974.

Chapter 7

Mixing and Mixing Enhancement in Supersonic Reacting Flowfields

J. P. Drummond* and M. H. Carpenter†
NASA Langley Research Center, Hampton, Virginia
and
D. W. Riggins‡
University of Missouri-Rolla, Rolla, Missouri

Abstract

Research has been undertaken to achieve an improved understanding of physical phenomena present when a supersonic flow undergoes chemical reaction. A detailed understanding of supersonic reacting flows is necessary to develop successfully advanced propulsion systems now planned for use late in this century and beyond. In order to explore such flows, a study was begun to create appropriate physical models for describing supersonic combustion and to develop accurate and efficient numerical techniques for solving the governing equations that result from these models. For this work, a computer program was written to study reacting flows. The program was constructed to consider the multicomponent diffusion and convection of important chemical species, the finite-rate reaction of these species, and the resulting interaction of the fluid mechanics and the chemistry. The program was first used to study a spatially developing and reacting mixing layer, and the results were analyzed to draw conclusions regarding the structure and growth of the evolving layer. The mixing layer provides a basic reacting flow-

*Senior Research Scientist.
†Research Scientist.
‡Assistant Professor.

field problem that can be studied, but it also serves as an excellent physical model for the mixing and reaction processes that take place in a scramjet combustor. Following the study of mixing layer development and growth, several techniques were considered to enhance the fuel-air mixing and growth of that layer to improve its overall combustion efficiency. Several alternate fuel injector configurations that developed from this study increased the degree of mixing and combustion that could be achieved.

Nomenclature

A_j = reaction rate constant for jth reaction
b_i = body force of species i
C_i = concentration of species i
$\dot{C}_i$ = time rate of change of C_i
c_p = specific heat at constant pressure
D_{ij}, D_T = binary and thermal diffusion coefficients, respectively
E = total internal energy; activation energy
$\vec{E}, \vec{F}, \vec{G}$ = flux vectors in x-, y-, and z- coordinate directions
f_i = mass fraction of species i
g_j, G_R = Gibbs energy of species i and of reaction, respectively
h = height of channel or duct
h_i, h_i^o = enthalpy and reference enthalpy of species i, respectively
$\vec{H}$ = source vector
K_b, K_f = backward and forward rate constants, respectively
K_{eq} = equilibrium constant
M = Mach number
M_i = molecular weight of species i
n_i = moles of species i
ns, nr = number of chemical species and reactions, respectively
p = pressure
$\dot{q}$ = heat flux
R^o = universal gas constant
T = temperature
T_R = reference temperature = 298 K
T_ϵ = effective temperature
t = time
Δt = time step
$\vec{U}$ = dependent variable vector
u, v = streamwise and transverse velocity, respectively

$\tilde{u}_i$ = streamwise diffusion velocity of species i
$\tilde{v}_i$ = transverse diffusion velocity of species i
$\tilde{V}_i$ = diffusion velocity vector of species i
$\dot{w}_i$ = species production rate of species i
x, y = streamwise and transverse coordinates, respectively
X_i = mole fraction of species i
λ = second viscosity coefficient
γ_{ij} = stoichiometric coefficient; species i, reaction j
γ = ratio of specific heats
δ = Kronecker delta function
ρ = density
σ = normal stress
σ_{ij} = effective collision diameter
τ = shear stress
μ_i = laminar viscosity of species i
μ = mixture laminar viscosity
ξ = computational streamwise coordinate
η = computational transverse coordinate
Ω_D = diffusion collision integral

Introduction

Research is currently under way, both in the United States and abroad, to develop advanced aerospace propulsion systems now planned for use late in this century and beyond. One such program is being carried out at the NASA Langley Research Center to develop a hydrogen-fueled supersonic combustion ramjet engine, also known as a scramjet, capable of propelling a vehicle at hypersonic speeds in the atmosphere. One phase of this research has been directed toward gaining a detailed understanding of the complex flow field present in the engine over a range of flow conditions. Numerical modeling of various regions of the engine flow has been shown to be a valuable tool for gaining insight into the nature of these flows. This approach has been used in conjunction with an ongoing experimental program to develop an effective analysis capability[1] .

The flow field in a scramjet engine is governed by the Navier-Stokes equations coupled to a system of equations describing each of the species present initially and produced by chemical reaction. The governing equations were solved in prior analyses using either explicit or implicit finite-difference techniques, with the chemical reaction process modeled by an ideal (mixing controlled) reaction model. Using these approaches, analyses of various ramjet and scramjet configurations have been carried out, and trends that were established by experiments have been predicted [2, 3].

Chemical reaction is not mixing-controlled throughout a scramjet combustor, however. Although chemical reaction may equilibrate in the rearward region of a well-designed combustor, chemistry in the forward portions of the combustor is certainly kinetically controlled. Finite-rate kinetics is, in fact, a critical issue in the design of flameholders in the engine, and this phenomenon must be considered along with the effects of molecular and turbulent fuel-air mixing to develop an accurate engine flow model. It is for this reason that attention has turned in the present work to a more basic and detailed analysis of chemically reacting flow fields. The long-term purpose of the present research is to develop detailed models for fuel-air mixing and reaction in an engine flow field, as well as accurate and efficient numerical methods for solving the equations governing reacting flow that result from these models.

Because of computer resource limitations, however, detailed modeling of the complete engine problem cannot be considered at the present time. A more tractable problem that relaxes only the complexities introduced by engine geometry is posed by the spatially developing, primarily supersonic, chemically reacting two-dimensional mixing layer. A major portion of the chemical reaction taking place in a supersonic combustor occurs in mixing layers. All of the difficulties introduced by the fluid mechanics, combustion chemistry, and interactions between these phenomena can be retained in the reacting mixing layer, making it an ideal problem for the detailed study of supersonic reacting flow.

Earlier Studies on Mixing Layers

Prior studies on supersonic reacting mixing layers have been quite limited. A fair amount of the work has been carried out, however, on nonreacting mixing layers, both supersonic and subsonic. Even without combustion, the results of these studies provided a significant amount of useful information for understanding reacting layers. Carpenter [4] studied the development of a laminar, free-shear layer behind steps and blunt bodies over a Mach number range of 0 to 10. He concluded that the development of the layer could best be understood in terms of vorticity transfer. The effect of compressibility was to increase the diffusion process in the layer, leading to more rapid development toward asymptotic conditions with increasing Mach number. Brown and Roshko[5] studied the subsonic mixing layer that developed between nitrogen and helium streams and found that the layer was dominated by large-scale coherent vortical structures. They found that these structures tended to convect at a nearly constant speed and that the size of the structures and the space between them changed discontinuously with movement downstream by the joining of those structures with their neighbors. Results of their experiment "suggested that turbulent mixing and entrainment was a process of entanglement on the scale of the large structures." They also found that very large changes in the density ratio (up to 49), measured transversely across the mixing layer, had only a small effect on the spreading of the layer. The authors concluded, therefore, that the significant reduction in supersonic mixing-layer growth rate with increasing Mach number was due primarily to compressibility effects rather than density effects, as had been thought in the past.

The role of coherent structures in turbulent processes in mixing layers was studied further by Roshko[6] . He found that the size of the coherent structures and the spacing between them increased with increasing downstream distance. The vortices were found to travel at a constant speed of $(u_1 + u_2)/2$, where u_1 and u_2 are the freestream velocities of the two streams making up the layer. Each vortex also had a finite life span that began and ended abruptly. Coincident with two or more of these endings, a new lifespan began with two or more vortices coalescing to form a new larger vortex. As noted earlier, each of these vortices was observed to move at a nearly constant speed, resulting in a fairly constant spacing between one vortex and its neighbor as they moved downstream during their lifetime. Developed mixing layers are self-similar, however, requiring that the spacing between vortices should increase linearly in the mean with increasing downstream distance. Roshko resolved this contradiction by reasoning that changes in the layer must occur discontinuously and irregularly along the layer such that the scale of the structure grew smoothly and linearly in the mean. Roshko further found that, in the transition region of the layer, there was only one spacing distance between neighboring vortices and that this spacing represented the most stable wavelength selected by the laminar portion of the layer. In this region, the scales had not yet become dispersed, as they did further downstream in the turbulent regime. Also, three-dimensional effects had not come into play in the transition region. Finally, Roshko noted that mixing layer growth probably occurred near or during the pairing event. Entrainment brought together "pieces" of fluid from either side of the layer, also enhancing the mixing process. Between each of these pairing/entrainment events, the vortices appeared to convect in an apparently passive fashion.

Ferziger and McMillian[7], in studies of the structure in turbulent shear flows, also noted the presence of coherent structures and pairing in a developing mixing layer. They went on to discuss the importance of a tearing mechanism by which vortices tended to be torn apart by shearing and then redistributed in parts to their neighboring vortices. They also pointed out the importance of three-dimensional effects in destabilizing the layer. The coherent structures present in the mixing layer tended to be unstable to three-dimensional perturbation that destroyed the spanwise coherence of the structures. Finally, Ferziger and McMillian noted that three-dimensional effects could also be introduced by streamwise vorticity produced by the stretching of vortical structures.

There has been additional work in the literature describing important structures present in developing mixing layers, but the authors of that literature have gone on to seek specific mechanisms leading to the production of the structures and their effect on the flow. Several of these authors have dealt particularly with mechanisms associated with retardation of mixing in the supersonic development of layers. Oh[8] hypothesized that when the local Mach number exceeded 1, some fraction of the turbulence energy in the flow was generated by shocks that formed about the eddies (eddy shocks). These shocks were quite weak, differing little from Mach waves, but having finite strength. Some of the eddies in the flow were decelerated by passing through these shocks, and the resulting disturbances produced pressure fluctuations. These fluctuations appeared to correlate well with velocity and density fluctuations in the flow. Favorable correlations in fluctuations of pressure and velocity gradient gave rise to values of the pressure dilation term $p'\partial u_j'/\partial x_j$ that acted as a source or sink of turbulent kinetic energy in the flow. This term vanished in incompressible flows and in low-speed mixing flows in which there was a large density variation. The term took on larger values, however, in high Mach number free-mixing layers and acted as a turbulent kinetic energy sink when gradients of mean Mach number and density had the same sign. Therefore, Oh reasoned that the pressure dilation term could act to reduce the turbulent shear level in high-Mach-number mixing layers, thereby slowing the growth of the layer relative to the incompressible case. This effect agreed with the results cited earlier in this chapter. Oh then carried out calculations by using these ideas, which appeared to validate his hypothesis. Papamoschou and Roshko[9] also observed that the spreading rate of compressible mixing layers was significantly reduced over that of incompressible layers, and they attributed that difference to compressibility effects. They deduced from their studies of large-scale structures in the layer that it was appropriate to define a natural coordinate system that moved with these structures. With this system, an alternative Mach number, termed the convective Mach number M_c, was defined as $M_c = (u - u_c)/a$, where u is the freestream velocity, u_c the convection velocity of the large-scale structures, and a the local speed of sound. The reduction in mixing-layer spreading rate (by approximately a factor of 3 or 4) was shown to correlate well with $M_c \approx 0.5$ and leveling off for $M_c > 1.0$. Reduced spreading therefore seemed to Papamoschou and Roshko to be due to a stabilizing effect of the convective Mach number.

Hussaini, et al.[10] offered a possible explanation for the correlation of mixing-layer spreading rate with convective Mach number. Their explanation was tied to the formation of the eddy shocklets that were described earlier. Hussaini and co-workers studied numerically the behavior of an eddy convecting subsonically, relative to a locally supersonic flow, with a convective Mach number greater than 1. Such flows could therefore support transient shock structures associated with the eddies. As the eddy accelerated in the supersonic flow, an eddy shocklet formed that tended to distort the eddy. As this process continued, an eddy bifurcation occurred, resulting in the formation of a vortex of opposite circulation. Additionally, the length scale of the original vortex was reduced. Therefore, it was seen that eddy shocklets could reduce turbulent mixing through both the production of counterfluctuating vorticity and the reduction of turbulence scale. Hussaini et al. stated that the mechanism for these effects resulted from the instantaneous inviscid pressure field induced about the front of the eddy. They further noted that the induced pressure field would always counter the initial vortex circulation over a portion of its contour and that, for long enough times and weak enough eddies, the formation of counter vorticity and consequent eddy splitting would occur, resulting in a significant alteration of the mixing-layer structure.

Many, if not all, of the important features described earlier for nonreacting subsonic and supersonic mixing layers also occurred in reacting layers. A majority of the studies on reacting mixing layers were carried out at subsonic rather than supersonic speeds, however. Yule et al.[11] found that, in a manner consistent with nonreacting flow, many combusting flows contained coherent burning structures that interacted as they were convected downstream. They termed the burning region that was associated with an eddy, and moving with it, a "flamelet" and found that the flamelet formed in only part of an eddy. They found that a range of eddy types existed in a diffusion flame (which occurred in a nonpremixed reacting mixing layer). Initially, there existed unstable laminar flow that contained an unstable laminar diffusion flame. That region was followed by one containing sheets of vortex rings with smooth tongues of flame at the interfaces between the vortices and unburned reactants. This region was followed by a zone of other orderly vortex structures, including helical vortices, which also produced relatively smooth tongues of flame. This zone contained the characteristics of transition observed in nonreacting flow. Here, viscous forces have a stabilizing influence on the flow. As the viscous forces became less important and inertial forces predominated further downstream, the authors found that the orderliness of the eddies decreased and the flow became increasingly unstable and three-dimensional. With the introduction of three-dimensional effects, randomly moving cell-like flamelets also appeared. Even further downstream, this process evolved into a fully turbulent flow, with eddies containing coherent ragged regions of burning, forming islands that were completely separated from the main flame. Yule et al.[11] also examined the structure of a single eddy containing a flamelet in a simple gas diffusion flame. The basic structure of a transitional eddy before it interacted with other eddies is given in Fig. 1[11]. The eddy contained separate regions of fuel and air that rolled up into the vortex, as well as a viscous core containing a mixture of fuel, oxidants, and products. A flame existed along the interface region where large transverse gradients of temperature and species concentration occurred. The local thickness of this region depended on the residence time and strength of the vortex, the local diffusion coefficients, and chemical kinetics. The molecular mixing required before fuel and air react was enhanced in the eddy by stretching

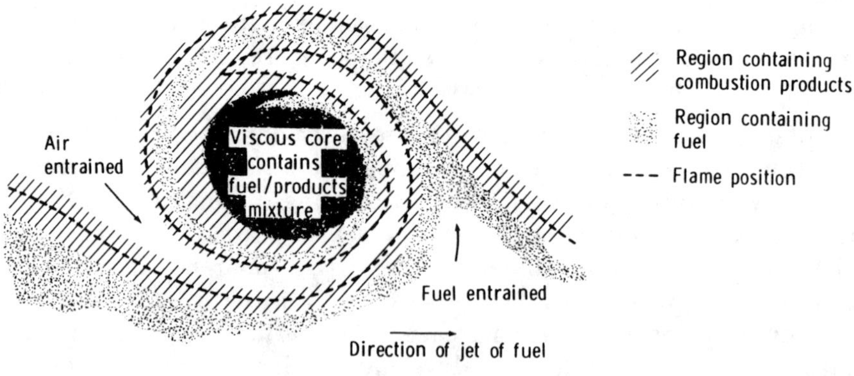

Fig. 1 Cross section of transitional vortex eddy in gas diffusion flame (Ref. 11).

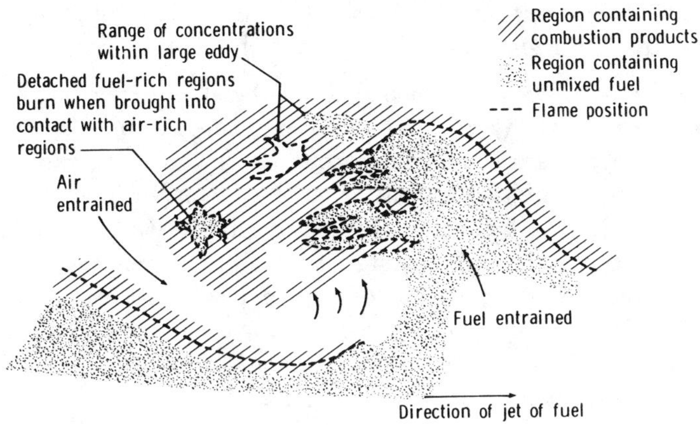

Fig. 2 Cross section of large turbulent eddy in gas diffusion flame (Ref. 11).

of the fuel/air interface due to the vorticity that the eddy contained. Preheating of fuel and air then took place primarily along the interface zone where mixing was taking place on a molecular scale. Combustion then occurred in the interface at or near-stoichiometric conditions. During these processes, the vortex continued to convect downstream, and the induced velocity within the eddy due to its vorticity continued to produce valleys and an increase in vortex dimensions. This eddy growth resulted in further entrainment of fuel and air, producing flame and mixing-layer growth.

Yule et al.[11] then went on to discuss the evolution of turbulent eddies from transitional eddies. The structure is pictured in Fig. 2[11] . The eddy has now taken on a three-dimensional structure, and it has begun to lose the circumferential coherence about its associated flamelet. Additionally, there now existed an irregular vorticity distribution within the eddy, which was interpreted to be due to the presence of smaller eddy scales now existing within the main eddy. Mixing down to molecular levels was still produced by vortical stretching, and the process appeared, in fact, to be more pronounced in the turbulent eddy. In addition, the irregularity of the structure also produced a range of flamelet structures, resulting in a "ragged" flame front trailing the eddy. Yule and co-workers concluded their study of large coherent structures in reacting flow by noting that such structures could lead to overall reduced combustion efficiency because of unmixedness. Unmixedness occurred when fuel and air could not effectively mix because each gas was bound up in vortical structures during its passage through a combustion region. Yule et al. did suggest that large eddies could be broken up by increasing the shear stresses in the flow in regions of steep velocity gradient or by the imposition of swirl into the flow.

Masutani and Bowman[12] also studied the structure of a chemically reacting plane mixing layer. They examined the reaction in the mixing layer between a stream of dilute nitrous oxide and a stream of dilute ozone and observed behavior similar to that seen by Yule et al. They found that the mixing layer had three streamwise states. First, fingers of unmixed free-stream fluid existed that sometimes reached entirely across the layer. Next, a region of mixed fluid appeared in a finite-thickness interfacial diffusion zone that bordered parcels of unmixed fluid. Finally, the layer consisted of regions of mixed fluid or nearly homogeneous composition in a global sense.

Keller and Daily[13] conducted an experimental study of a gaseous, two-stream, reacting-mixing layer flow fueled by propane, with one stream made of hot combustion products and the other stream containing cold unburned reactants. They found that the mixing-layer structure was qualitatively unaffected by heat release for the range of conditions that they studied. Mungal, et al.[14] experimentally studied the reacting mixing layer created between a dilute hydrogen stream and a dilute fluorine stream over a wide range of conditions. They also observed the presence of large, hot, coherent structures in the layer that strongly influenced the mixing and entrainment of fuel and oxidant and the overall structure of the flowfield.

Hermanson, et al.[15] extended the work described earlier[14], but with significantly higher heat release. They found that, at the higher temperature resulting in this case, the flow still appeared to be dominated by large-scale structures that were separated by cold tongues of fluid that extending well into the layer. Thus, the structure did not appear to be altered by heat release and continued to be predominately two-dimensional. Hermanson and co-workers also found that, with significant heating and the resulting large density changes, the shear-layer thickness did not increase and, in fact, showed a slight decrease. This reduction in layer thickness with increasing heat release was further confirmed by the resulting velocity profiles which showed noticeably higher values of transverse velocity gradient with increased heating. Hermanson, et al. then went further to note that since the layer width did not increase with temperature and since the density of the layer was substantially reduced by heating, the volumetric entrainment rate of freestream fluid into the layer must also be greatly reduced by heat release. Pitz and Daily[16] carried out an experiment to study a turbulent propane-air mixing layer downstream of a rearward-facing step.

They also found that large-scale structures dominated the flow and that the growth of these eddies influenced the reaction zone. Reaction took place mainly in the eddies although the eddies were not confined to the velocity gradient region of the layer. Therefore, the resulting flame spread faster into the premixed reactants than did the mixing layer defined by the mean velocity. Thus, the region of the mixing layer defined by the velocity gradient did not coincide with the region of high chemical reaction and heat transfer. Broadwell and Dimotakis [17] surveyed a number of recent papers describing experiments of reacting mixing layers. Based on these papers and their experience, they then discussed the implications for modeling such flows. Their three principal conclusions were that molecular transport retained a significant role in turbulent mixing phenomena, even when the flow was fully developed; large-scale structures controlled entrainment, which then provided conditions for the subsequent mixing processes; and mixing layers remained unsteady at the largest temporal and spatial scales.

Reacting mixing-layer studies using analytical or numerical approaches have also been carried out. Carrier, et al.[18] used a singular perturbation technique to modify their Burke-Schumann thin flame solution for a more realistic finite- thickness reaction zone in a mixing layer. They studied the effect of fluid strain on the flame; their strain increased the interfacial exposure of fuel and oxidant, and convected additional reactant into the flame. Riley and Metcalfe [19] directly simulated a subsonic, temporally developing and reacting mixing layer by using a pseudospectral numerical method and a binary single-step irreversible reaction with no heat release. Using the approach, they were able to consider the effect of the turbulence field on chemical reaction. Their results were shown to be consistent with similarity theory and in approximate agreement with experimental data. McMurtry, et al.[20] extended the preceding work to consider the effect of chemical heat release on a subsonic, temporally developing mixing layer. They solved both the compressible form of the governing equations as well as a more computationally efficient form of the equations valid for low Mach numbers. Reaction was again modeled with a binary, single-step, irreversible reaction. McMurtry and co-workers found with their simulations that the thickness of the mixing layer and the amount of mass entrained into the layer decreased when the heat-release rate due to exothermic reaction was increased. Likewise, the resulting product formation also decreased as the heat-release rate increased.

Menon, et al.[21] studied the stability of a laminar, premixed, spatially developing, supersonic mixing layer undergoing chemical reaction. They introduced an infinitesimal disturbance into the layer and examined its spatial stability for both reacting and nonreacting flows. Chemical reaction was shown to have a significant effect on flow stability. Menon and co-workers found that, with reaction, the disturbance amplification rate was higher and the wave speed lower as compared to nonreactive cases. Also, the freestream Mach number was shown to have little effect on stability when the flow was reacting.

Current Studies on Supersonic Mixing Layers

In this study, a numerical model has been developed for describing general two- or three-dimensional, high subsonic or supersonic, chemically reacting flows. This model was then adapted to a supersonic, chemically reacting mixing layer. Reaction in many practical devices takes place in mixing layers, so that the problem chosen, while geometrically simple, still retained the fluid mechanical and chemical complexities under consideration. Computer programs have been developed that numerically solve the governing equations resulting from the model. The programs used either a hybrid MacCormack-Householder technique[22] or a family of high-order, compact finite-difference techniques developed by Carpenter[23] to solve the Navier-Stokes and species continuity equations that describe multiple species undergoing chemical reaction. Momentum, heat, and mass transfer with reaction were defined with a multicomponent finite-rate scheme and a real gas thermodynamics model was employed.

Using the computer programs developed in this work, detailed studies of the supersonic, spatially developing and reacting mixing layer were undertaken. Because of their importance in subsonic layers, consideration was given first to the existence of vortical structures in a supersonic reacting mixing layer. The effects of such structures on the development of the layer were then explored and compared with the literature cited earlier. Particular emphasis was given in this study to the mixing of fuel and oxidant in the layer, the resulting chemical reaction, the effect of chemical heat release on mixing, and the existence of supersonic unmixedness. The existence of unmixedness next lead to a study of fuel-air mixing enhancement to improve the degree of chemical reaction that could be achieved in a mixing layer. Findings from this study were finally applied to actual scramjet engine configurations. The overall combustion efficiency of these configurations was significantly improved using the new techniques.

Before beginning our analysis of the spatially developing and reacting mixing layer, it is appropriate to review the theory and numerical methods on which the analysis is based. They are discussed in the following section.

Theory

Governing Equations

The Navier-Stokes, energy, and species continuity equations governing multiple species undergoing chemical reaction have been derived by Williams.[24] The terms used in these and subsequent equations are defined in the Nomenclature at the beginning of this chapter. The governing equations are given by

Continuity:

$$\frac{\partial \rho}{\partial t} + \nabla \cdot (\rho \vec{V}) = 0 \tag{1}$$

Momentum:

$$\frac{\partial (\rho \vec{V})}{\partial t} + \nabla \cdot (\rho \vec{V} \vec{V}) = \nabla \cdot \tau + \rho \sum_{i=1}^{ns} f_i \vec{b}_i \tag{2}$$

Energy:

$$\frac{\partial (\rho E)}{\partial t} + \nabla \cdot (\rho \vec{V} E) = \nabla \cdot (\tau \cdot \vec{V}) - \nabla \cdot \vec{q} + \rho \sum_{i=1}^{ns} f_i \vec{b}_i (\vec{V} + \tilde{V}_i) \tag{3}$$

Species continuity:

$$\frac{\partial (\rho f_i)}{\partial t} + \nabla \cdot (\rho \vec{V} f_i) = \dot{w}_i - \nabla \cdot (\rho f_i \tilde{V}_i) \tag{4}$$

where

$$\tau \equiv \tau_{ij} = -\delta_{ij} p + \mu \left(\frac{\partial u_i}{\partial x_j} + \frac{\partial u_j}{\partial x_i} \right) + \delta_{ij} \lambda \frac{\partial u_k}{\partial x_k} \tag{5}$$

and

$$\vec{q} = -k \nabla T + \rho \sum_{i=1}^{ns} h_i f_i \tilde{V}_i + R^o T \sum_{i=1}^{ns} \sum_{j=1}^{ns} \left(\frac{X_j D_{T_i}}{M_i D_{ij}} \right) (\tilde{V}_i - \tilde{V}_j) \tag{6}$$

Also,

$$E = \sum_{i=1}^{ns} h_i f_i - \frac{p}{\rho} + \sum_{i=1}^{3} \frac{u_i^2}{2} \tag{7}$$

$$h_i = h_i^o + \int_{T_r}^{T} c_{p_i} dT \qquad i = 1, 2, ..., ns \tag{8}$$

$$p = \rho R^o T \sum_{i=1}^{ns} \frac{f_i}{M_i} \tag{9}$$

Radiation heat transfer is not included in Eq. (6). The diffusion velocities are found by solving

$$\nabla X_i = \sum_{j=1}^{ns} \frac{X_i X_j}{D_{ij}} (\tilde{V}_j - \tilde{V}_i) + (f_i - X_i) \frac{\nabla p}{p} + \frac{\rho}{p} \sum_{j=1}^{ns} f_i f_j (\vec{b_i} - \vec{b_j}) + \sum_{j=1}^{ns} \frac{X_i X_j}{\rho D_{ij}} \left(\frac{D_{Tj}}{f_j} - \frac{D_{Ti}}{f_i} \right) \frac{\nabla T}{T} \tag{10}$$

Note that if there are ns chemical species, then $i = 1, 2, ..., (ns - 1)$ and $(ns - 1)$ equations must be solved for the species f_i. The final species mass fraction f_{ns} can then be found by conservation of mass since $\sum_{i=1}^{ns} f_i = 1$.

Thermodynamics Model

To calculate the required thermodynamic quantities, the specific heat for each species is first defined by a fourth-order polynomial in temperature,

$$\frac{c_{p_i}}{R} = A_i + B_i T + C_i T^2 + D_i T^3 + E_i T^4 \tag{11}$$

The coefficients are found by a curve fit of the data tabulated in Ref. 25. When the specific heat of each species is known, the enthalpy of each species is then found from Eq. (8) and the total internal energy is computed from Eq. (7).

To determine the equilibrium constant (required in the next section) for each chemical reaction being considered, the Gibbs energy of each species must first be found. For a constant pressure process, c_p/T from Eq. (11) is first integrated over temperature to define the entropy of the species, and then the resulting expression is integrated again over temperature to obtain a fifth-order polynomial in temperature for the Gibbs energy of each species.

$$\frac{g_i}{R} = A_i(T - T \ln T) + \frac{B_i}{2} T^2 + \frac{C_i}{6} T^3 + \frac{D_i}{12} T^4 + \frac{E_i}{20} T^5 + F_i - G_i T \tag{12}$$

The coefficients F_i and G_i are again defined in Ref. 25. The Gibbs energy of reaction is then calculated as the difference between the Gibbs energy of product and reactant species.

$$\Delta G_{R_j} = \sum_{i=1}^{ns} \gamma''_{ji} g_i - \sum_{i=1}^{ns} \gamma'_{ji} g_i \qquad j = 1, 2, ..., nr \tag{13}$$

The equilibrium constant for each reaction can then be found from[26]

$$K_{eq_j} = \left(\frac{1}{R^o T}\right)^{\Delta n} \exp\left(\frac{-\Delta G_{R_j}}{R^o T}\right) \tag{14}$$

where Δn is the change in the number of moles when going from reactants to products.

Chemistry Models

The rate of chemical reactions is often defined by using the Arrhenius law. A modified form of the Arrhenius law is usually employed when modeling supersonic combustion. It is given by

$$K_{f_j} = A_j T^{N_j} \exp\left(\frac{-E_j}{R^o T}\right) \tag{15}$$

The values of the pre-exponential constant A, power constant N, and activation energy E have been determined for a number of reaction schemes. Unfortunately, there is a great deal of uncertainty for many chemical reactions. One of the best understood mechanisms, however, is the hydrogen-air reaction system. This is not the reason that hydrogen fuel was chosen for several scramjet concepts, but it has proved convenient for its combustor analysts. Values for A, N, and E for a typical hydrogen-air mechanism are given in Table 1. With the forward rate known, the reverse rate is then given by

$$K_{b_j} = \frac{K_{f_j}}{K_{eq_j}} \tag{16}$$

Once the forward and reverse reaction rates have been determined, the production rates of the species are found from the law of mass action. For the general chemical reaction,

$$\sum_{i=1}^{ns} \gamma'_{ji} C_i \overset{K_{f_j}}{\rightleftharpoons} \sum_{i=1}^{ns} \gamma''_{ji} C_i \qquad j = 1, 2, ..., nr \tag{17}$$

the law of mass action states that the rate of change of concentration of species i by reaction j is given by [24]

Table 1 Fine-rate chemistry model and Arrhenius rate coefficients for each reaction

Reaction Number	Reaction	A	N	E,kJ/g-mole
1.	$H_2+O_2=OH+OH$	0.1700E+14	0.00	201.5
2.	$H+O_2=OH+O$	0.1420E+15	0.00	68.6
3.	$OH+H_2=H_2O+H$	0.3160E+08	1.80	12.78
4.	$O+H_2=OH+H$	0.2070E+15	0.00	57.5
5.	$OH+OH=H_2O+O$	0.5500E+14	0.00	29.3
6.	$H+OH=H_2O+M$	0.2210E+23	-2.00	0.0
7.	$H+H=H_2+M$	0.6530E+18	-1.00	0.0
8.	$H+O_2=HO_2+M$	0.3200E+19	-1.00	0.0
9.	$HO_2+OH=H_2O+O_2$	0.5000E+14	0.00	4.2
10.	$HO_2+H=H_2+O_2$	0.2530E+14	0.00	2.9
11.	$HO_2+H=OH+OH$	0.1990E+15	0.00	7.5
12.	$HO_2+O=OH+O_2$	0.5000E+14	0.00	4.2
13.	$HO_2+HO_2=H_2O_2+O_2$	0.1990E+13	0.00	0.0
14.	$HO_2+H_2=H_2O_2+H$	0.3010E+12	0.00	78.2
15.	$H_2O_2+OH=HO_2+H_2O$	0.1020E+14	0.00	7.9
16.	$H_2O_2+H=OH+H_2O$	0.5000E+15	0.00	41.9
17.	$H_2O_2+O=OH+HO_2$	0.1990E+14	0.00	24.7
18.	$M+H_2O_2=OH+OH$	0.1210E+18	0.00	190.4
For the	single-step reaction	0.5510E+15	0.00	30.2

$$(\dot{C}_i)_j = (\gamma''_{ji} - \gamma'_{ji})[K_{f_j} \prod_{i=1}^{ns} C_i^{\gamma'_{ji}} - K_{b_j} \prod_{i=1}^{ns} C_i^{\gamma''_{ji}}] \qquad i = 1, 2, ..., ns \quad (18)$$

The net rate of change in concentration of species i by reaction j is then found by summing the contributions from each reaction,

$$\dot{C}_i = \sum_{j=1}^{nr} (\dot{C}_i)_j \quad (19)$$

Finally, the production of species i can be found by multiplying its rate of change of concentration by its molecular weight.

$$\dot{w}_i = \dot{C}_i M_i \quad (20)$$

The source terms in Eq. (4) are now determined as a function of the dependent variables.

Molecular Diffusion Models

The coefficients governing the molecular diffusion of momentum, energy, and mass are determined from models based on kinetic theory. The set of models that is often used is now described. Individual species viscosities are computed from Sutherland's law,

$$\frac{\mu}{\mu_o} = \left(\frac{T}{T_o}\right)^{1.5} \frac{T_o + S}{T + S} \tag{21}$$

where μ_o and T_o are reference values and S is Sutherland's constant. These constants are tabulated for many species in Refs. 27 and 28. Once the viscosity of each species has been determined, the mixture viscosity is found from Wilke's law [29]

$$\mu_m = \sum_{i=1}^{ns} \frac{\mu_i}{1 + \frac{1}{X_i} \sum_{j=1, j\neq i}^{ns} X_j \phi_{ij}} \tag{22}$$

where

$$\phi_{ij} = \frac{[1 + (\frac{\mu_i}{\mu_j}\frac{\rho_j}{\rho_i})^{0.5} (\frac{M_i}{M_j})^{0.25}]^2}{\frac{4}{\sqrt{2}}(1 + \frac{M_i}{M_j})^{0.5}} \tag{23}$$

Species thermal conductivities are also computed from Sutherlands's law

$$\frac{k}{k_o} = \left(\frac{T}{T_o'}\right)^{1.5} \frac{T_o + S'}{T + S'} \tag{24}$$

with different values of the reference values k_o and T_o' and the Sutherland's constant S'. These values are also tabulated for a number of species in Refs. 27 and 28. The mixture thermal conductivity is computed using conductivity values for the individual species and Wassilewa's formula [30]

$$k_m = \sum_{i=1}^{ns} \frac{k_i}{1 + \frac{1}{X_i} \sum_{j=1, j\neq i}^{ns} X_j \phi_{ij}'} \tag{25}$$

where $\phi_{ij}' = 1.065\ \phi_{ij}$ and ϕ_{ij} is taken from Eq. (23).

For dilute gases, Chapman and Cowling used kinetic theory to derive the following expression for the binary diffusion coefficient D_{ij} between species i and j [27] :

$$D_{ij} = \frac{0.001858T^{1.5}(\frac{M_i+M_j}{M_iM_j})^{0.5}}{p\sigma_{ij}^2\Omega_D} \tag{26}$$

Here, the diffusion collision integral Ω_D is approximated by

$$\Omega_D = \bar{T}^{-0.145} + (\bar{T} + 0.5)^{-2} \tag{27}$$

where $\bar{T} = T/T_{\epsilon_{ij}}$. Values of the effective temperature T_ϵ and the effective collision diameter σ are taken to be averages of the separate molecular properties of each species, giving

$$\sigma_{ij} = 0.5(\sigma_i + \sigma_j) \tag{28}$$

and

$$T_{\epsilon_{ij}} = (T_{\epsilon_i}T_{\epsilon_j})^{0.5} \tag{29}$$

For most molecules, the thermal diffusion coefficient is generally small when compared with the binary diffusion coefficient and, therefore, the thermal diffusion coefficient can be neglected. This is fortunate since values of the thermal diffusion coefficient are generally not known for most species. For low-molecular-weight molecules such as hydrogen, though, the thermal diffusion coefficient can be important. A set of relationships for the thermal diffusion coefficient of species having a molecular weight less than 5 has been developed by Kee et al.[31] . The reader is referred to Ref. 31 for further information and some numerical details for computing thermal diffusion coefficients of light molecules.

Once the binary and thermal diffusion coefficients for all species combinations are known, the diffusion velocities of each species can be computed from Eq. (10). The diffusion velocity is the velocity induced on each species by all diffusion processes that are present in the flow. The solution of Eq. (10) requires solving a simultaneous equation system, with the number of equations equivalent to the number of species present for each component of the diffusion velocity. It should be noted that for i species, however, the system of i equations defined by Eq. (10) is not linearly independent. One of the equations must be replaced by the constraint $\sum_{i=1}^{ns} \rho f_i \tilde{V}_i = 0$ to make the system linearly independent. The resulting simultaneous system of equations must then be solved for the diffusion velocities.

The process of solving for the diffusion velocities can be computationally quite expensive. A coupled system of equations must be solved for each of the three components of the diffusion velocity at each computational grid point. This process can require as much time as solving the Navier-Stokes equations for the three components of the convection velocities. Alternately, for hydrogen-air chemistry, where large amounts of nitrogen are present, it is sometimes assumed that each species is present as a "trace" in a mixture with N_2 [32]. Then each species is assumed to diffuse only into N_2 with that process defined by its binary diffusion coefficient with N_2. Finally, for engineering calculations, it is often further assumed that the diffusivities of each chemical species present in the flow are the same. Then the diffusion of each species into the remaining species varies only with its respective concentration gradient. The diffusion velocities then decouple, and Eq. (10) reduces to

$$\tilde{V}_{i,j} = -\frac{D}{f_i}\frac{\partial f_i}{\partial x_j} \tag{30}$$

where $\tilde{V}_{i,j}$ is the diffusion velocity vector of the ith species in the jth coordinate direction (j = [x,y,z]) and D is the binary diffusion coefficient. If the binary diffusion with N_2 is not used, the value of D is determined by choosing an appropriate value of the Schmidt number Sc since $D = \mu/\rho Sc$. The mixture viscosity μ is determined as before from Wilke's law. When the binary diffusion assumption is invoked, it is often further assumed that the mixture thermal conductivity can be defined by $k = C_p/\mu Pr$ after an appropriate value of the Prandtl number Pr has been chosen.

Turbulent Diffusion Models

Although the techniques for defining the molecular diffusion of momentum, heat, and mass are reasonably well established in a supersonic reacting flow, the same statement cannot be made for our ability to describe the turbulent diffusion of these quantities. Work to develop methods for modeling turbulent supersonic combustion is now in its early stages. Conventional approaches, which are employed in the calculations that follow, have included the use of algebraic eddy-viscosity models or differential transport models. Several eddy-viscosity models have been used, in particular the Cebeci-Smith [33] and the Baldwin-Lomax models [34]. The differential transport models include the k/ϵ turbulent kinetic energy model and its variants,[35] a modified k/ϵ model that includes a supersonic flow compressibility correction,[36, 38] and a multiple dissipation length scale (k/multiple ϵ) model with a compressibility correction[36, 37] that addresses the existence of multiple dissipation length scales existing in the energy cascade of a turbulent flow. In addition to these differential transport models, the algebraic Reynolds stress models of Rodi[39] and Sindir[40] have also been considered for use in modeling turbulent supersonic reacting flows. A review of all of these models has been given by Sindir[40]. In that review, Sindir also has critically compared the models against several nonreacting flow experiments prior to using the models for studying flows with reaction. He concluded that forms of the algebraic Reynolds stress model that he considered produced the best agreement with nonreacting data. He also found that the multiple dissipation length scale model did not offer any advantage over the basic k/ϵ model.

All of the turbulence models just described have a major disadvantage when applied to reacting flowfields: they fail to account for the important coupling between the fluid mechanics and the chemistry. Turbulent fluctuations in the fluid mechanic variables have a direct effect on the species production rates. The coupling between these two fields occurs through the Arrhenius rate expression, Eq. (15), and the law of mass action, Eq. (18). The Reynolds averaging process applied to the governing equations eliminates the direct effect of temperature and species fluctuations on species production rates. For example, a positive temperature fluctuation would cause a decrease in the size of the exponential argument of the Arrhenius rate expression, with a corresponding increase in the forward kinetic rate of a particular reaction. This would, in turn, produce an increase in the time rate of change of the products of that reaction. More important, if the reaction were at a critical point, where perhaps a small increase in temperature would cause a reaction to enter an ignition stage, the entire species distribution of the flowfield downstream could be changed.

Two promising ways of accounting for the effects of fluid and species fluctuations on chemical reaction would be through probability density functions or direct numerical simulation. The application of the probability density function approach to a reacting flow has been covered by Stephen Pope in a recent book,[41] and so this subject will not be further addressed here. Direct numerical simulation offers another attractive approach for modeling a turbulent reacting flow. The method has been used for several years to accurately model lower-speed reacting flows[42, 19, 43]. With this approach, the Navier-Stokes and species continuity equations are resolved down to the smallest scale features of the flowfield. The size of those scales goes inversely with the Reynolds number of the flowfield. Clearly, for the high Reynolds numbers that occur in typical supersonic reacting flows, the smallest scales can become quite small, necessitating a very fine computational grid to resolve them. Also, when high-speed flow undergoes chemical reaction, additional scales are introduced by the combustion process. Herein lies the principal difficulty in applying direct simulation to a high-speed flow. The difficulty is not so much one of numerical algorithms as it is of computer power. Highly accurate numerical algorithms are required, but appropriate high-order finite-difference methods, finite-volume methods, or spectral methods have been developed that satisfy that requirement. The large number of computational grid points required to resolve the smallest scales in the flow requires, however, large computer storage and, therefore, meaningful calculations can be carried out only on large-memory machines. Currently, direct numerical simulations have been made for nonreacting flows with Reynolds numbers up to about 10,000 on a Cray 2 computer[44] . Work is proceeding to simulate directly a chemically reacting flow of a similar Reynolds number.

As an alternative to direct numerical simulation, with its intensive memory requirements, it is possible to model, rather than compute, the smallest scales. In this approach, termed large-eddy simulation, the larger scales above a chosen wavelength are still computed. The smaller scales below the cutoff wavelength are modeled, however, using a subgrid scale model. Large-eddy simulation is an attractive alternative because only the larger-scale effects are computed, lessening the computer memory requirements for higher-Reynolds-number flows. Subgrid scale models must be constructed, though, that give an accurate rendering of the physics of small-scale phenomena. This is a difficult task. Work is under way to develop subgrid scale models for nonreacting flow, for example the early work of Schumann[45] and later work described by Speziale et al.[46] . Large-eddy simulation is an attractive technique for modeling high speed reacting flows. Little has been done so far with this technique, but it warrants serious attention in the future.

Discretization of the Governing Equations

Once the governing equations and the required modeling are in hand, the equations can be discretized and integrated in space and time. We begin this process by expressing the governing equations (1) in vector form. In that form they become,

$$\frac{\partial \mathbf{U}}{\partial t} + \frac{\partial \mathbf{E}}{\partial x} + \frac{\partial \mathbf{F}}{\partial y} + \frac{\partial \mathbf{G}}{\partial z} = \mathbf{H} \tag{31}$$

where $\mathbf{U}$ is the vector of dependent variables, $\mathbf{E}, \mathbf{F}$, and $\mathbf{G}$ are flux vectors containing convective and diffusive terms, and $\mathbf{H}$ is the source term containing body forces and the chemistry production terms. The temporally discrete form of Eq. (31) can then be written as

$$\mathbf{U}^{n+1} = \mathbf{U}^{n} - \Delta t[\frac{\partial \mathbf{E}^{n}}{\partial x} + \frac{\partial \mathbf{F}^{n}}{\partial y} + \frac{\partial \mathbf{G}^{n}}{\partial z} - \mathbf{H}^{n+1}] \tag{32}$$

where n is the old time level and $n+1$ is the new time level. The flux terms are written at the old time level in this study because the equations are advanced in real time at the smallest fluid time scale. As an option, the chemical source term is written implicitly[47, 48] to allow for stiffness in the governing equations when chemistry time scales become small compared to fluid time scales early in a calculation or when the system approaches chemical equilibrium. Otherwise, solution times would become prohibitively long.

Once the temporal discretization of Eq. (32) is established, the spatial derivatives must also be discretized. The earliest work employed the MacCormack-Householder scheme. In this approach, the unsplit, spatially second-order MacCormack finite- difference scheme[49] was used along with the Householder matrix solver[50] required to solve the block diagonal matrix resulting from the implicit source terms. Details of the method are given in Ref. 22.

To achieve a higher level of numerical accuracy, several methods with higher spatial accuracy were introduced into the computer codes by Carpenter. These included the statially fourth-order accurate method of Gottlieb and Turkel[51], and the family of compact fourth-order schemes of Carpenter[53]. In the latter case, the tedious algebra required to develop the compact schemes necessitated the use of the symbolic manipulator MACSYMA. The compact schemes offered increased spatial accuracy and applicability at the grid point next to the boundary, but have more restrictive stability requirements than does the Gottlieb-Turkel scheme. All the schemes were dissipative in nature, and could be used to capture weak discontinuities without artificial damping. The specific problem being addressed dictated which algorithm should be used, since each offered specific advantages relative to others.

For calculations where only the steady-state solution was desired, a new scheme emerged as an alternative to the standard MacCormack formulation. The ideas of Abarbanel and Kumar[52] for developing compact steady-state algorithms for the Euler equations, were extended by Carpenter[23] into MacCormack-like schemes. The resulting scheme was a second-order temporal and second-order spatial scheme during the transient calculation, but degenerated to yield fourth-order spatial accuracy at steady-state. The new scheme offered additional accuracy, and did not lose the robustness inherent in MacCormack-like schemes.

Mixing-Layer Results

Following the development of the theory and the solution procedure described here, the combustion code was checked by comparison with a standard test case, namely the known solution for a spatially developing mixing layer. Lock[27] solved the incompressible boundary-layer equations for a laminar mixing layer with upper stream velocity U_1 and lower stream velocity U_2. The configuration is shown in Fig. 3 along with Lock's solution for air with a velocity ratio $U_2/U_1 = 0.5$. The solution is similar and is plotted in terms of a similarity variable η vs the nondimensional streamwise velocity u/U_1. The definition of y was modified using the Howarth-Dorodnitzyn transformation

$$y = \int_0^y \frac{\rho}{\rho_e} dy \tag{33}$$

to allow comparison with the solution to the same problem based on the compressible Navier-Stokes equations. In Eq. (33), y is the transverse coordinate, and ρ and ρ_e are the field and edge densities, respectively. The definition for η reduces for incompressible flow to that used by Lock. Results from the computation for $U_2/U_1 = 0.5$ are also shown in Fig. 3. The calculations were made on a computational grid with 51 nodes in the streamwise direction, 51 nodes in the spanwise direction, and 51 nodes in the transverse direction across the layer. The grid was highly compressed in the transverse direction, with a minimum grid spacing of 0.1 mm. The results become similar at a value of x/L equal to or greater than 0.1, where x is the streamwise coordinate and L is the streamwise length of the physical domain. The comparison of the computation with the exact solution of Lock is excellent for all values of η. The transformation coordinate η is highly stretched relative to the physical coordinate y. It is therefore clear from the comparison that the large velocity gradient across the mixing layer is well resolved by the chosen grid. Other validation cases for the code can be found in the literature[22, 54, 23, 53, 55].

Following checkout of the program with several test cases, the chemically reacting flowfield in a nonpremixed, laminar, supersonic, spatially developing mixing layer was simulated with the code using both the

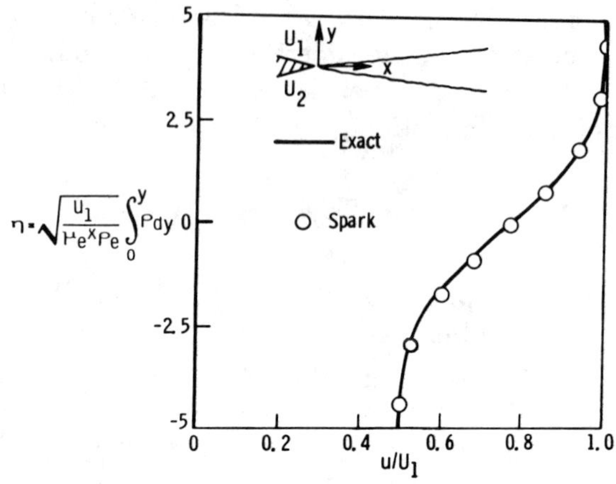

Fig. 3 Comparison of exact and numerical solution of a supersonic shear layer.

second-order and fourth-order algorithms. The case that was chosen involved a hydrogen-air mixing layer initially separated by a finite-thickness splitter plate. The configuration is described schematically in Fig. 4. The overall domain is 5 cm high and 5 cm long. The height chosen places the boundaries well into the freestream, and the length allows initial development of the mixing layer. Initially, hydrogen fuel and air are separated by a 0.5-cm-long splitter plate that is 0.02 cm thick and centered at $y = 2.5$ cm. Downstream of the plate, the fuel and air mix and ignition occurs at some further distance downstream of the plate base. After ignition, chemical reaction between the fuel and the air takes place. For the problem being considered, cold gaseous hydrogen is introduced above the plate at Mach 1.5, a Reynolds number of 3700 based on plate thickness, a temperature of 293 K, and a

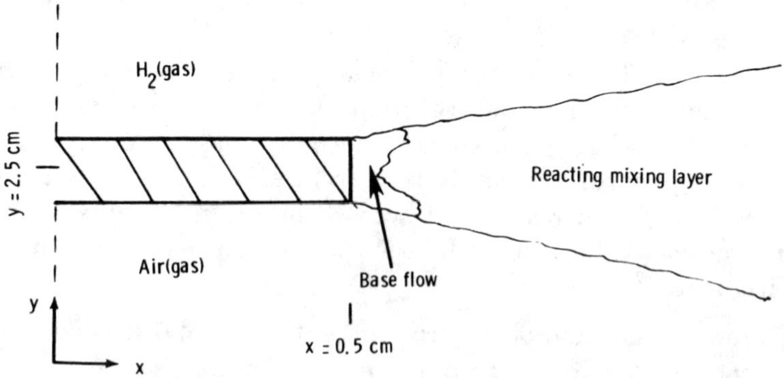

Fig. 4 Reacting mixing-layer schematic.

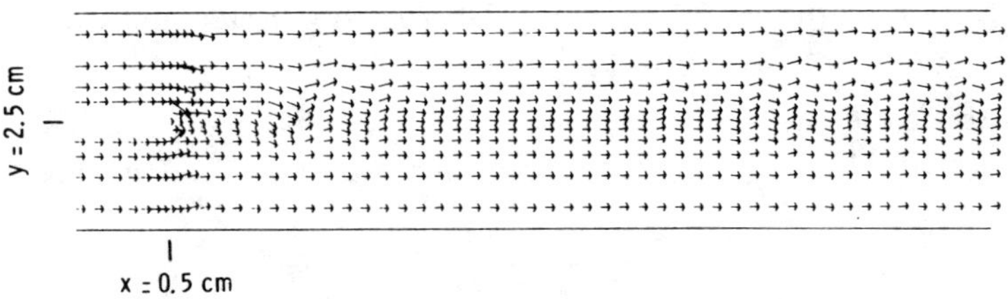

Fig. 5 Mixing-layer velocity vector field; every fourth vector plotted (0.1 ms).

pressure of 0.101 MPa (1 atm). Hot air is introduced below the plate at Mach 1.5, a Reynolds number of 731, a temperature of 2000 K, and a pressure of 0.101 MPa. These conditions result in an initial hydrogen velocity of 1953 m/s and an air velocity of 1297 m/s, yielding a hydrogen-to-air velocity ratio of 1.5.

By using the configuration and conditions just described, the mixing-layer flowfield is marched in time from the specified initial conditions to the conditions existing at 0.1 ms. The solution is obtained on a spatial grid, with 219 nodes in the streamwise direction and 51 nodes in the transverse direction. The grid is compressed in x near the trailing edge of the plate and highly compressed in y in the region of the mixing layer, using a distribution identical to that used in the validation case described earlier. The resulting flowfield is described in figures 5-23, which give pictures of the flow at an instant in time. Figure 5 shows a velocity vector field plot of the flow close to, and on either side of, the splitter plate and the developing mixing layer downstream of the base of the splitter plate. (Velocity vectors are shown for only every four streamwise and transverse grid points in this region.) Expansions of both streams through Prandtl-Meyer fans can be observed at the trailing edge of the plate. The higher-velocity hydrogen stream and the lower-velocity airstream are apparent, as is the wake flow downstream of the plate. The development of the mixing layer with streamwise distance can also be seen. Two regions of instability are also apparent in Fig. 5. The first region lies just downstream of the splitter plate, approximately 1.0 cm beyond the initial station. The second region lies well downstream, at approximately 4.0 cm from inflow. The flow is relatively quiescent between these two regions. The first instability is associated with an unstable separation bubble on the trailing edge of the splitter plate separating hydrogen and air, and the second instability is a Kelvin-Helmholtz instability that occurs downstream in a mixing layer with a sufficient initial flow disturbance. The instabilities can also be seen in Fig. 6 which shows a plot at 0.1 ms of streamwise velocity vs streamwise coordinate at three constant transverse stations located well within the mixing layer.

To examine the effect of chemical heat release on the instabilities in the flowfield, the identical calculation was repeated without chem-

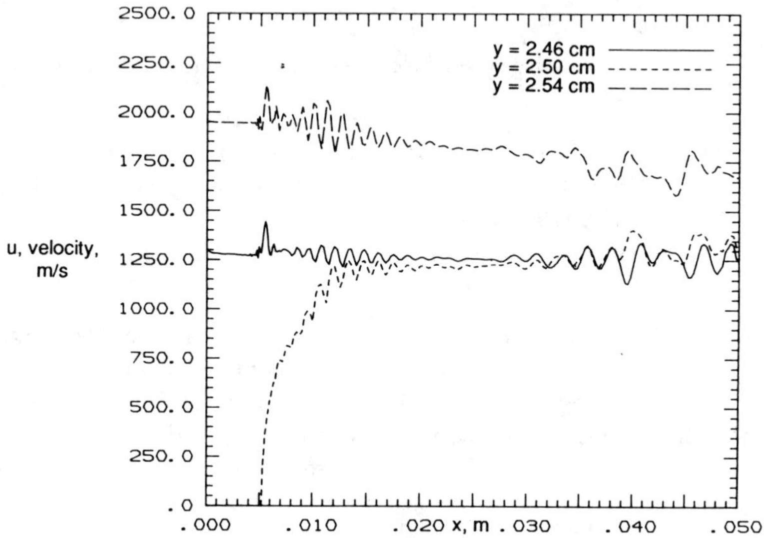

Fig. 6 Streamwise velocity vs x at y locations for reacting flow (0.1 ms).

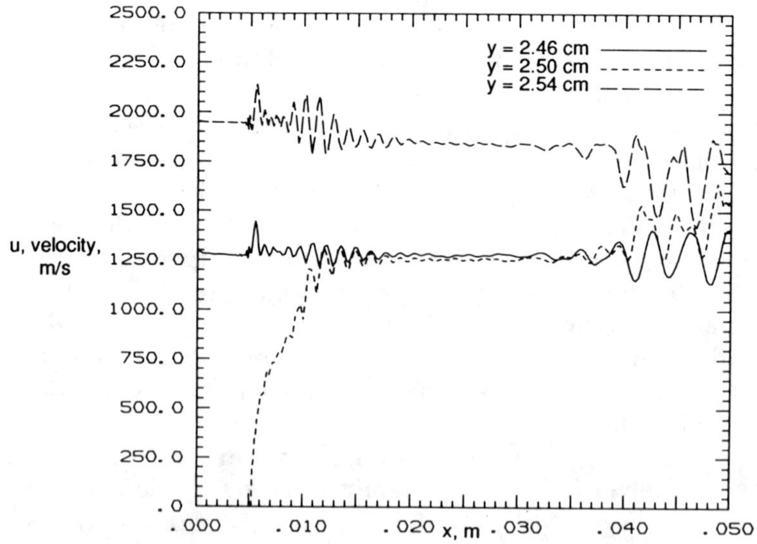

Fig. 7 Streamwise velocity vs x at y locations for nonreacting flow (0.1 ms).

ical reaction. The resulting plot of streamwise velocity vs streamwise coordinate at 0.1 ms is shown in Fig. 7. Comparison of Fig. 6 and 7 suggests that the upstream disturbance is essentially unaffected but that chemical reaction is small and endothermic in that region. Well downstream, however, heat release is large and exothermic, and that heat release reduces the amplitude of the instability. This result is consistent with the findings of Refs. 15 and 20 for subsonic flow, which showed mixing retarded by heat addition in an unconfined flow.

Contour plots of velocity and static temperature in the streamwise direction along the mixing layer are shown in Fig. 8 and 9. By comparing these figures, it can be seen that the temperature profiles at each x-station are consistently broader than the streamwise velocity profiles. This result is consistent with the discussion and experimental observations described earlier from Refs. 1 through 20, and in particular Ref. 16. Vortical structures are present in the mixing layer, and the

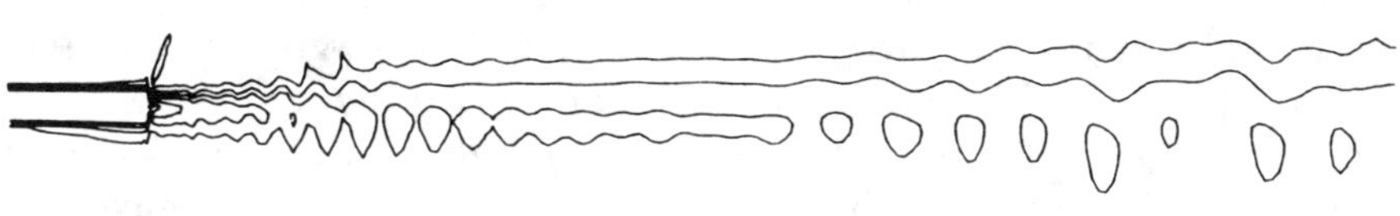

Fig. 8 Streamwise velocity contours in mixing layer.

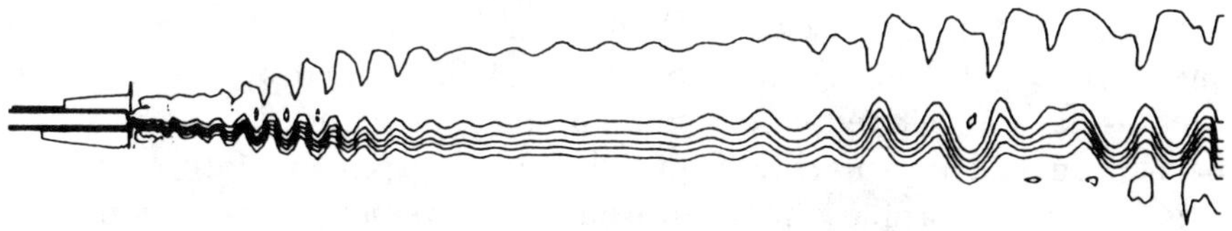

Fig. 9 Temperature contours in mixing layer.

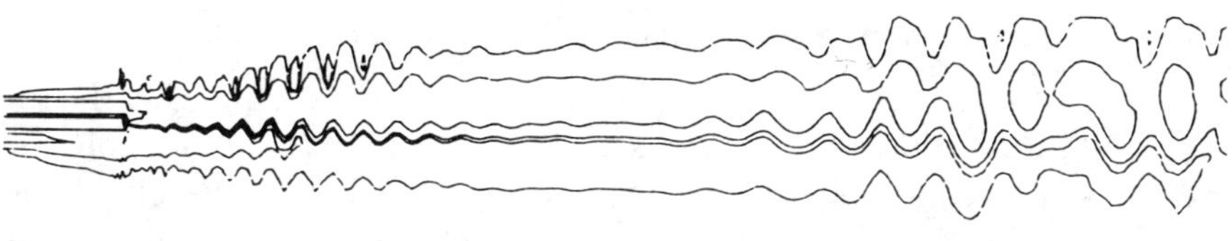

Fig. 10 Vorticity contours in mixing layer.

existence and growth of these vortices influence the growth and reaction in the mixing layer. The vortical character can be seen in Fig. 10, which gives the vorticity distribution in the mixing layer. Chemical reaction takes place not only in the interior of the mixing layer but also in the eddies on the edges of the layer. These eddies lie outside the high- velocity gradient region of the layer, as can be seen by comparing Fig. 8 with Fig. 10. Therefore, the resulting flame spreads transversely faster into the unreacted species than did the mixing layer defined by the high-velocity gradient zone. Thus, the region of the mixing layer defined by the velocity gradient is not as transversely wide as the flame zone defined by temperature gradient in the mixing layer, in agreement with Ref. 16.

Figures 11-17 show plots at seven streamwise stations (x = 0.51, 0.58, 1.0, 2.0, 3.0, 4.0, and 5.0 cm) of the major chemical species (H_2, O_2, and H_2O) and minor chemical species (OH, H, O, HO_2, and H_2O_2). Contour plots giving the two-dimensional distribution of the species are given in Fig. 18-23. Initially, at $x = 0.51$ cm (fig. 11b), fuel and air have just begun to mix, and no significant amount of water has yet formed. A very narrow band of hydrogen peroxide (H_2O_2) is present just above the splitter plate center, and a very small amount of hydroperoxyl (HO_2) lies just below that spike. At $x = 0.58$ cm (Fig. 12), the hydrogen and oxygen profiles begin to broaden, but no water has yet appeared in the layer. The hydrogen peroxide spike is still the most predominant and, although the profile has not broadened, the peak has increased. (Note that the ordinate in Fig. 12b has been rescaled.) A small amount of hydroperoxyl still lies below the hydrogen peroxide peak, and small amounts of atomic hydrogen (H) and atomic oxygen (O) have appeared there. At $x = 1.0$ cm, as described in Fig. 13, the hydrogen and oxygen profiles have developed, and a small amount of water (8% by mass) has been produced in a narrow profile below the splitter plate centerline. The H_2 and O_2 profiles are appropriately depressed in the region of the water peak. Noticeably increased profiles of H, O, and OH also appear at this station just below the splitter plate centerline ($y = 2.5$ cm). The O and OH profiles lie slightly below the water peak; this is consistent with the spatial distribution of reactant species. Small amounts of HO_2 and H_2O_2 still remain at, and just above, the plate centerline.

Figure 14 presents the species profiles at $x = 2.0$ cm. The H_2, O_2, and H_2O profiles have broadened significantly more at this station, and the water peak has risen to approximately 23% by mass. The minor species profiles have also broadened significantly, with atomic oxygen peaking at 3.0%, hydroxyl peaking at 2.0%, and atomic hydrogen peaking at 0.8%, all by mass. Small amounts of hydroperoxyl and hydrogen

a) b)

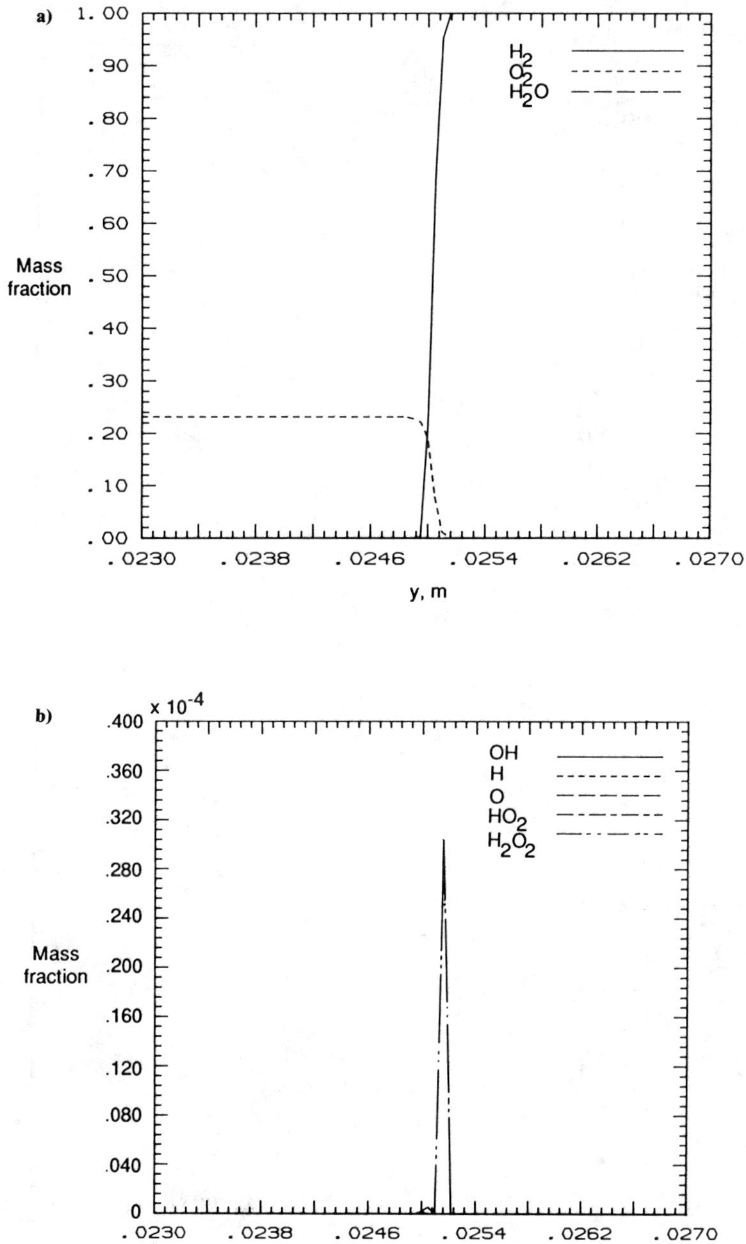

Fig. 11 Mass fraction vs y at $x = 0.51$ cm: a) major species; b) minor species.

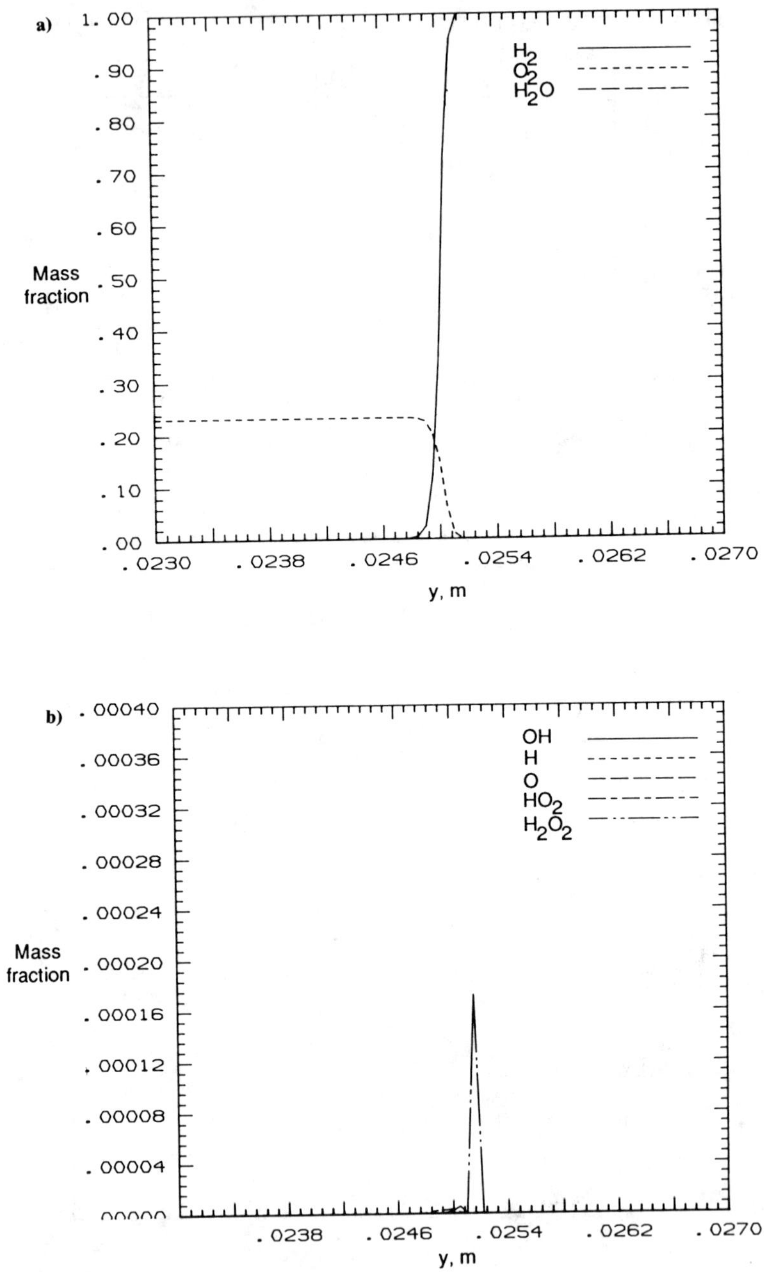

Fig. 12 Mass fraction vs y at $x = 0.58$ cm: a) major species; b) minor species.

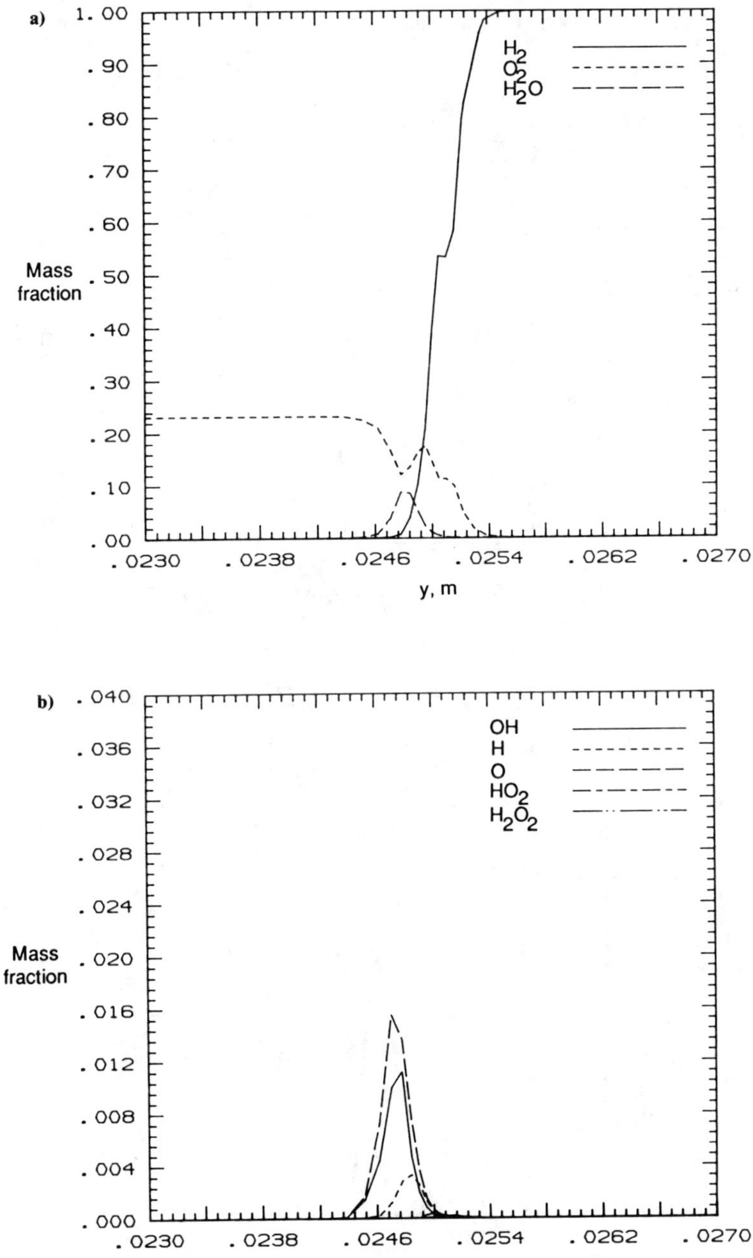

Fig. 13 Mass fraction vs y at $x = 1.0$ cm: a) major species; b) minor species.

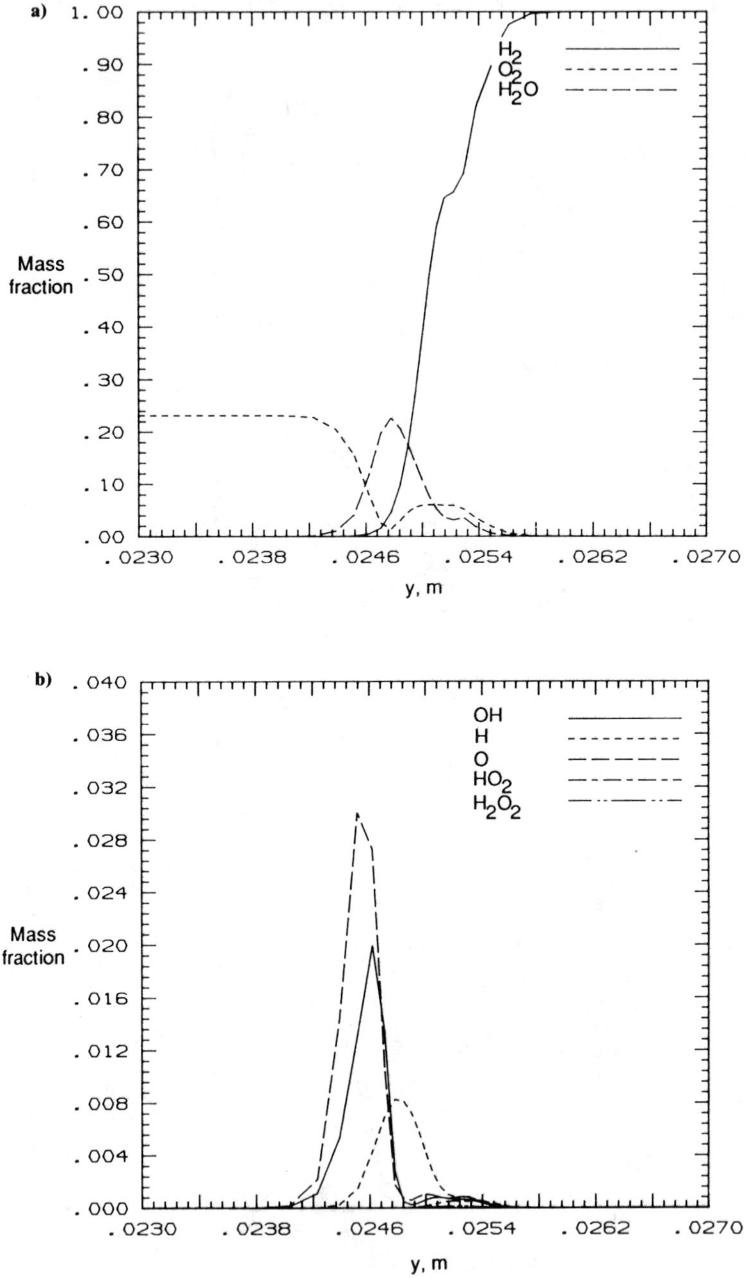

Fig. 14 Mass fraction vs y at x = 2.0 cm: a) major species; b) minor species.

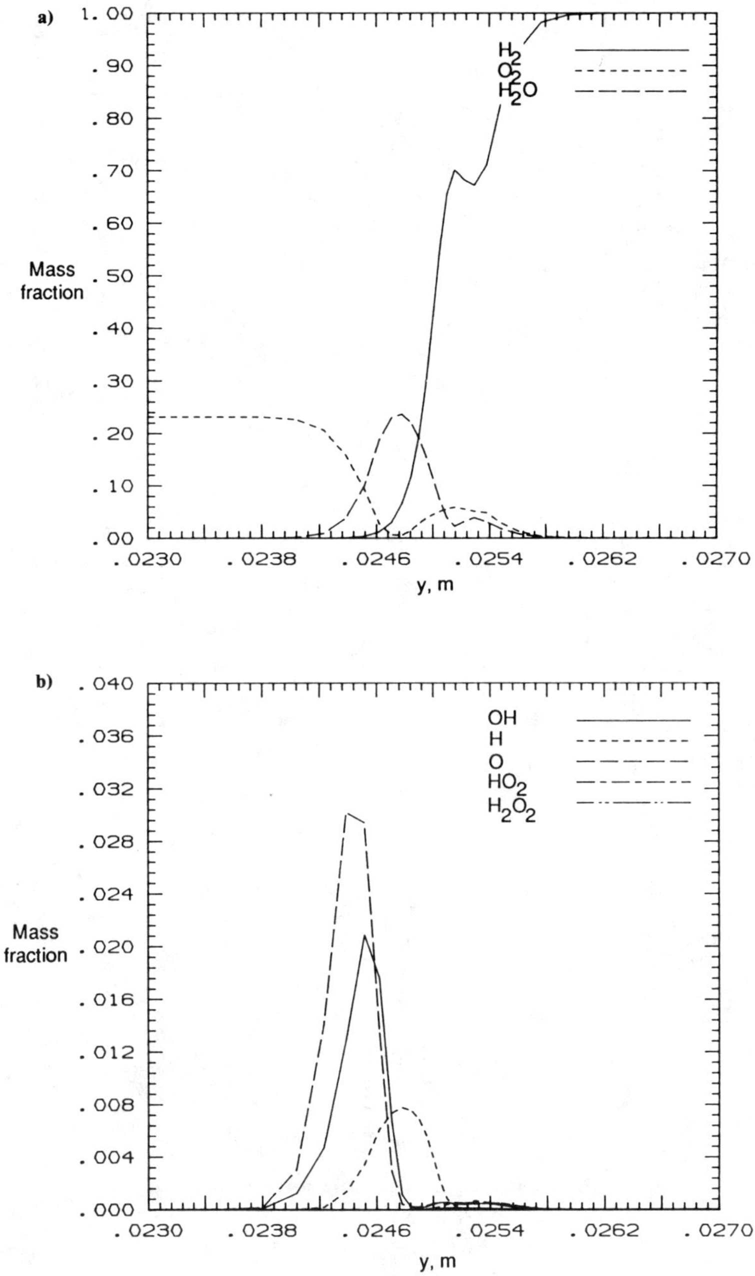

Fig. 15 Mass fraction vs y at $x = 3.0$ cm: a) major species; b) minor species.

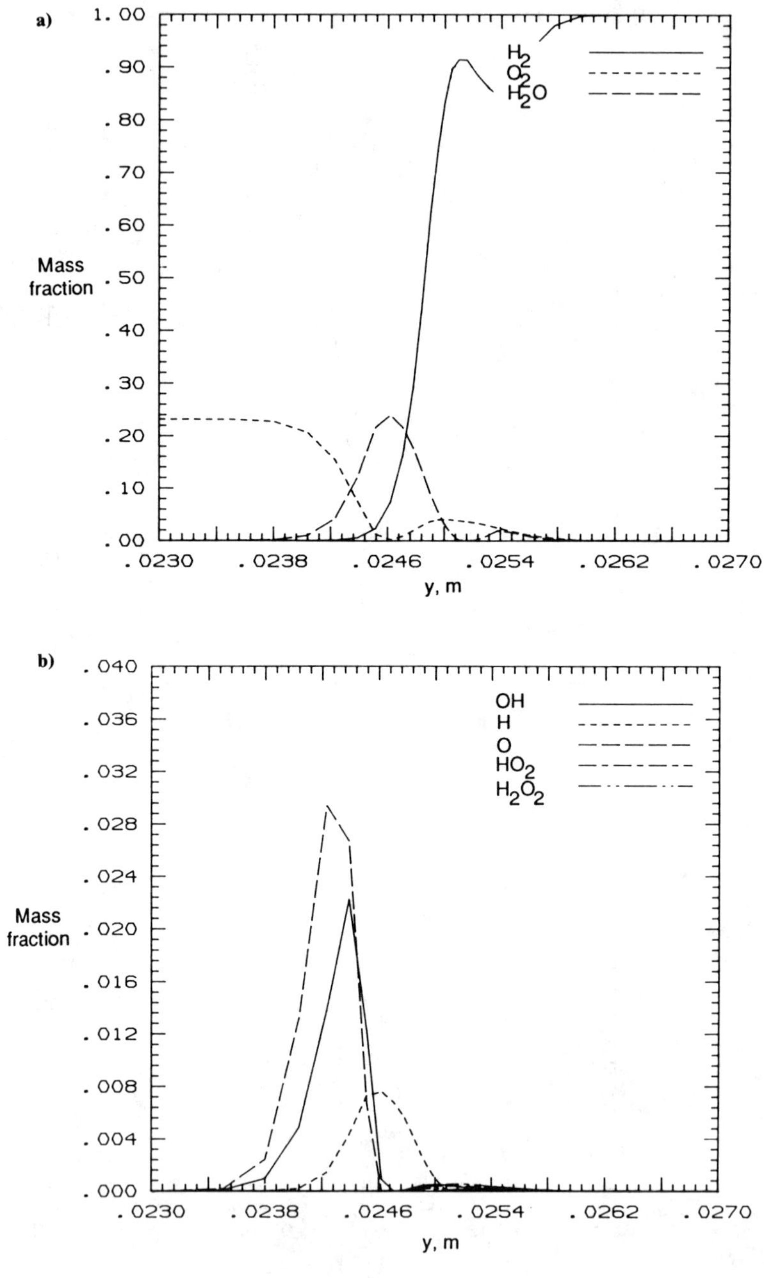

Fig. 16 Mass fraction vs y at $x = 4.0$ cm: a) major species; b) minor species.

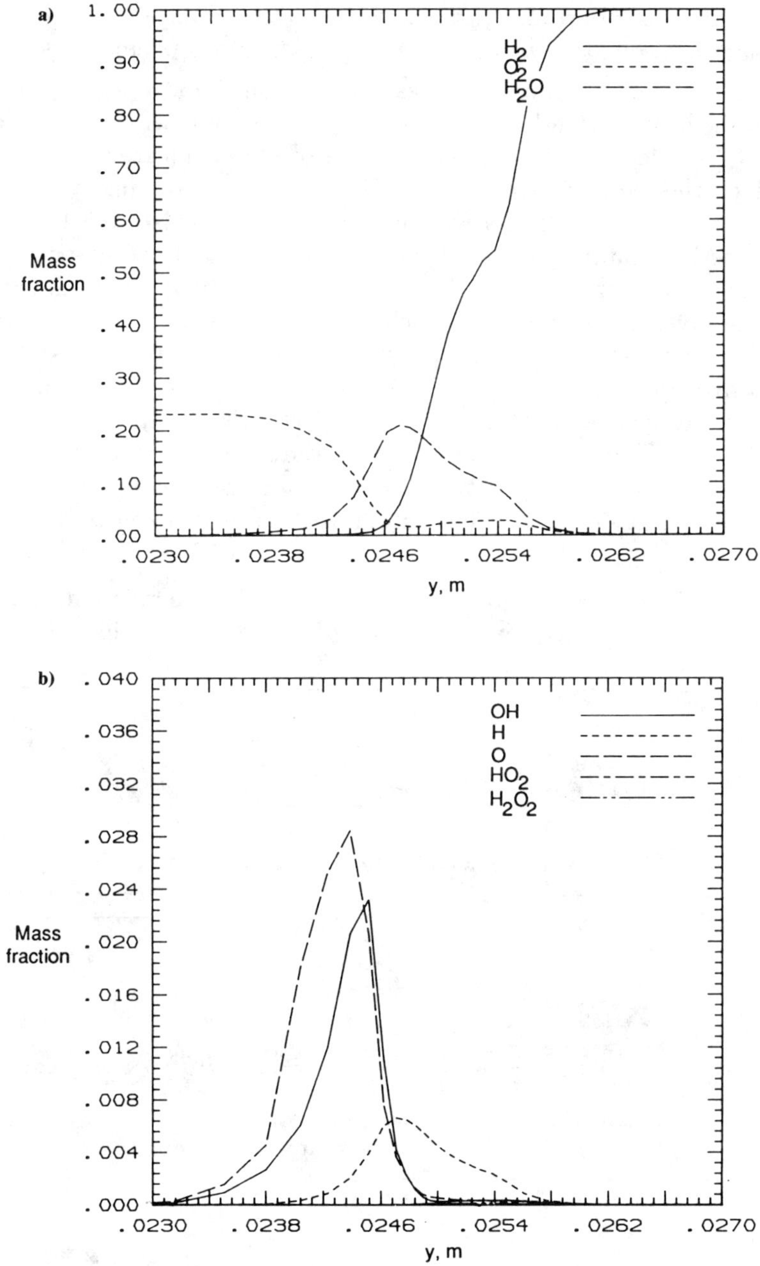

Fig. 17 Mass fraction vs y at $x = 5.0$ cm: a) major species; b) minor species.

peroxide are still present just above the splitter plate centerline. With further movement downstream at $x = 3.0$ and 4.0 cm (Figs. 15 and 16), the major and minor species profiles continue to develop, increasing both in width and in peak values. There are distinct distortions in the H_2 profiles in both figures because of eddies located on the upper edge of the mixing layer. There is also a general migration of each profile to lower values of y with increasing streamwise coordinate. The increase in product species along the lower edge of the mixing layer is a direct result of preferential burning in this region of the layer. The mixing layer is most nearly stoichiometric there, and the temperature reaches values that favor rapid ignition and combustion. At the last streamwise station given in Fig. 17, $x = 5.0$ cm, the major and minor species profiles broaden considerably further and shift to even small values of y. The noticeable increase in the rate of spread of the species profiles is associated with the second instability present in the mixing layer in this region and is consistent with transitioning to a turbulent state.

Two-dimensional contour plots of the species are given in Fig. 18-23. The resulting structure as the mixing layer develops, described previ-

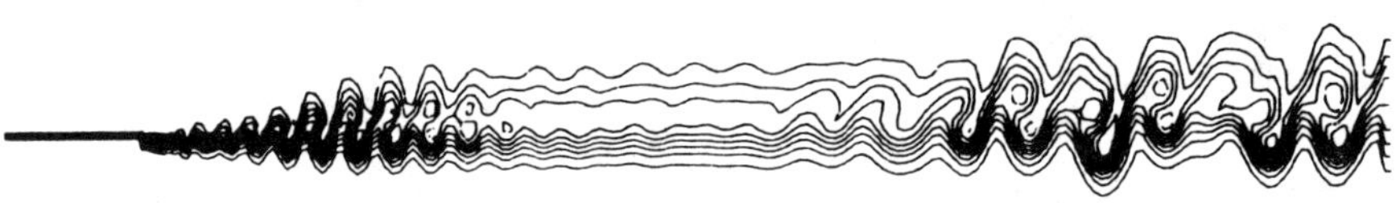

Fig. 18 Hydrogen mass fraction contours in mixing layer.

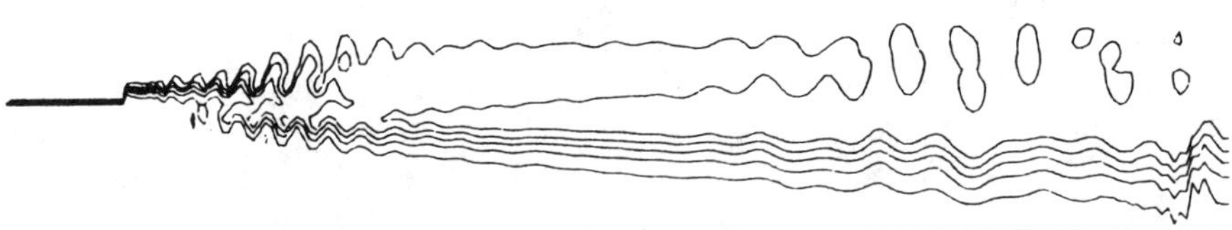

Fig. 19 Oxygen mass fraction contours in mixing layer.

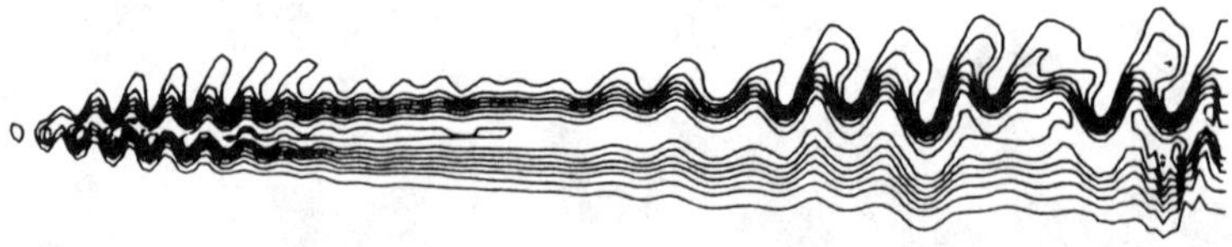

Fig. 20 Atomic hydrogen (H) mass fraction contours in mixing layer.

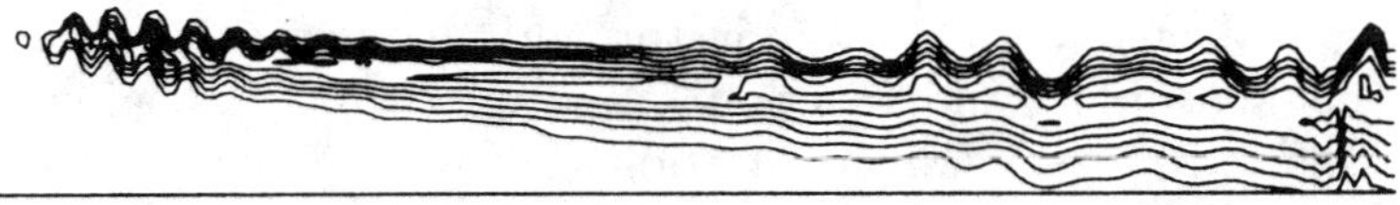

Fig. 21 Atomic oxygen (O) mass fraction contours in mixing layer.

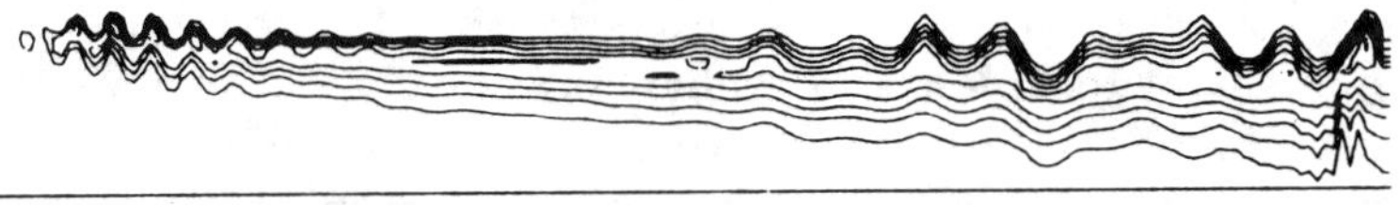

Fig. 22 Hydroxyl (OH) mass fraction contours in mixing layer.

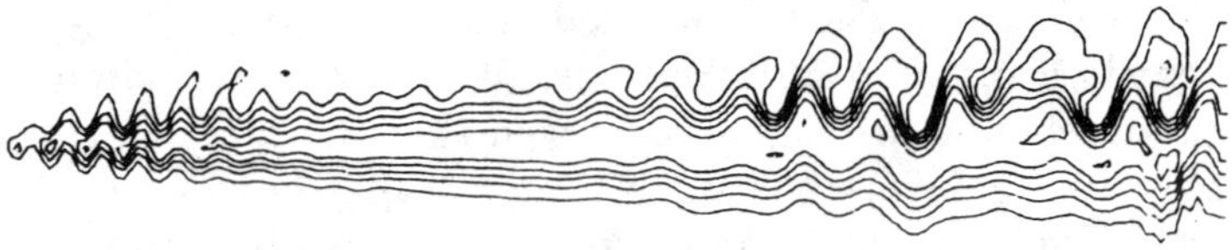

Fig. 23 Water mass fraction contours in mixing layer.

ously in Fig. 11-17, can be seen clearly in these figures. The first and second regions of instability are apparent for each species shown. The more rapid transverse spread of each species in the latter third of the layer can also be seen. A quiescent region between the two instabilities also occurs for each species, as expected. Additionally, there is a general downward migration of each of the product species with increasing streamwise coordinate. The structure of the product species, typified by water, in the downstream region of the layer is also interesting. The vortical nature of the flow, seen earlier in Fig. 10, results in regions of unreacted hydrogen gas being captured by regions (or "folds") of product water. Once captured, the regions of hydrogen have difficulty mixing with oxygen so that they can ultimately react. This phenomenon, often termed "unmixedness," reduces the overall level of reaction that can be achieved and contributes to a reduction in the efficiency of combustion. Under these conditions, mixing enhancement is required to achieve an acceptable degree of mixing and combustion efficiency.

Mixing Enhancement

Significant research has recently been directed toward the optimization of the scramjet combustor and, in particular, the efficiency of fuel-air mixing and reaction taking place in the engine. In the very high speed vehicle configurations currently being considered, achieving a high combustor efficiency becomes particularly difficult. With increasing combustor Mach number, the degree of fuel-air mixing that can be achieved through natural convective and diffusive processes is reduced, leading to an overall decrease in combustion efficiency and thrust.

Because of these difficulties, attention has now turned to the development of techniques for enhancing the rate of fuel-air mixing in the combustor. As discussed earlier in this chapter, Brown and Roshko [5] showed that the spreading rate of a supersonic mixing layer decreased with increasing Mach number, exhibiting a factor of 3 decrease in spread rate as compared with an incompressible mixing layer with the same density ratio. They concluded that the reduced spread rate was primarily a function of compressibility. It was also noted earlier that Papamoschou and Roshko[9] observed that the spreading rate of compressible mixing layers was significantly reduced over that of incompressible layers. The reduction in mixing- layer spreading rate, by approximately a factor of 3 or 4, was shown in their experiment to correlate well with increasing convective Mach number. Papamoschou and Roshko therefore concluded that the reduced spreading rate was attributable to a stabilizing effect of the convective Mach number. Ragab and Wu [56] analyzed the spatially developing supersonic mixing layer using linear spatial stability theory. They, too, found that the decreased spreading rate of the mixing layer correlated well with convective Mach number, and their predictions agreed well over most of the Mach number range with the results of Papamoschou and Roshko. The work presented in Refs. 5-56 thus showed, through both experiment and theory, the difficulty in achieving a high degree of mixing in unenhanced supersonic mixing layers.

Several authors, faced with this challenge, have numerically examined potential techniques for enhancing the mixing rates in supersonic mixing layers. Guirguis et al. [57] showed that the spreading rate of a confined mixing layer could be improved if the pressure of the two streams was different. Encouraged by this result, Guirguis[58] inserted a bluff body at the base of the splitter plate separating the two streams. The body produced an instability further upstream in the layer and resulted in a more rapid rate of spread. Kumar et al. [59] discussed a number of mixing problems that might exist in scramjet combustors. Several techniques for enhancing turbulence and mixing in combustor flowfields were suggested, and one enhancement technique that employed an oscillating shock was studied numerically. In this case, a premixed stoichiometric hydrogen-air flow was processed through a spatially and temporally oscillating shock wave, and the resulting flow was studied without and with chemical reaction. The oscillating shock was shown to increase the level of turbulence in the flowfield, and the degree of turbulence enhancement was seen to increase with a decreasing frequency of shock oscillation. Chemical reaction as defined by an equilibrium model was shown to have little effect relative to the nonreacting results. Drummond and Mukunda[60] studied fuel-air mixing and reaction in a supersonic mixing layer and applied several techniques for enhancing mixing and combustion in the layer. They found that when the mixing layer, with its large gradients in velocity and species, was processed through a shock with strong curvature, vorticity was produced. The vorticity then interacted with the layer and significantly increased the degree of mixing and reaction. Based on the results of the study, an alternate fuel injector configuration that used interacting parallel and transverse fuel injectors was designed by Drummond and Mukunda, and that configuration significantly increased the amount of fuel-air mixing and combustion over a given combustor length. This work will be reviewed later in this section.

Experimental studies have also been carried out to evaluate mixing-enhancement techniques. Menon[61] studied the interaction between a weak shock wave and a supersonic mixing layer. He found that significantly increased spreading of the shear layer occurred downstream of the shock/shear-layer interaction region. King et al.[62] studied the enhancement of mixing through the interaction of parallel and transverse jets. They found that the mixing rate of the injectors could be significantly increased by the combined parallel and transverse injection compared to parallel injection alone. Sullins et al.[63] studied the shear layer between a Mach 2-3 airstream and a sonic airstream. They observed, using schlieren photography and velocity profile measurements, a 60% increase in shear-layer growth rate when the static pressure of the supersonic stream reached approximately 1.3 times that of the sonic stream. The observations of Sullins et al. were therefore consistent with the numerical results of Guirguis.

In this section, we will apply the theory and computer program developed in the preceeding section to several candidate mixing- enhancement techniques. We will begin with a study of two-dimensional mixing layers and jets and examine several means for improving the extent of mixing and combustion in these flows. Based on the results of this study, an alternate fuel injector configuration will be numerically designed that significantly improves the amount of fuel-air mixing and combustion that can be achieved. We will then move ahead to three dimensions and model the flowfield in a generic supersonic combustor. Those flowfield results will indicate a less than acceptable degree of fuel-air mixing in the combustor. A new fuel injector configuration will again be designed to improve the level of mixing efficiency.

Mixing Enhancement of Mixing Layers and Jets

Once the computer code developed in this effort was validated for nonreacting flow, it was then applied to several reacting flow cases. The purpose of this study was to assess several candidate configurations for enhancing fuel-air mixing and reaction in a mixing layer which would then lead to a better understanding for achieving improved mixing and combustion in the supersonic combustor of a scramjet engine. The first three cases involved a supersonic, spatially developing, and chemically reacting mixing layer. The first of these cases, shown in Fig. 24, served as a benchmark calculation in that it contained no enhancement mechanism. Case 1 involved a mixing layer developing between a fuel stream and an airstream that were initially separated by an infinitely thin splitter plate. The fuel stream was made up of a mixture of 10% hydrogen and 90% nitrogen, introduced above the plate at a velocity of 2672 m/s, a static temperature of 2000 K, and a static pressure of 0.101 MPa (1 atm). Nitrogen gas was included to reduce the speed of sound of the fuel mixture. Air was introduced below the plate at a velocity of 1729 m/s, a static temperature of 2000 K, and a static pressure of 0.101 MPa. These conditions resulted in a Mach number of 2 for both streams. The physical domain considered in this case was 0.1 m long and 0.1 m high. The computational grid was identical to that used in the validation case described earlier. The simulation of this case was

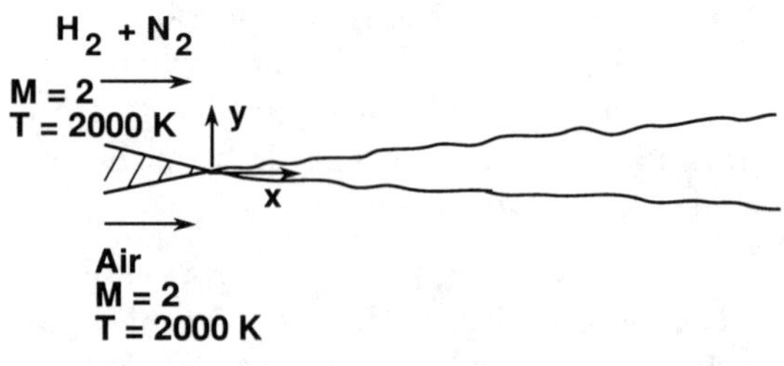

Fig. 24 Schematic of the supersonic reacting mixing layer in case 1.

begun at $t=0$ with static conditions ($u=0$, $v=0$) in the flow domain. At this time, fuel flows and airflows were initiated off the trailing edge of the splitter plate (in a method analogous to opening two valves) at the conditions given earlier, and the gases then proceeded downstream. The mixing layer then evolved between the two gases in both space and time. The calculation was then advanced in a real-time sense until an integration time of approximately 0.1 ms was reached. This time represented 14 computational sweeps of the flowfield and allowed a periodic solution to develop. Chemical reaction of the hydrogen and air took place after a sufficient degree of mixing had occurred. The chemical reaction in this case (and the next two cases) was modeled using the one-step hydrogen-air model described in Table 1.

Results for case 1 are shown in Fig. 25. The first result given is a contour plot of the velocity field. The mixing layer is seen to develop slowly in a smooth laminar fashion. A Kelvin-Helmholtz instability begins downstream at an x/L of approximately 0.8, but there is an insufficient length in the region of interest for the instability to evolve significantly. A similar result, showing temperature contours in the layer, is also given in Fig. 25. The temperature is also seen to rise smoothly in the layer as a result of both viscous heating and chemical reaction. The instability is also apparent downstream, and increased water production and an associated temperature rise to 2353 K result because of increased reaction in the vortical structures that evolve in this region. This can be seen more clearly in the final plot in the fig-

Fig. 25 Velocity, temperature, and water mass fraction contours in case 1.

ure, a contour plot of water mass fraction in the layer. Water begins appearing at an x/L of about 0.1, then evolves at a slowly increasing rate until an x/L of 0.8 is reached, and the rate increases somewhat in the instability reaching a peak value of 2.4% by mass. Even with the increased water production that occurs in the region of the instability, however, the overall degree of reaction is still quite limited in this case. This difficulty is compounded even further by the limited transverse spread of the layer in the y-coordinate direction with increasing values of the streamwise coordinate x. The mixing layer must exhibit significantly more transverse spread within the limit of the streamwise coordinate if an acceptable level of mixing and combustion efficiency is to be achieved.

In an attempt to improve the level of mixing and reaction in the benchmark mixing-layer case just considered, two mixing-enhancement mechanisms were employed. The first approach (denoted as case 2) is described in Fig. 26. Here, the mixing layer considered in case 1 is processed through two shocks entering the flow domain from the upper and lower boundaries. Each shock is set at an angle of 10 deg by choosing appropriate boundary conditions along the upper and lower boundaries beyond the shock that are consistent with a 10-deg shock in a Mach 2 flow of hydrogen/nitrogen and air, respectively. The two shocks then propagate from the upper and lower boundaries into the flow and across the mixing layer. The resulting flowfield, taken again at about 0.1 ms, is shown in Fig. 27. Inspection of the velocity field shows no marked enhancement in mixing-layer spread rate because of the stationary shocks. The instability appears slightly further upstream at $x/L = 0.7$, just behind the location of shock interaction with the layer. The amplitude of the instability is not increased, however, and the viscous region of the layer as defined by the velocity gradient is not thickened relative to the previous case. The temperature contours given in Fig. 27 show identical trends. Temperatures reach a peak value of

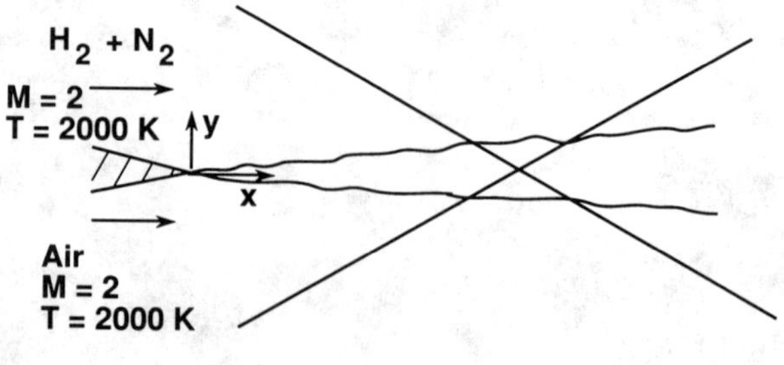

Fig. 26 Schematic of the supersonic reacting mixing layer interacting with two shocks in case 2.

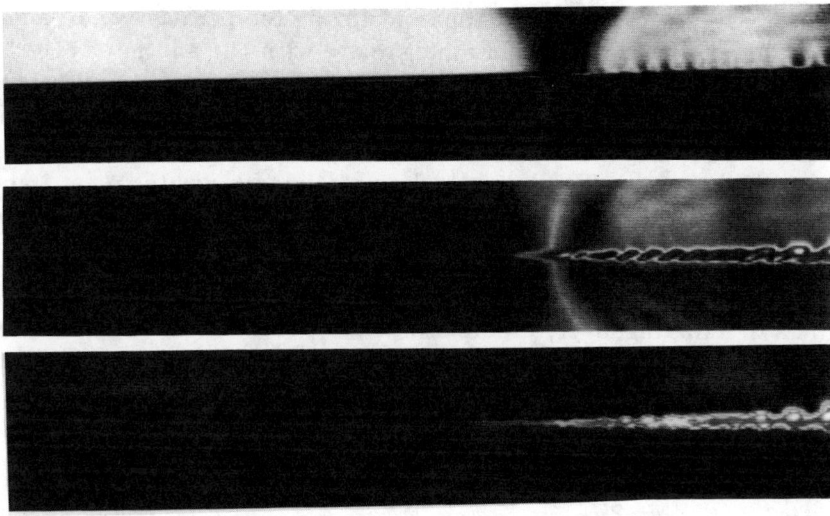

Fig. 27 Velocity, temperature, and water mass fraction contours in case 2.

3178 K in the center of the layer near $x/L = 1.0$, but the increase is due primarily to the temperature rise through the shocks and some small increase in water production resulting from the higher temperature of the reactants. Even with this further increase in water production, giving peak values of about 6% by mass, the overall degree of reaction and the amount of product that is produced is still quite low. This can be seen by viewing the water contour plot in Fig. 27 which shows a layer thickness defined by water that is not any greater than that observed in the previous unenhanced case.

The second enhancement study, designated as case 3, is described in Fig. 28. Conditions are again the same as in the previous two studies. In this case, a small square cylinder is placed along the fuel/air interface at $x/L = 0.2$. The body is 0.0012 m high and 0.002 m long. When the body is placed in the flow, it results in the formation of a bow shock just ahead of the body. The shock has strong curvature in the immediate neighborhood of the body. When the high-velocity gradient region of the mixing layer is processed by this curved shock, vorticity is produced. The vorticity is then convected downstream where it produces enhanced macromixing of fuel and air. This effect can be seen in the results given in Fig. 29. The velocity field can be seen to become unstable near the trailing edge of the interference body, and the thickness of the layer as defined by the velocity gradient grows rapidly with increasing streamwise coordinate. Rapid growth of the layer can also be seen in the temperature contour plot given in figure 29. A significant amount of vortical structure is also apparent in the temperature field, and the individual vortices grow and amalgamate with one another as the layer develops with increasing streamwise distance. A peak temperature of 2500 K was observed at the center of several of the vortical structures. The water mass fraction also peaked at the centers with values as high as 3.4% by mass. The layer thickness defined by water mass fraction that is shown in the figure is also significantly greater at all values of x relative to the previous two cases in this study.

Although spread rate gives a good visual indication of the development of a mixing layer, a more useful and practical indication of mixing is given by the mixing efficiency of the layer. The mixing efficiency is

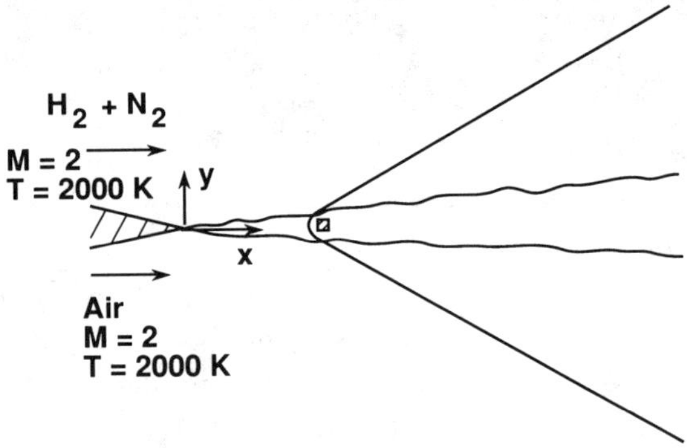

Fig. 28 Schematic of the supersonic reacting mixing layer interacting with a curved shock in case 3.

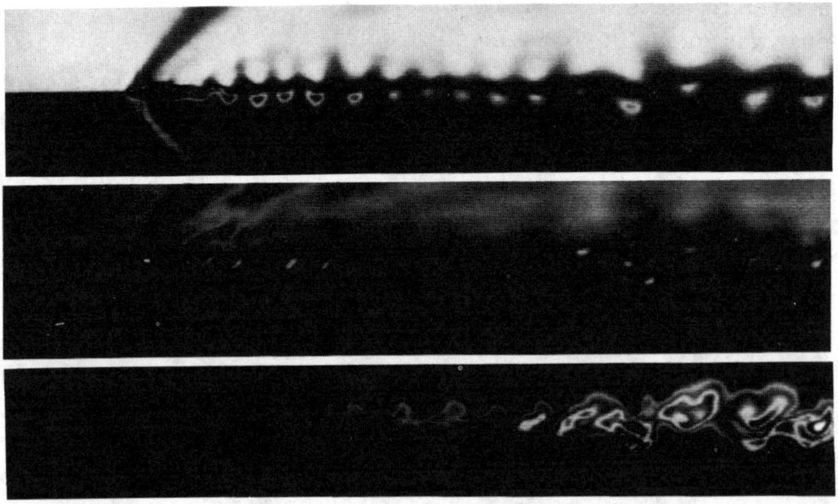

Fig. 29 Velocity, temperature, and water mass fraction contours in case 3.

defined as a number between 0 and 1 that specifies the amount of fuel that could react at any x station if chemical reaction was taking place. Therefore, if all fuel could be consumed at a given x station, the mixing efficiency at that station would be unity. The mixing efficiency plotted as a function of the streamwise coordinate for cases 1-3 is given in Fig. 30. The efficiency values for the three cases should be viewed in a relative sense because there are significant regions of fuel and air that can never mix since they are located at large transverse distances away from the mixing layer. What is important in this comparison, then, is the relative degree between the cases to which efficient mixing takes place. All three cases show a similar streamwise development of mixing efficiency to the 0.02-m station. Beyond that location, cases 1 and 2 exhibit a similar slow development in efficiency, whereas case 3 experiences a significant growth in mixing efficiency. The more rapid spread of the mixing layer due to the higher level of induced vorticity in case 3, which was observed in the earlier results, translates directly into a significantly higher level of mixing efficiency. Near the outflow station, the peak mixing efficiency of case 3 is approximately four times that of case 1 and three times that of case 2. The oscillatory nature of the mixing efficiency plots for the three cases is related to the instantaneous structure present in each of the mixing layers at the time the results are plotted. The structure of the plots can be directly tied to the representation of layers given in Fig. 25, 27, and 29. At that instant in time (0.1 ms), the highest rate of chemical reaction and the largest amount of product at a given x station are present in the neighborhood of the largest vortical structures, whereas less product is present at stations where the layer pinches between the vortices.

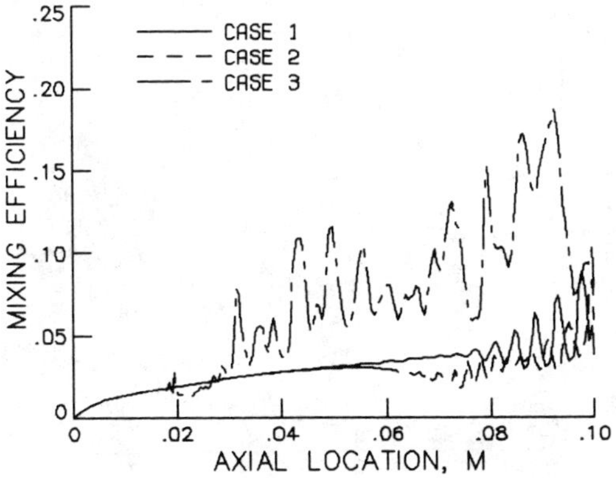

Fig. 30 Mixing efficiency vs streamwise station for cases 1-3.

Once fuel and air have mixed, the degree to which they then react is defined at any streamwise station by the combustion efficiency. The combustion efficiency is again a number between 0 and 1 that defines the degree of chemical reaction that has taken place. The combustion efficiency that results for cases 1-3 is given in Fig. 31. Comparison among the three cases should again be made in a relative sense because of the geometry of the mixing layer. In addition, the one-step chemistry used to model reaction for cases 1, 2, and 3 underestimates water production relative to the more complete model used later in this study. The model is still useful for this comparison, however. The combustion efficiency results given in Fig. 31 follow trends quite similar to those observed for mixing efficiencies given in Fig. 30. All three cases exhibit similar increases in combustion efficiency up to the 0.02-m station, but then case 3 shows a significant increase over the other two cases. This trend continues until the 0.08-m station, where the combustion efficiency for case 2 increases rapidly. Re-examination of the mixing efficiency for case 2 shows, however, that this increase is not related to improved mixing, but rather to an increased reaction rate following a significant temperature rise through the shocks present in case 2. Overall, combustion efficiencies for case 3 are nearly 3 times higher than case 1 and 1.5 times higher than those for case 2. It is also important to note that the higher levels of mixing efficiency achieved in case 3 occur well upstream in the layer relative to efficiency increases in the other two cases. Therefore, enhancement techniques of the type used in case 3 may be an effective means of shortening the overall combustor length while still retaining a high degree of chemical reaction and the associated thrust.

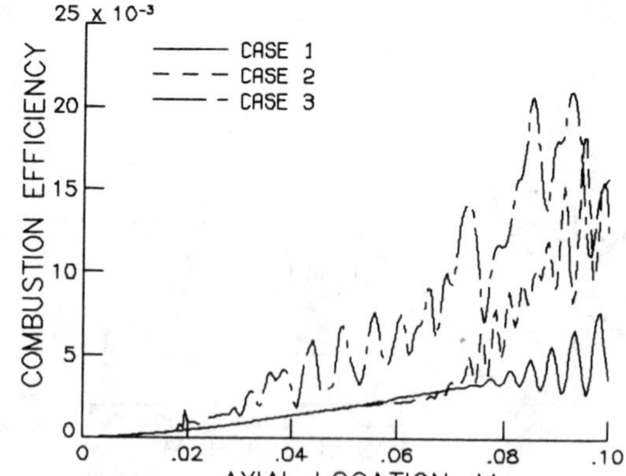

Fig. 31 Combustion efficiency vs streamwise station for cases 1, 2, and 3.

Table 2 Statistical features of cases 1 and 3

Case Configuration	(1) M=2, Fuel, 2000 K  M=2, Air, 2000 K	(3) M=2, Fuel, 2000 K 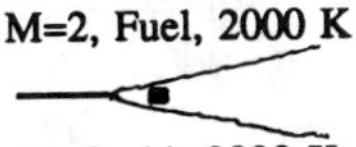 M=2, Air 2000 K
δ_w/x	0.04 - 0.044	0.056 - 0.058
$\sqrt{u'^2}/U$	3.5%	10%
$\sqrt{T'^2}/T$	3.6%	6.7%
v/U_o	~0.4%	~1%
$\sqrt{v'^2}/v$	3.5%	12%
$\sqrt{\rho'^2}/\bar{\rho}$	17%	34%
$\sqrt{f'^2_{H_2O}}/f_{H_2O}$	2.5%	4.7%
$\overline{\rho u'v'}/U_o^2\rho_o^2$	9.1 x 10^4	1.84 x 10^{-3}

To better understand the success of enhancement on fuel-air mixing, statistics of the resulting flowfield in case 3 were next examined. The results of the case 1 benchmark analysis were also included in this study to allow comparison with case 3. To extract these statistics from the simulations, values of the flow variables were collected at selected spatial locations over 50 time steps representing approximately two computational sweeps of the flowfield. The resulting statistical features of the flow are summarized in Table 2. Included in the table are peak values of root-mean-square total velocity, streamwise velocity, temperature, density, and water mass fraction normalized by their respective mean values. Also included are peak values of the mean total velocity, the peak Reynolds stress normalized by the freestream dynamic pressure, and the layer vorticity thickness normalized by streamwise distance. In addition, a spectral analysis of the fluctuating variables was also conducted. Results of the analysis were correlated in terms of a frequency nondimensionalized by the ratio of local velocity and local vorticity thickness. Significant energy was found to exist in the upstream fluctuating velocity field at frequencies of 0.01, 0.06, 0.09, and 0.12. Well downstream in the layer, the energy spectrum was quite broad, and there were no significant local peaks. This distribution indicated that the flow was transitioning, thus justifying the collection of statistical data. The tabulated values of vorticity thickness in Table 2 indicate that the mean growth rate is enhanced by about 40% in case 3 as compared to case 1. The fluctuating quantities also exhibit significantly higher values for case 3. The Reynolds stress, and the fluctuating values of temperature, density, and water mass fraction are approximately twice as large in the enhanced case compared to the unenhanced case, and the fluctuating total velocity is nearly four times as high. The statistical results, therefore, also indicate a significant improvement in mixing and combustion in case 3.

The success achieved with the mixing-enhancement technique employed in case 3 motivated the authors to apply this approach to a more realistic configuration. The fuel-injection strut of a conventional scramjet engine was chosen for study. The fuel-injection struts provide locations for instream injection of gaseous hydrogen fuel into the air coming from the inlet of the engine. Fuel is injected in both a parallel and transverse direction to the incoming airstream. Transverse injection predominates over parallel injection when the engine is operating in the high-Mach-number regime to speed fuel-air mixing and combustion. At lower Mach numbers, more parallel injection is used to slow the mixing process and achieve a heat-release schedule similar to that achieved at higher Mach numbers. Figure 32 is a schematic of

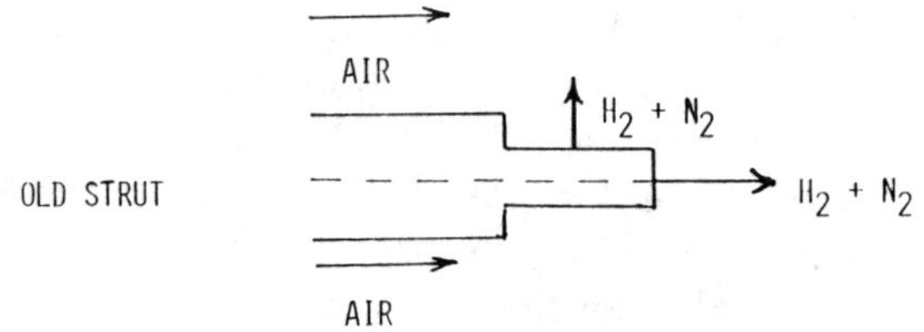

Fig. 32 Schematic of conventional fuel-injector strut configuration.

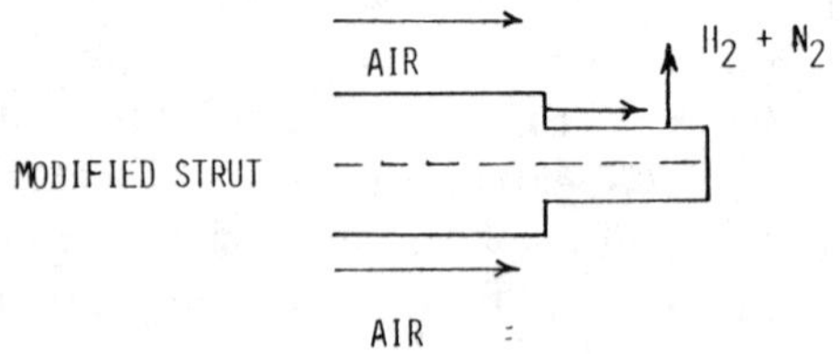

Fig. 33 Schematic of modified fuel-injector strut configuration.

a typical injector configuration on the trailing edge of a fuel injection strut. Transverse injection takes place following a rearward-facing step that provides improved flameholding, and parallel injection occurs at the strut base. The results achieved in case 3 suggested that a reorientation of the injectors might improve the rate of fuel-air mixing and chemical reaction that could be achieved downstream of the strut. That change is reflected in Fig. 33. In the new design, the parallel injector has been moved from the base of the strut to the vertical wall of the rearward-facing step. The transverse injector now produces a curved bow shock that interacts with the high- velocity gradient of the jet from the parallel injector, resulting in vorticity production. The transverse jet thus serves a function similar to that provided by the interference body employed in case 3. The present design is more practical, however. The solid interference body would produce significant losses in an engine, and aerodynamic heating would also pose a problem. The transverse injector is present in both strut designs, however, and so it introduces no significant losses in the new design compared to the old one.

To assess the new strut design, a computational study was again performed. The calculation was begun at the rearward-facing step. A parallel slot fuel injector, located on the face of the step, injected a mixture of 10% hydrogen and 90% nitrogen gas at a velocity of 2672 m/s, a temperature of 2000 K, and a pressure of 0.101 MPa (1 atm). The injector was 0.064 cm high and located 0.032 cm above the strut wall. A transverse slot fuel injector was located 0.26 cm downstream of the step. An identical hydrogen-nitrogen mixture was injected there at 2672 m/s, a temperature of 2000 K, and a pressure of 0.505 MPa. The slot was sized to be one-fifth the width of the parallel jet so that the same amount of fuel was introduced from each injector. The strength of the transverse jet insured that it would produce a shock of sufficient strength to result in significant vorticity production.

With the new strut now configured, two cases were considered. The first case (identified as case 4) considered operation of only the parallel injector. Case 5 involved operation of both the parallel and transverse injectors. Both cases were computed on a computational grid of 218 streamwise points and 51 transverse points. The grid was highly compressed in the transverse direction about the parallel injector and highly compressed in the streamwise direction about the transverse injector. Chemistry was modeled in both cases using the 9-species, 18-reaction, hydrogen-air scheme described in Table 1. This model provided a more realistic description of reaction than the one-step model, but it was

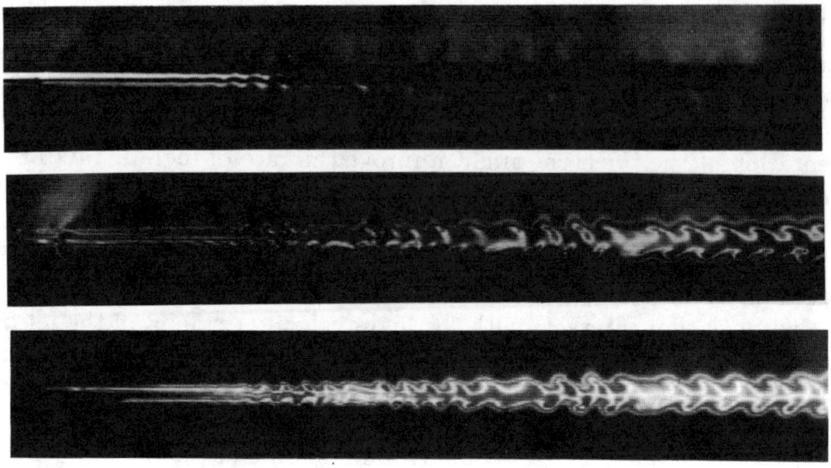

Fig. 34 Velocity, temperature, and water mass fraction contours in case 4 with only the parallel injector.

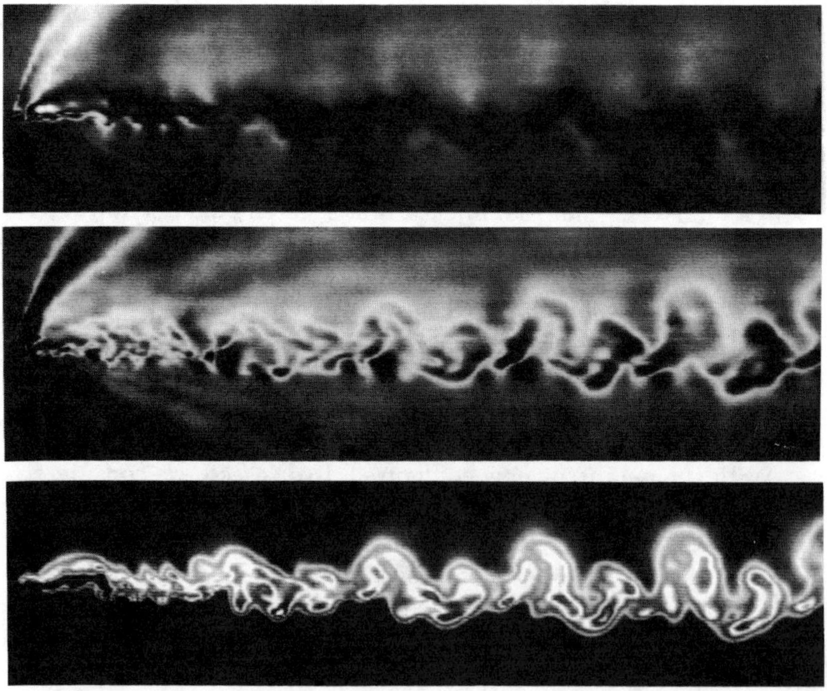

Fig. 35 Velocity, temperature, and water mass fraction contours in case 5, with interaction of the parallel and transverse injectors.

computationally more expensive. However, it seemed appropriate to apply the more detailed model to these practical cases for the most accurate representation of product production and heat release in the simulation. Results for case 4 at a time of 0.1 ms are shown in Fig. 34. The unenhanced parallel jet behaves much like the unenhanced mixing layer in case 1. The velocity contour plot exhibits a mild instability further upstream in this case, but there is no significant growth of the jet. Lack of growth can also be observed in the temperature contours in Fig. 34. The thermal layer is somewhat thicker than the velocity thickness of the previous plot because of burning, heat release, and a resulting temperature rise on the edges of the layer. A peak temperature of 2636 K is reached near the end of the jet. Water contours, which result from the complex reaction process, are shown in the final plot. As expected, the water contours closely resemble the temperature contours, and only a moderate degree of spread in the water field is observed. Peak values of around 22% by mass of water are achieved near the end of the layer, however, indicating that a significant degree of chemical reaction does occur where fuel and air are able to mix sufficiently.

To improve the degree of fuel-air mixing, the transverse fuel injector was activated so that it might interact with the parallel injector. The resulting flowfield, again at a time of 0.1 ms, is shown in Fig. 35. The degree of mixing enhancement induced when the parallel and transverse jets interact is significant. The interaction of the parallel jet with the curved bow shock ahead of the transverse jet again produces vorticity, resulting in increased mixing. The bow shock can be seen in both the velocity and temperature contour plots. The more rapid development of the fuel jets can also be seen in those plots, along with a significant increase in the spread of the jet. A peak temperature of 2705 K is reached downstream in the reaction zone. The water contour plot in Fig. 35 also shows markedly more jet development due to the interaction. The jet spreads much more rapidly than in the previous case, and water mass fractions as high as 24% occur across appreciable portions of the jet.

A more quantitative comparison of the last two cases is again made by examining their mixing and combustion efficiencies. A comparison of mixing efficiencies for cases 4 and 5 is given in Fig. 36. The mixing efficiency of case 5 increases much more rapidly than that of case 4 because of enhancement. Efficiencies of around 90% are achieved in only 40% of the solution domain length, whereas the unenhanced case 4 requires 75% of the solution length to achieve the same level of mixing. This is especially noteworthy since twice as much fuel is being injected in case

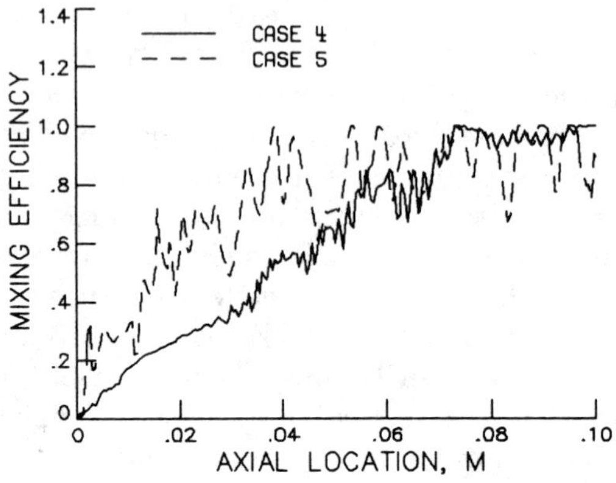

Fig. 36 Mixing efficiency vs streamwise station for cases 4 and 5.

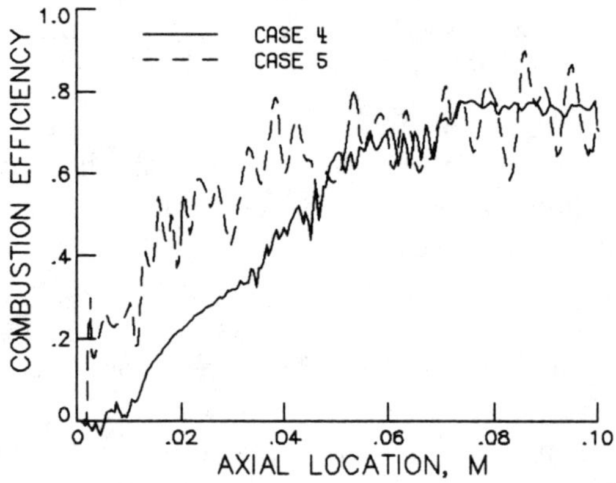

Fig. 37 Combustion efficiency vs streamwise station for cases 4 and 5.

5 compared to case 4. Similar results are also observed when combustion efficiencies for the two cases are compared in Fig. 37. Combustion efficiencies of about 80% are obtained for the enhanced case within the first 40% of the solution length, whereas the unenhanced case requires 75% of the solution length to achieve the same level of reaction. The higher combustion efficiencies that occur at earlier streamwise stations are again achieved with twice the amount of injected fuel. It is clear, then, that the simple enhancement technique employed in this study is quite effective in achieving a higher level of fuel-air mixing and combustion in shorter streamwise distances compared to more conventional unenhanced approaches. The same or similar approaches should there-

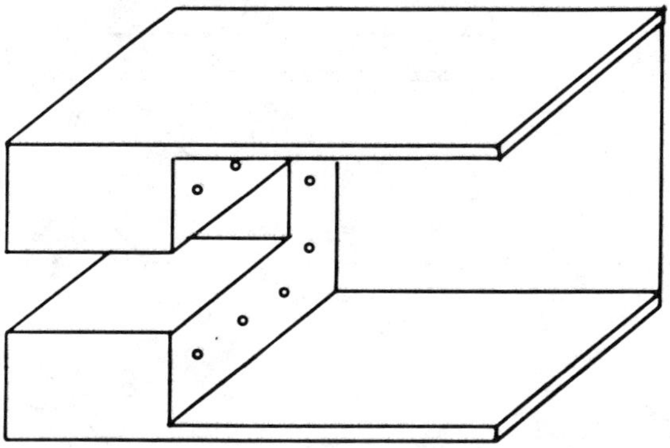

Fig. 38 Generic scramjet combustor.

fore offer an efficient means for achieving improved levels of combustion efficiency with a shorter overall combustor length.

Mixing Enhancement in Scramjet Engine Flow Fields

We will now move from two-dimensional to three-dimensional analyses and describe a numerical study of fuel-air mixing in a model generic supersonic combustor. The generic scramjet combustor that was considered is shown in Fig. 38. Results from the study showed that the amount of fuel penetration and fuel-air mixing that could be achieved by the present design was limited below that required for acceptable combustor performance. To improve the degree of fuel-air mixing, two injector ramp enhancement configurations, developed in an experimental effort by Northam, et al.[64], were chosen for study using numerical simulation. One of these configurations was shown to improve the degree of fuel-air mixing significantly, and appeared to be an excellent candidate for use in actual combustor configurations.

Generic Scramjet Combustor

Following the two-dimensional mixing-enhancement studies discussed in the previous section, a three-dimensional analysis was undertaken to study the generic model scramjet combustor described in Fig. 38. Each of the four combustor walls contains a step 2.0 cm high located 1.5 cm downstream of the inlet. Four hydrogen fuel injectors are located on each step. One wall has been removed in the figure so that the internal structure can be seen. Air enters the combustor from an upstream inlet at Mach 2.5, a static temperature of 2000 K, and a pressure of

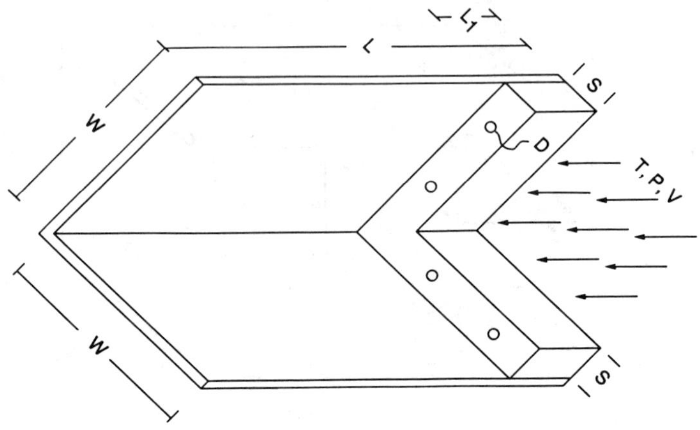

Fig. 39 Computational domain of generic scramjet combustor.

0.101 MPa (1 atm). Gaseous hydrogen fuel is injected under choked conditions at Mach 1.0 at a temperature of 250 K and a pressure of 0.203 MPa. It can be seen from Fig. 38 that the combustor is symmetric about transverse and spanwise planes through its centerline. Therefore, only one quadrant of the combustor was considered in the analysis. That region is diagrammed in Fig. 39.

The combustor flowfield was solved using the fourth-order compact scheme described earlier in the chapter. The computational grid used in the combustor study had 61 points in the streamwise direction and 45 points in both the transverse and spanwise directions. In the streamwise direction, the points were concentrated at the step, and they were exponentially stretched away from the step. In both the transverse and spanwise directions, the points were compressed at the walls and at the steps. An exponentially expanding grid was then connected to the inner grid to span the region from the step to the combustor centerline. The entire grid was also adjusted to account for transverse and spanwise growth of the fuel jets.

To specify boundary conditions, the wall velocities were set to zero and the wall temperature was set to 1100 K. The boundary-layer assumption on pressure was also employed to specify zero pressure gradients on all walls. The walls were also assumed to be noncatalytic such that the gradient of each chemical species vanished at the walls. Symmetry conditions were used at the two symmetry planes. The inflow plane was supersonic, and so the velocities, temperature, pressure, and species distributions were specified and held fixed at that location. At the outflow plane, the flow was again supersonic, and so extrapolation of velocity, temperature, pressure, and species from upstream values was employed.

The turbulence field in the combustor calculation was modeled using the Baldwin-Lomax turbulence model[34]. Following the work of Riggins[66], the turbulent Schmidt number was assumed to be 0.5. The calculation, carried out initially without chemical reaction, required approximately 3 hours on the NASA Langley Cray 2 computer. The resulting hydrogen mass fraction contours at six successive streamwise stations, beginning near the steps and ending near the combustor outflow boundary, are shown as a contour plot in Fig. 40. The contour levels range from 0.0 to 1.0 in increments of 0.1. The flow direction is from right to left. The contour levels on the initial plane show high hydrogen concentrations in the vicinity of the four fuel injectors. The hydrogen from each injector mixes with the inlet air with increasing streamwise distance until the near-wall region has nearly uniform levels of hydrogen. It is important to note that the hydrogen from all the injectors remains in the region within one step height from each wall. The shear layer and expansion around the step confines the injected hydrogen to this region, and no penetration into the freestream is achieved. It is also interesting to note that the injectors located near the corner are effectively swept by the strong vortical flow moving into the corner. The hydrogen is mixed very effectively with inlet air in that region, but the vorticity also tends to hold the mixture in the corner. The injectors away from the corner are not affected by the corner vorticity, and they mix much more slowly.

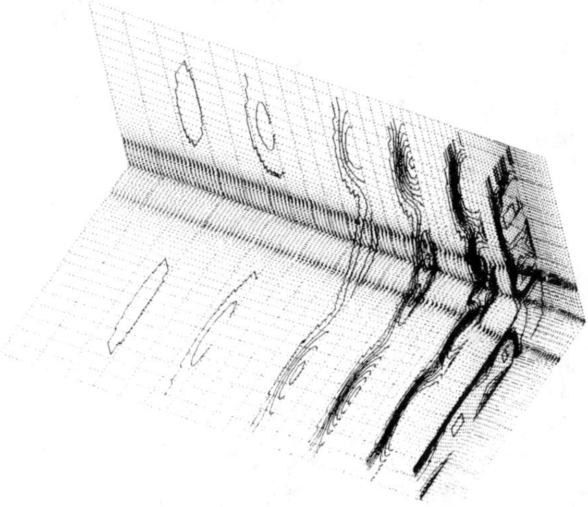

Fig. 40 Hydrogen mass fraction contours in generic scramjet combustor with four fuel injectors.

The combustor flowfield was next studied with chemical reaction of the hydrogen fuel and air. Only the two injectors near the corner were considered in this study to make the problem more tractable computationally. The two injectors away from the corner were eliminated. Otherwise, the combustor geometry and flow conditions remained unchanged. The 9-species, 18-reaction chemistry model described in Table 1 was used to model combustion processes. The governing equations were solved on a grid with 41 points in the streamwise direction and 25 points in the transverse and spanwise directions. This grid appeared adequate to resolve important convective effects. Further study on finer grids is planned to resolve more adequately fine-scale mixing and chemical reaction. Solution times were approximately one hour on the Cray 2.

Results from the reacting flow study are shown in Fig. 41-45. The results are represented in the same manner as those shown in Fig. 40. In this case, the flow direction is chosen to be from left to right to represent the contours more clearly. Contours of streamwise velocity are shown in Fig. 41. The velocity ranges from 1177 to 2504 m/s, and the contours are divided into 10 levels. The flow separates in the corner through the fifth cross-stream plane and reattaches beyond that point. Boundary layers form and grow along each wall, with typical thickening in the corner region. The temperature field in the combustor is shown in Fig. 42. Temperature is contoured between 142 and 2617 K with 10 contour levels. As before, cold hydrogen fuel is pumped into the corner by the corner vortices. Contour structure in the corner representing the cold hydrogen can be seen in the second data plane.

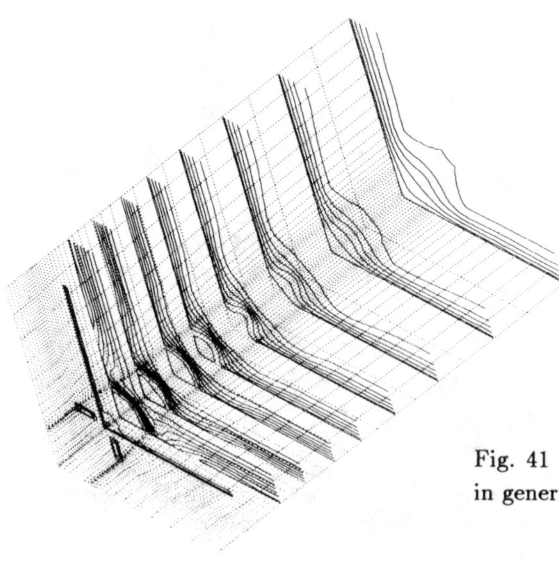

Fig. 41 Streamwise velocity contours in generic scramjet combustor.

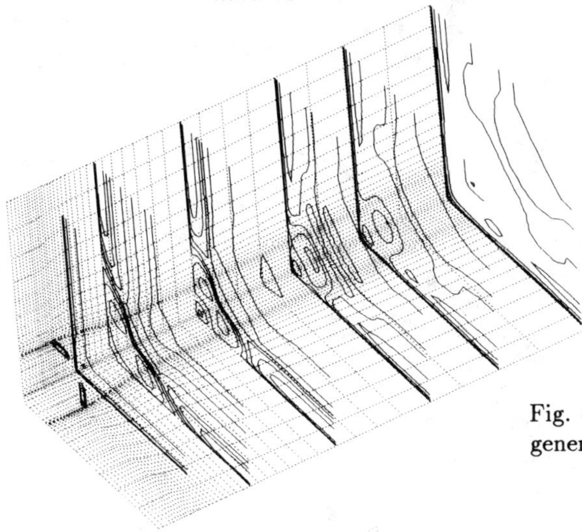

Fig. 42 Temperature contours in generic scramjet combustor.

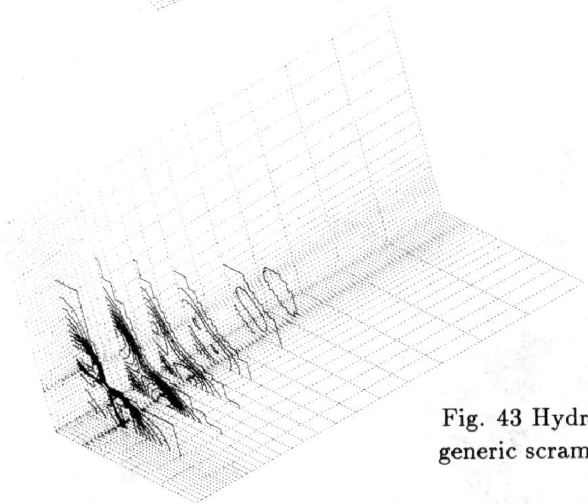

Fig. 43 Hydrogen mass fraction contours in generic scramjet combustor.

Further downstream, the hydrogen begins to mix with hotter air at the third data plane, the fuel-air mixture ignites, and reaction occurs in the corner region from that point downstream. The hydrogen mass fraction distribution is shown in Fig. 43. The hydrogen mass fraction ranges between 0.1 and 1.0 over 10 contour levels. As noted earlier, hydrogen is forced by the corner vortices into the corner. The fuel then begins mixing with air and, by the seventh data station, it is completely consumed to form water. The resulting water mass fraction is shown in Fig. 44. Water mass fraction is plotted between 0.05 and 0.30 over 10 contour levels. Small amounts of water begin to form a small distance downstream of the injectors well away from the corner on the edge of the fuel jets. Somewhat further downstream, a small amount of water begins to appear in the immediate vicinity of the corner that is convected upstream by the recirculation in the separated flow. By

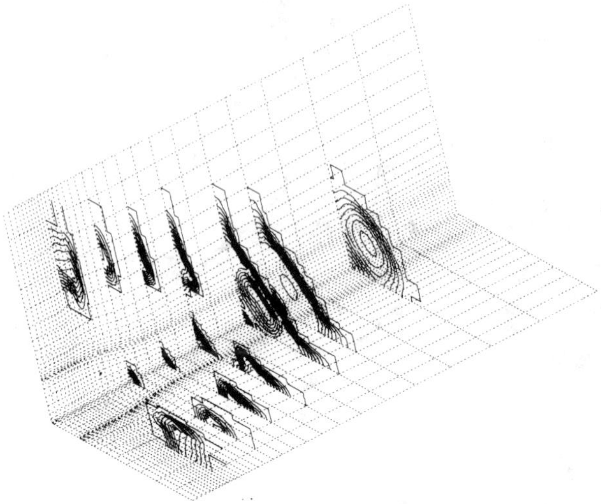

Fig. 44 Water mass fraction contours in generic scramjet combustor.

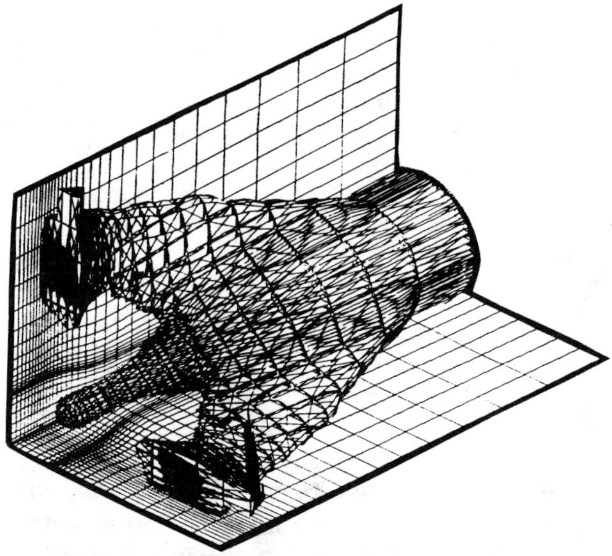

Fig. 45 Volume rendering of water mass fraction contours in generic scramjet combustor.

the fifth data plane in the figure, fuel and air have become well mixed, and large amounts of water are formed. Reaction is nearly complete at the last data plane shown in the figure.

A more global picture of the water distribution can be seen in the volume rendering given in Fig. 45. A volume rendering represents the water mass fraction levels by small volumes rather than contour lines. Color levels are typically used to represent the value of the mass fraction, but they cannot be shown in this chapter. The volume rendering is still useful here, however, to show a continuous path of the water mass fraction. The water that is forced upstream to the step by flow recirculation is evident in the figure. It also becomes quite clear in the figure how the inlet flow and corner vortices force the fuel-air mixture into the corner, not allowing sufficient penetration of the mixture into the primary combustor flow. It is very evident from this figure that enhancement of fuel-air mixing must be provided if an acceptable degree of combustion efficiency is to be obtained from this generic combustor. Mixing enhancement is the subject of the following subsection of this chapter.

Fuel-Air Mixing Enhancement

A number of approaches have been suggested for enhancing the mixing of high-speed fuel-air flows. Several of these approaches were discussed in the Introduction. A particularly attractive option has been suggested by Northam, et al.[64] in their experimental study of wall-mounted parallel injector ramps used to enhance the relatively slow mixing of fuel and air normally associated with parallel fuel injection. Parallel injection may be useful at high speeds to extract energy from hydrogen that has been used to cool the engine and the airframe of a hypersonic cruise vehicle. The ramp injector configurations were intended to induce vortical flow and local recirculation regions similar to the rearward-facing step that has been used for flameholding in reacting supersonic flow.

Two ramp configurations were considered in the experiment. They are both shown in Fig. 46, which was taken from Ref. 64. In both configurations, hydrogen gas was injected at Mach 1.7 from conical nozzles in the base of the two ramps which were inclined at 10.3 deg to the combustor wall. The injector diameters were 0.762 cm. The sidewalls of the unswept ramps were aligned with Mach 2.0 streamwise airflow from a combustion facility, whereas the swept ramps were swept at an angle of 80 deg. Each ramp was 7.0 cm long and ended in a nearly square base, 1.52 cm on a side. Both ramp designs were chosen to induce vorticity to enhance mixing and base flow recirculation to provide flameholding. The swept ramp injector, becuse of its delta shape, was intended to induce higher levels of vorticity and, therefore, higher levels of mixing. Hydrogen injection occurred at a streamwise velocity of 1747 m/s, a transverse velocity of 308 m/s, and a static temperature and pressure of 187 K and 325,200 Pa, respectively. The facility air crossed the leading edge of the wedges at a streamwise velocity of 1300 m/s, a static temperature of 1023 K, and a static pressure of 102,000 Pa. The air was vitiated following heating by a burner with oxygen, nitrogen, and water mass fractions of 0.2551, 0.5533, and 0.1818, respectively.

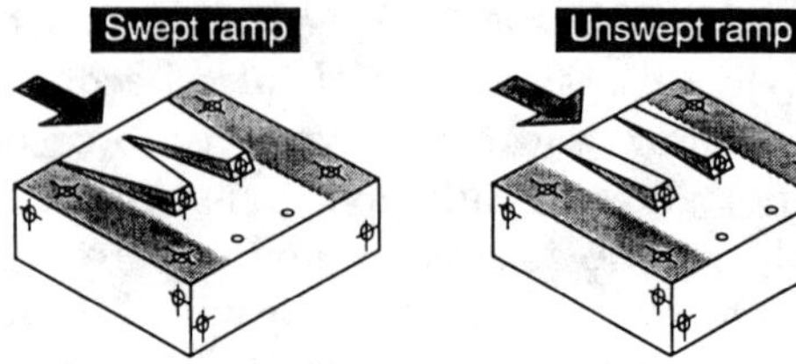

Fig. 46 Swept and unswept ramp fuel-injector configurations.

The overall fuel-air equivalence ratio was 0.6. Other experimental and facility details are given in Ref. 64.

Both the unswept and swept parallel injector ramps were studied computationally. Only fuel-air mixing was considered in this study. The fourth-order-accurate compact algorithm in the program was again used. The facility test section surrounding the ramps and considered in the computation was 13.97 cm long and 3.86 cm high. Symmetry planes were chosen to pass transversely through each fuel injector to define the spanwise computational boundaries. That region was spanned by a computational grid with 43 points in the streamwise direction and 61 points in the transverse and spanwise directions, respectively. The grid was compressed such that each injector was represented by 220 points. No-slip boundary conditions were specified on the lower facility wall and on the ramp walls. Free-slip boundary conditions were employed on the upper facility wall. Initial calculations were carried out assuming only laminar diffusion.

Results from the computational study for both the unswept and swept injector ramps are shown in Figs. 47-55. Pressure contours through a cross plane through the center of the injectŏr are given for the unswept and swept ramps in Figs. 47a-47b, respectively. The ramp

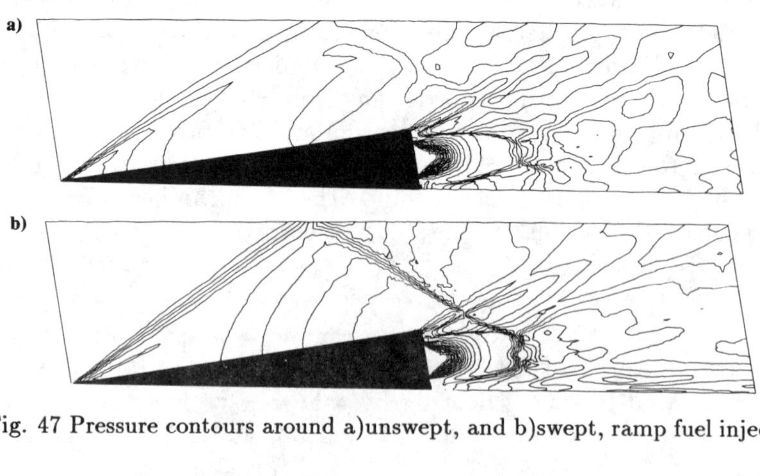

Fig. 47 Pressure contours around a)unswept, and b)swept, ramp fuel injector.

shock, the expansion fan at the end of the ramp, and the plume of the fuel jet can be seen in both cases. The swept ramp exhibits a greater shock angle compared to the unswept ramp because flow more readily spills further upstream about the unswept wedge as a result of its narrower spanwise dimension. The predicted shock angle in both cases agrees with the experiment.

Figures 48-50 show the cross-stream velocity vectors for the unswept and swept cases at three successive downstream planes ($x = 6.60$, 8.06, and 13.20 cm) oriented perpendicular to the test section walls. Part a of the figures displays the unswept ramp results, and part b the swept ramp results. The planar cut extends from the lower to the upper wall of the test section, and it slices through the center of the right fuel jet. The left boundary is located halfway between the two ramps. At the $x = 6.6$-cm station, which lies just ahead of the end of the ramps, a streamwise vortex has formed at the edge of each ramp. The vortex formed by the swept ramp is considerably larger, however, and it persists well into the flow above the ramp and to the ramp centerline. This condition continues at the $x = 8.06$-cm station located around 1 cm downstream of the end of the ramps. The swept ramp vortex has now begun to interact with the hydrogen fuel jet, enhancing its penetration into the airstream. The pluming of the higher-pressure

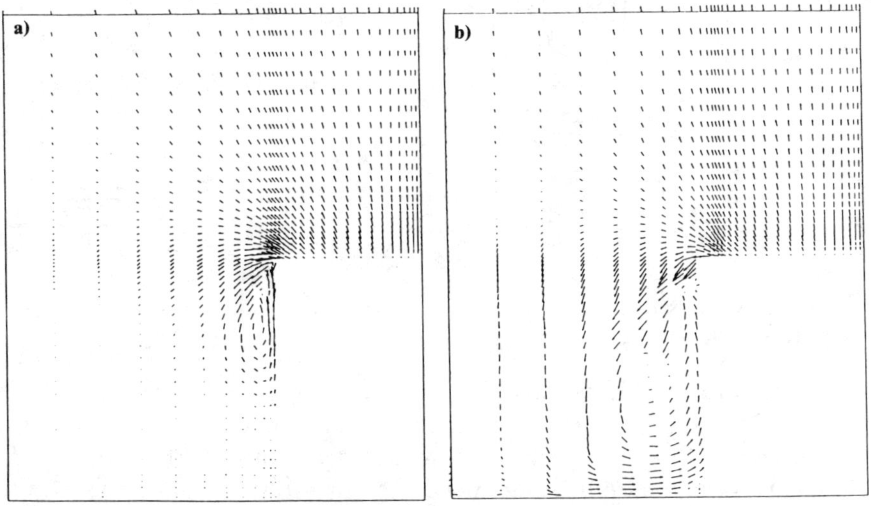

Fig. 48 Cross-stream velocity vectors for a)unswept, and b)swept, ramp at $x = 6.60$ cm.

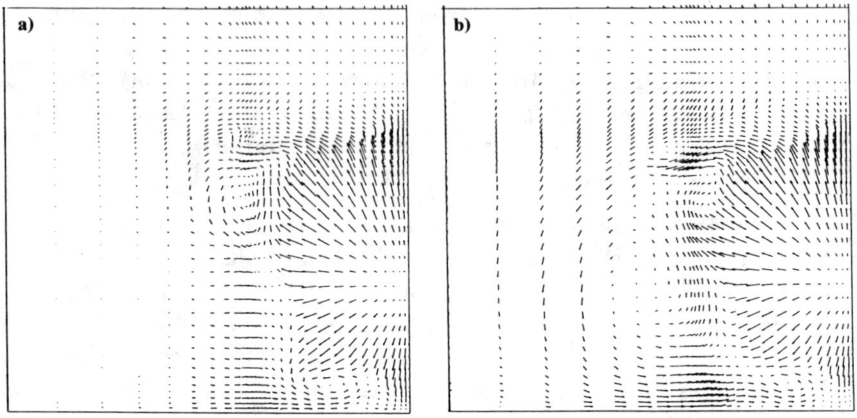

Fig. 49 Cross-stream velocity vectors for a)unswept, and b)swept, ramp at $x = 8.06$ cm.

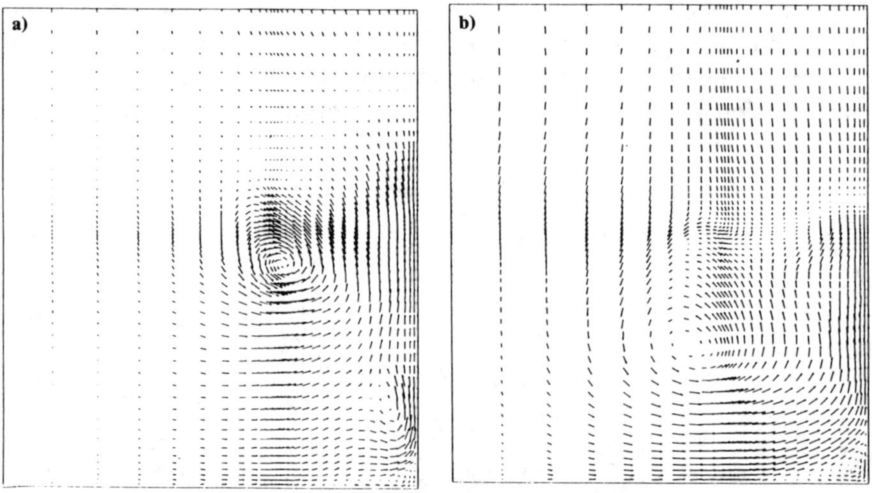

Fig. 50 Cross-stream velocity vectors for a)unswept, and b)swept, ramp at $x = 13.2$ cm.

fuel jets into the air can also be seen in both cases. At the $x = 13.20$-cm station, located 6.2 cm beyond the end of the ramps, the swept ramp vortex has continued to grow and has moved further toward the jet centerline. There is pronounced fuel-air mixing enhancement as the vortex spreads across the test section, pumping hydrogen fuel into the airstream. Some enhancement is also provided by the unswept ramp, but it is not nearly as pronounced as that provided by the swept ramp.

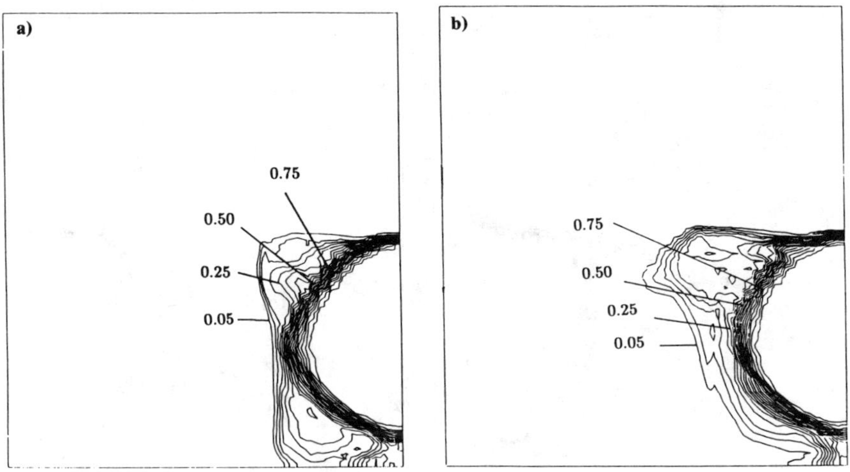

Fig. 51 Cross-stream hydrogen mass fraction contours for a)unswept, and b)swept, ramp at $x = 7.30$ cm.

The transport of hydrogen fuel into the airstream can be seen more clearly by studying the location of hydrogen mass fraction contours in several test section cross planes that are plotted with increasing streamwise distance. Figures 51-55 show the hydrogen mass fraction contours at five successive downstream planes ($x = 7.3$, 8.06, 9.6, 11.3, and 13.2 cm), again oriented perpendicular to the test section walls. As before, part a of the figures displays the unswept ramp results, and part b the swept ramp results. The results in Fig. 51 occur 0.3 cm downstream of the end of the ramp. With the swept ramp, the larger streamwise vortex has already begun to sweep the hydrogen fuel across into the airstream and away from the lower wall. The smaller streamwise vortex of the unswept ramp also begins to transport hydrogen away from the jet, but not nearly as much as does the swept ramp. As a result, more hydrogen is transported toward the lower wall boundary layer in the unswept case. The same trends continue at the $x = 8.06$- and 9.6-cm stations as shown in Figs. 52 and 53, respectively. At $x = 11.3$ cm, as shown in Fig. 54, the swept ramp enhancer has lifted the fuel jet almost completely off the lower wall. Significant amounts of hydrogen have also been carried across the test section. On the other hand, the unswept ramp enhancer still allows a large amount of hydrogen to be transported along the lower wall, and the spanwise transport is not nearly as great. At $x = 13.2$ cm, the final streamwise station shown in Fig. 55, the spanwise spread of the fuel jet enhanced by the swept ramp is 46% greater than the spanwise spread because of the unswept

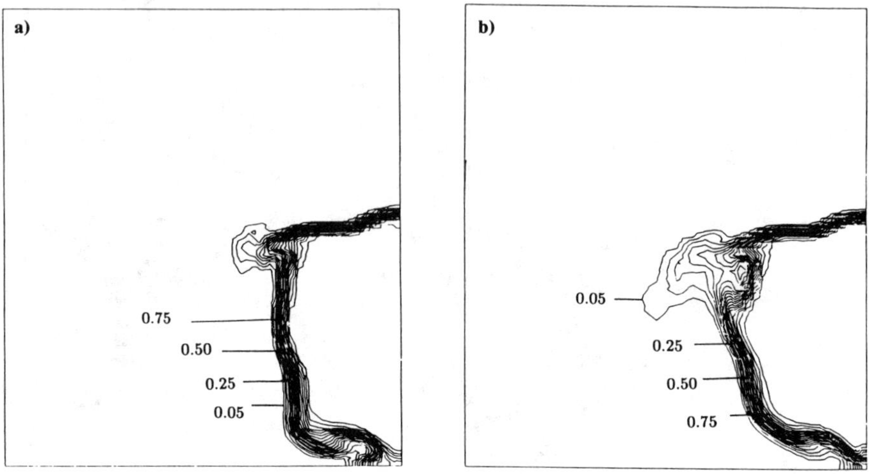

Fig. 52 Cross-stream hydrogen mass fraction contours for a)unswept, and b)swept, ramp at $x = 8.06$ cm.

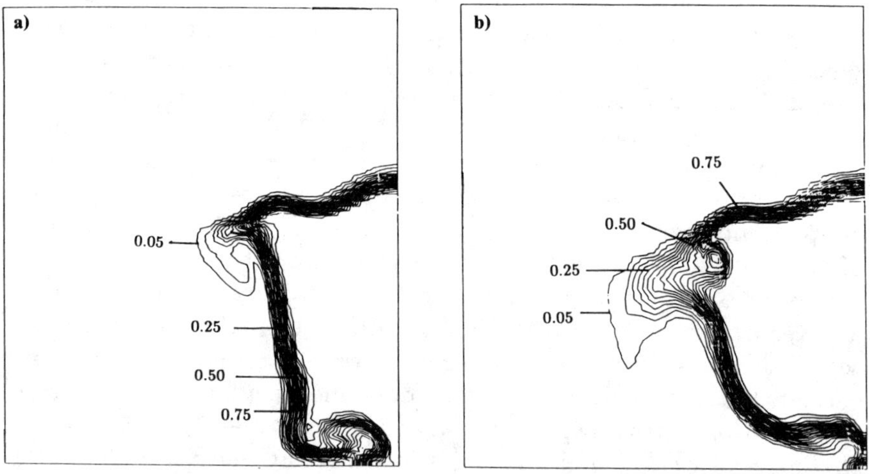

Fig. 53 Cross-stream hydrogen mass fraction contours for a)unswept, and b)swept, ramp at $x = 9.60$ cm.

ramp. In addition, the swept enhancer has resulted in the fuel jet being transported completely off the lower wall. Finally, an eddy of hydrogen has broken completely away from the primary hydrogen jet, increasing the fuel-air interfacial area even further. Clearly then, the swept ramp enhancer significantly increases the overall spread and mixing of the hydrogen fuel jets.

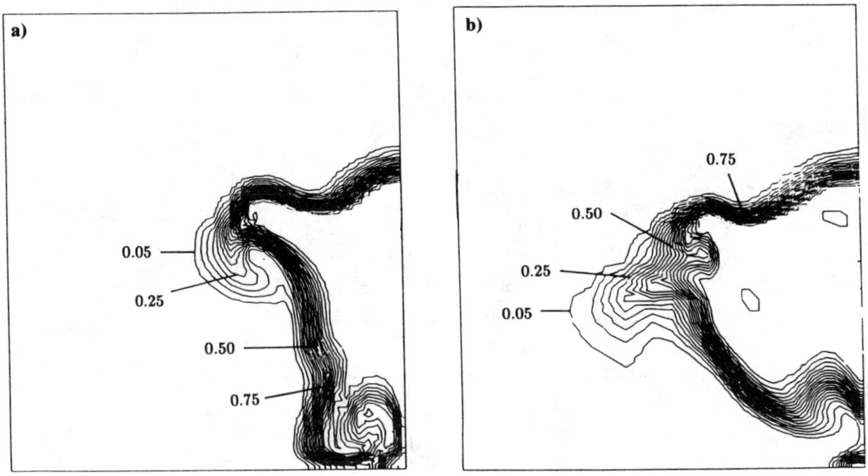

Fig. 54 Cross-stream hydrogen mass fraction contours for a)unswept, and b)swept, ramp at $x = 11.3$ cm.

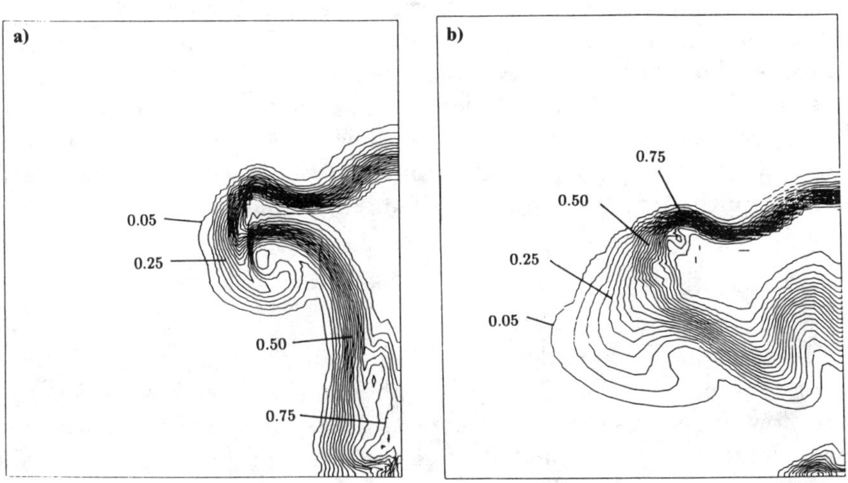

Fig. 55 Cross-stream hydrogen mass fraction contours for a)unswept, and b)swept, ramp at $x = 13.2$ cm.

A final study of both the unswept and swept ramp enhancers was undertaken with turbulent flow. As before, turbulence was modeled using the Baldwin-Lomax turbulence model. Results are given in Fig. 56 at $x = 13.2$ cm, the final downstream station. The trends observed in the laminar calculation are also observed here for both cases, with approxi-

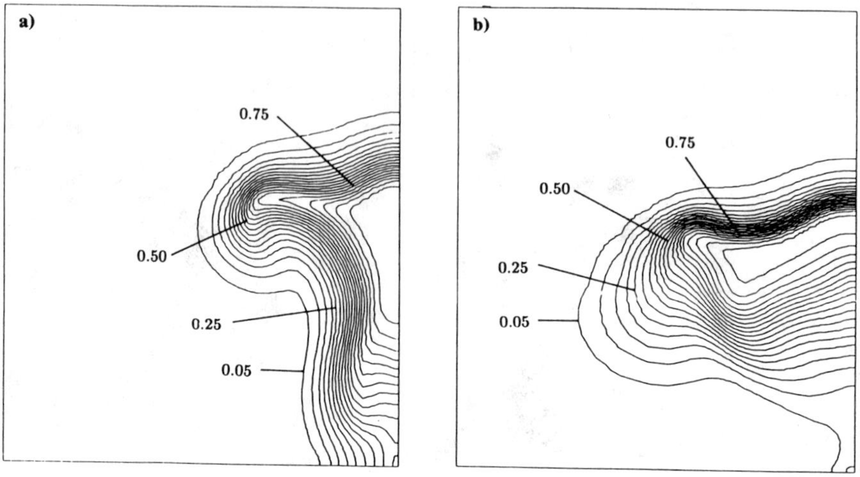

Fig. 56 Cross-stream hydrogen mass fraction contours with turbulent flow for a)unswept, and b)swept, ramp at $x = 13.2$ cm.

mately a 43% improvement in spanwise spread of the fuel jet enhanced by the swept ramp. The turbulent processes diffuse the interior region of each jet, however, compared to the laminar result. Because gradients are somewhat relieved by turbulence, the swept ramp no longer bursts the fuel jet as occurred with laminar flow. Therefore, with turbulent flow as described by the present turbulence model, the swept ramp still remains quite attractive as a fuel-air mixing enhancer.

Concluding Remarks

Research has been undertaken in this study to achieve an improved understanding of important physical phenomena present when a supersonic flow undergoes chemical reaction. To explore the behavior of such flows, detailed physical models of convective and diffusive mixing and finite-rate chemical reaction in supersonic flow were developed. Numerical algorithms were then constructed to solve the equations governing supersonic chemically reacting flow that resulted from these models. Computer programs were written using the algorithms, and each program was used to study a spatially developing and reacting supersonic mixing layer. The results obtained from these studies were then analyzed, and conclusions were drawn concerning the structure of the reacting mixing layer. Those conclusions are now summarized.

Supersonic reacting flows exhibited many of the same features observed for subsonic reacting and nonreacting flows. As in subsonic flows, exothermic heat release in unconfined supersonic flows retarded fuel-air mixing. The vortical structure of the flow, noted in much of the subsonic nonreacting flow literature, was shown to be quite predominant in supersonic reacting flow as well. In agreement with the earlier reacting subsonic literature, the vortical structure had a marked effect on chemical reaction in supersonic flow. Significant burning took place in the eddies on the edges of the mixing layer, broadening the reaction zone relative to the layer thickness defined by the velocity gradient. In addition, the vortical flow resulted in the rollup of unburned reactants inside a layer of partially or fully burned products. This phenomenon, often termed "unmixedness" in subsonic flows, prohibited the reaction of captured reactants and reduced the overall efficiency of the combustion process. Unmixedness was thus shown to be a potential problem in reducing the efficiency of supersonic, as well as subsonic combustion, and techniques are needed to overcome its negative effects on mixing and combustion efficiency.

Because of this need for mixing enhancement in supersonic reacting flows, techniques for enhancing fuel-air mixing and combustion were also explored in this chapter. The initial studies were limited to two dimensions. A supersonic, spatially developing, and chemically reacting mixing layer well represents the early stages of mixing and reaction that take place in a scramjet combustor. The mixing layer was therefore considered as a model problem for the initial phases of this study. Once this problem was chosen, a simulation of a reacting mixing layer without enhancement was first performed using the computer code to serve as a benchmark for the enhancement studies that followed. Two calculations employing enhancement were then carried out, the first employing planar shocks and the second a shock with curvature. The second enhancement approach proved more attractive. The curved shock produced vorticity when it interacted with the high-velocity gradient in mixing layer, and a higher degree of mixing and reaction then resulted. Based on the results of this study, an alternate fuel-injector configuration employing interacting jets was computationally designed. That configuration significantly increased the amount of fuel-air mixing and reaction over a given combustor length.

The study was then extended to three dimensions and a technique for enhancing fuel-air mixing and reaction in a scramjet combustor was explored. The computer program was first used to study fuel-air mixing in a generic scramjet combustor. The design of that combustor did not promote a sufficient degree of fuel penetration and fuel-air mixing. Two parallel injector ramp enhancement configurations, one swept and the other without sweep, were then considered to assess their ability to provide a sufficient degree of fuel-air mixing and mixing efficiency. The swept injector ramp configuration induced a large degree of streamwise vorticity and significantly increased the amount of fuel-air mixing that was achieved compared to the unswept configuration. The swept injector ramp therefore appeared to be an attractive option for mixing and combustion enhancement in a scramjet combustor.

References

[1] Drummond, J. P., Rogers, R. C., and Evans, J. S., Combustor Modelling for Scramjet Engines. *Combustor Modelling,* AGARD-CP-275, Feb. 1980, pp. 10-1–10-30.

[2] Drummond, J. P., and Weidner, E. H., "Numerical Study of a Scramjet Engine Flowfield," *AIAA Journal,* Vol. 20, Sept. 1982, pp. 1182-1187.

[3] Drummond, J. P., "Numerical Study of a Ramjet Dump Combustor Flowfield," *AIAA Journal,* Vol. 23, April 1985, pp. 604-611.

[4] Carpenter, P. W., "A Numerical Investigation Into the Effects of Compressibility and Total Enthalpy Difference on the Development of a Laminar Free Shear Layer," *Journal of Fluid Mechanics,* Vol. 50, Pt. 4, Dec. 29, 1971, pp. 785-799.

[5] Brown, G. L., and Roshko, A., "On Density Effects and Large Structure in Turbulent Mixing Layers," *Journal of Fluid Mechanics,* Vol. 64, Pt. 4, July 24, 1974, pp. 775-816.

[6] Roshko, A., "Progress and Problems in Understanding Turbulent Shear Flows," *Turbulent Mixing in Nonreactive and Reactive Flows,* edited by S. N. B. Murthy, Plenum Press, New York, 1975, pp. 295-316.

[7] Ferziger, J. H., and McMillian, O. J., *Studies of Structure and Modeling in Turbulent Shear Flows.* NEAR TR 335, Nielsen Engineering and Research, Inc., Mountain View, CA, Dec. 1984.

[8] Oh, Y. H., "Analysis of Two-Dimensional Free Turbulent Mixing," AIAA Paper 74-594, June 1974.

[9] Papamoschou, D., and Roshko, A., "Observations of Supersonic Free Shear Layers," AIAA Paper 86-0162, Jan. 1986.

10 Hussaini, M. Y., Collier, F., and Bushnell, D. M., "Turbulence Alteration Due to Shock Motion," *Turbulent Shear-Layer/Shock-Wave Interactions,* edited by J. Délery, Springer-Verlag, New York, 1986, pp. 371-381.

11 Yule, A. J., Chigier, N. A., and Thompson, D., "Coherent Structures in Combustion," *Symposium on Turbulent Shear Flows,* Pennsylvania State Univ., University Park, PA, 1977, pp. 7.41–7.50.

12 Masutani, S. M., and Bowman, C. T., *The Structure of a Chemically Reacting Plane Mixing Layer.* Stanford Univ. paper presented at the Western States Section/Combustion Institute, 1984 Spring Meeting, Boulder, CO, April 2-3, 1984.

13 Keller, J. O., and Daily, J. W., "The Effects of Large Heat Release on a Two Dimensional Mixing Layer," AIAA Paper 83-0472, Jan. 1983.

14 Mungal, M. G., Dimotakis, P. E., and Hermanson, J. C., "Reynolds Number Effects on Mixing and Combustion in a Reacting Shear Layer," AIAA Paper 84-0371, Jan. 1984.

15 Hermanson, J. C., Mungal, M. G., and Dimotakis, P. E., "Heat Release Effects on Shear Layer Growth and Entrainment," AIAA Paper 85-0142, Jan. 1985.

16 Pitz, R. W., and Daily, J. W., "Combustion in a Turbulent Mixing Layer Formed at a Rearward-Facing Step," *AIAA Journal,* Vol. 21, Nov. 1983, pp. 1565-1570.

17 Broadwell, J. E., and Dimotakis, P. E., "Implications of Recent Experimental Results for Modeling Reactions in Turbulent Flows," AIAA Paper 84-0545, Jan. 1984.

18 Carrier, G. F., Fendell, F. E., and Marble, F. E., "The Effect of Strain Rate on Diffusion Flames," *SIAM Journal of Applied Mathematics,* Vol. 28, Mar. 1975, pp. 463-500.

19 Riley, J. J., and Metcalfe, R. W., "Direct Simulations of Chemically Reacting Turbulent Mixing Layers," AIAA Paper 85-0321, Jan. 1985.

20 McMurtry, P. A., Jou, W.-H., Riley, J. J., and Metcalfe, R. W., "Direct Numerical Simulations of a Reacting Mixing Layer With Chemical Heat Release," AIAA Paper 85-0143, Jan. 1985.

21 Menon, S., Anderson, J. D., Jr., and Pai, S. I., "Stability of a Laminar Premixed Supersonic Free Shear Layer With Chemical Reactions," *International Journal of Engineering Science,* Vol. 22, No. 4, 1984, pp. 361-374.

22 Drummond, J. P., "A Two-Dimensional Numerical Simulation of a Supersonic, Chemically Reacting Mixing Layer," NASA TM-4055, 1988.

23 Carpenter, M. H., "Three-Dimensional Computations of Cross-Flow Injection and Combustion in a Supersonic Flow," AIAA Paper 89-1870, June 1989.

24 Williams, F. A., *Combustion Theory.* Addison-Wesley, Reading, MA, 1965, pp. 358-429.

25 McBride, B. J., Heimel,S., Ehlers, J. G., and Gordon, S., "Thermodynamic Properties to 6000 K for 210 Substances Involving the First 18 Elements," NASA SP-3001, 1963.

26 Kanury, A. M., *Introduction to Combustion Phenomena.* Gordon and Breach, New York, 1982, pp. 363-371.

27 White, F. M., *Viscous Fluid Flow.* McGraw-Hill, New York, 1974, pp. 28-36.

28 Suehla, R. A., "Estimated Viscosities and Thermal Conductivities of Gases at High Temperature," NASA TR R-132, 1962.

29 Wilke, C. R., "A Viscosity Equation for Gas Mixtures," *Journal of Chemistry and Physics,* Vol. 18, No. 4, pp. 517-519, 1950.

30 Berman, H. A., Anderson, J. D., and Drummond, J. P., "Supersonic Flow Over a Rearward Facing Step with Transverse Nonreacting Hydrogen Injection," *AIAA Journal,* Vol. 21, Dec. 1983, pp. 1701-1713.

31 Kee, R. J., Warnatz, J., and Miller, J. A., "A Fortran Computer Code Package for the Evaluation of Gas-Phase Viscosities, Conductivities, and Diffusion Coefficients," Sandia Rep. SAND83-8209, March 1983.

32 Schetz, J. A., Billig, F. S., and Favin, S., "Flowfield Analysis of a Scramjet Combustor with a Coaxial Fuel Jet," *AIAA Journal,* Vol. 20, Sept. 1982, pp. 1268-1274.

33 Cebeci, T., and Smith, A. M. O., *Analysis of Turbulent Boundary Layers.* Academic, New York, 1974.

34 Baldwin, B. S., and Lomax, H., "Thin Layer Approximations and Algebraic Model for Separated Turbulent Flows," AIAA Paper 78-257, Jan. 1978.

35 Jones, W. P., and Launder, B. E., "The Prediction of Laminarization with a Two-Equation Model of Turbulence," *International Journal of Heat and Mass Transfer,* Vol. 15, 1972, p. 301.

36 Fabris, G. , Harsha, P. T., and Edelman, R. B., "Multiple-Scale Turbulence Modeling of Boundary Layer Flows for Scramjet Applications," NASA CR-3433, 1981.

37 Hanjalic, K., and Launder, B. E., "Sensitizing the Dissipation Equation to Irrotational Strains," *Journal of Fluids Engineering, Transactions of ASME,* Vol. 102, 1980, pp. 34-40.

38 Hanjalic, K., Launder, B. E., and Schiestel, R., "Multiple-Time Scale Concepts in Turbulent Transport Modeling," Second Symposium on Turbulent Shear Flows, Imperial College, London, July 1979.

39 Rodi, W., "The Prediction of Free Turbulent Boundary Layers by Use of a Two-Equation Model of Turbulence," Ph.D. Thesis, Univ. of London, 1972.

40 Sindir, M. M., "Numerical Study of Turbulent Flows in Backward-Facing Step Geometries, Comparison of Four Models of Turbulence," Ph.D. Thesis, Univ. of California, Davis, CA, June 1982.

41 Oran, E., and Boris, J., *Recent Advances in Numerical Methods for Chemically Reacting Flows.* American Institute of Astronautics and Aeronautics, Washington, D. C., 1991.

42 Riley, J. J., Metcalfe, R. W., and Orszag, S. A., "Direct Numerical Simulations of Chemically Reacting Turbulent Mixing Layers," *Physics of Fluids,* Vol. 29, No. 2, 1986, pp. 406-422.

43 McMurtry, P. A., Jou, W.-H., Riley, J. J., and Metcalfe, R. W., "Direct Numerical Simulations of a Reacting Mixing Layer with Chemical Heat Release," *AIAA Journal,* Vol. 24, 1986, pp. 962-970.

[44] Erlebacher, G., and Hussaini, M.Y., "Stability and Transition in Supersonic Boundary Layers," AIAA Paper 87-1416, June 1987.

[45] Schumann, U., "Subgrid Scale Model for Finite Difference Simulations of Turbulent Flows in Plane Channels and Annuli," *Journal of Computational Physics,* Vol. 18, 1975, pp. 376-404.

[46] Speziale, C. C., Erlebacher, G., Zang, T. A., and Hussaini, M. Y., "On the Subgrid-Scale Modeling of Compressible Turbulence," NASA CR-178420, 1987.

[47] Bussing, T. R. A., and Murman, E. M., "A Finite Volume Method for Calculation of Compressed Chemically Reacting Flows," AIAA Paper 85-0311, Jan. 1985.

[48] Widhopf, G. F., and Victoria, K. J., "On the Solution of the Unsteady Navier-Stokes Equations Including Multicomponent Finite Rate Chemistry," *Computers and Fluids,* Vol. 1, 1973, pp. 159-184.

[49] MacCormack, R. W., "The Effect of Viscosity on Hypervelocity Impact Cratering," AIAA Paper 69-354, April 1969.

[50] Householder, A. S., *The Theory of Matrices in Numerical Analysis,* Dover, New York, 1964, pp. 122-140.

[51] Gottlieb, D., and Turkel, E., "Dissipative Two-Four Methods for Time-Dependent Problems," *Mathematics of Computation,* Vol.30, No. 136, 1976, pp. 703-723.

[52] Abarbanel, S., and Kumar, A., "Compact High-Order Schemes for the Euler Equations," *Journal of Scientific Computing,* Vol. 3, No. 3, 1988.

[53] Carpenter, M. H., "A High-Order Compact Numerical Algorithm for Supersonic Flows," Proceedings of the 12th International Conference on Numerical Methods in Fluid Dynamics, Springer-Verlag, New York, 1990.

[54] Drummond, J. P., Carpenter, M. H., Riggins, D. W., and Adams, M. S., "Mixing Enhancement in a Supersonic Combustor," AIAA Paper 89-2794, July 1989.

[55] Riggins, D. W., Drummond, J. P., and McClinton C. R., "Mixing Enhancement in a Scramjet Engine," AIAA Paper 90-0203, Jan. 1990.

56 Ragab, S. A., and Wu, J. L., "Instabilities in the Free Shear Layer Formed by Two Supersonic Streams," AIAA Paper 88-0038, Jan. 1988.

57 Guirguis, R. H., Grinstein, F. F., Young, T. R., Oran, E. S., Kailasanath, K., and Boris, J. P., "Mixing Enhancement in Supersonic Shear Layers," AIAA Paper 87-0373, Jan. 1987.

58 Guirguis, R. H., "Mixing Enhancement in Supersonic Shear Layers: III. Effect of Convective Mach Number," AIAA Paper 88-0701, Jan. 1988

59 Kumar, A., Bushnell, D. M., and Hussaini, M. Y., "A Mixing Augmentation Technique for Hypervelocity Scramjets," AIAA Paper 87-1182, June 1988.

60 Drummond, J. P., and Mukunda, H. S., "A Numerical Study of Mixing Enhancement in Supersonic Reacting Flow Fields," AIAA Paper 88-3260, July 1988.

61 Menon, S., "Shock-Wave-Induced Mixing Enhancement in Scramjet Combustors," AIAA Paper 89-0104, Jan. 1989.

62 King, P. S., Thomas, R. H., and Schetz, J. A., "Combined Tangential-Normal Injection Into a Supersonic Flow," AIAA Paper 89-0622, Jan. 1989.

63 Sullins, G. A., Lutz, S. A., Carpenter, D. A., and Taylor, M. A., "Experimental Investigation of a Shear Layer in Supersonic Flow," 1989 JANNAF Propulsion Meeting, Cleveland, OH, May 1989.

64 Northam, G. B., Greenberg, I., and Byington, C. S., "Evaluation of Parallel Injector Configurations for Supersonic Combustion," AIAA Paper 89-2525, July 1989.

65 Schlichting, H., *Boundary-Layer Theory,* McGraw-Hill, New York, 1968.

66 Riggins, D. W., Mao, M., Bittner, R. D., McClinton, C. R., and Rogers, R. C., "Numerical Modeling of Normal Fuel Injection: Effect of Turbulent Schmidt Number," NASP CR-1043, April 1989.

Chapter 8

Study of Combustion and Heat Exchange Processes in High-Enthalpy Short-Duration Facilities

V. K. Baev,* V. V. Shumsky,† and M. I. Yaroslavtsev†
Institute of Theoretical and Applied Mechanics, USSR Academy of Sciences, Novosibirsk, USSR

Abstract

Aerodynamic facilities modelling the maximally high parameters of free stream stagnation and T are required to study gasdynamic models with combustion (GMC). The use of aerodynamic facilities with the operation duraction of a few seconds and more for such investigations considerably complicates the solution of the problem because of the following reasons: a high cost of facilities as well as the conducting of tests, need in thermal protection of gasdynamic tunnel circuit and the model, air pollution with errosion products of premix chamber walls and other tunnel units. Moreover, to obtain a drive gas with parameters $P_0 \sim 100$ MPa, $T_0 \sim (2000\text{-}3000)$K there are required energetic powers whose realization is not possible in a number of cases.

Therefore, the high-enthalpy aerodynamic facilities with operation regime duration (0.2-1)s are preferable when studying GMC. The advantages of such facilities as compared with those of long duration (> 1s) are as follows:

- there is no need in thermal protection of tunnel units, models and primary transducers;
- the presence of a possibility to model the maximally high P_0 and T_0 as well as to change them within wide limits;

*Deputy Director
†Senior Scientist

- the gas heating at practically constant density ρ_0 permits to get high Re providing the turbulent boundary layer 100-150 mm in length, that favourably affects the start-up and functioning of the air intake of the model;

- the thermal measurements on the model are simplified because there is no need in stepwise introduction of the model into gas stream.

In the present paper, some aspects of tests of GMC in high-enthalpy hot-shot tunnel of the Institute of Theoretical and Applied Mechanics are reflected.

I. Introduction

In experimental studies on models in hypersonic airflows with fuel injection, ignition and burning as well as complex interaction of the incoming flow with the chemical reaction zone, not only must the similarity creteria characteristic of an aerodynamic experiment (the M, Re, St, We, etc. numbers) hold but, in addition, the natural values of the pressure p and temperature T must be reproduced, since the conditions of ignition (especially self-ignition), the rates of chemical reactions and, therefore, the liberation of heat in time and space depend strongly on p and T [1-3].

The need to maintain high values of p and T, which are characteristic of hypersonic flows, in airflow around the model makes the experiment extremely complicated because the flow rates of high-enthalpy air are high and it is difficult to perform measurements at high temperatures. Both factors make it necessary to reduce the test time, which ultimately leads to the idea of employing high-enthalpy facilities operating for short times for tests of gasdynamic models with combustion. One of the first attempts in this direction was a test of a gasdynamic model with combustion of hydrogen in a shock tube with $M_{in} \approx 11$ [4]. In spite of the short operating time (~ 3 ms), a number of parameters were measured in the channel of the model in this work. Three milliseconds is, however, not enough time to study either the working process or, especially, the force characteristics, since several milliseconds are required to establish a boundary layer and flow in the separation zones,[5] whereas transient processes in models with combustion can last for tens of milliseconds[6].

The short operating time and the fundamental feature, associated with the decrease in the values of p and T during the experiment, of a number of high-enthalpy, pulsed facilities of the tunnel type with short operating times make it necessary to carefully analyse all questions regarding the technology of tests on gasdynamic models with combustion. The characteristics of the operation of a system for injecting combustible media into a model, measurements of the resistance X and other important parameters, including the distribution of the pressure and heat fluxes over the surfaces of gasdynamic models, were studied in Ref.7. It is shown in Ref.7 that the methods developed yield reliable quantitative information in tests of diverse models with combustion and short operating times. The methodological investigations performed in Refs.7-9 form the basis for further continuation of the work in order to obtain a final answer to the question of the possibility of using high-enthalpy short-lived facilities (test times of tens of milliseconds) for studying the working process in gasdynamic models with heat and mass inflow.

II. The Study of Elements of Working Process in the Model with Combustion

Studies of injection of ethyl alcohol, water, kerosene, and organoboron compounds performed in similar facilities[10,11] made it possible to extend the range of previous investigations of the penetration of jets into super- and hypersonic gas flows (see, for example, Ref.12) to high values of the stagnation parameters. The experiments performed in Ref.10 lead to the conclusion that, in the range of stagnation temperatures T_0 = 1500-2400° K (T = 400 - 1150 K for M = 2.4 - 3.7), the mechanism of spraying the liquids studied, sprayed normally into the airflow above a plate, is determined, as for low temperatures of the incoming air (T = 60 - 200° K), solely by the aerodynamic fragmentation of the jet. This is attributable to the fact that the temperatures T_0 < 2400° K are not high enough for evaporation of a significant fraction of the liquid droplets (at room temperature up to the moment of injection) in the starting section, which could reduce the penetration depth. The heating of droplets without their evaporation, however, does not affect the characteristics of penetration: in the zone T = 60 - 1150° K, the results on the penetration depth of a jet of liquid into a supersonic gas flow, the width of the spray plume, and the distance at which the spray emerges onto the asymptote agree well with one another and obey the laws established in Ref.12 for cold air.

Figure 1 shows the boundaries of self-ignition of hydrogen and liquid substances in a supersonic airflow [13,14]. For the case in which injection occurred through an opening on the surface of a 190 x 190-mm plate 80 mm from the front edge, self-ignition began immediately outside the orifice of the injector for both hydrogen and organoboron, although, for the values of p and T in the experiments and M = 2.4, the induction zone was 10 - 80 cm long. This indicates that, under these conditions, the separation zones, in particular, the zone of separation of the boundary layer in front of the point of injection, play the determining role.

To reduce the stagnation temperature T_{0s}, at which self-ignition of the fuel mixture occurs, an additional wedge was placed at a distance 10 mm along the flow in front of the point of injection, as a result of which a ledge 10 mm high was formed (see Fig. 1b). The value of T_{0s} decreased by 100° K for hydrogen and by 350° K for organoboron liquid. The formation, with the help of a shock, incident on the spray plume, not only of a zone of flow separation but also a local region with high values of p and T (owing to the wadges placed above the plate over its entire width as shown in Fig.1c), further reduces T_{0s}. It is evident from Fig.1 that although the values of T_{0s} for organoboron liquid in a supersonic flow over a plate are significantly higher than for hydrogen, the artificial formation of even small zones with a return flow and high values of p and T can appreciably improve the conditions for self-ignition of liquid.

When kerosene is injected through a plate with a step, self-ignition was not observed in any experiment up to T_0 = 2200° K. This is explained quantitatively by comparing the curves of

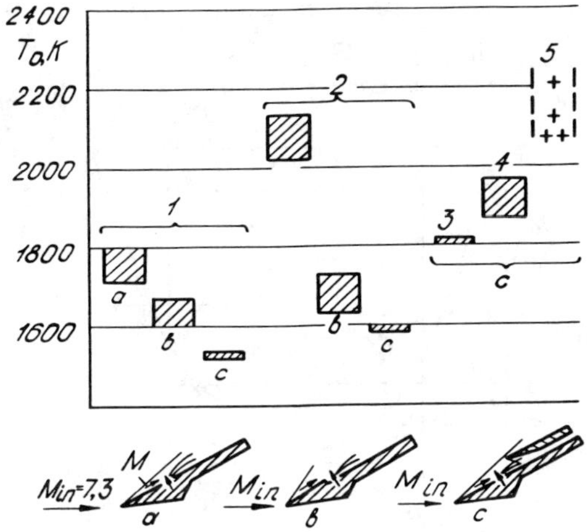

Fig. 1 Boundaries of self-ignition : a) plate, b) plate with step, c) plate with deflector; 1) hydrogen, 2) 100% organoboron liquid, 3) mixture of 50% organoboron liquid and 50% kerosene, 4) mixture of 25% organoboron liquid and 75% kerosene, 5) kerosene.

the induction time $T_u = f(T,p)$ for the combustible substances studied [14]. In a mixture of kerosene with organoboron liquid, the organoboron liquid self-ignites, so that when its concentration in the mixture is reduced, T_{0s} increases.

The existence of both separation zones and locations with a local increase in p and T and the possibility of the formation of shock waves (SW) in the outflowing fuel-air mixture are very important for insuring self-ignition [15]. In addition, the creation of conditions for the formation of such waves enables stable ignition of combustible gas at temperatures much lower than T_{0s} for the same gas in air. Thus, in Ref.15, the conditions under which hydrogen is spontaneously ignited in air with the initial temperatures of both gases equal to 20°C were studied. The physical picture of this "anomalous" self-ignition is as follows. After diaphragm 1 rapture (Fig.2), a SW propagates along chamber 2 and compresses the air in the chamber. In the region of the contact surface, conditions (pressure, temperature, and composition) sufficient for self-ignition of the hydrogen-air mixture can be created. The products of combustion forming in the chamber in the region of the contact surface ignite after flowing out

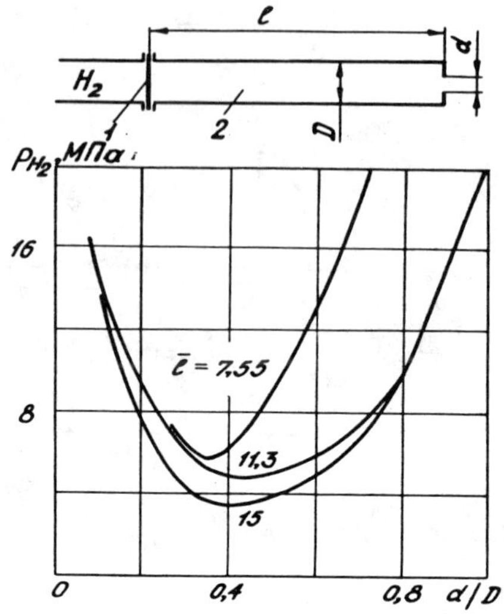

Fig. 2 Boundaries of self-ignition.

through the opening d, the hydrogen jet following them. Figure 2 shows the boundaries of self-ignition of hydrogen, flowing after the diaphragm is ruptered into a chamber of length l = l/D as a function of d/D for D = 8 mm. Here PH_2 is the hydrogen pressure in front of the diaphragm before it is ruptured. The region of self-ignition lies above the corresponding curve. One can see from Fig.2 that this system enables stable ignition of cold hydrogen in surrounding air for comparatively moderate values PH_2 = 6-8 MPa.

Figure 3 shows the specific heat fluxes q in the region of interaction of the boundary layer and the oblique shock in the case of airflow over a plate with M_{in} = 11. The value of Re were calculated based on the parameters in region 3 and a length of 0.14 m, from the tip of the plate to the starting point of the region of interaction. The value of q in the zone of the shock is 130 times higher than that with unperturbed flow over a plate. The experimental data illustrate the average intensity of the heat fluxes in the zones of separation of the boundary layer on the surfaces of compression of the air-intake units in a hypersonic airflow.

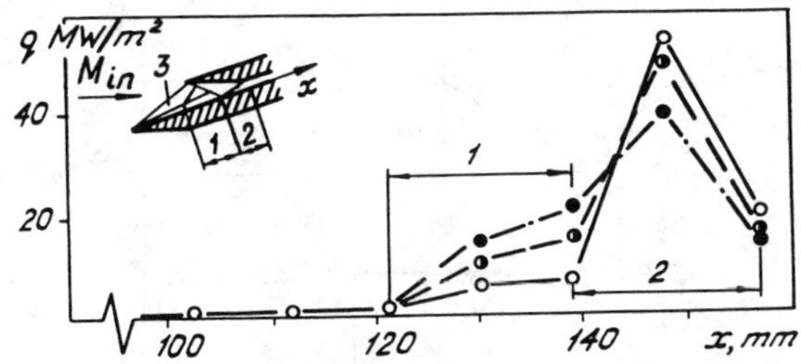

Fig. 3 Specific heat fluxes toward the surface of plate: 1) in region of separation of boundary layer, 2) in region of shock, 3) flow on wedge behind oblique shock, M = 5.4. $Re \cdot 10^{-6}$: ○) 1.2; ◐) 1; ●) 0.8.

III. The Choice of Geometrical Correlations by Developing Models with Combustion

Study of separate stages of the working process (injection, spraying, self-ignition) in high-enthalpy facilities operating for short times on comparatively simple models of the flat-plate type, employed in Refs. 8, 10, 11, 13, and 14, enabled analysis of all processes occurring in channels with a complicated shape.

Figure 4 shows a diagram of the gasdynamic channel of the model, as well as the dependence of the relative areas of this channel on the coordinate x. The experiments were performed for M_{in}= 7.7-7.9 [6,16,17]. The ranges of the values of the parameters of the air flowing onto the model were as follows: p_{0in} = 70 - 7 MPa, T_{0in} = 2250 - 970° K, p_{in} = 110 -20 hPa, T_{in} = 220 - 80°K. The change in the parameters of the air during the test with τ = 50-55 ms (typical of pulsed tunnels) is presentes in Ref.10. The model was built in the form of one-half an axisymmetric body [17] (when the working process was studied) or axisymmetric body (when the force characteristics were measured, Fig.5) [6,16]. At the same time, the geometry of the inner channel was left unchanged. Hydrogen was fed through the injector into zone a at the center (along the vertical) of the combustion chamber, either upstream with an air ratio α = 1.8-3 or downstream with α = 0.2-1.3.

An air-intake device without undercutting of the edge, i.e., with a large inner compression (Fig.5), was employed because, in a pulsed facility, firing occurs by

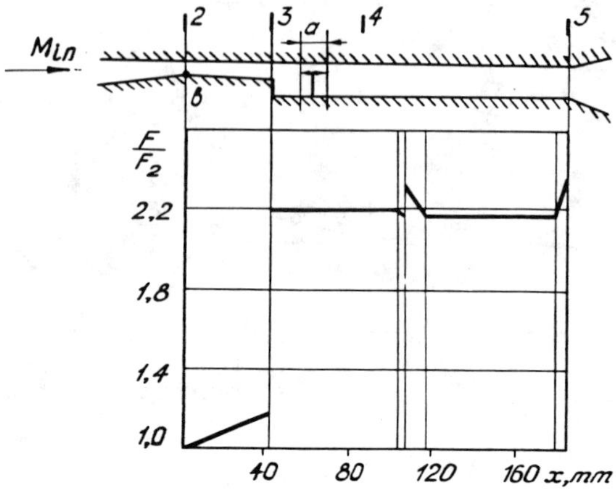

Fig. 4 Sketch of gasdynamic channel of the model.

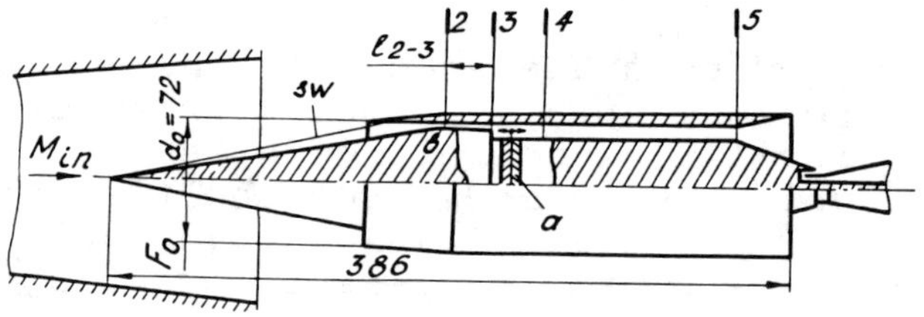

Fig. 5 Gasdynamic model for tests at M = 7.3-7.9: a - injector zone, SW - shock wave.

the wave mechanism [18]. Because of this, the edge can be made quite thin with a small wave drag, which is important when separating from the total force measured by the scales the particular part that is exerted only by the inner surfaces of the model, the internal thrust.

The air-intake unit with no undercutting and with a one- or two-shock central body operated in the fired state with a relative area $f_2 = F_2/F_0 > 0.18$-0.19, where F_2 is the area of the channel in section 2 and F_0 is the area of the inlet in the model. Analysis of the experimental data and the results of two-dimensional calculations of the flow in the channel of the model showed that two factors affected the operation of the air-intake unit in the ignited state: 1) the position of the point at which the bow shock, reflected from the edge, strikes the central body relative to the corner point b and 2) the decrease in the Re number in the flow regime. Both factors are related with the separation of the boundary layer on the central body in the region of the throat of the diffuser. The fact that the shock reflected from the edge falls above point b downstream and that Re decreases the values at which the boundary layer here becomes laminar lead to the following that the boundary layer does not withstand the pressure gradient on transition through a system of shocks, which sharply enlarges the separation zone in the region of the corner point. Passing into the throat, such separation covers the flow range section and disrupts airflow into the model. For the inlet configurations examined, this occurred for $f_2 \leqslant 0.18-0.19$. Since the smaller f_2, the larger the thrust characteristics of the inner channel of the model are,[19] further experiments with hydrogen combustion in the model were performed with the minimum possible value $f_2 = 0.19$ for the given configuration of the air-intake unit.

IV. The Working Process in Gasdynamic Model with Combustion

In the gasdynamic channels, illustrated in Figs. 4 and 5, a two-regime process of heat and mass inflow was realized. The essence of this process is as follows [20]. The ratio of the area of the transverse section at the outlet and inlet into the combustion chamber (CC), i.e., the degree of expansion of the combustion chamber, $\bar{F}_c = F_2/F_5 > 1$. If the relative heating of the working body T_{05}/T_{0in} in the channel of the model is quite large, then, at the initial section of the CC, a transition occurs from supersonic to subsonic flow, and heat flows on the average into the subsonic flow. If the ratio T_{05}/T_{0in} is such that the heat inflow cannot hold back the shock (pseudoshock) in the CC, then heat will, on the average, flow to the supersonic flow: two regimes of heat inflow can be realized in the same channel. The regime actually realized depends on the values of M_{in}, T_{05}/T_{0in}, and F_c and the hydraulic losses. In Refs. 6,16, and 17, the operation of a two-regime CC in the regime with heat inflow into the subsonic flow was studied.

For $M_{in} = 7.3-7.9$; at the inlet to the CC, $M_2 = 4 - 4.6$. For these Mach numbers, the transition in the channel from super- to subsonic flow occurs in the pseudoshock [21-24]. For gasdynamic models, the starting section of the combustion chamber, where the pseudoshock lies, is important from the viewpoint of its effect on the operation of the air-intake unit. If the inclrease in pressure in the CC is transferred upstream behind section 2 on the compression section, then it affects the operation of the model since the transfer of disturbances on the compression surface, as a rule, causes a disruption of airflow into the model. If, on the other hand, the disturbances are not transmitted above section 2, then the CC does not affect the operation of the model. Therefore, a correct choice of the length l_{2-3} of the CC initial section is determining from a point of view of the normal functioning of the inner model canal.

Figure 6 shows the distribution of the relative pressure $\bar{p} = p/p_{in}$ along the model. For experiments with combustion, the pressure distributions for different values of T_{05}/T_{0in} are presented: T_{0in} = 1170, 1100, 1075 for curves 3-1, respectively. When the relative heating increased, further air inflow was disrupted, the flow rate dropped sharply, and hydrogen combustion in the CC was disrupted.

Since the pseudoshock has a definite length, the length of the initial section of the CC l_{2-4} is important for the operation of the model: from section 2 to the termination of the intense rise in pressure in the channel (section 4 in Figs. 4 and 5). For a number of configurations of the initial section of the CC, it was impossible to burn hydrogen without disruption of airflow into the model. Taking only one measure, increasing l_{2-4} (thanks to which, the point of rapid rise in the pressure was shifted downstream from section 2), made it possible to burn hydrogen without disrupting airflow into the model. In reality, some disturbances are transferred to the compression surface without disrupt-

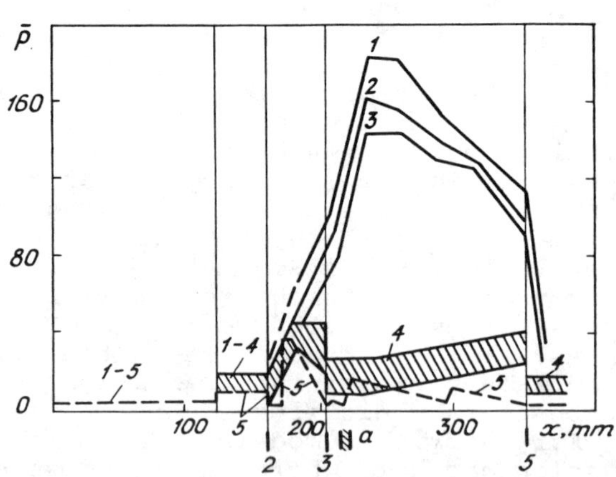

Fig. 6 Distribution of static pressure over central body of the model: a - zone of injectors. 1-3) α = 0.9; 4) without hydrogen inflow for different values of Re and $\bar{T}_w = T_w/T_{0in}$ = 0.13-0.32 ($T_w \approx 300$ K is the temperature of the wall); 5) two-dimensional calculation with $\varkappa$ = 1.4.

ion of inflow. This is also indicated by motion pictures of the flow around the model and by the fact that after the head of the pseudoshock moves to section 2 (with an increase in the realative heating of the working body), some increase in the relative heating without disruption is still possible (see curves 1 and 2 in Fig.6). In Ref.22 this effect is called fixation of the pseudoshock. However, the fact that this "margin" is not large with respect to the relative heating indicates that the value of l_{2-4} for operation in the region of large relative heatings of the working body must be at least as large as the length of the pseudoshock.

Figure 7 shows the distribution of heat fluxes: 1, 3, into a central body, 2,4, into the inlet cowling of the model inner canal; 1,2, in experiments with combustion (α = 0.9 - 0.53); 3,4, in experiments without combustion ($\alpha = \infty$); $q_i F_0/m_a h_{0in}$ is the specific heat flux, where h_{0in} is the specific stagnation enthalpy of the air freestream. Physically, the value $\bar{q}_i$ is the specific heat flux q_i at the corresponding point of the inner cannel referred to the stagnation enthalpy $m_a h_{0in}/F_0$ transferred by the air freestream in 1 through 1 m^2 of the cross section; or $\bar{q}_i$ is the heat that disappeared at the corresponding point of the inner canal through the surface with the area F_0 referred to the stagnation enthalpy $m_a\ h_{0in}$ enetering into a model with the freestream.

The stagnation on the initial section of the CC is extremely nonuniform over the height of the channel. Me-

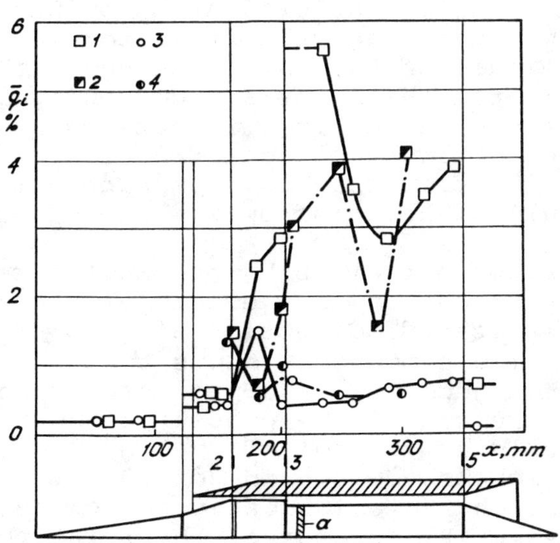

Fig. 7 Heat fluxes toward the inner channel of the model.

asurement of the heat flows to the central body and the edge showed that the stagnation of the airflow in the region of the central body begins closer to section 2 than at the edge. Traces of soot on the walls show that the flow is also very nonuniform along the periphery of the initial section: stagnation on it does not correspond to the classical picture of flow in pseudo-shocks, occurring in channels with circular or square cross sections, but rather is closer to the flow (nonuniform) observed in rectangular channels with a large ratio of the sides[24,25].

For $\alpha < 0.5$, the pressure gauges record in section l_{2-4} an oscillatory process; this is explained by the interaction of the separation zone in the region of the corner point b and the pseudo-shock "pushed" forward out of the combustion chamber as a result of heat inflow. The reasons for the appearance of oscillations with low and intermediate frequency in channels with a step are examined in a series of studies, for example, in Refs. 26 and 27.

Although the disruption of airflow into the model in different experiments occurred for different values of α, τ and T_{0in}, the value of the complex T_{05}/T_{0in} at the moment of disruption, was the same for different experiments $\beta = (m_{H_2} + m_a)/m_a$, where m_{H_2} and m_a are the mass flow rates of hydrogen and air. For $\bar{F}_c = 1.88$-2.2, the value of this complex at the moment of disruption was significantly lower than the value this complex should have in the case of thermal blocking of the CC. This indicated that, for all CC configurations tested, thermal blocking was not responsible for disruption. Disruption occurred because the zone in which the flow changed from super- to subsonic as a result of heat inflow was too close to section 2.

Thus, when fuel is burned in the regime of subsonic combustion in the initial section of the CC, a complicated collection of gasdynamic and thermodynamic processes occurs that is very sensitive to the geometric and process parameters.

Measurement of the pressures and heat fluxes along the inner channel of the models confirmed the flow scheme adopted in Ref.20 for calculating the thrust characteristics in a two-stage CC: for the values of $\bar{F}_c$ occurring, the relative heatings of the working body, and hydraulic losses at the outlet from the CC, a flow with $M_5 \approx 1$ was realized; heat flowed into the subsonic flow with a minimum value of the number $M \approx 0.5$, which corresponded to the point of maximum pressure, below the injectors downstream.

When hydrogen is burned, approximately 20% of the heat is lost in wall of the inner channel of model for $T_w \approx 300°$ K. In addition, in the range α = 1.3-0.5, T_{0in} = 1600-1000° K, the relative amount of heat lost in the wall of the model is independent of both and T_{0in}, indicating that 1) when hydrogen is burned, the losses in the CC are the largest component of the heat losses in the wall of the inner channel, and 2) the completeness of combustion is approximately the same for the entire range α = 0.5-1.3.

Figure 8 shows the completeness of combustion in the channel of the model, determined by two independent methods. Both methods give close and quite high values, ξ = 0.8 for T_{in} = 80-140° K (taking into account the heat lost in the wall of the model).

The first portions of hydrogen flowed into the model when a supersonic flow was already established in the CC. Heat inflow owing to combustion of hydrogen changes the supersonic flow in the chamber to subsonic flow. The time required for this equaled 20-28 ms (in many experiments characterized by high values of T_{0in} and, therefore, low values of $T_{05}/T_{0in} = f(\tau)$, the regime of heat inflow did not change). Further experiments showed, however, that most of this transient process is associated with the drop in the physical parameters of the air flowing into the model during the operation of the facility. For constant parameters of the inflowing air, which was insured by using a booster in the high-enthalpy facility, as discussed later, the time of the transient process equaled only several milliseconds, i.e., the protracted transition process, observed in experiments on a pulsed facility, was a consequence of the peculiarities of the facilities and not of the physical phenomena accompanying the change in the conditions of heat inflow. This indicated that the pressure and the temperature have a strong effect on the conditions of hydrogen combustion in the model of the self-ignition regime. This is also indicated by the fact that the flame is extinguished immediately after airflow is disrupted, which is accompanied by an almost stepped drop in the pressure in the CC by a factor of ~ 4.

Experiments in which, with α = 1.7-3, the thrust of the model is measured and the internal thrust is separated from it showed that the experimental force characteristics of the inner channel correspond to those calculated employing the procedure of Ref.20.

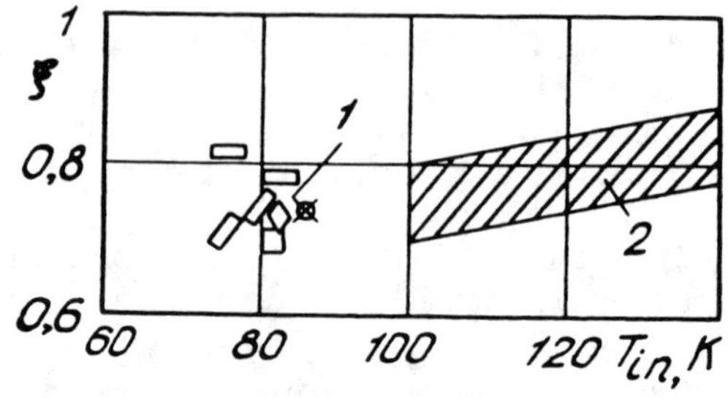

Fig. 8 Completeness of combustion: 1) based on measurement of pressure and heat fluxes (α = 0.5-1.3), 2) from results of weight measurements.

V. The Study of Micromodels

The successful results obtained with M_{in} = 7.3-7.9 in the IT-301 high-enthalpy facility in gasdynamic models with hydrogen combustion raised the question of whether M_{in} can be reduced through the dimensions of the model. Such a construction with a diameter of the inlet into the air intake d_0 = 23 mm was tested for M_{in} = 4.9 [28] in the IT-301 facility. The results of these studies showed that all characteristics of the working process and the force characteristics obtained in the experiments with M_{in} = 7.3-7.9 are also observed with M = 4.9 on the model d_0 = 23 mm.

VI. Facilities with Constant Parameters of a Working Body

Experiments in the IT-301 facility with M_{in} = 4.9-7.9 also showed the negative aspects of tests of gasdynamic models with combustion for falling values of the parameters of the inflowing air. First of all, the transient process by which the heat inflow regimes change becomes protracted (and, in many experiments for high values of T_{0in}, a change in the heat flow regime is impossible). Second, when the fuel consists of substances with kinetic and reaction characteristics that are worse than those of hydrogen, there are difficulties in burning these substances. For sufficiently high values of p and T occurring in the first few milliseconds of operation of the pulsed facility, it was possible to realize self-ignition of these substances. When the stagnation parameters decrease in the course of the process, new values of p and T are no longer adequate for efficient

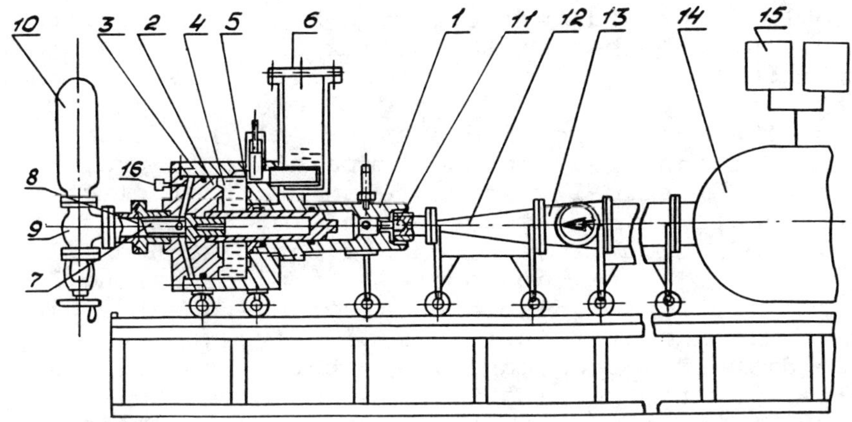

Fig. 9 Sketch of high-enthalpy facility with booster.

combustion of these substances. In this case, parameters of the inflowing air that are constant in time are required. To this end, a facility in which p_{0in} and T_{0in} are maintained constant with the use of a booster was developed.[29]

A sketch of the facility is shown in Fig. 9. The main elements of the facility are premix chamber - I, two-step piston - 2, casing of pressure multiplier - 3, damping liquid - 4, controlled drain holes - 5, drain tank - 6, circular plunger - 7, diaphragm - 8, cut-off plate - 9, bottle with pushing gas - 10, diaphragm - 11, profiled nozzme - 12, test section - 13, exhaust vacuum capacity - 14, vacuum pumps - 15, indicator of the piston location - 16.

Before the start-up of the facility, exhaust vacuum capacity, test section, nozzle are pumped with preevacuation pumps up to the pressure 10^{-1} - 10^{-2} mm Hg. A diaphragm placed in front of the critical nozzle section isolates the premix chamber volume from the evacuated section of the facility. The premix chamber is filled with a drive gas up to initial pressure. Thereat the two-step piston moves away to the left up to the contact with circular plunger touching a diaphragm that isolates the volume with drive gas from the multiplier cavity. In this location of the piston, the gap between a wall of the multiplier casing lid and a large piston step accounts for ~ 0.5 - 1 mm and is fixed with the help of the piston travel indicator. In order to prevent untimely diaphragm breakdown by circular plunger, the pressure in multiplier cavity is maintained at the level providing the forces balance affecting the piston both from the premix chamber side and from that of the large piston step. After the cut-off plate opens, the drive gas remains isolated from the multiplier body cavity only by diaphragm.

The gas pressure and temperature in premix chamber drastically increase on start of electro-arch heater. Thereat the balance of forces affecting the piston is violated and the piston displaces by 0.5 - 1 mm up to the multiplier casing lid. Circular plunger displaces also and cuts the diaphragm. Under the influence of drive gas pressure, the diaphragm opens and gas enters into the multiplier cavity, putting the piston in motion. The running speed of piston is set by the value of drive gas pressure and the size of controlled drain holes for damping liquid. A main assignment of the damping liquid is preventing of self-oscillatory piston motions when the pressure affects it stepwise. Depending on the flow rate of high-temperature gas in premix chamber, the speed of the piston may vary from several centimetres up to several metres per second by using different relations of pressure and sizes of drain holes.

The above principle of the start-up of stabilization system of the flow parameters is employed in the case when the drive gas pressure exceeds 70 atm and the thickness of diaphragm isolating the volume with drive gas exceeds (1-1.5)mm. While facility operates at small stagnation parameters (the pressure in premix chamber after discharge is about ~ 50.0 MPa) and the drive gas pressure does not exceed 7.0 MPa, the diaphragm failure is made with the help of hydroimpulsive device whose principle of functioning is based on the effect of hydroimpact in liquid.

Either means of the start-up of a stabilizing system provide the time of reaching the computed speed of motion equal to 1-2 ms.

Such a scheme of hot-shot wind tunnel has been chosen as a result of computational studies that enabled us to analyse dynamics of the piston running and to determine geometrical parameters of all units of the system. Their typical values for a forechamber volume $V_f(0) = 9.3$ cm³ are presented in Fig. 10. Here Re_m is the Re number per 1 m; w_{in} is the velocity of the incident flow. The plateau exists for 120-130 ms. In gene-

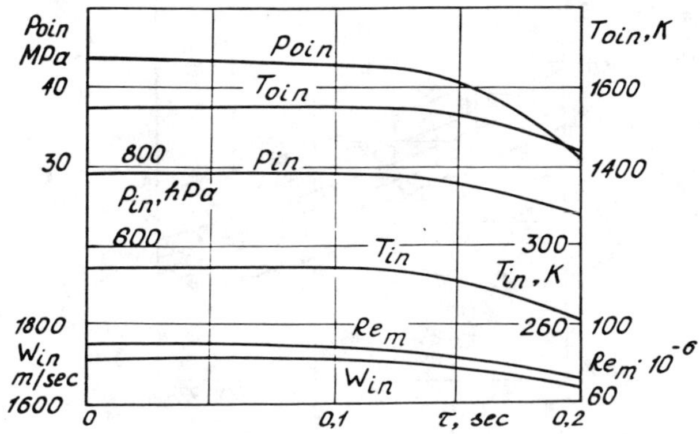

Fig. 10 Parameters of the air flowing onto the model; $V_f(0) = 9.3\ dm^3$, $d_{cr} = 10$ mm, $M_\infty = 5$.

ral, however, by combining the pressure under which the air is pumped into the forechamber, the values of Q, d_{cr}, and the pressure on the side of the large piston of the booster p_b, the flow parameters can vary during the process over a wide range, or they can be fixed beforehand.

VII. The Results of Tests of Models with Combustion in Hot-Shot Facility with Constant Parameters of Working Medium

In this facility, the thrust-cost characteristics of the inner channel of the model were studied for $M_{in} = 5$.[28]

Fig. 11 shows the inner thrust factor of the model as a function of T_{in} and $\bar{F}_c$ for hydrogen with $\alpha = 1$. The solid lines show the calculation following Ref. 20, and the symbols indicate the experimental values. For $\bar{F}_c = 2.85$, two computed curves are presented: curve 6 corresponds to the burning of hydrogen in the model with $\xi = 0.7$, and curve 5 corresponds to $\xi = 0.8$. One can see that the experimental data fell between curves 5 and 6, i.e., heat flow in the experiments corresponded to $\xi = 0.75$-0.8. If one keeps in mind that 15-20% of the heat liberated on combustion with $\alpha = 1$

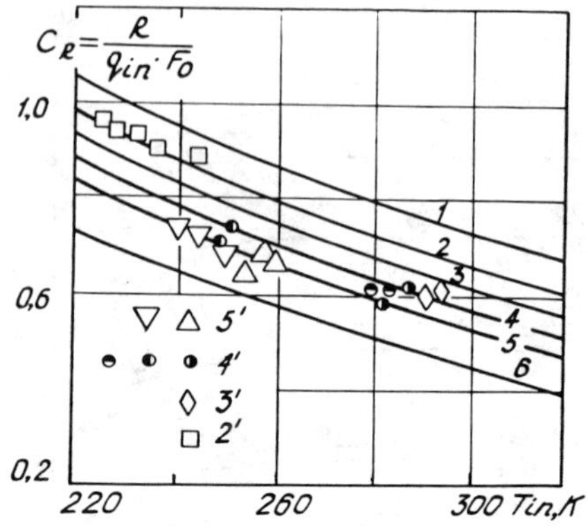

Fig. 11 Inner thrust factor $\bar{F}_C$: 1) 1.54; 2, 2') 1.96; 3, 3') 2.24; 4, 4') 2.57; 5, 5', 6) 2.85.

and ξ = 1 was lost in the inner channel of the model for the test conditions in a facility with constant parameters of the inflowing air, then the physical completeness of the combustion of hydrogen is quite high, at least 0.0-0.95. The higher values of ξ in these experiments with M_{in} = 5 as compared with that of ξ in the model in Fig. 4 with M_{in} = 7.3-7.9 are attributable to two factors: 1) the more uniform distribution of hydrogen over the transverse cross section of the CC and 2) the higher pressure in the CC. In experiments with M_{in} = 7.3-7.9, the pressure in the combustion chamber equaled 0.05-0.5 MPa and, in experiments with M_{in} = 5, the pressure equaled 1.8-4 MPa, depending on $\bar{F}_C$. The high physical completeness of combustion (ξ = 0.9-0.95) occurred for the entire range $\bar{F}_C$ = 2.85-1.96 tested in the experiments.

Curve 1 in Fig. 11 corresponds to the minimum possible degree of expansion of the combustion chamber $\bar{F}_C$ = 1.54 under the following conditions: T_{in} = 220° K, ξ = 0.8; the coefficient of restoration of the total pressure related only to the hydraulic losses in the CC on section 2-5, σ = 0.9 Thus, curve 1 characterizes the limiting values of C_R, which can be obtained in tests of this model in a high-enthalpy facility with a loss in the wall of 15-20% of the heat released by the combustion of hydrogen. It is obvious that already, for $\bar{F}_C$ = 1.96, the values of C_R are close to the maximum possible values.

Experiments with $\bar{F}_c$ = 2.85 and 2.57 were performed with all of the hydrogen injected in the first zone (the angle of inclination of the injectors relative to the axis φ_1 = 135°).[28] For $\bar{F}_c$ = 2.24, however, this injection scheme did not provide normal operation: the flow of air in the model was disrupted immediately after the injection of the first portions of hydrogen. For this reason, for $\bar{F}_c$ = 2.24, hydrogen was injected through the zones with $\varphi_1 = \varphi_2$ = 135° ; ~1/3 of the hydrogen in the first zone, and the remaining hydrogen was injected in the second zone. For this injection scheme, the model operated with α = 1 throughout the entire operating time of the facility, with an inner thrust corresponding to curve 3 in Fig. 11. But for $\bar{F}_c$ = 1.96, this scheme is also unsatisfactory: for $\bar{F}_c$ = 1.96 and 135° all of the hydrogen musy be injected into the second zone. In this case, however, ξ = 0.5-0.6, since the remaining length of the CC was already insufficient for efficient combustion of hydrogen. Transfer to hydrogen injection through the first zone with φ_1 = 45° enabled, in this case F_c = 1.96, operation with ξ = 0.8 (see curve 2 in Fig.11). For this injection method, however, in a series of experiments with $\bar{F}_c$ = 1.96, stalling of the input unit was observed during some part of the regime.

Thus, the physical picture of the flow in the initial section of the CC, revealed for M_{in} = 7.3-7.9 on the model with d_0 = 72 mm, was manifested fully in experiments with M_{in} = 5 with a reduction of $\bar{F}_c$. The significant rise in the pressure in the CC accompanying the decrease in F_c was the main factor responsible for the disruption of air inflow accompanying movement of pseudoshock forward along the flow. Until the head of the pseudoshock (or the separation zones displaced by it) emerge onto the surface of compression of the air intake unit, the inflow of air into the model proceeds as computed. The redistribution of fuel injection along the CC with a decrease in $\bar{F}_c$ prevented the pseudoshock from emerging onto the compression surface, thereby insuring that the pseudoshock is located in the initial section of the CC.

In the case $\bar{F}_c$ = 1.69, normal operation of the model was achieved only if all of the fuel was injected into the second zone with φ_2 = 45° But, in this case, ξ = 0.5-0.6.

The experiments performed showed that, for operation of the CC in the regime with heat inflow into the subsonic flow, it is, in principle, possible to reduce $\bar{F}_c$ to a minimum value, determined by the gas-thermodynamic flow. For this it is necessary to provide conditions under which the pseudoshock is located in the initial section of the CC so that the pseudoshock does not start on the surface of compression.

VIII. Conclusion

In this review, we gave an incomplete description of the most important processes in the implementation of the working process in gasdynamic channels with heat and mass inflow under conditions for which high-enthalpy airflow around the model occurs over a short period of time. A more detailed exposition of some investigations, as well as an entire series of other interesting phenomena, that were not discussed in this review because of space limitations is given in the references discussed earlier.[6-10, 13-17, 20, 28, 29]

These investigations showed that pulsed wind tunnels are a completely reliable tool for studying gasdynamic models with heat and mass inflow. The working process in the gasdynamic channel (compression of air in an air intake unit, interaction of the chemical-reaction zone with the supersonic airflow) follows the varying parameters of the flow moving toward the model, the characteristic times of the constituent elements of the working process in gasdynamic models are significantly shorter than the characteristic times over which the parameters of the working regime of the tunnel change. The main result of the investigations carried out on diverse models with combustion for M_{in} = 5-8 is the experimental demonstration of the possibility of organizing a highly efficient working process in complicated, short channels with hypersonic velocities of the air flowing around the model, making it possible to obtain quite high characteristics of the inner channel of the models, as well as to compare computational methods with experimental results.

This also shows that high-enthalpy facilities operating in a short-time regime are promising tools for studying diverse processes in gasdynamic models with combustion.

References

[1]Kurziner, R. I., Jet Engines for High Supersonic Flight Velocities , Mashinostroeniye, Moscow, 1977 (in Russian).

[2]Baev, V. K., Golovichev, V. I., Tretyakov, P. K., Konstantinovsky, V.A., Garanin, A.F., Yasakov, V.A., "Combustion in Supersonic Flow", Nauka, Novosibirsk, 1984.

[3]Shchetinkov, E. S., "Physics of Combustion of Gases", Nauka, Moscow, 1965.

[4]Osgerby, I. T., Smithson, H. K., and Wagner, D. A., "Development of a Double-Oblique-Shock Scramjet Model in a Shock Tunnel", Journal of Aircraft, Vol. 8, Sept. 1970, pp. 1703-1705.

[5]Holden, M. S., "Establishment Time of Laminar Separated Flows", AIAA Journal, Vol. 9, November 1971, pp. 2296-2298.

[6]Baev, V. K., Shumsky, V. V., and Yaroslavtsev, M. I., "Study of Two-Regime Combustion Chamber Operation by Subsonic Heat Delivery", Gas Dynamics of Flows in Nozzles and Diffusers, Novosibirsk, 1982.

[7]Baev, V.K., Shumsky, V.V., and Yaroslavtsev, M. I., "Methodical Questions of Aerodynamic Model Tests with Combustion in High-Enthalpy Facilities of Short-Time Regime", Izvestiya Sibirskogo Otdeleniya Akademii Nauk SSSR, Seriya Tekhnicheskikh Nauk, Vol. 1, No. 4, 1984, pp. 68-77.

[8]Shumsky, V.V., "Methodical Questions of Flow past a Plate by Conical Stream", Third All-Union School on Methods of Aerophysical Research, Sec., Novosibirsk, 1982.

[9]Belousov, S. V., Golod, V. V., Pronin, Yu. A., Yaroslavtsev, M. I., "Weight Tests of Heavy Models in Pulsed Regime at Hypersonic Velocities", Methods and Technology of Aerophysical Research, Novosibirsk, 1978.

[10]Baev, V. K., Boshenyatov, B.V., Pronin, Yu. A., Shumsky, V. V., "Study of Liquid Injection in Supersonic Flow of High-Enthalpy Gas", Fizika Goreniya i Vzryva, Vol. 17, No. 3, May-June 1981, pp. 72-76.

[11]Gladyshev, M. K., Gorelov, V. A., and Korolev, V. S., "Experimental Study of Liquid Jet - Hypersonic Stream Interaction", Uchenye Zapiski TsAGI, Vol. 9, No. 1, 1978, pp.115-120.

[12]Prudnikov, A. G., Volynskii, M. S., and Sagalovich, V. N., Mixing and Combustion Processes in Air-Jet Engines, Mashinostroeniye, Moscow, 1971.

[13]Baev, V. K., Boshenyatov, B. V., Pronin, Yu. A., and Shumsky, V. V., "Experimental Study of Hydrogen Ignition Injected into Supersonic Hot Air Flow", Gas Dynamics of Combustion in Supersonic Flow, Novosibirsk, 1979, pp. 53-64.

[14]Baev, V. K., Pronin, Yu. A., and Shumsky, V. V., "Self-Ignition of Liquid Substances in Supersonic Air Flow", Fizika Goreniya i Vzryva, Vol. 18, No. 4, 1982, pp. 22-26.

[15]Baev, V. K., Shumsky, V. V., and Yaroslavtsev, M. I., "Self-Ignition of a Combustible Gas Escaping into a Medium of Gaseous Oxidizer", Fizika Goreniya i Vzryva, Vol. 19, No. 5, 1983, pp. 73-80.

[16]Baev, V. K., Shumsky, V. V., and Yarpslavtsev, M. I., "Study of Gas Dynamics of a Model with Coùbustion in Shock Tube", Zhurnal Prikladnoi Mekhaniki i Tekhnicheskoi Fiziki, Vol. 24, No. 6, Nov.-Dec. 1983, pp. 58-66.

[17]Baev, V. K., Shumsky, V. V., and Yaroslavtsev, M. I., "Pressure Distribution and Heat Transfer in Gasdynamic Model with Combustion Streamlined by High-Enthalpy Air Stream", Zhurnal Prikladnoi Mekhaniki i Tekhnicheskoi Fiziki, Vol. 26, No. 5, Sept.-October 1985, pp. 56-65.

[18] Lashkov, A. I., and Nikolsky, A.A., "Wave Start-Up of Supersonic Diffuser", Inzhenerhyi Zhurnal, Vol. 2, No. 1, 1962, pp. 11-16.

[19]Zatoloka, V. V., Zvegintsev, V. I., and Shumsky, V. V., "The Influence of Compression Process in Air Intake on Specific Thrust Characteristics of Scramjet", Izvestiya Sibirskogo Otdeleniya Akademii Nauk SSSR, Seriya Tekhnicheskikh Nauk, Vol. 2, No. 8, 1978, pp. 3-12.

[20]Saren Y. A., and Shumsky V. V., "Characteristics of Scramjet with Two-Regime Combustion Chamber", Gas Dynamics of Flows in Nozzles and Diffusers, pp. 73-85.

[21]Emmons, G., Foundations of Gas Dynamics, IL, Moscow, 1963 (Russian translation).

[22]Gurylev, G. V., and Trifonov, A. K., "Transition of Supersonic Flow into Subsonic in Tunnel with Expanding Initial Section", Uchenye Zapiski TsAGI, Vol. 11, No. 4, 1980, pp. 80-89.

[23]Zimont, V. L., and Ostras', V. N., "Deceleration in Pseudoshock by Supersonic Flow in Channel", Ideas of F. A. Tsander and Development of Rocket and Space Science and Technology, Naukan Moscow, 1983, pp. 93-107.

[24]Kuz'min, V. A. , "Deceleration of Supersonic Flow in Rectangular Channel", Gas Dynamics of Aircraft Engines, No. 1, Kazan, 1978, pp. 31-37.

[25]Ktalkherman, M. G., Mal'kov, V. M., and Ruban, N.A., "Slowing Down of a Supersonic Flow in Rectangular Channel of Constant Cross Section", Zhurnal Prikladnoi Mekhaniki i Tekhnicheskoi Fiziki, Vol. 25, No. 6, Nov.-December 1984, pp. 48-57.

[26]Clark, W. H., "An Experimental Investigation of Pressure Oscillations in a Side Dump Ramjet Combustor", AIAA Paper 80-1117, 1980.

[27]Smith, D. A., and Zukoski, E.E., "Combustion Instability by Unsteady Vortex Combustion", AIAA Paper 85-1248, 1985.

[28]Baev, V. K., Shumsky, V. V., and Yaroslavtsev, M.I., "Force Characteristics and Flow Parameters in Combustion Models", Zhurnal Prikladnoi Mekhaniki i Tekhnicheskoi Fiziki, Vol. 25, No. 1, Jan.-Feb. 1984, pp. 103-109.

[29]Antonov, A. S., Baev, V. K., Dmitriev, V. A., Pronin, Y. A., Puzyrev, L. N., Shumsky, V. V., and Yaroslavtsev, M. I.,"Gasdynamic Facility IT-302M", The IV-th All-Union School on Methods of Aerophysical Studies, Novosibirsk, 1986.

Chapter 9

Facility Requirements for Hypersonic Propulsion System Testing

M. G. Dunn,* J. A. Lordi,† C. E. Wittliff,‡ and M. S. Holden‡

Calspan Advanced Technology Center, Buffalo, New York

Abstract

Facility requirements for hypersonic propulsion system testing and the capabilities or limitations of selected hypersonic testing facilities are briefly reviewed. The specific advantages and limitations of the shock tunnel for hypersonic propulsion testing are discussed, and areas are identified where shock-tunnel research can supplement other ground tests and flight experiments in the development of air-breathing propulsion at hypersonic speeds. Other chapters in this volume will describe several other relevant facilities in detail, so the discussion in this chapter will be limited to a response to many of the perceived limitations associated with the short-duration approach to obtaining fundamental data applicable to high-enthalpy environments. Shock-tunnel testing of both scramjet combustor and exhaust nozzle configurations at conditions corresponding to flight at Mach numbers of 10-12 at altitudes of 100,000 ft and above are considered. Sample computations are presented to demonstrate that real-gas effects, and in particular, the departure from vibrational and chemical equilibrium in the facility nozzle expansion do not significantly impact the ability to simulate the desired test flows. Rather, it is the facility limitations in producing the flow enthalpies and test time that impact the simulation of flight conditions for NASP-like configurations.

I. Introduction

The design of new vehicles for hypersonic flight requires advances not only in structural materials, thermal protection

---

*Vice President, Research Fellow.

†Department Head, Physical Chemical Sciences.

‡Principal Aeronautical Engineer.

systems, propulsion systems, and control systems but also in test facilities. In the late 1950s, through the following decade, and until the mid-1970s extensive hypersonic-related research was reported throughout the world in the unrestricted literature. By the mid-1970s interests changed; hypersonics research funding faded but did not disappear completely. It can be inferred that Soviet Union research remained relatively active during this period. The U.S. Air Force remained active at basically a maintenance level in the hypersonics field because of the obvious importance of hypersonics to many military missions. A few research laboratories in the United States were active in order to support this interest of the Air Force but, in general, the natural result of drastically reduced funding and military interest was that many of the groups and almost all of the universities that had been doing hypersonics research now found other areas of research to pursue. Further, the development of new facilities at both government and private laboratories was reduced drastically and, in most cases, ceased entirely. Many of the researchers involved found careers in other research areas, and many of the hypersonic facilities were dismantled. Fortunately, however, the achievements and the difficulties associated with the active hypersonics area of the 1950s and 1960s were reasonably well documented in generally available literature, and selected research facilities and personnel remained active.

In 1985, the U.S. government embarked on a program to restart hypersonic awareness within the universities by initiating a multiyear program titled "Training and Research in Hypersonics." This was followed in 1986 by the announcement of a joint NASA and Defense Department program to build the National Aerospace Plane (NASP), or the X-30 flight research vehicle, which has launched a renewed interest in hypersonics in the United States Prior to the NASP announcement, the French (HERMES), British (HOTOL), and West Germans (Sanger II) were under way with their own hypersonic programs, as were the Japanese and the Soviets. It is important to note that these vehicle configurations and the flight regimes associated with current hypersonic vehicles are very different from those that were studied in the 1950-1970 era. In the 1960s and 1970s there was a high level of confidence in designing and operating a very blunt (Gemini, Mercury, and Apollo) type of entry body, slender (either blunt or sharp-nosed) weapon type of entry body and the winged, rocket-powered X-15, which operated at Mach numbers lower than those of current interest.

In order to address the test facility requirements associated with the newer generation of hypersonic vehicles, it is important to revisit the achievements of the past 30 years and the reasons for some of the failures, but it is essential that we recognize that some of the problems and the facility requirements of the 1980s and 1990s are very different from those that were previously faced.

During the past few years, work has continued on current problems even though the optimum tools, both computational and experimental, have not been available. The research tools available can be utilized to extract the maximum amount of information possible while, in parallel, developing the required new facilities, computational codes, and flight programs. It is important to recognize that there has been and will continue to be a funding limitation requiring that a priority list be established that is consistent with national goals in hypersonics. Section II provides a review of past and current hypersonic facility studies. It will briefly review one of the extensive efforts to establish facility requirements for the hypersonic problems of the 1970 era, followed by a similar brief review of just completed studies that were intended to define facility requirements for the solution of problems associated with the next generation of hypersonic vehicles. Section III describes some of the many problems currently associated with wing-body hypersonic airplanes and several currently operational ground-based facilities or facilities in the development stage that can be used now and in the future to obtain relevant data.

Section IV has been devoted to a closer look at facility limitations in general and the short-duration shock tunnel in particular. Problem areas are discussed wherein this device can make significant contributions to the type of unified computational, ground-test, and flight-experiment program that will be necessary in order to resolve the complex issues associated with the development of either a single-stage-to-orbit vehicle or an air-breathing/rocket-assist-to-orbit vehicle. In the course of this discussion, an attempt has been made to respond to some of the perceived difficulties with these facilities while trying also to present a realistic assessment of their limitations.

II. Facility Studies and Requirements

This section will be devoted to a review of past and current studies of hypersonic facility requirements. In Sec. A, a few selected studies are described and, in Sec. B a very brief description of current hypersonic flow problem areas and facility requirements is given.

A. Brief Review of Hypersonic Facility Studies

In 1970, McDonnell Aircraft Company reported the results of a one-year hypersonic facility (HYFAC) program study that attempted to identify high-priority research required for future hypersonic cruise aircraft, to evaluate research potential and cost of candidate research facilities, and to assess the usefulness of these facilities in support of other aerospace systems. The similarities between these requirements and those of current problems facing the hypersonic community are striking. The results of the study are described in Refs. 1-8 and represent one group's analysis of the requirements of both ground-based and flight research facilities required in order to develop an air-breathing winged-body vehicle capable of operating at hypersonic speeds. The material presented in these reports was focused on a 300-ft, a 200-ft, or a 100-ft-long vehicle and the associated mission requirements as they were understood in 1970, which are reasonably consistent with those of today. Further, the vehicle configurations sketched in these reports and the propulsion systems described were in many ways similar to current designs, but details of the test requirements were very different. The specific subjects addressed included: 1) gasdynamic facilities, 2) turbo- machinery/ramjet facility, 3) scramjet engine research facilities, 4) structural/fluid system facility, and 5) material research facility. A detailed listing of the research objectives that the recommended facilities were to satisfy is given in Ref. 2. This list contains 102 objectives ranging from the desire to improve fundamental knowledge of hypersonic boundary-layer behavior to the development and integration of engine components into complete scramjet systems. In a later phase of the study, reported in Ref. 1, the list was reduced to 10 primary objectives.

The McDonnell study was performed in three phases: Phase I was the preliminary analysis during which the 102 objectives noted earlier were formulated and 54 possible facilities were examined; the study reduced the number of candidate facilities to 11. For Phase II, these 11 facilities were evaluated on the basis of technical advantages at minimum cost, which resulted in the selection of 5 candidate facilities. Phase III involved an improved definition of the 5 candidate facilities and a more detailed estimate of the construction and development costs. The 5 selected facilities were: 1) GD20 polysonic blowdown wind tunnel, 2) GD7 hypersonic impulse gas dynamic research facility, 3) E20 compound turbomachinery test facility, 4) E9 dual-mode ramjet engine research facility, and 5) S20 structures research facility. The authors of Refs. 1-8 felt that a ground-based facility would serve their needs if they could produce one-fifth of the maximum flight Reynolds number over the full Mach number range at a dynamic pressure of 2000 psf. (for a 310-ft-long airplane). Preliminary design drawings for all of these facilities were completed, but none of them was constructed.

A more recent review of facility requirements for hypersonic propulsion testing is the one presented in 1987 by Smith et al.[9] These authors have also addressed the problem of hypersonic winged-body vehicles as did the study reported in Refs. 1-8, and they list a number of testing issues and "enabling technologies" that must be developed in order to obtain solutions to the problems of interest. Many of the testing issues and technology questions are consistent with the earlier McDonnell study, but Smith et al. have put them in perspective with current hypersonic problems. After summarizing the state of the art in 1987, the authors recommend that the proper approach would be to modify a number of facilities that they list, to evaluate the potential of other proposed modifications which are also listed, and to construct a new HYFAC at AEDC. Implicit in the discussion of Ref. 9 is the suggestion that the appropriate course of action would be construction of an advanced arc heater (of the 400-mw class) and the development of a magneto hydro dynamic facility to achieve the high flow velocities.

Solutions to problems associated with the hypersonic flight of an air-breathing wing-type vehicle require the use of several different experimental facilities, since no one facility can handle all of the problems. One of us (C.E.W.) has completed a survey of existing hypersonic ground-test facilities in North America,[10] with emphasis on hypersonic aerodynamic and aerothermal testing. This survey includes a description of current status, test section size, Mach number capability, Reynolds number capability, and stagnation pressure and temperature capability for continuous-flow tunnels, intermittent-flow (air, nitrogen, CF_4, or helium) tunnels, shock tunnels, and gun tunnels. In addition, the survey provides a comparison of several current flight vehicle Reynolds number vs Mach number requirements with the capabilities of existing facilities.

Hallion[11,12] has recently edited a two-volume report§ that is interesting reading for those involved in the hypersonics areas. Volume 1,[11] titled "The Hypersonic Revolution, from Max Valier to Project Prime, 1924-1967," is available to anyone interested. Volume II,[12] titled "The Hypersonic Revolution, From Scramjet to the National Aero-Space Plane, 1964-1986," is controlled, with availability only to U.S. government person-

§These volumes can be obtained by writing to: ASD/HO, Bldg. 16, Area B WPAFB, OH 45433-6503, Att: Albert Misenko.

nel and their contractors. The two volumes consist of eight case studies in the history of hypersonic technology. The specific cases studies in Volume I are:

1) Transiting From Air to Space: The North American X-15
2) Strangled Infant: The Boeing X-20A Dyna-Soar
3) The Development of Winged Reentry Vehicles: An Essay from the NACA-NASA Perspective, 1952-1963
4) ASSET: Pioneer of Lifting Reentry
5) Project PRIME: Hypersonic Reentry From Space

The specific cases studied in Volume II are:

6) Confronting Scramjet: The NASA Hypersonic Ramjet Experiment
7) The Piloted Lifting Body Demonstrators: Supersonic Predecessors to Hypersonic Lifting Reentry
8) Space Shuttle: Fulfillment of a Dream

All of these case studies are written by people who were intimately involved in the effort, and the authors have made an attempt to document their material carefully. Both of these reports put into perspective a major, far-reaching technical development comparable to the NASP effort and provide the basic grounds for evaluation of current progress in this difficult undertaking.

It is beyond the scope of this paper to review the many publications of the 1960s and 1970s that summarized the facility requirements for hypersonic propulsion system testing. All of this work can easily be found in a conventional library search. It is more appropriate for us to move on to the more recent studies of the 1980s and then to present a few of the requirements from our perspective. Four prestigious panels have been commissioned by various agencies during the past few years to study and make recommendations on various aspects of the problems encountered in high-speed flight. We noted earlier in this chapter that the common purpose of these panels was to evaluate the existing capability with respect to current and future problems. Each of the panels addressed a very specific, but different, aspect to the overall problem of high-velocity flight in the Earth environment and spent long hours evaluating information and formulating recommendations. All four of the studies have now been completed, and the printed version of their findings are available in Refs. 13-16.

The Smelt Committee report[13] is specific in recommending facilities that should be developed to support subsonic/transonic/supersonic needs and in establishing a priority for their recommendations. In general, this committee was probably more interested than the other three panels in the lower portions of the velocity range. The committee also recommends actions that should be followed with respect to new hypersonic facilities. With respect to hypersonic facilities, the committee concludes that "no ground-based facility exists, or can be expected to be built in the near future, of sufficient size to test a large model of an aerospace vehicle with engines operating at any hypersonic Mach number. Furthermore, CFD is not likely to provide answers over all hypersonic Mach numbers of interest. Thus, there will continue to be a requirement for flight testing as a tool in design validation." The report provides a rather brief discussion of the application of short-duration (they call them "impulse") facilities to the problems of hypersonic flight. They do not recommend any further development of short-duration facilities. It is our opinion that this committee over emphasizes the potential difficulties that may be caused by nonequilibrium chemistry, which can be associated with the expansion of a gas from very hot reservoirs. Often, some of the community has associated this potential difficulty only with short-duration facilities. As will be demonstrated in this chapter, this is not a limitation over a wide range of simulated conditions and, furthermore, can be avoided if the proper analysis is performed in designing the experiment.

The Kerrebrock Committee[14] had a different charter than the other committees mentioned here. Their purpose was to evaluate the existing analysis and experimental state of the art as it pertains to providing reasonable prediction capability for current military missions involving hypersonic flight of vehicles. This committee described what they felt to be the solid points and the soft spots in existing hypersonic technology. From the viewpoint of this paper, the committee findings and recommendations in the general area of aerodynamics are most appropriate. They break the flight regime into two regions: 1) low hypersonic speeds ($6<M<10$) and 2) high hypersonic speeds ($M>10$). For region 1 they assign the highest priority to the design, construction, and operation of a "quiet" wind tunnel. For the high hypersonic speeds, the suggestion of this committee is that the field of low-density flows should be given special consideration. They are also very concerned about reaction rate coefficient information for the high-temperature environment associated with this flight regime and suggest that work needs to be initiated in this area.

The charter of the Buchanan Committee[15] was to address the requirements for hypersonic test facilities. This committee has defined hypersonics as "flight in or through the atmosphere at speeds greater than or equal to Mach 5." Draft copies of Refs. 13, 14, and 16 were available to the Buchanan Committee during their study. Reference 15 is a comprehensive document that concludes by presenting recommendations for action that should be taken in the technologies (aerodynamics, propulsion, stability and control, structures, and materials and avionics) and for a core program for ground-test facilities. The Buchanan Committee report provides an assessment of the current needs for hypersonic test facilities. Reference 16 is basically an in-depth review and status report (as of 1987) of the major NASA wind-tunnel facilities. References 14-16 do a thorough job of documenting their work. To gain an appreciation for the findings of the respective committees, the interested person should take the time necessary to read these committee reports carefully.

In Ref. 17, the effect of test flow nonequilibrium on the simulation of hypersonic flows is examined, together with simulation requirements for current vehicles of interest. The design of hypersonic shock-tunnel experiments for the study of hydrogen-burning scramjet combustors is also described. Nonequilibrium effects and facility limitations that lead to relaxing the requirements for full duplication of the flight conditions are described. It is illustrated that the effects of shock-tunnel nonequilibrium on hypersonic flow measurements and, specifically, on scramjet combustor experiments can be minimized, if not eliminated, if the shock-tunnel test conditions are properly selected and, moreover, representative combustor inlet conditions can be achieved for a range of relevant flight conditions. It is certainly true that existing facilities cannot provide full duplication, but it is possible in many cases to relax the full duplication requirement and to duplicate the flow velocity and one of the thermodynamic variables corresponding to atmospheric conditions at a given altitude. If the flow is expanded from an equilibrium reservoir to the desired velocity and the density corresponding to a given altitude, then, in general, the static temperature will be higher than the static temperature at that altitude. For hypersonic flows, velocity and density-altitude duplication preserve a match in the total enthalpy, or stagnation temperature, and in the dynamic pressure. In Sec. IV of this paper, some specific examples (which include nonequilibrium chemistry) will be treated as they relate to hypersonic propulsion measurements in short-duration facilities. Measurements performed in these facilities have in the past been shown to be very valuable in designing flight vehicles and in interpreting flight data.

Another facility requirement study that should be mentioned here (the results cannot be described in any detail because the report distribution is limited to U.S. government agencies and their contractors) is the study recently completed at AEDC and reported in Ref. 18. The suggested facility is intended to provide a test capability for hypersonic propulsion system development and materials testing. This report is the one mentioned in Ref. 9, and the primary findings of the study are given there.

The Smelt Committee and the Buchanan Committee both find that CFD, ground-based experiments and flight experiments will all be required to obtain satisfactory performance of hypersonic vehicles. Further, the Buchanan Committee report notes that many different ground-based facilities will be required to obtain the data necessary to validate CFD codes for conditions representative of flight situations. Section III of this paper is devoted to a discussion of a few ground-based facilities and their advantages and disadvantages for hypersonic research and testing. It is suggested here that there is a definite need for flight experiments and, further, that a well-coordinated set of ground-based measurements can increase the probability of success of the flight experiments, ultimately resulting in a significant cost savings.

B. Summary of Hypersonic Flow Problem Areas

With the development of powerful numerical techniques to obtain detailed solutions to the full or reduced time-averaged Navier-Stokes equations, the role of experiment has been increasingly one of examining in detail the fundamental flow mechanisms that must be modeled by the Navier-Stokes equations rather than aiding in the construction of gross models of the flows of interest. Because of the enormous energy levels involved for flows above Mach 8, it is unrealistic to expect the possibility of duplicating the environment around a full-size vehicle or engine in a "continuous" ground-test facility. Our efforts, therefore, must be directed to isolating the key aerothermal phenomena of interest and developing an experimental environment to obtain a full or partial simulation from which we can perform the measurements necessary to develop and evaluate models to be used in the solution of some form of Navier-Stokes equations. Although the major modeling problems are associated with the construction of models of turbulence and chemical kinetics for boundary layers and other shear layers subjected to strong pressure gradients and influenced by chemical nonequilibrium and combustion, there are a number of key problem areas in which modeling must be addressed if we are to predict overall vehicle performance over the complete flow regime.

In the high-Mach-number, high-temperature, low-density regime, there are important modeling problems associated with the catalytic surface/reacting gas interaction. Whereas research in low density flows has continued at a steady pace in low-enthalpy flows, there has been almost no experimental research during the past two decades in hypervelocity (V > 10,000 fps or 3.05 km/s) flows. One of the areas of greatest interest for small re-entry vehicles such as the French Hermes is the transitional flow regime between free-molecular and continuum flow. Here the short-duration facility, in one of its various modes of operation, has been used successfully to generate the high-purity hypervelocity flows necessary to obtain the required measurements. Flowfield and surface diagnostics have been developed to obtain measurements in short-duration flows (~1 ms), typically generated in these facilities when operated at high-temperature conditions. Thus, there is every reason why further studies should be conducted to contribute significantly to an increased understanding in this flow regime.

One of the most important areas for research in hypersonic laminar boundary-layer flows is associated with the understanding of chemical nonequilibrium and combustion effects on the size and characteristics of separated flows developed in regions of shock-wave/boundary-layer interaction. Whereas solutions to the full Navier-Stokes equations have been shown to successfully describe separated regions in non-reacting hypersonic flows, nonequilibrium air chemistry or combustion in the recirculation regions provides problems that can be resolved only with combined experimental and numerical studies. In the experimental study of these flows, flow duration becomes an important parameter, and test times of one or more milliseconds may be required for establishment of complex interacting flows. For this application, larger reflected shock tunnels are currently in existence that would be acceptable facilities to use. Construction of such a facility is a low-risk venture because of the existing experience factor. The influence of nonequilibrium chemistry on hypersonic ramjet performance in laminar flows is an area for which experimental research is essential to provide the answers to critical questions. Likewise, one can have little confidence in numerical solutions to laminar combusting hypersonic flows without experimental verification.

Predicting boundary-layer transition to turbulent flow and the characteristics of transitional flows presents one of the formidable problems in hypersonic flows. Whereas such flows are difficult to predict at any Mach number, at hypersonic speeds, they are so extensive and have such large effects on aerothermal loads that they must be adequately described. Since the basic mechanisms of transition are not understood, only a careful combination of measurements in ground and flight tests will provide the insight required to validate the modeling of these flows. Semi-empirical transition prediction techniques such as the $(Re)^N$ method requires validation on both the fundamental and correlational level. Measurements are required to define the basic instability modes associated with transition in hypersonic flow, and the mechanisms involved with the gross breakdown of the laminar flow and the transition to fully developed turbulent flow. Extensive regions of low-Reynolds-number turbulent flows in and downstream of the transition can dominate the aerothermal characteristics of hypersonic vehicles and the performance of air-breathing engines in the transitional flow regime.

The structure of turbulent boundary layers in regions of shock-wave/turbulent-boundary-layer interaction, and shear-layer mixing regions in hypersonic flows are strongly influenced by compressibility effects. How compressibility influences shock/turbulence interaction, flow unsteadiness and eddy shocklets, and flow structure remains to be determined. However, it is clear that accurately predicting the size and structure of turbulent flows in strong pressure gradients will require detailed insight from experimental research. The shear-layer development in mixing regions between dissimilar gases, in regions of jet injection, and also film or transpiration cooling at hypersonic speeds require extensive experimental research. Clearly, adding nonequilibrium flow chemistry and combustion to the problem, as they occur in a scramjet engine, presents a situation well beyond the current state of the art in understanding and computation.

Because of the intrinsic integrated structure of the airframe and engine for a hypersonic air-breathing vehicle, all of the problems mentioned previously are important in the design of the engine of an air-breathing vehicle. The nonequilibrium flow developed in the leading-edge flowfield will influence the flow development on the compression ramps and possibly into the inlet. Earlier experimental studies have shown that strong distortions to the inlet flows result from viscous/inviscid interaction and crossflow over compression ramps. Flow distortions can arise from the compression ramp and cowl shocks separating the sidewall and cowl boundary layer. The sidewall boundary layer will be intrinsically transitional in nature, and predicting the flows in the presence of interacting shock waves and flow separation will be extremely difficult. Regions of shock/shock interaction on the cowl lip and leading-edge of the sidewalls will induce large leading edge heating rates as well as significant flow disturbances inside the engine. The blunt leading edges of the sidewalls and cowl will induce flow chemistry that may exert significant effects on combustion chemistry and the regions of shock-wave/boundary layer interaction inside the engine. Two- and three-dimensional separated flows may be induced on the floor, the sidewalls, and the cowl of the engine. The characteristics of these flows must be carefully simulated in experiment and/or predicted in numerical computations. It is in these flows that there are currently no successful turbulence models. Unquestionably, various forms of film and transpiration cooling must be employed to maintain the engine integrity. Fuel injection from the engine walls and possibly struts will generate complex interacting flowfields, which again must be carefully simulated numerically and experimentally. Here combustion and nonequilibrium turbulence and flow chemistry are all important. Without experimental verification, current models of turbulent mixing in these flows are seriously in question. Again, flow chemistry and viscous/inviscid flow interaction control the flow in, and hence the performance of, the nozzle. In these flows, relaminarization may also play an important, but poorly understood, role, which must be explored in experimental simulations.

III. Ground-Based Hypersonic Facilities

In 1985, NASA published an Aeronautical Facility Catalog[19,20] that updated and supplemented previous surveys of ground-test facilities. Reference 19 surveyed wind tunnels over the subsonic to hypersonic speed range, and Ref. 20 covered air-breathing propulsion and flight simulation facilities. Information on foreign (non-U.S.) facilities was also included. More recently, Wittliff[10] surveyed existing hypersonic ground-test facilities in North America. That survey, however, was restricted to hypersonic aerodynamic facilities. Concurrent with the survey in Ref. 10, Wendt reported a survey of European hypersonic wind tunnels[21] that also was restricted to aerodynamic facilities. Taken together, these three surveys provide a relatively complete description of the hypersonic ground-test facilities in North America and Europe as of 1987. In addition, Ref. 19 describes a number of propulsion test facilities that are applicable to hypersonic air-breathing vehicles. Not included in these surveys are the free-piston driver shock tunnels in Australia (or the shock tunnel at the University of Sheffield, in the United Kingdom). Also, a number of new facility developments have been undertaken since

1987. Therefore, an updated discussion of current facilities will be presented here that includes these facilities.

In this section, the discussion will center on how the various types of facilities can be used to obtain partial solutions to some of the many problems described in Secs. I and II. It is paramount to recognize that no existing facility is capable of full duplication of the flow about a large-scale model of a NASP vehicle for Mach numbers above 10. The following discussion will concern itself first with the roles of some of the more or less conventional hypersonic wind tunnels and various propulsion test facilities. Then, attention will be given to high-enthalpy short-duration facilities such as shock tunnels, including the free-piston driver type, and the expansion tube/tunnel. Finally, a scramjet combustion study performed in the Calspan 96- inch (2.4-m) hypersonic shock tunnel will be described in Sec. IIID as an illustration of the application of shock tunnels to this type of research.

A. Hypersonic Wind Tunnels

Even though the emphasis of this paper is on facility requirements for hypersonic propulsion system testing, one cannot overlook the fact that, for hypersonic vehicles like the NASP, the airframe and propulsion system cannot be treated separately. The forebody of the vehicle will provide the external compression associated with the inlet, whereas the afterbody is an integral part of the engine nozzle. The airframe and the engine are truly an integrated system. Nevertheless, it should be possible to study the forebody inlet flowfield separately from the propulsion system. Similarly, the flow over the complete vehicle without the propulsion system can be studied for the purpose of providing a data base for CFD code validation as well as the effects of vehicle configuration on aerodynamic forces. It is in this sense that many existing hypersonic wind tunnels can contribute to the NASP program even though they can not provide full duplication of flight conditions for a complete model.

For flight velocities up to about 20,000 fps (or 6.1 km/s), the flight path of a hypersonic air-breathing vehicle can be typified by a freestream dynamic pressure in the range 1000-2000 psf (48-96 kPa); see Fig. 1. Along such flight paths, real-gas effects behind a normal shock wave begin at a velocity of about 9000 fps (2.7 km/s), where 10% of the oxygen is dissociated. Initially, this phenomenon is confined to the flow near a blunt-nose cowl or leading edge. If the vehicle is slender and at a low angle of attack, much of the flow continues to behave as an ideal gas until higher flight velocities

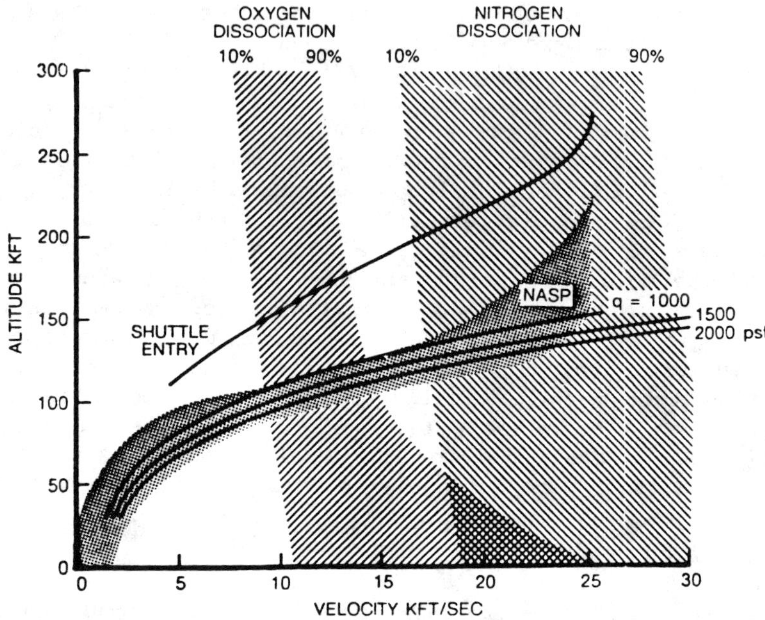

Fig. 1 NASP flight regime and normal shock wave conditions.

are attained. This situation is favorable to the use of the conventional hypersonic wind tunnels, which can produce flow Mach numbers up to 18 but at velocities not much greater than 7000 fps (2.1 km/s). Essentially, these are ideal-gas wind tunnels, a fact that is even more evident in a number of them where nitrogen or even helium is used as the test gas. These facilities can be useful in studying flow phenomena that can be characterized by only the Mach number, Reynolds number, specific heat ratio γ, and wall-to-inviscid flow temperature ratio. Two of the larger facilities in this category are the 5-foot-diameter (1.5-m) Naval Surface Weapons Center (NSWC) tunnel #9 using nitrogen, and the NASA ARC 3.5-ft. (1.1-m) hypersonic wind tunnel. They can produce relatively high Reynolds numbers in the Mach number range 5-14. Another important large hypersonic tunnel is the NASA LaRC 8-ft (2.4-m) high-temperature tunnel. Currently, this tunnel operates in the Mach number range 5.8-7.2, and the test medium is the product of combustion of methane and air. It has been used primarily for aerostructural heating research. This facility is to be upgraded with Mach 4 and 5 nozzles and a new heater with oxygen replenishment that will allow it to be used for propulsion testing as well as aerothermal and structural testing.

These hypersonic wind tunnels and others listed in Refs. 10 and 19 can be useful for aerodynamic and aerothermodynamic studies of the NASP and similar vehicles provided the Reynolds numbers are sufficiently close to the flight values.

B. Propulsion Test Facilities

There are four propulsion test facilities listed with the hypersonic wind tunnels of Ref. 19. These are the electric-arc-heated NASA LaRC scramjet test facility (M = 4.7 to 6.0), the GASL high temperature storage heater propulsion wind tunnel (M = 0.1 to 12), the GASL vitiated-air heater (VAH) propulsion wind tunnel (M = 2.7 to 8.0) and the GASL high mass flow storage heater propulsion wind tunnel (HPB) (M = 0.1 to 7.0). They have maximum total temperature capabilities of 4000, 2000, 4500 and 1700°R, respectively. Two other significant propulsion test facilities are the NASA LaRC Mach 4 scramjet test facility[22] and the AEDC APTU facility (M = 4.5)[19]. These latter two facilities have maximum total temperatures of 2200 and 2000°R, respectively, and have vitiated-air combustion heaters.

All of these facilities, plus a number of other hypersonic propulsion test facilities are listed in Ref. 23. Also included in that listing are storage heater test cells at Johns Hopkins/Applied Physics Laboratory (JHU/APL), vitiated-air combustion heated facilities at Marquardt, NWC China Lake, JHU/APL and Aerojet, and electric-arc-heated facilities at NASA ARC and JHU/APL.

As part of the NASP program, Marquardt[24] and Aerojet[25] have upgraded an existing facility and built a new facility, respectively, for scramjet engine testing at conditions simulating Mach 8 and 100,000-ft (30.5-km) altitude. These facilities will be able to test full-scale engine modules. At NASA LaRC, test chamber PSL-4 of the Propulsion Systems Laboratory has been modified to provide a Mach 3 to 6 direct-connect test capability[26]. These three new or upgraded facilities all use vitiated-air combustion heaters. In addition, NASA LaRC has proposed that the hypersonic tunnel facility (HTF) at Plum Brook be reactivated[26]. The HTF is a nonvitiated, blowdown, free-jet facility capable of testing large-scale engines at Mach numbers of 5,6, and 7.

All of the propulsion test facilities, plus the upgraded NASA LaRC 8-ft (2.4-m) high-temperature tunnel, are useful for studying scramjet engines or combustors up to full-scale hardware. However, the limitations of these facilities for NASP testing are that they operate at total temperatures too low to simulate flight velocities of 10,000 fps (3.05 km/s) or higher.

C. High-Enthalpy Facilities

For both aerodynamic and propulsion testing of hypersonic air-breathing vehicles, facilities having a stagnation enthalpy greater than 2000 BTU/lb (4.17 J/kg) are required in order to include real-gas effects. Another facility requirement is high stagnation pressure capability to simulate the altitudes or Reynolds numbers of interest. Current facilities having high total pressure capability are electric-arc-heated facilities, shock tunnels and the expansion tube/tunnel. Figure 2 is a simple sketch of several of these facilities.

Electric-Arc Facilities. Although there are a number of arc-heater facilities at McDonnell Aircraft Company, AEDC, and the various NASA laboratories (ARC, JSC, and LaRC), they are used primarily for materials testing. One exception is the NASA ARC aerodynamic heating facility[27], which can produce Mach numbers up to 15 with a total enthalpy up to 15,000 BTU/lb (3.49×10^7 J/kg) but has a maximum stagnation pressure of only 5 atm. The NASA LaRC scramjet test facility, mentioned in the previous section, has an electric-arc heater and is used

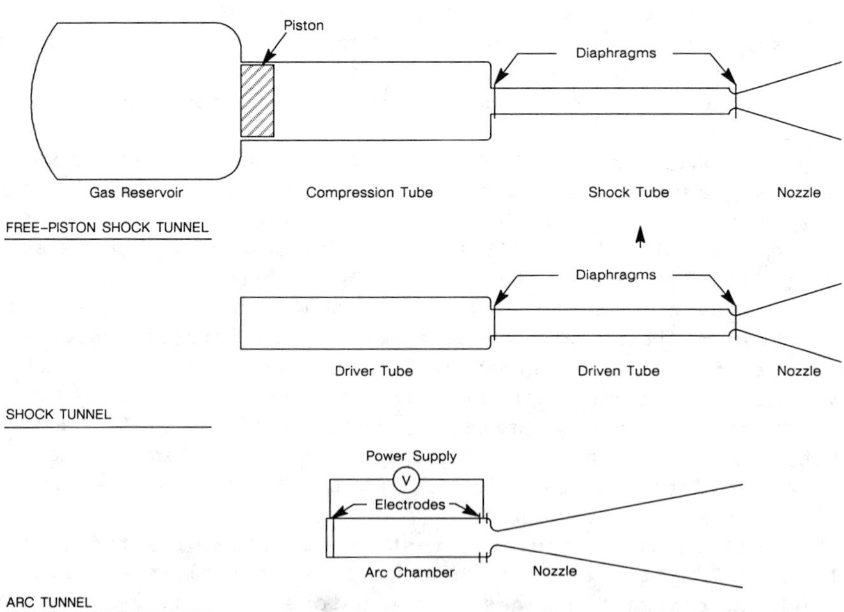

Fig. 2 NASP flight regime and normal shock-wave conditions.

for propulsion research. However, this facility has a maximum stagnation pressure of 40 atm and a stagnation enthalpy of about 1000 BTU/lb (2.33×10^6 J/kg).

Shock Tunnels. Although shock tunnels have steady flow test times measured in milliseconds, they are capable of operating at both high stagnation pressures and enthalpies. Shock tunnels can operate in either a reflected shock-wave mode or a nonreflected shock mode[28]. The currently active shock tunnels in the United States contributing to the NASP program are the 48- and 96-in. hypersonic shock tunnels (HST) at Calspan ATC, the reactivated 30-in. shock tunnel at Boeing, and the reactivated shock tunnel that is a part of the NASA ARC aerodynamic hypervelocity free-flight facility[29]. All three shock tunnels have reported maximum enthalpies of about 6000 BTU/lb (1.4×10^7 J/kg). The Calspan 96-in. HST and the Boeing shock tunnel have maximum stagnation pressures of 1350 and 450 atm, respectively. The Calspan 48-in. HST, which is not rated for using hydrogen, currently has a maximum stagnation pressure of 270 atm and stagnation enthalpy of 1850 BTU/lb (4.3×10^6 J/kg).

The shock tunnel at the University of Sheffield, in the United Kingdom has been used for turbulent mixing[30] and supersonic combustion[31] research. This combustion-driven shock tunnel has been used to produce airflows with a stagnation condition of 10,800°R and 200 atm corresponding to a total enthalpy of 4450 BTU/lb (1.03×10^7 J/kg).

In Australia, there are four free-piston shock tunnels that were developed by Stalker[32] at the Australian National University and the University of Queensland. In these shock tunnels, a free piston, driven by high-pressure air, is used to compress and heat the helium driver gas to 8100°R (4500°K). Mach 22 shock waves can be generated in the driven tube. These shock tunnels can be operated in either reflected shock or nonreflected shock mode. With such strong shock waves in the driven tube, total enthalpies of 10,000 BTU/lb (2.33×10^7 J/kg) have been achieved in the reflected shock mode. At these enthalpies, the useful test time is on the order of 100 μs or less. These free piston shock tunnels are being used to study real gas and hydrogen combustion[33] phenomena. Currently, there are two free-piston shock tunnels under construction in the United States in support of the NASP program. The Rocketdyne Division of Rockwell International is building a large tunnel that will have a 72-inch diameter test section[34]. Operating in the reflected shock mode, this facility is designed to produce stagnation pressures of about 1500 atm at total enthalpies of 13,000 BTU/lb (3.02×10^7 J/kg). This would provide flow velocities up to 25,000 fps. At these very high enthalpies, the duration of uniform flow (test time) becomes a very important consideration. There are some measurements that can be performed within 50–100 μs, but there are others that cannot. This point will be addressed more fully in Sec. IVA. A smaller free-piston shock tunnel is being built by GALCIT for real-gas investigations.

Expansion Tube/Tunnel. The NASA LaRC expansion tube/tunnel[35] was deactivated in 1983 and was moved to GASL as part of the NASP program, where it was reactivated in 1989 as the Hypersonic Pulse (HYPULSE) facility[36]. A critical examination of its performance at LaRC has been described by Miller[37]. He reported airflow velocities up to 18,000 fps (total enthalpy of 6900 BTU/lb (1.60 x 10^7 J/kg) and uniform flow test times of approximately 200 μs. At LaRC, this facility was used for aerodynamic studies. At GASL, it is planned to use this facility to study scramjet combustion. The early experiments reported by GASL were performed with a helium driver gas at room temperature. This restricts the performance to lower total enthalpies than those achieved at LaRC. However, it is planned to extend the performance capability of the facility by the addition of a free-piston driver[36] that could heat the helium driver gas to 4000 K.

D. Scramjet Combustion Studies in Shock Tunnels

Swithenbank and his colleagues were studying supersonic combustion in a shock tunnel in the late 1960's.[31,38] These were direct-connect experiments wherein hydrogen was injected into a Mach 3 or 4 airstream. As a part of the NASP program, a series of semi-direct-connect tests have been made in the Calspan 96-in. shock tunnel with a scramjet combustor model designed and fabricated by GASL. Typically, the shock tunnel was operated at a total pressure of 1000 atm and a stagnation temperature of over 6000°R (total enthalpy of approximately 2000 BTU/lb (4.65 x 10^7 J/kg)). The tunnel freestream conditions were M = 9.45, static pressure of 0.26 psia, and static temperature of 428°R. The model inlet ramp was mounted at an angle of 20 deg to the freestream flow so that the combustor entrance conditions were typically Mach 4.3, with a static pressure of 6.5 psia and a static temperature of 1800°R. These conditions closely simulated Mach 10 flight at about 100,000-ft altitude. The design of the GASL model allowed for variation in the combustor height and for installation of a wide variety of hydrogen injectors using hydrogen. The results clearly demonstrated the applicability of the shock tunnel to scramjet combustion studies in that combustion was achieved and the effectiveness of various injectors was measured.

IV. Real Gas Effects and Facility Limitations on the Simulation of Hypersonic Flows

A. Effects of Test Flow Nonequilibrium and Facility Limitations on the Simulation of Hypersonic Flowfields

The development of ground-test facilities to simulate the flow about hypersonic vehicles is very challenging because of the high total enthalpies required. This challenge is being addressed again after many years of inactivity in response to the NASP effort.

The stagnation enthalpies encountered in flight at hypersonic speeds result in flowfield temperatures high enough to dissociate the chemical species in air (see Fig. 1). The reservoir state in a test facility that then expands the air to high velocity is, of course, at comparable enthalpies and so, in the reservoir, the test gas is also dissociated. In the subsequent expansion to hypersonic speeds, the flow can depart from thermal and chemical equilibrium, and the conditions of the gas in the freestream flow of the test section differ from those of the freestream gas for flight in the atmosphere. The static pressures and temperatures differ, and the air contains oxygen atoms and nitric oxide molecules. The latter species is relatively easily ionized, and so electrons are also present. At enthalpies corresponding to flight velocities above orbital speed, nitrogen would also dissociate, and other ionized species would appear in significant concentration.

In addition to the very high stagnation enthalpies, extreme reservoir pressures are needed to simulate hypersonic flows. Indeed, the reservoir pressures associated with the high degrees of expansion in hypersonic tunnels place a more stringent limitation on the flight conditions that can be simulated than does the total enthalpy. The lack of duplication of flight conditions due to facility limitations is coupled to any lack of duplication due to nonequilibrium effects, simply because the degree of nonequilibrium varies with reservoir conditions. Both nonequilibrium and facility limitations lead to relaxing the requirements for full duplication of the flight conditions. In this section, the impact of nonequilibrium on the lack of full simulation is emphasized, but both factors are addressed.

The hypersonic flight regimes of several vehicles of current interest are shown in Fig. 3. The altitude-velocity range shown there includes Shuttle and ballistic missile re-entry flight paths that bracket the flight paths of interceptor missiles. The aero-assisted orbital transfer vehicle (AOTV) trajectories and a representative flight envelope for NASP, which

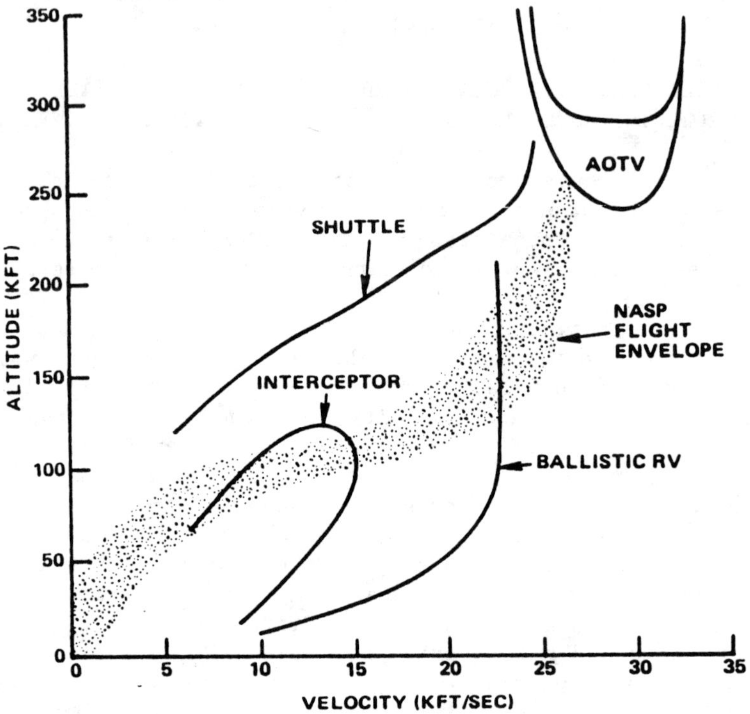

Fig. 3 Altitude–Velocity map of flight regime of interest.

falls within the flight corridor for air-breathing vehicles, also are shown.

In order to relate the flight regimes shown in Fig. 3 to hypersonic facility requirements, it is assumed, for the moment, that the test flow can be expanded from a reservoir in equilibrium. By using the thermodynamic state of the atmosphere at any altitude and the desired flow velocity, the equilibrium properties of air can be used to define the reservoir conditions from which expansion would produce a test flow duplicating the chosen flight conditions. The equilibrium reservoir conditions corresponding to a portion of the flight regime of interest are shown in Fig. 4. There the temperatures and pressures are shown for a stagnated, equilibrium reservoir condition for each altitude and velocity. Reservoir pressures greater than 4000 atm are not shown because such conditions are well beyond the capability of existing facilities and, furthermore, represent thermodynamic states for which the properties of air are subject to considerable departure from the ideal-gas equation of state.

A curve is shown on Fig. 4 that represents the extension of the full flow duplication region that is possible by using a nonreflected shock tunnel, i.e., by not stagnating the test gas. It should be noted that such operation entails a significant decrease in test time, which is already very limited at these high flow enthalpies.

In order to attain full duplication at lower altitudes or at higher flow velocities than the nonreflected shock-tunnel curve, some type of advanced facility concept, such as an MHD-augmented shock-tunnel facility would be required. The discussion in this section draws heavily on a study of similitudes for inviscid and viscous hypersonic flows that was done for such a facility, the high-rho facility, over 20 years ago[39]. In that study, the full duplication boundary shown in Fig. 4 was assumed. With this concept for further accelerating the flow over and above the velocity achievable by a nozzle

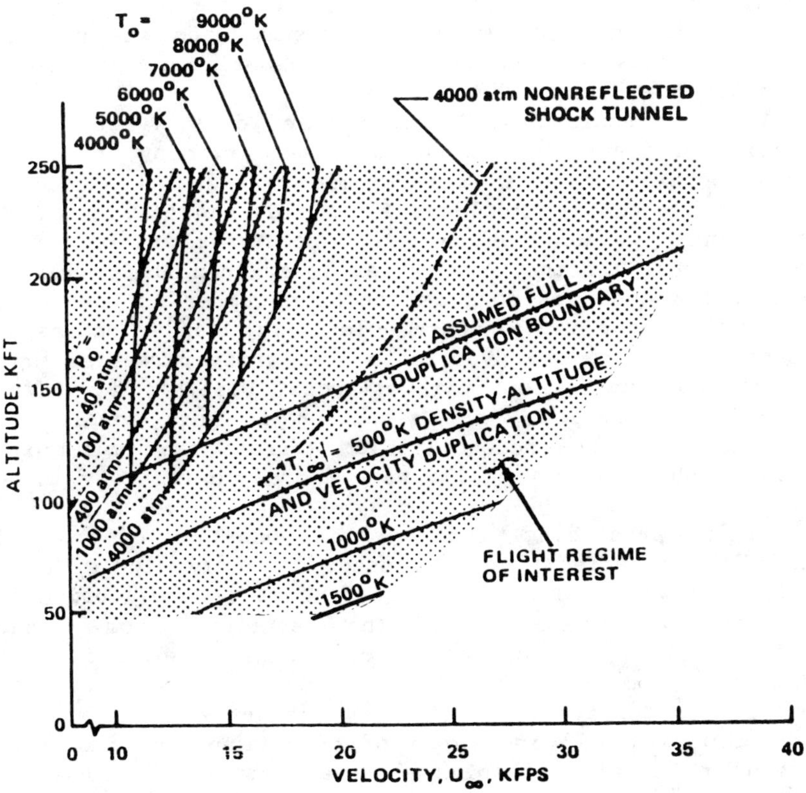

Fig. 4 Altitude velocity map showing hypersonic tunnel performance.

expansion from equilibrium stagnation conditions, flight conditions at higher velocities and lower altitudes can be duplicated. In other words, test flows can be produced that correspond to effectively higher reservoir temperatures and pressures.

It is possible to relax the full duplication requirement and to duplicate the flow velocity and one of the thermodynamic variables corresponding to atmospheric conditions at a given altitude. Often, the flight Mach number is duplicated in hypersonic flows by expanding the test flow to lower velocity and static temperature. Alternatively, for stagnation pressures corresponding to altitudes below the full duplication boundary, if the flow is expanded from an equilibrium reservoir to the desired velocity and the density corresponding to a given altitude, then, in general, the static temperature will be higher than the static temperature at that altitude. The curves shown in Fig. 4 give the static temperatures that result from expansion to the plotted velocities and the densities at the associated altitude. The curves for density-altitude duplication shown in Fig. 4 result from the effective stagnation pressures and temperatures associated with the full duplication capability shown there.

For hypersonic flows, velocity and density-altitude duplication preserve a match in the total enthalpy, or stagnation temperature, and in the dynamic pressure. Consequently, for blunt bodies, the real-gas flowfields can be simulated to a good approximation. Although the freestream static temperature is higher and the Mach number lower than the corresponding flight values, as long as the Mach number is high enough (>5), Mach number independence holds. For slender bodies with thickness ratio τ, the hypersonic similarity parameter $M\tau$ can be duplicated under some circumstances. However, for blunted slender configurations, the situation is more complex. The issues of test flow nonequilibrium effects on simulation of the flows about such vehicles were addressed recently in Ref. 17.

Before proceeding, the important question of test flow nonequilibrium effects is addressed. In early studies of hypersonic flows, several investigations of nonequilibrium nozzle flows were reported.[40–42] The departure from chemical equilibrium in hypersonic expansions of air results in a test gas that is composed not only of molecular oxygen and nitrogen, but also of atomic oxygen and nitrogen as well as nitric oxide (NO). Over a wide range of conditions, roughly corresponding to total enthalpies associated with orbital velocities and below, N_2 is essentially equilibrated in nozzle expansions. Even though the N concentration is negligible in the element balance, it must be included in the chemical model because of its role in the shuffle reactions, which control the NO concentration:

$$O + N_2 \leftrightarrows NO + N \tag{1}$$

$$N + O_2 \leftrightarrows NO + O \tag{2}$$

Over the same range of flow velocities, NO^+ is the dominant source of electrons in air.

The results of a sample calculation of the quasi-one-dimensional expansion of air are shown in Fig. 5. Fig. 5a illus-

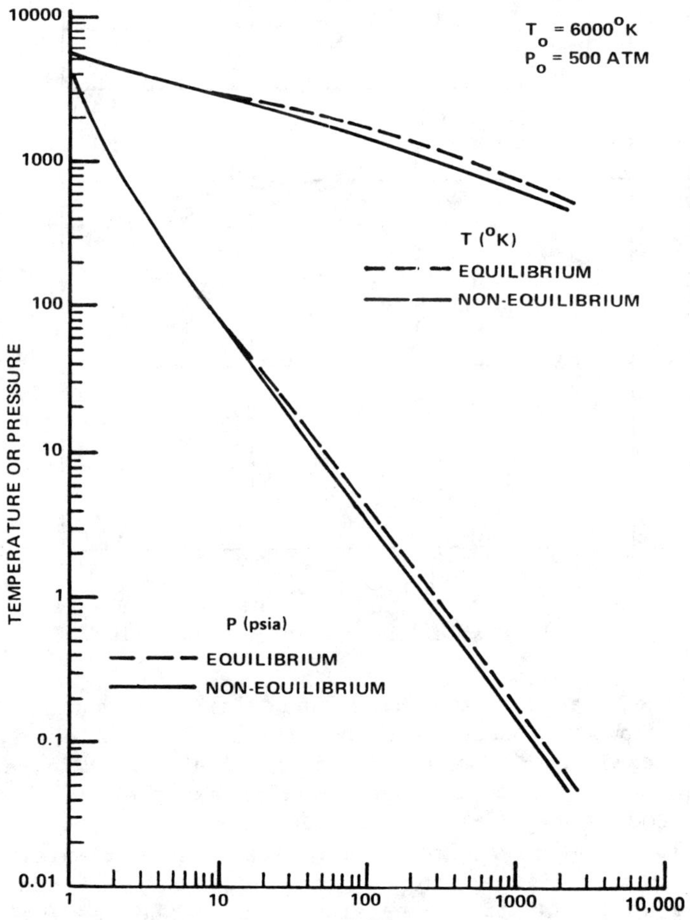

Fig. 5a Nozzle flow gas dynamic properties.

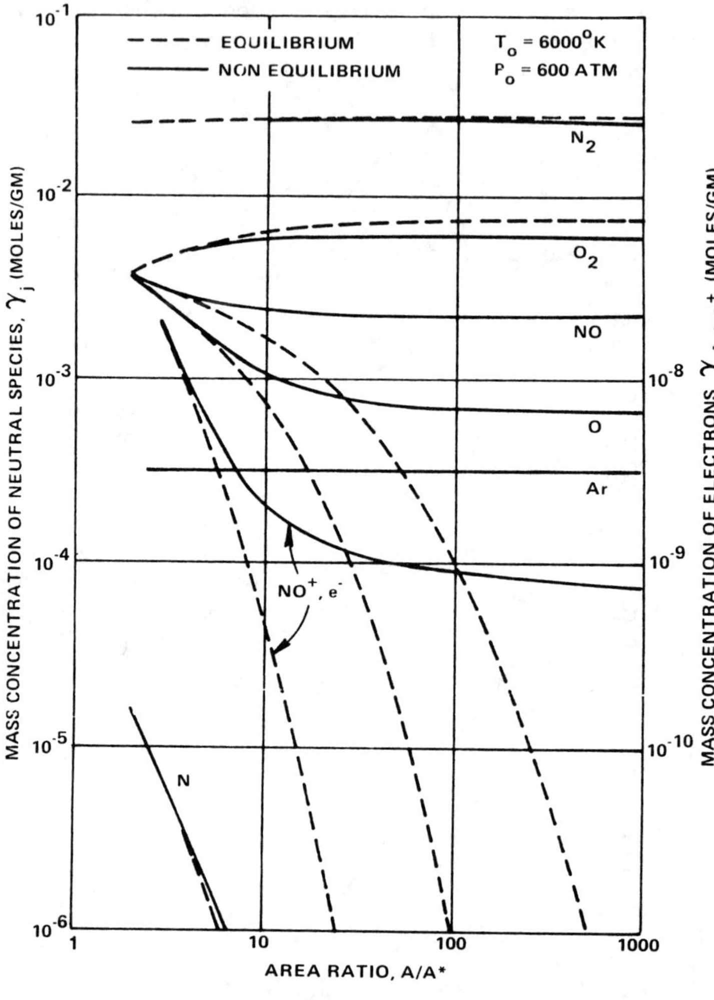

Fig. 5b Nozzle flow species concentration.

trates the static temperature and static pressure, whereas Fig. 5b shows the species concentrations for an expansion in the "D" nozzle of Calspan's 96-in. shock-tunnel. The reservoir or reflected shock conditions for this example are a temperature of 6000 K and a pressure of 500 atm, which produce a test flow velocity at high expansion ratios of approximately 14,000 fps. From the figure, the expected lag can be seen in the decrease in the static temperature due to the chemical energy not returned to the nonequilibrium flow. The static pressure also falls below the value for an equilibrium expansion; the density is affected very little by the nonequilibrium effects.

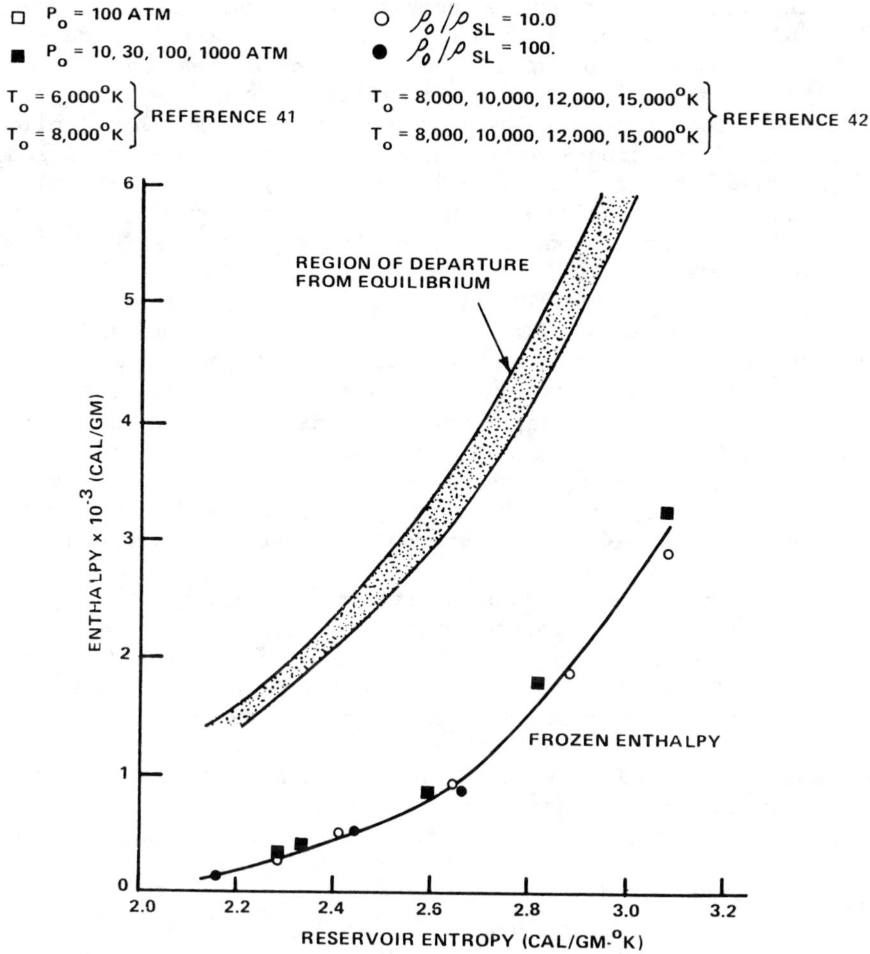

Fig. 6 Correlation of frozen enthalph data for nonequilibrium nozzle flow solutions.

The species concentrations are shown in mass concentrations units (moles/g). This removes the density dependence from the distributions and highlights the freezing process, i.e., when frozen, γ_j = constant. For this case, the NO concentration is about 5% of the total and exceeds the "frozen" O-atom concentration. The results of calculations similar to this example are used to examine the effects of test flow nonequilibrium on the interpretation of experiments in hypersonic facilities. In the remainder of this section, some generally useful results in correlating nonequilibrium effects are briefly reviewed.

In Refs. 41 and 42, a number of nonequilibrium nozzle flow calculations were made for expansions of air from reservoir temperatures up to 15,000°K and densities up to 100 amagats (standard atmosphere densities, ϱ/ϱ_0). The calculations were done for an axisymmetric nozzle having a hyperbolic variation of the cross-section area with axial distance. Based on the sudden freezing approximations for hypersonic expansion of simpler gases, it was found that the frozen enthalpy in nonequilibrium expansions of air could be correlated with the reservoir entropy. When the frozen enthalpy determined from the above computations for air was plotted against reservoir entropy, the results collapse to a single curve, as shown in Fig. 6. The correlation with reservoir entropy is a consequence of the small entropy production due to chemical nonequilibrium effects. The limiting cases of frozen and equilibrium flows are both isentropic. The entropy values in the correlation can then be associated with that for the reservoir conditions and can be obtained from an equilibrium computation of the reservoir entropy or a Mollier diagram for air.

The frozen species concentrations in hypersonic expansions of air also correlate with reservoir entropy. Fig. 7, which was reported in Ref. 43, is based on the solutions reported in Refs. 41 and 42. The correlation with reservoir entropy is valid only for a specific nozzle geometry. However, the behavior of the results for the hyperbolic nozzle are representative and so are useful in defining the expected trends in the results for other facilities.

The nonequilibrium effect on velocity at high area ratios (>1000) can be related to the frozen enthalpy by

$$U_{neq}/U_{eq} = (1 - H_f/H_0)^{0.5}$$

With the preceding correlations and the facility performance assumed in defining the full flight duplication region of Fig. 4, it was concluded[39] that nonequilibrium effects on flow velocity and concentration are negligible outside that region of facility performance. Whereas the simplification that nonequilibrium effects are confined to the velocity-altitude region in which full duplication of flight conditions would otherwise be possible is convenient, that situation may be optimistic because of the performance of advanced facilities considered in Ref. 39. In general, the situation is more complex, and both facility limitations and nonequilibrium effects must be considered in the simulation of hypersonic flows. In some cases, special similitudes like binary scaling[44] may be appealed to in order to achieve some flexibility in facility

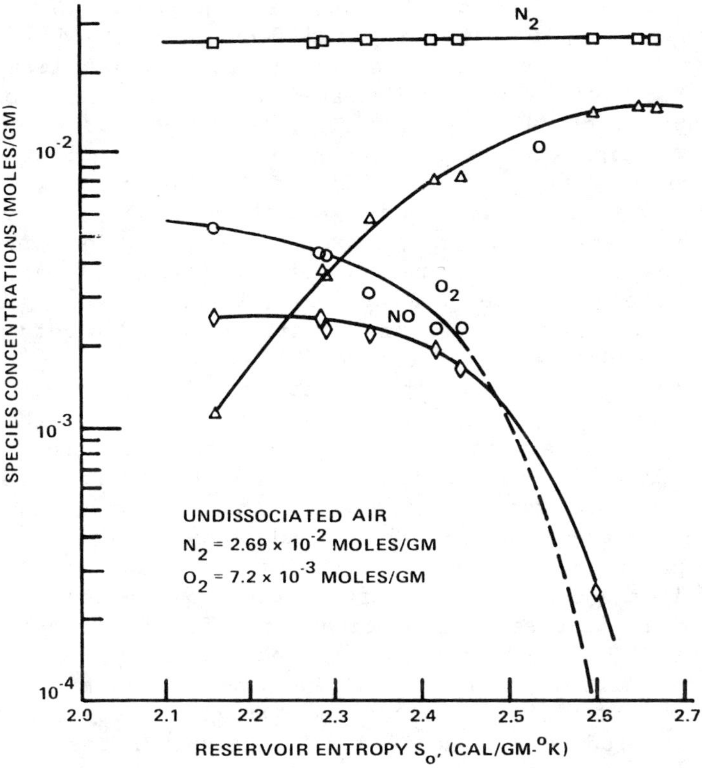

Fig. 7 Correlations of nozzle flow results for frozen species concentrations.

requirements and model scale. However, in other instances, it is possible to operate the test facility at high enough enthalpy to produce the physical phenomena being studied and to use a judicious combination of measurements and models that contain the appropriate physics to interpret the experiment.[17]

In Ref. 17, a specific example of simulating the ionized flowfields about blunt-nosed slender bodies was presented. This example was used to demonstrate that electron density profiles could be produced in a model flowfield, which were essentially the same as those in the corresponding flight vehicle flowfield despite the lack of full duplication of the flight conditions. The successful reproduction of flight vehicle electron density profiles was shown to be critically dependent on model scales. The example chosen was at a flight or test flow velocity of 14,000 fps in the Calspan 96-in. shock tunnel and a simulated density altitude of about 150 kft. These test conditions correspond to the reservoir conditions (T_0 = 6000°K, p_0 = 500 atm) for which the computed nonequilibrium effects on the nozzle flow were presented earlier. In the tunnel case, the velocity and density altitude were the same as the flight case, but the Mach number was lower, and the nonequilibrium nozzle expansion resulted in non-zero freestream electron densities. Despite these differences between the flight and tunnel cases, the electron density profiles could be simulated at relevant test model scales. Such an approach is being followed in current and previous experimental programs in which plasma interactions in hypersonic flowfields have been measured. It should be re-emphasized that Langmuir probe measurements are an integral part of the experiments in order to confirm that the expected electron density flowfields are indeed produced.

Comparing the NASP flight corridor shown in Fig. 3 to the full-duplication boundary of Fig. 4, it can be seen that, for flow velocities between 10 and 20 kfps, this facility limitation boundary lies within the altitude velocity region of the flight corridor. Recall that, along this boundary, expansion from the corresponding reservoir conditions is subject to negligible nonequilibrium effects (<1% of the total enthalpy frozen out in the expansion). Consequently, for the hypervelocity wind tunnels that seek to duplicate the freestream conditions along the NASP flight corridor, the nonequilibrium effects are not large. Although measurable, the nonequilibrium effects in the flow properties will not limit the capability of the facility to produce the desired test velocity. The nonequilibrium effects on freestream temperatures and species concentrations are examined in the subsequent subsection.

B. Real-Gas Effects on the Simulation of Scramjet Flows

Aside from some specific examples, the discussion of facility limitations and nonequilibrium effects applies to all hypervelocity tunnels, regardless of whether the reservoir conditions are produced by electrical resistance heating, combustion, arc-discharge, or shock-wave compression. In this subsection, the nonequilibrium effects on shock-tunnel expansions are examined for selected examples at which NASP-related experiments are under way and, in some cases, have already been completed. In particular, real-gas effects on the test flow at which scramjet combustor experiments being done in the 96-in. shock tunnel at Calspan are evaluated, and finite-reaction-rate nozzle expansion computations follow for this shock tunnel with the $D^* = 1.6$ in. throat configuration.

As a rule of thumb, the 96-in. HST facility is capable of fully duplicating the freestream conditions corresponding to a flight Mach number of 10 at an altitude of 100,000 ft with a test time of about 1 ms. These conditions, which correspond to freestream velocity of nearly 10,000 fps, are those for which the scramjet combustor experiments have been run. Here we concentrate on the real-gas effects on the shock-tunnel test flow conditions, which correspond to reservoir conditions of T_0 = 3500°K and P_0 = 1000 atm. Real-gas effects on the expansion from a somewhat higher reservoir condition (T_0 = 4000°K) are also evaluated, not only to illustrate the sensitivity of the results to total temperature but also because an upgrade of the shock-tunnel capability at Calspan that will produce somewhat higher enthalpies is in progress. A more important aspect of this upgrade, which is designed around an existing 6-in. diameter shock tube, is that the test time can be nearly doubled over that available in the 96-in. shock tunnel.

At reservoir conditions of temperatures between 3500 and 4000°K and pressures of 1000 atm, real-gas effects in air include both dissociation effects and intermolecular force effects. At these high pressures and temperatures, the dissociated products of air can no longer be described by an ideal gas equation of state. The departure from an ideal gas due to the pair-wise interactions between the constituent atoms and molecules must be accounted for in evaluating the properties of the gas mixture. In discussions of hypersonic facilities, nonequilibrium effects are often cited as limiting the ability to define the test flow conditions. However, at these reservoir density levels, intermolecular forces can be even more important in the initial part of the nozzle expansion and are seldom mentioned. Indeed, the effect of intermolecular forces on the compressibility in the reservoir is about 10%, whereas the dissociation effects on this quantity are only about 1%. The properties of a chemical equilibrium expansion for the case T_0 = 4000°K and p_0 = 1000 atm are tabulated in Table 1. These values were computed using the tables of equilibrium properties of air presented in Ref. 45. The compressibility factors in the equation of state, as defined in Ref. 45, are Z and Z*, where

$$P = Z\varrho\frac{R_0}{M}T = ZZ^*\varrho\frac{R_0}{M_0}T$$

where p, ρ, and T are the mixture pressure, density and temperature, respectively, M_0 is the undissociated molecular weight, and M is the molecular weight of the dissociated mixture. Thus, Z represents the contribution of the intermolecular force effects to the compressibility and Z* the contribution of dissociation effects.

Table 1 Properties of equilibrium expansion of air (T_0 = 4000°K, P_0 = 1000 atm)

	Reservoir	Throat	A/A^a = 10	A/A^a = 100	A/A^a = 1000
T, k	4000.0	3598.9	1606.3	734.8	247.2
P, atm	976.8	535.8	10.08	0.4715	0.0165
ρ, kq/m3	78.59	49.51	2.197	0.2131	0.0195
Z	1.0956	1.0600	1.0023	1.0002	1.000
Z^a	1.0086	1.0051	1.0000	1.0000	1.0000
u, ft/s	—	3895	8855	10020	10498
X_{NO}	0.0988	0.0782	0.0022	—	—
X_O	0.0181	0.0109	0.0000	—	—
X_{O_2}	0.1481	0.1630	0.2084	—	—
X_{N_2}	0.7241	0.7373	0.7795	—	—
$(\Sigma v)_{N_2}/H_o$	0.0981	0.0852	0.0196	—	—

a Conditions at sonic throat

From the results for the equilibrium expansion shown in Table 1, it can be seen that, before the flow has expanded to an area ratio of 10 in the supersonic section of the nozzle, both contributions to the compressibility have decreased to less than 1%. Consequently, the dominant real-gas effects present in the reservoir and the initial portion of the nozzle expansion are the intermolecular force effects on the equations of state for air. The practical implications of this conclusion is that such effects must be included when inferring the test flow properties. For the shock-tunnel example, the reservoir conditions often are inferred from measurements of the wave speed in the shock tube and of the pressure in the reflected shock reservoir. Computation of the remaining reservoir properties can be in error by failing to account for those real-gas effects.

For the total or reservoir temperatures in the range that can be obtained in shock tunnels, the dominant internal modes of excitation and chemical composition effects include: vibrational excitation of nitrogen molecules (N_2) and the formation of oxygen atoms (O) and nitric oxide (NO) from the dissociation of N_2 and O_2. In hypersonic nozzle expansions from the reservoir conditions being considered here, these quantities remain very close to their equilibrium values into the supersonic portion of the nozzle flow. Computations of finite-rate effects on the latter stages of the air expansion from the reservoir conditions of T_0 = 3500°K and T_0 = 4000°K at p_0 = 1000 atm have been made. In the ensuing discussion, the results of nozzle computations to obtain the test section properties corresponding to these shock tunnel conditions are described in detail.

The potential effects of nozzle nonequilibrium on hypersonic testing have been the subject of extensive earlier studies and were reviewed in a recent paper by the authors.[17] In order to successfully carry out the nozzle-flow computations from these reservoir conditions, a scheme for the numerical solution of stiff equations is required. The stiffness of the chemical species rate equations for high-pressure, moderately high-temperature air results from the behavior of the so-called shuffle reactions, Eqs. (1) and (2). These reactions have much larger rates than the net rate of production of NO. Reaction (3) tends to be dominated by the reverse rate and reaction (4) by the forward rate, so that NO is consumed by reaction (3) and produced by reaction (4), resulting in nearly a stationary state or constant concentration as follows:

$$N + NO \rightarrow N_2 + O \tag{3}$$

$$N + O_2 \rightarrow NO + O \tag{4}$$

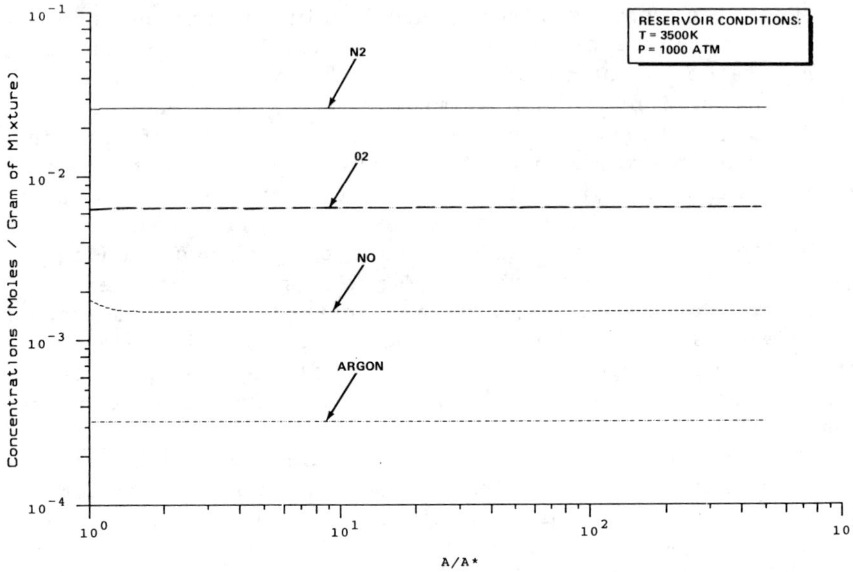

Fig. 8a Nozzle species concentrations versus area ratio.

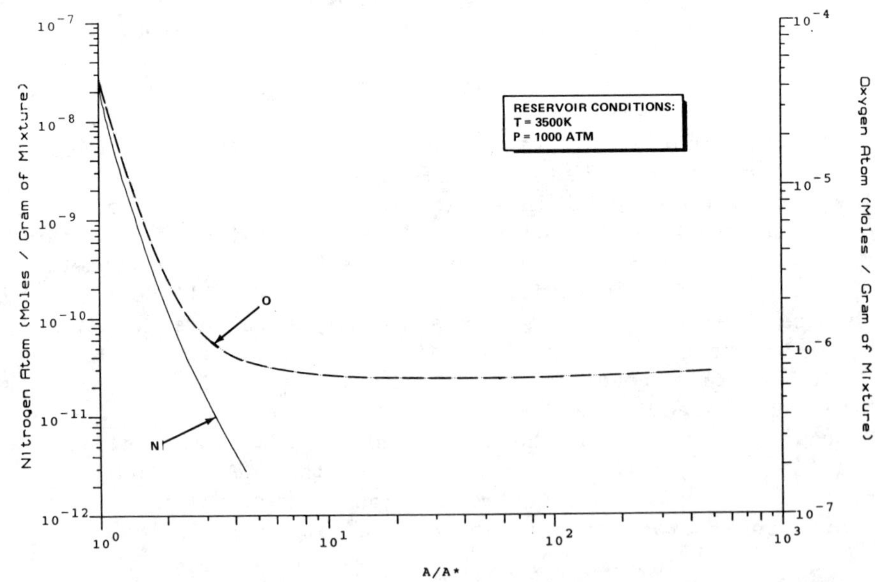

Fig. 8b Nozzle species concentrations versus area ratio.

Nitrogen atoms, while a trace species at the temperatures of interest here, are consumed by both reactions, thereby decreasing the N concentration even further. The O atoms are produced by both reactions but are consumed by the three-body recombination process until collisional freezing occurs.

The results of the nozzle calculations for the reservoir conditions of 3500°K, 1000 atm indicate that the freestream conditions are essentially ideal-gas, undissociated air in the shock-tunnel test section. The neutral species concentration variations along the nozzle expansion are plotted in Figs. 8a and 8b. There, the behavior of the neutral species chemistry is graphically illustrated, with the NO concentration reaching a nearly constant, stationary-state value not far below the reservoir concentration. The computed concentration of NO in the test section is small, being less than 5 mole %. As illustrated on Fig. 8b, the frozen O-atom concentration is much lower than the NO concentration which, with the N-atom concentration (which is extremely small to begin with), is consumed very rapidly in the expansion at almost the rate corresponding to chemical equilibrium. At the 4000°K reservoir condition, the test section NO concentration is somewhat lower than that for the 3500°K case, being about 2.5%.

The N_2 vibrational temperature defined by the ratio of the first vibrational level is plotted, along with the translational/rotational temperature in Fig. 9. The lower vibrational levels are essentially in a Boltzmann distribution at the frozen vibrational temperature of 1800°K indicated in the figure. This vibrational temperature value is well above the translational temperature, but the fraction of flow enthalpy tied up in this internal degree of freedom is only about 2% at this stage in the expansion (see Table 1).

There are two aspects of the departure of the flowfield from undissociated, low-temperature air in the test section of the shock tunnel. The first is the species concentrations being other than N_2 and O_2. In a previous paper, Ref. 17, we reviewed the potential effect of NO on scramjet combustor testing and, based on that study, the above-mentioned amount of NO in the test flow was concluded to have no effect on the supersonic combustor studies. The use of the shock tunnel for external flow studies was discussed in Sec. III and the effects of test flow nonequilibrium on plasma studies were addressed in Sec. IVA. Shock-tunnel studies can provide an important source of aerothermal data on NASP configurations as well. Here we also note that the computed effects of nonequilibrium on the test section Mach number based on the frozen sound speed is very small.

Unless vehicle scale and all the freestream properties are duplicated, then the flowfield about complex configura-

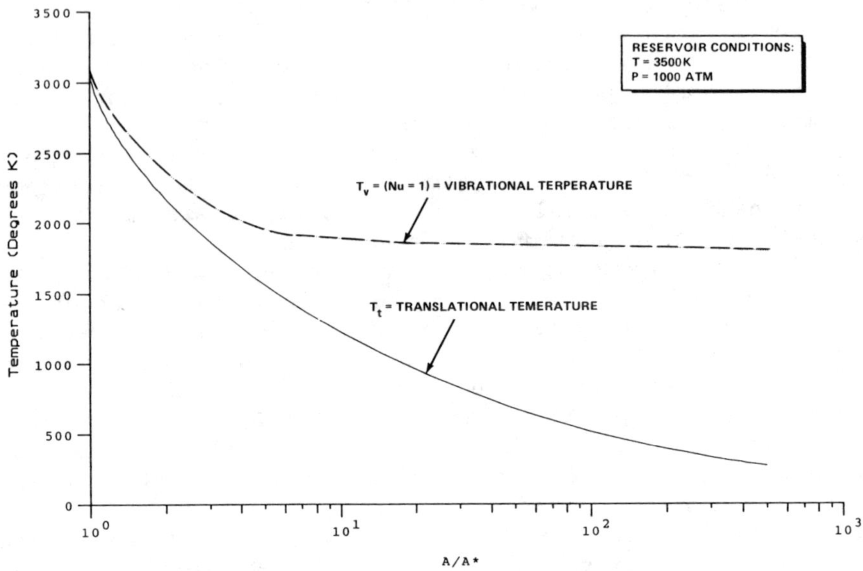

Fig. 9 Nozzle temperature distribution versus area ratio.

tions will differ to some degree from the flight flowfields. The discussions here demonstrate that facility limitations and nonequilibrium effects prevent full duplication. However, meaningful measurements can still be made that provide validation data for CFD codes and that can guide the design and reduce the risk of flight experiments. The approach taken here is that, if carefully designed, ground-test experiments can simulate some of the major physical effects of nonequilibrium hypersonic flowfields. Such experiments require predictive techniques to design the experiments and specific flowfield measurements to verify that the design conditions of the experiment have been met. Measurements of freestream conditions, including the nonequilibrium vibrational temperatures and species concentrations, should also be part of such measurements.

Although short-duration test facilities can make significant contributions to NASP vehicle design, a major question associated with such facilities is the issue of test time. This facility limitation is taken up in the next subsection.

C. Test-Time and Flow-Establishment Time in Short-Duration Facilities

The question of test time in short-duration facilities is an important one that must be addressed. The available test time must be significantly greater than the response time of the instrumentation, the time for establishment of fully developed flow and, in the case of combustion, the time for mixing and combustion to go to completion. The time required to establish fully developed flow will depend on the model configuration and the flow variables (Mach number, Reynolds number). For model configurations that do not require establishment of separated-flow regions, the flow establishment time is very short, corresponding to a few flow lengths for the flow conditions associated with these hypersonic flows. A significant portion of this flow establishment time comes from the nozzle starting time, which is not useful for testing anyway. The time required for mixing and combustion to go to completion depends on the model configuration, the injection configuration, the flow variables, and the combustion gases. In the case of hydrogen/air combustion (where local separation occurs near the injector), it has been shown experimentally that the time for mixing and combustion to be established for the flow conditions used in this chapter and described in Ref. 17 is on the order of 50–80 μm. The test time required to perform the combustion experiments is at least 80 μm. For the model size used in the experiments referred to earlier, a test time corresponding to several flow lengths was felt to be necessary, which translates into a required test time on the order of 500–800 μm.

In Ref. 46, the subjects of available test time, flow establishment time, and information provided by different diagnostic tools in short-duration facilities have been addressed. The facilities we are talking about here are generally composed of a driver tube that contains the driver gas, another tube (the driven-gas tube) in which the test gas is shocked by the driver gas, and a test section in which the shock-processed gas is directed. It is very important that the quality of the flow be completely understood and well documented for each of these subsections of the device, especially in the driven-gas tube and the test section. In the strict sense, the test time is defined as the time during which the shock-processed gas is uncontaminated by driver gas. For the short-duration facilities being discussed in this chapter, it is not always necessary to demand absolutely uncontaminated test gas because the total pressure and the velocity in the expanding gas stream are not influenced to a significant degree by small amounts of driver-gas contaminant. The type of measurements and the amount of driver-gas contamination permitted in the test-gas slug can be divided into three general groups: 1) force and moment measurements, for which a significant amount of driver-gas contamination can be permitted; 2) heat-transfer and pressure measurements associated with aerothermal studies, which can tolerate a reasonable amount of driver-gas contamination before the measurements are influenced; and 3) spectroscopic, chemical-kinetic, electromagnetic, and combustion measurements, which can tolerate very little driver-gas contamination in the test-gas slug because even a small amount of driver-gas contamination has a significant influence on temperature. Supersonic combustion studies would fall in group 3 because such studies would obviously involve the kinetics of the fuel-air mixture as well as spectroscopic diagnostics.

Regardless of whether one is interested in group 1, 2, or 3, it is very important to determine experimentally the available test time in the driven tube, the starting process in the test section, and the flow establishment time for the particular model, even though such experiments are expensive and difficult. Ref. 46 presents the details of such an experimental determination that was performed for one of the Calspan short-duration facilities and illustrates how the driven tube and the test section wave diagrams can be constructed from experimental data; the results of this construction were then compared to the predictions obtained using available analytical tools. Further, comparisons are also presented of the results obtained using several different diagnostic tools, all looking at the same physical event. It is demonstrated that stagnation-point heat transfer and pitot-pressure measurements are both very poor indicators of test time. However, both techniques are reasonable indicators of the starting process in the test section flow. If one is interested in performing spectroscopic, chemical-kinetic, electromagnetic, or combustion measurements, then the requirements on the purity of the test gas (very little driver-gas contamination can be permitted) are severe. It is demonstrated in Ref. 45 that, for the case of driven-tube Mach number in excess of about 9, the test section test time is approximately equal to the test time behind the incident shock in the driven tube. The implication of this is that the tailored interface or equilibrium interface concept[47-49] that is often assumed to be applicable for force and moment measurements or heat-transfer and pressure measurements may be compromised by real-gas effects in this case. It is important to look carefully at what one is doing when making this assumption.

The obvious starting point for the type of measurement program described in the previous paragraph is to utilize the many analytical tools available to estimate what one should expect. Calculations of the ideal wave diagram for the driven-gas tube can be initiated using the material presented in either Ref. 50 or 51. Unfortunately, for the high-enthalpy conditions generally of interest in hypersonic aerodynamics, the gas does not always behave in an ideal way, and corrections must be made to the test time estimates for both a laminar and a turbulent boundary layer using the techniques presented by Mirels[52, 53]. An estimate of whether the driven-tube boundary layer is laminar or turbulent can be obtained using the result presented in Ref. 54. Several authors[55–57] have investigated the driven-tube test time and compared the result with the predictions of Refs. 52 and 53. If the facility operates in a mode for which the reflected shock from the tube endwall is used to increase the temperature and pressure of the reservoir of gas to be directed into the test section, then the physics of the flow in this reflected-shock region becomes important. Many authors[56–59] have studied the complex flow in the reflected-shock region.

For these short-duration facilities, the test section is generally evacuated in order to minimize the starting time. Many authors have studied the unsteady starting process, but the work of Smith[60, 61] is particularly useful for estimating the time required to achieve uniform flow at any axial location in the expanding flow. Out of Smith's work comes the easiest way of estimating the earliest time at which uniform flow can be anticipated, which is simply the integral of dx/(u-a) at the particular location. This technique has been shown to be very accurate.[57, 61–63]

The time required to establish uniform flow either within or over a complex model for which there may be boundary-layer flow separation is relatively difficult to predict, especially if the boundary layer is turbulent. Holden[64–67] has obtained a large body of data relevant to establishing separated flows. There has been some success in calculating flow establishment times using full Navier-Stokes solvers for the case of laminar boundary-layer separation. However, the turbulent boundary-layer case has been more difficult to calculate.

In the 1960s and early 1970s it was common to perform an analysis and experimental program like that described in the previous paragraphs in order to insure that one understood the test time, nozzle starting time, flow establishment time, and freestream conditions associated with a particular facility and even with a particular experiment to be performed within that facility. These studies were a good idea then, and they remain so now.

D. The Use of Short-Duration Facilities to Study Scramjet Exhaust Nozzle Flows

In addition to being a valuable tool for the study of scramjet combustor flows, short-duration facilities can be applied to the study of scramjet exhaust nozzle flows. The results of a preliminary study having the objective of designing such experiments are summarized in this subsection. The degree of chemical reaction in the nozzle affects the wall pressure distribution sufficiently to impact the nozzle exhaust/afterbody contribution to the thrust/drag level for the integrated vehicle. Scramjet nozzle expansions subject to the effect of nonequilibrium chemistry may be studied by measurement of the wall pressure and heat-transfer distribution. Moreover, the development of modern, nonintrusive diagnostics permits the detailed profiles of the nozzle expansion, including the concentrations of radical species such as OH, to be probed. Such detailed measurements are key to understanding the fundamental physics and chemistry of such complex reacting flowfields.

Another important aspect of this simulation that should not go unnoticed is that of scale size, which plays a prominent role so that the appropriate properties of the boundary layers and shear layers within the exhaust nozzle flowfield are allowed to develop. Facilities and techniques are available that will permit combustor/nozzle exhaust measurements to be performed at the appropriate flight conditions and for model sizes sufficiently large that the viscous effects are properly represented.

Three different techniques have been considered for producing realistic nozzle inlet conditions, among them the appropriate products of combustion, in simulations of scramjet nozzle exhausts. These include:

1) Direct connect experiments using a shock tube to produce the inlet flow.
2) Semi-direct connect using a shock tunnel to produce the inlet to a combustor (similar to the scramjet combustor experiments described earlier).
3) Gaseous-equivalent combustion simulation to provide the inlet flow to a nozzle.

Each of these techniques has its own set of advantages and disadvantages. The experience base that has been established in such measurement programs, as well as nozzle computational codes, are being applied to evaluate the optimum method. The basic approach being taken in this effort is to make detailed measurements on subscale, but sufficiently large nozzle components to examine the effects on finite-rate chemistry on the exhaust. The flow times will not be duplicated but, since the model scale is smaller, the degree of nonequilibrium is greater, and so the "measurability" of the effect (e.g., on wall-pressure distribution) should be greater in the model experiments. Computations can be done to insure that the flow is not frozen and, moreover, nonintrusive diagnostics can be employed to monitor quantities such as the OH concentration, as well as to measure the properties at the nozzle wall.

In the application of shock tunnels to studies of NASP propulsion problems, development of nonintrusive diagnostics plays an important role in the measurement of combustor and nozzle flowfield properties as well as freestream test conditions.

V. Conclusions

Facility requirements and capabilities for hypersonic propulsion system testing have been reviewed with emphasis on short-duration test facilities. The significant contributions of selected previous and current studies on the subject have been put in perspective. With this review completed there are several points that can be made:

1) Many of the current hypersonic flow problems are identical to those identified 20 years ago.
2) Because of the enormous energy associated with the vehicles and trajectories of interest, it will be impossible to build a continuous-flow ground-test facility to test a full-size NASP-type vehicle.
3) The new problems associated with modern configurations will require integrated CFD, ground-test, and flight-test efforts to obtain satisfactory solutions.

In this regard, it is important to emphasize that analytical tools are available to design laboratory experiments that can provide required data. In many cases, full duplication of flight conditions can be produced in component tests in ground test facilities. In other cases, where departure from full duplication occurs, experiments can be designed in which the essential physical phenomena are reproduced and numerical models are employed to scale the results to flight conditions. In such cases the freestream conditions and the test environment must be determined as part of the experiment.

Several different ground-test facilities will be required to address the wide range of problem areas. Short-duration facilities can provide the high energy levels for a significant portion of the trajectory and are currently being used to obtain important data, especially in the investigation of high Mach number supersonic combustion.

Real gas effects on nozzle expansions of air from the reservoir conditions required to produce the high enthalpy flows corresponding to simulation of NASP flight conditions were examined. Sample computations were reported for nozzle expansions from reservoir conditions at which scramjet combustor experiments have been performed at Calspan and which fully duplicate the freestream conditions corresponding to flight at M_∞ = 10 and h = 100,000 ft. The results of these calculations demonstrate that:

1) For these conditions, real-gas effects on the reservoir and the initial portion of the expansion include significant effects of intermolecular forces on the compressibility of air. Such effects must be considered when determining the flow properties in high enthalpy facilities.
2) Nozzle expansions of air in hypervelocity tunnels at these conditions remain in vibrational and chemical equilibrium well into the supersonic region. Nonequilibrium effects are not large enough to limit the facility capability to duplicate the velocity and altitude conditions along the NASP corridor. However, nonequilibrium effects on the nozzle expansion can be significant for species-specific measurements.
3) The freestream concentrations of NO and the nonequilibrium N_2 vibrational temperatures at the enthalpy levels corresponding to flight at M = 10 are measurable but do not appear to affect scramjet combustor experiments. In general, however, it is recommended that such quantities as the free stream temperature, the N_2 vibrational temperature, and the NO concentration be measured in hypervelocity facilities and that specific experiments be designed to determine what, if any, impact nonequilibrium effects on the test flow case have on experiments at higher enthalpies.

Acknowledgments

The authors would like to thank Dr. Charles E. Treanor for his comments during the preparation of this manuscript. They would also like to thank John Moselle, and Michael Stanton for performing the calculations described in this paper.

References

[1]Staff, "Hypersonic Research Facilities Study, Vol. 1, Summary," NASA CR 114322, Oct.

[2]Staff, "Hypersonic Research Facilities Study, Vol. II, Part 1, Research Requirements and Ground Facility Synthesis," NASA CR-114323, 1970.

[3]Staff, "Hypersonic Research Facilities Study, Vol. II, Part 2, Flight Vehicle Synthesis," NASA CR-114324, 1970.

[4]Staff, "Hypersonic Research Facilities Study, Vol. III, Part 1, Research Requirements and Ground Facility Synthesis," NASA CR-114325, 1970.

[5]Staff, "Hypersonic Research Facilities Study, Vol. III, Part 2, Flight Vehicle Synthesis," NASA CR-114326, 1970.

[6]Staff, "Hypersonic Research Facilities Study, Vol. IV, Part 1, Flight Research Facilities," NASA CR-114327, 1970.

[7]Staff, "Hypersonic Research Facilities Study, Vol. IV, Part 2, Ground Research Facilities," NASA CR-114328, 1970.

[8]Staff, "Hypersonic Research Facilities Study, Vol. IV, Part 3, Research Requirements Analysis and Facility Potential," NASA CR-114329, 1970.

[9]Smith, V.K., Keel, L.C., and Boudreau, A.H., "Ground Testing Facilities Requirements for Hypersonic Propulsion Development," AIAA Paper 87-1884 July 1987.

[10]Wittliff, C.E., "A Survey of Existing Hypersonic Ground Test Facilities - North America," Aerodynamics of Hypersonic Lifting Vehicles, AGARD Conference Proceedings 428, April 1987.

[11]Hallion, R.P. (ed.), "The Hypersonic Revolution, From Max Valier to Project Prime, Volume 1, 1924-1967," Special Staff Office, Aeronautical Systems Division, Wright-Patterson Air Force Base, OH, 1987.

[12]Hallion, R.P. (ed.), "The Hypersonic Revolution, From Scramjet to the National Aero-Space Plane, Volume II, 1964-1986," Special Staff Office, Aeronautical Systems Division, Wright-Patterson Air Force Base, OH, 1987.

[13]Smelt, R., "Review of Aeronautical Wind Tunnel Facilities," Committee Report sponsored by NASA, and conducted by National Research Council; available from National Academy Press, Washington, DC, 1988.

[14]Kerrebrock, J., "Hypersonic Technology for Military Applications," Committee Report sponsored by U.S. Air Force Systems Command and conducted by the U.S. Air Force Studies Board; available from National Academy Press, Washington, DC, 1989.

[15]Buchanan, L., "SAB Ad Hoc Committee on Requirements for Hypersonic Test Facilities"; Volume I Study Report and Volume II Appendices, available from the United States Air Force Scientific Advisory Board, May 1989.

[16]Wind Tunnel Study Task Team Report, "Assessment of NASA's Major Wind Tunnel Facilities," available from the National Aeronautics and Space Administration, Washington, DC, August 1987.

[17]Lordi, J.A., Boyer, D.W., Dunn, M.G., Smolarek, K.K., and Wittliff, C.E., "Description of Nonequilibrium Effects on Simulation of Flows About Hypersonic Vehicles," AIAA Paper 88-0476, 1988.

[18]Roepke, R.G., "Hypersonic Facility Study," AEDC-TMR-88-P6, Jan. 1988.

[19]Penaranda, F.E., and Freda, M.S. (eds.), "Aeronautical Facilities Catalogue, Vol. 1, Wind Tunnels," NASA RP-1132, 1985.

[20]Penaranda, F.E. and Freda, M.S. (eds.), "Aeronautical Facilities Catalogue, Vol. 2, air-breathing Propulsion and Flight Simulators," NASA RP-1133, 1985.

[21]Wendt, J.F., "European Hypersonic Wind Tunnels," Aerodynamics of Hypersonic Lifting Vehicles, AGARD CP-428, April 1987.

[22]Andrews, E.H., Torrence, M.G., Anderson, G.Y., Northam, G.B. and Mackley, E.A., "Langley Mach 4 Scramjet Test Facility," NASA TM-86277, 1985.

[23]Billig, F.S. Waltrup, P.J., Gilreath, H.E., White, M.E., Van Wie, D.M., and Pandolfini, P.P., "Proposed Supplement to Propulsion System Management Support Plan," JHU/APL-NASP-86-1, July 1986.

[24]Aviation Week and Space Technology, Sept. 5, 1988, pp. 211, 215.

[25]Aviation Week and Space Technology, June 12, 1989, p. 263.

[26]Haas, J., Chamberlin R., and Dicus, J., "New Hypersonic Facility Capability at NASA-Lewis Research Center," AIAA Paper 89-2534, July 1989.

[27]Snyder, C.T., and Presley, L.L., "Current Wind Tunnel Capability and Planned Improvements at NASA Ames Research Center," AIAA Paper 86-0729-CP, March 1986.

[28]Wittliff, C.E., Wilson, M.R., and Hertzberg, A., "The Tailored-Interface Hypersonic Shock Tunnel," Journal of the Aerospace Sciences, April 1959.

[29]Pirrello, C.J., Hardin, R.D., Heckart, M.V. and Brown, K.R., "An Inventory of Aeronautical Ground Research Facilities, Vol. 1 - Wind Tunnels," NASA CR-1874, Nov. 1971.

[30]Swithenbank, J., Eames, I.W., Chin, S.B., Ewan, B.C.R., Yang, Z.Y., Cao, J. and Zhao, X., "Turbulent Mixing in Supersonic Combustion Systems," AIAA Paper 89-0260, Jan. 1989.

[31]Swithenbank, J., and Parsons, R.J., "Combustion Research in a Shock Tunnel," New Experimental Techniques in Propulsion and Energetics Research, AGARD CP-38, edited by D. Andrews and J. Surugue, Oct. 1970.

[32]Stalker, R.J., "Shock Tunnels for Real Gas Hypersonics," Aerodynamics of Hypersonic Lifting Vehicles, AGARD CP-428, April 1987.

[33]Stalker, R.J., and Morgan, R.G., "Parallel Hydrogen Injection into ConstantArea, High-Enthalpy, Supersonic Airflow," AIAA Journal, Vol. 20, Oct. 1982.

[34]Aviation Week and Space Technology, Jan. 30, 1989, p. 65.

[35]Miller, C.J., and Jones, J.J., "Development and Performance of the NASA Langley Research Center Expansion Tube/Tunnel, A Hypersonic-Hypervelocity Real-Gas Facility," Shock Tubes and Waves, Proceedings of the 14th International Symposium, Ed. R.D. Archer and B.E. Milton, Sydney Shock Tube Symposium Publishers, Sydney, 1983.

[36]Rizkalla, O. Bakos, R.J., Chinitz, W., Pulsonetti, M.V. and Erdos, J.I., "Use of An Expansion Tube to Examine Scramjet Combustion at Hypersonic Velocities," AIAA Paper 89-2536, July 1989.

[37]Miller, C.G., "A Critical Examination of Expansion Tunnel Performance," AIAA 10th Aerodynamic Testing Conference, AIAA Paper 78-768 April 1978.

[38]Wood, M.P., "The Shock Tunnel as a Supersonic Combustion Test Facility," Fuel Society Journal, 1966.

[39]Wittliff, C.E., Sundaram, T.R., Rae, W.J. and Lordi, J.A., "Study of HighDensity Hypervelocity Flows and Similitudes," AEDC TR-67-72, April 1967.

[40]Lordi, J.A., Mates, R.E., and Moselle, J.R., "Computer Program for the Numberical Solution of Nonequilibrium Expansion of Reacting Gas

Mixtures," Cornell Univ., Ithaca, NY, Cornell Aeronautical Laboratory Rep. AD-1689-A-6, Oct. 1965.

[41]Eschenroeder, A.Q., Boyer, D.B., and Hall, J.G., "Exact Solutions for Nonequilibrium Expansions of Air with Coupled Chemical Reactions," Cornell Univ., Ithaca, NY, Cornell Aeronautical Laboratory Rep. AF-1413-A-1, May 1961.

[42]Lordi, J.A., and Mates, R.E., "Nonequilibrium Expansions of High-Enthalpy Airflows," AIAA Journal, Vol. 3, Oct. 1965, pp. 1972-1974.

[43]Harris, C.J., "Nonequilibrium Effects on High-Enthalpy Expansion of Air," AIAA Journal, Vol. 4, June 1966, pp. 1148-1149.

[44]Gibson, W.E., and Marrone, P.V., "A Similitude for Nonequilibrium Phenomena in Hypersonic Flight," The High Temperature Aspects of Hypersonic Flow, edited by W.C. Nelson, AGARDograph 68, Macmillan, New York, 1964, pp. 105-132.

[45]Hilsenrath, J., and Klein, M., "Tables of Thermodynamic Properties of Air in Chemical Equilibrium Including Second Virial Corrections from 1500 K to 15,000 K," AEDC-TR-65-58, March 1965.

[46]Dunn, M.G., Moller, J.C., and Steele, R.C., "Development of a New HighEnthalpy Shock Tunnel," AIAA Paper 88-2782, June 1988.

[47]Copper, J.A., Miller, H.R., and Hameetman, F.H., "Correlation of Uncontaminated Test Durations in Shock Tunnels," Fourth Hypervelocity Techniques Symposium, Univ. of Denver, Boulder, CO, Nov. 1965, pp. 274-310.

[48]Wittliff, C.E., Wilson, M.R., and Hertzberg, A., "The Tailored-Interface Hypersonic Shock Tunnel," Journal of Aerospace Sciences, Vol. 26, No. 4, April 1959, pp. 219- 228.

[49]Copper, J.A., "Experimental Investigation of the Equilibrium Interface Technique," Physics of Fluids, Vol. 5, July 1962, pp. 844-849.

[50]Glass, I.I., and Hall, J.G., "Handbook of Supersonic Aerodynamics, Section 18; Shock Tubes," NAVORD Rept. 1488, (Vol. 6), Dec. 1959.

[51]Gaydon, A.G. and Hurle, I.R., The Shock Tube in High Temperature Chemical Physics, Reinhold, New York, 1963.

[52]Mirels, H., "Test Time in Low-Pressure Shock Tubes," Physics of Fluids, Vol. 6, Sept. 1963, pp. 1201-1214.

[53]Mirels, H., "Shock-Tube Test-Time Limitations Due to Turbulent-Wall Boundary Layer," AIAA Journal, Vol. 2, Jan. 1964, pp. 84-92.

[54]Hartunian, R.A., Russo, A.L., and Marrone, P.V., "Boundary-Layer Transition and Heat Transfer in Shock Tubes," Journal of the Aerospace Sciences, Vol. 27, Aug. 1960, pp. 587-594.

[55]Dunn, M.G., "Experimental Shock-Tube Investigations of Conditions Behind Incident and Reflected Shocks in Air for Shock Mach Between 8.5 and 16.5," Fifth Hypervelocity Techniques Symposium, Univ. of Denver, Boulder, CO, Vol. 1, March 1967.

[56]Fuehrer, R.C., "Measurements of Incident-Shock Test Time and Reflected Shock Pressure at Fully Turbulent Boundary Layer Test Conditions," Seventh International Symposium on Shock Tube and Waves, Toronto, Canada, University of Toronto Press, June 1969.

[57]Dunn, M.G., "Experimental Study of High-Enthalpy Shock-Tunnel Flow. Part I: Shock-Tube Flow and Nozzle Starting Time," AIAA Journal, Vol. 7, Aug. 1969, pp. 1553-1560.

[58]Mark, H., "The Interaction of a Reflected Shock Wave with the Boundary Layer in a Shock Tube," NACA, TM 1418, March 1958.

[59]Davies, L., "The Interaction of the Reflected Shock with the Boundary Layer in a Shock Tube and Its Influence on the Duration of Hot Flow in the Reflected Shock Tunnel," ARC 27 110-Hyp. 505, 1965, Aeronautical Research Council.

[60]Smith, C.E., "An Analytic Study of the Starting Process in a Hypersonic Nozzle," Rept. SUDAER 135, Sept. 1962, Stanford Univ., Stanford, CA; also Proceedings of the 1964 Heat Transfer and Fluid Mechanics Institute, Stanford Univ. Press, Stanford, CA.

[61]Smith, C.E., "The Starting Process in a Hypersonic Nozzle," Rept. 1000, July 1965, Univ. of Oxford, Oxford, U.K.

[62]Dunn, M.G., "Experimental Study of High-Enthalpy Shock-Tunnel Flow, Part II: Nozzle-Flow Characteristics," AIAA Journal, Vol. 7, Sept. 1969, pp. 1717-1724.

[63]Boyer, D.W., and Andre, S., "Theoretical and Experimental Studies of Microwave Interaction Effects in a Hypersonic Air PLASMA," Sandia Corp., Albequerque, NM, Rpt. SC-CR-72 3164, Oct. 1972.

[64]Holden, M.S., "The Establishment Time of Laminar Separated Flows," Cornell Univ., Ithaca, NY, Cornell Aeronautical Laboratory Rept. CAL 179, March 1971; also AIAA Journal, Vol. 9, Nov. 1971, pp. 2296-2298

[65]Holden, M.S., "Three-Dimensional Supersonic and Hypersonic Flows Including Separation," Von Karman Institute 1988-1989 Lecture Series, Brussels, Belgium, May 8-12, 1989.

[66]Holden, M.S., "Studies of the Structure of Attached and Separated Regions of Viscous/Inviscid Interaction and the Effects of Combined Surface Roughness and Flow in High Reynolds Number Hypersonic Flows," Calspan-UB Research Center, CUBRC Rept. 88682, Dec. 1988.

[67]Holden, M.S., Havener, A.G., and Lee, C.H., "Shock Wave/Turbulent Boundary Layer Interaction in High-Reynolds-Number Hypersonic Flows," Calspan-UB Research Center, Buffalo, New York, CUBRC Rept. 86681, July 1987.

Author Index

PROGRESS IN ASTRONAUTICS AND AERONAUTICS SERIES VOLUMES

*1. **Solid Propellant Rocket Research** (1960)
Martin Summerfield
Princeton University

*2. **Liquid Rockets and Propellants** (1960)
Loren E. Bollinger
Ohio State University
Martin Goldsmith
The Rand Corp.
Alexis W. Lemmon Jr.
Battelle Memorial Institute

*3. **Energy Conversion for Space Power** (1961)
Nathan W. Snyder
Institute for Defense Analyses

*4. **Space Power Systems** (1961)
Nathan W. Snyder
Institute for Defense Analyses

*5. **Electrostatic Propulsion** (1961)
David B. Langmuir
Space Technology Laboratories, Inc.
Ernst Stuhlinger
NASA George C. Marshall Space Flight Center
J.M. Sellen Jr.
Space Technology Laboratories, Inc.

*6. **Detonation and Two-Phase Flow** (1962)
S.S. Penner
California Institute of Technology
F.A. Williams
Harvard University

*7. **Hypersonic Flow Research** (1962)
Frederick R. Riddell
AVCO Corp.

*8. **Guidance and Control** (1962)
Robert E. Roberson, Consultant
James S. Farrior
Lockheed Missiles and Space Co.

*9. **Electric Propulsion Development** (1963)
Ernst Stuhlinger
NASA George C. Marshall Space Flight Center

*10. **Technology of Lunar Exploration** (1963)
Clifford I. Cummings
Harold R. Lawrence
Jet Propulsion Laboratory

*11. **Power Systems for Space Flight** (1963)
Morris A. Zipkin
Russell N. Edwards
General Electric Co.

*12. **Ionization in High-Temperature Gases** (1963)
Kurt E. Shuler, Editor
National Bureau of Standards
John B. Fenn, Associate Editor
Princeton University

*13. **Guidance and Control—II** (1964)
Robert C. Langford
General Precision Inc.
Charles J. Mundo
Institute of Naval Studies

*14. **Celestial Mechanics and Astrodynamics** (1964)
Victor G. Szebehely
Yale University Observatory

*15. **Heterogeneous Combustion** (1964)
Hans G. Wolfhard
Institute for Defense Analyses
Irvin Glassman
Princeton University
Leon Green Jr.
Air Force Systems Command

16. **Space Power Systems Engineering** (1966)
George C. Szego
Institute for Defense Analyses
J. Edward Taylor
TRW Inc.

17. **Methods in Astrodynamics and Celestial Mechanics** (1966)
Raynor L. Duncombe
U.S. Naval Observatory
Victor G. Szebehely
Yale University Observatory

18. **Thermophysics and Temperature Control of Spacecraft and Entry Vehicles** (1966)
Gerhard B. Heller
NASA George C. Marshall Space Flight Center

*Out of print.

***19. Communication Satellite Systems Technology** (1966)
Richard B. Marsten
Radio Corporation of America

***20. Thermophysics of Spacecraft and Planetary Bodies: Radiation Properties of Solids and the Electromagnetic Radiation Environment in Space** (1967)
Gerhard B. Heller
NASA George C. Marshall Space Flight Center

21. Thermal Design Principles of Spacecraft and Entry Bodies (1969)
Jerry T. Bevans
TRW Systems

22. Stratospheric Circulation (1969)
Willis L. Webb
Atmospheric Sciences Laboratory, White Sands, and University of Texas at El Paso

23. Thermophysics: Applications to Thermal Design of Spacecraft (1970)
Jerry T. Bevans
TRW Systems

24. Heat Transfer and Spacecraft Thermal Control (1971)
John W. Lucas
Jet Propulsion Laboratory

25. Communication Satellites for the 70's: Technology (1971)
Nathaniel E. Feldman
The Rand Corp.
Charles M. Kelly
The Aerospace Corp.

26. Communication Satellites for the 70's: Systems (1971)
Nathaniel E. Feldman
The Rand Corp.
Charles M. Kelly
The Aerospace Corp.

27. Thermospheric Circulation (1972)
Willis L. Webb
Atmospheric Sciences Laboratory, White Sands, and University of Texas at El Paso

28. Thermal Characteristics of the Moon (1972)
John W. Lucas
Jet Propulsion Laboratory

29. Fundamentals of Spacecraft Thermal Design (1972)
John W. Lucas
Jet Propulsion Laboratory

30. Solar Activity Observations and Predictions (1972)
Patrick S. McIntosh
Murray Dryer
Environmental Research Laboratories, National Oceanic and Atmospheric Administration

31. Thermal Control and Radiation (1973)
Chang-Lin Tien
University of California at Berkeley

32. Communications Satellite Systems (1974)
P.L. Bargellini
COMSAT Laboratories

33. Communications Satellite Technology (1974)
P.L. Bargellini
COMSAT Laboratories

34. Instrumentation for Airbreathing Propulsion (1974)
Allen E. Fuhs
Naval Postgraduate School
Marshall Kingery
Arnold Engineering Development Center

35. Thermophysics and Spacecraft Thermal Control (1974)
Robert G. Hering
University of Iowa

36. Thermal Pollution Analysis (1975)
Joseph A. Schetz
Virginia Polytechnic Institute
ISBN 0-915928-00-0

37. Aeroacoustics: Jet and Combustion Noise; Duct Acoustics (1975)
Henry T. Nagamatsu, Editor
General Electric Research and Development Center
Jack V. O'Keefe, Associate Editor
The Boeing Co.
Ira R. Schwartz, Associate Editor
NASA Ames Research Center
ISBN 0-915928-01-9

38. Aeroacoustics: Fan, STOL, and Boundary Layer Noise; Sonic Boom; Aeroacoustics Instrumentation (1975)
Henry T. Nagamatsu, Editor
General Electric Research and Development Center
Jack V. O'Keefe, Associate Editor
The Boeing Co.
Ira R. Schwartz, Associate Editor
NASA Ames Research Center
ISBN 0-915928-02-7

39. Heat Transfer with Thermal Control Applications (1975)
M. Michael Yovanovich
University of Waterloo
ISBN 0-915928-03-5

40. Aerodynamics of Base Combustion (1976)
S.N.B. Murthy, Editor
J.R. Osborn,
Associate Editor
Purdue University
A.W. Barrows
J.R. Ward,
Associate Editors
Ballistics Research Laboratories
ISBN 0-915928-04-3

41. Communications Satellite Developments: Systems (1976)
Gilbert E. LaVean
Defense Communications Agency
William G. Schmidt
CML Satellite Corp.
ISBN 0-915928-05-1

42. Communications Satellite Developments: Technology (1976)
William G. Schmidt
CML Satellite Corp.
Gilbert E. LaVean
Defense Communications Agency
ISBN 0-915928-06-X

43. Aeroacoustics: Jet Noise, Combustion and Core Engine Noise (1976)
Ira R. Schwartz, Editor
NASA Ames Research Center
Henry T. Nagamatsu,
Associate Editor
General Electric Research and Development Center
Warren C. Strahle,
Associate Editor
Georgia Institute of Technology
ISBN 0-915928-07-8

44. Aeroacoustics: Fan Noise and Control; Duct Acoustics; Rotor Noise (1976)
Ira R. Schwartz, Editor
NASA Ames Research Center
Henry T. Nagamatsu,
Associate Editor
General Electric Research and Development Center
Warren C. Strahle,
Associate Editor
Georgia Institute of Technology
ISBN 0-915928-08-6

45. Aeroacoustics: STOL Noise; Airframe and Airfoil Noise (1976)
Ira R. Schwartz, Editor
NASA Ames Research Center
Henry T. Nagamatsu,
Associate Editor
General Electric Research and Development Center
Warren C. Strahle,
Associate Editor
Georgia Institute of Technology
ISBN 0-915928-09-4

46. Aeroacoustics: Acoustic Wave Propagation; Aircraft Noise Prediction; Aeroacoustic Instrumentation (1976)
Ira R. Schwartz, Editor
NASA Ames Research Center
Henry T. Nagamatsu,
Associate Editor
General Electric Research and Development Center
Warren C. Strahle,
Associate Editor
Georgia Institute of Technology
ISBN 0-915928-10-8

47. Spacecraft Charging by Magnetospheric Plasmas (1976)
Alan Rosen
TRW Inc.
ISBN 0-915928-11-6

48. Scientific Investigations on the Skylab Satellite (1976)
Marion I. Kent
Ernst Stuhlinger
NASA George C. Marshall Space Flight Center
Shi-Tsan Wu
University of Alabama
ISBN 0-915928-12-4

49. Radiative Transfer and Thermal Control (1976)
Allie M. Smith
ARO Inc.
ISBN 0-915928-13-2

50. Exploration of the Outer Solar System (1976)
Eugene W. Greenstadt
TRW Inc.
Murray Dryer
National Oceanic and Atmospheric Administration
Devrie S. Intriligator
University of Southern California
ISBN 0-915928-14-0

51. Rarefied Gas Dynamics, Parts I and II (two volumes) (1977)
J. Leith Potter
ARO Inc.
ISBN 0-915928-15-9

52. Materials Sciences in Space with Application to Space Processing (1977)
Leo Steg
General Electric Co.
ISBN 0-915928-16-7

53. Experimental Diagnostics in Gas Phase Combustion Systems (1977)
Ben T. Zinn, Editor
Georgia Institute of Technology
Craig T. Bowman, Associate Editor
Stanford University
Daniel L. Hartley, Associate Editor
Sandia Laboratories
Edward W. Price, Associate Editor
Georgia Institute of Technology
James G. Skifstad, Associate Editor
Purdue University
ISBN 0-015928-18-3

54. Satellite Communications: Future Systems (1977)
David Jarett
TRW Inc.
ISBN 0-915928-18-3

55. Satellite Communications: Advanced Technologies (1977)
David Jarett
TRW Inc.
ISBN 0-915928-19-1

56. Thermophysics of Spacecraft and Outer Planet Entry Probes (1977)
Allie M. Smith
ARO Inc.
ISBN 0-915928-20-5

57. Space-Based Manufacturing from Nonterrestrial Materials (1977)
Gerard K. O'Neill, Editor
Brian O'Leary, Assistant Editor
Princeton University
ISBN 0-915928-21-3

58. Turbulent Combustion (1978)
Lawrence A. Kennedy
State University of New York at Buffalo
ISBN 0-915928-22-1

59. Aerodynamic Heating and Thermal Protection Systems (1978)
Leroy S. Fletcher
University of Virginia
ISBN 0-915928-23-X

60. Heat Transfer and Thermal Control Systems (1978)
Leroy S. Fletcher
University of Virginia
ISBN 0-915928-24-8

61. Radiation Energy Conversion in Space (1978)
Kenneth W. Billman
NASA Ames Research Center
ISBN 0-915928-26-4

62. Alternative Hydrocarbon Fuels: Combustion and Chemical Kinetics (1978)
Craig T. Bowman
Stanford University
Jorgen Birkeland
Department of Energy
ISBN 0-915928-25-6

63. Experimental Diagnostics in Combustion of Solids (1978)
Thomas L. Boggs
Naval Weapons Center
Ben T. Zinn
Georgia Institute of Technology
ISBN 0-915928-28-0

64. Outer Planet Entry Heating and Thermal Protection (1979)
Raymond Viskanta
Purdue University
ISBN 0-915928-29-9

65. Thermophysics and Thermal Control (1979)
Raymond Viskanta
Purdue University
ISBN 0-915928-30-2

66. Interior Ballistics of Guns (1979)
Herman Krier
University of Illinois at Urbana-Champaign
Martin Summerfield
New York University
ISBN 0-915928-32-9

***67. Remote Sensing of Earth from Space: Role of "Smart Sensors"**(1979)
Roger A. Breckenridge
NASA Langley Research Center
ISBN 0-915928-33-7

68. Injection and Mixing in Turbulent Flow (1980)
Joseph A. Schetz
Virginia Polytechnic Institute and State University
ISBN 0-915928-35-3

69. Entry Heating and Thermal Protection (1980)
Walter B. Olstad
NASA Headquarters
ISBN 0-915928-38-8

70. Heat Transfer, Thermal Control, and Heat Pipes (1980)
Walter B. Olstad
NASA Headquarters
ISBN 0-915928-39-6

71. Space Systems and Their Interactions with Earth's Space Environment (1980)
Henry B. Garrett
Charles P. Pike
Hanscom Air Force Base
ISBN 0-915928-41-8

72. Viscous Flow Drag Reduction (1980)
Gary R. Hough
Vought Advanced Technology Center
ISBN 0-915928-44-2

73. Combustion Experiments in a Zero-Gravity Laboratory (1981)
Thomas H. Cochran
NASA Lewis Research Center
ISBN 0-915928-48-5

74. Rarefied Gas Dynamics, Parts I and II (two volumes) (1981)
Sam S. Fisher
University of Virginia
ISBN 0-915928-51-5

75. Gasdynamics of Detonations and Explosions (1981)
J.R. Bowen
University of Wisconsin at Madison
N. Manson
Université de Poitiers
A.K. Oppenheim
University of California at Berkeley
R.I. Soloukhin
Institute of Heat and Mass Transfer, BSSR Academy of Sciences
ISBN 0-915928-46-9

76. Combustion in Reactive Systems (1981)
J.R. Bowen
University of Wisconsin at Madison
N. Manson
Université de Poitiers
A.K. Oppenheim
University of California at Berkeley
R.I. Soloukhin
Institute of Heat and Mass Transfer, BSSR Academy of Sciences
ISBN 0-915928-47-7

77. Aerothermodynamics and Planetary Entry (1981)
A.L. Crosbie
University of Missouri-Rolla
ISBN 0-915928-52-3

78. Heat Transfer and Thermal Control (1981)
A.L. Crosbie
University of Missouri-Rolla
ISBN 0-915928-53-1

79. Electric Propulsion and Its Applications to Space Missions (1981)
Robert C. Finke
NASA Lewis Research Center
ISBN 0-915928-55-8

80. Aero-Optical Phenomena (1982)
Keith G. Gilbert
Leonard J. Otten
Air Force Weapons Laboratory
ISBN 0-915928-60-4

81. Transonic Aerodynamics (1982)
David Nixon
Nielsen Engineering & Research, Inc.
ISBN 0-915928-65-5

82. Thermophysics of Atmospheric Entry (1982)
T.E. Horton
University of Mississippi
ISBN 0-915928-66-3

83. Spacecraft Radiative Transfer and Temperature Control (1982)
T.E. Horton
University of Mississippi
ISBN 0-915928-67-1

84. Liquid-Metal Flows and Magnetohydrodynamics (1983)
H. Branover
Ben-Gurion University of the Negev
P.S. Lykoudis
Purdue University
A. Yakhot
Ben-Gurion University of the Negev
ISBN 0-915928-70-1

85. Entry Vehicle Heating and Thermal Protection Systems: Space Shuttle, Solar Starprobe, Jupiter Galileo Probe (1983)
Paul E. Bauer
McDonnell Douglas Astronautics Co.
Howard E. Collicott
The Boeing Co.
ISBN 0-915928-74-4

86. Spacecraft Thermal Control, Design, and Operation (1983)
Howard E. Collicott
The Boeing Co.
Paul E. Bauer
McDonnell Douglas Astronautics Co.
ISBN 0-915928-75-2

87. Shock Waves, Explosions, and Detonations (1983)
J.R. Bowen
University of Washington
N. Manson
Université de Poitiers
A.K. Oppenheim
University of California at Berkeley
R.I. Soloukhin
Institute of Heat and Mass Transfer, BSSR Academy of Sciences
ISBN 0-915928-76-0

88. Flames, Lasers, and Reactive Systems (1983)
J.R. Bowen
University of Washington
N. Manson
Université de Poitiers
A.K. Oppenheim
University of California at Berkeley
R.I. Soloukhin
Institute of Heat and Mass Transfer, BSSR Academy of Sciences
ISBN 0-915928-77-9

89. Orbit-Raising and Maneuvering Propulsion: Research Status and Needs (1984)
Leonard H. Caveny
Air Force Office of Scientific Research
ISBN 0-915928-82-5

90. Fundamentals of Solid-Propellant Combustion (1984)
Kenneth K. Kuo
Pennsylvania State University
Martin Summerfield
Princeton Combustion Research Laboratories, Inc.
ISBN 0-915928-84-1

91. Spacecraft Contamination: Sources and Prevention (1984)
J.A. Roux
University of Mississippi
T.D. McCay
NASA Marshall Space Flight Center
ISBN 0-915928-85-X

92. Combustion Diagnostics by Nonintrusive Methods (1984)
T.D. McCay
NASA Marshall Space Flight Center
J.A. Roux
University of Mississippi
ISBN 0-915928-86-8

93. The INTELSAT Global Satellite System (1984)
Joel Alper
COMSAT Corp.
Joseph Pelton
INTELSAT
ISBN 0-915928-90-6

94. Dynamics of Shock Waves, Explosions, and Detonations (1984)
J.R. Bowen
University of Washington
N. Manson
Université de Poitiers
A.K. Oppenheim
University of California at Berkely
R.I. Soloukhin
Institute of Heat and Mass Transfer, BSSR Academy of Sciences
ISBN 0-915928-91-4

95. Dynamics of Flames and Reactive Systems (1984)
J.R. Bowen
University of Washington
N. Manson
Université de Poitiers
A.K. Oppenheim
University of California at Bereley
R.I. Soloukhin
Institute of Heat and Mass Transfer, BSSR Academy of Sciences
ISBN 0-915928-92-2

96. Thermal Design of Aeroassisted Orbital Transfer Vehicles (1985)
H.F. Nelson
University of Missouri-Rolla
ISBN 0-915928-94-9

97. Monitoring Earth's Ocean, Land, and Atmosphere from Space — Sensors, Systems, and Applications (1985)
Abraham Schnapf
Aerospace Systems Engineering
ISBN 0-915928-98-1

98. Thrust and Drag: Its Prediction and Verification (1985)
Eugene E. Covert
Massachusetts Institute of Technology
C.R. James
Vought Corp.
William F. Kimzey
Sverdrup Technology AEDC Group
George K. Richey
U.S. Air Force
Eugene C. Rooney
U.S. Navy Department of Defense
ISBN 0-930403-00-2

99. Space Stations and Space Platforms — Concepts, Design, Infrastructure, and Uses (1985)
Ivan Bekey
Daniel Herman
NASA Headquarters
ISBN 0-930403-01-0

100. Single- and Multi-Phase Flows in an Electromagnetic Field: Energy, Metallurgical, and Solar Applications (1985)
Herman Branover
Ben-Gurion University of the Negev
Paul S. Lykoudis
Purdue University
Michael Mond
Ben-Gurion University of the Negev
ISBN 0-930403-04-5

101. MHD Energy Conversion: Physiotechnical Problems (1986)
V.A. Kirillin
A.E. Sheyndlin
Soviet Academy of Sciences
ISBN 0-930403-05-3

102. Numerical Methods for Engine-Airframe Integration (1986)
S.N.B. Murthy
Purdue University
Gerald C. Paynter
Boeing Airplane Co.
ISBN 0-930403-09-6

103. Thermophysical Aspects of Re-Entry Flows (1986)
James N. Moss
NASA Langley Research Center
Carl D. Scott
NASA Johnson Space Center
ISBN 0-930403-10-X

104. Tactical Missile Aerodynamics (1986)
M.J. Hemsch
PRC Kentron, Inc.
J.N. Nielsen
NASA Ames Research Center
ISBN 0-930403-13-4

105. Dynamics of Reactive Systems Part I: Flames and Configurations; Part II: Modeling and Heterogeneous Combustion (1986)
J.R. Bowen
University of Washington
J.-C. Leyer
Université de Poitiers
R.I. Soloukhin
Institute of Heat and Mass Transfer, BSSR Academy of Sciences
ISBN 0-930403-14-2

106. Dynamics of Explosions (1986)
J.R. Bowen
University of Washington
J.-C. Leyer
Université de Poitiers
R.I. Soloukhin
Institute of Heat and Mass Transfer, BSSR Academy of Sciences
ISBN 0-930403-15-0

107. Spacecraft Dielectric Material Properties and Spacecraft Charging (1986)
A.R. Frederickson
U.S. Air Force Rome Air Development Center
D.B. Cotts
SRI International
J.A. Wall
U.S. Air Force Rome Air Development Center
F.L. Bouquet
Jet Propulsion Laboratory, California Institute of Technology
ISBN 0-930403-17-7

108. Opportunities for Academic Research in a Low-Gravity Environment (1986)
George A. Hazelrigg
National Science Foundation
Joseph M. Reynolds
Louisiana State University
ISBN 0-930403-18-5

109. Gun Propulsion Technology (1988)
Ludwig Stiefel
U.S. Army Armament Research, Development and Engineering Center
ISBN 0-930403-20-7

110. Commercial Opportunities in Space (1988)
F. Shahrokhi
K.E. Harwell
University of Tennessee Space Institute
C.C. Chao
National Cheng Kung University
ISBN 0-930403-39-8

111. Liquid-Metal Flows: Magnetohydrodynamics and Applications (1988)
Herman Branover, Michael Mond, and Yeshajahu Unger
Ben-Gurion University of the Negev
ISBN 0-930403-43-6

112. Current Trends in Turbulence Research (1988)
Herman Branover, Michael Mond, and Yeshajahu Unger
Ben-Gurion University of the Negev
ISBN 0-930403-44-4

113. Dynamics of Reactive Systems Part I: Flames; Part II: Heterogeneous Combustion and Applications (1988)
A.L. Kuhl
R & D Associates
J.R. Bowen
University of Washington
J.-C. Leyer
Université de Poitiers
A. Borisov
USSR Academy of Sciences
ISBN 0-930403-46-0

114. Dynamics of Explosions (1988)
A.L. Kuhl
R & D Associates
J.R. Bowen
University of Washington
J.-C. Leyer
Université de Poitiers
A. Borisov
USSR Academy of Sciences
ISBN 0-930403-47-9

115. Machine Intelligence and Autonomy for Aerospace (1988)
E. Heer
Heer Associates, Inc.
H. Lum
NASA Ames Research Center
ISBN 0-930403-48-7

116. Rarefied Gas Dynamics: Space-Related Studies (1989)
E.P. Muntz
University of Southern California
D.P. Weaver
U.S. Air Force Astronautics Laboratory (AFSC)
D.H. Campbell
University of Dayton Research Institute
ISBN 0-930403-53-3

117. Rarefied Gas Dynamics: Physical Phenomena (1989)
E.P. Muntz
University of Southern California
D.P. Weaver
U.S. Air Force Astronautics Laboratory (AFSC)
D. Campbell
University of Dayton Research Institute
ISBN 0-930403-54-1

118. Rarefied Gas Dynamics: Theoretical and Computational Techniques (1989)
E.P. Muntz
University of Southern California
D.P. Weaver
U.S. Air Force Astronautics Laboratory (AFSC)
D.H. Campbell
University of Dayton Research Institute
ISBN 0-930403-55-X

119. Test and Evaluation of the Tactical Missile (1989)
Emil J. Eichblatt Jr.
Pacific Missile Test Center
ISBN 0-930403-56-8

120. Unsteady Transonic Aerodynamics (1989)
David Nixon
Nielsen Engineering & Research, Inc.
ISBN 0-930403-52-5

121. Orbital Debris from Upper-Stage Breakup (1989)
Joseph P. Loftus Jr.
NASA Johnson Space Center
ISBN 0-930403-58-4

122. Thermal-Hydraulics for Space Power, Propulsion and Thermal Management System Design (1989)
William J. Krotiuk
General Electric Co.
ISBN 0-930403-64-9

123. Viscous Drag Reduction in Boundary Layers (1990)
Dennis M. Bushnell
Jerry N. Hefner
NASA Langley Research Center
ISBN 0-930403-66-5

124. Tactical and Strategic Missile Guidance (1990)
Paul Zarchan
Charles Stark Draper Laboratory, Inc.
ISBN 0-930403-68-1

125. Applied Computational Aerodynamics (1990)
P.A. Henne
Douglas Aircraft Company
ISBN 0-930403-69-X

126. Space Commercialization: Launch Vehicles and Programs (1990)
F. Shahrokhi
University of Tennessee Space Institute
J.S. Greenberg
Princeton Synergetics Inc.
T. Al-Saud
Ministry of Defense and Aviation Kingdom of Saudi Arabia
ISBN 0-930403-75-4

127. Space Commercialization: Platforms and Processing (1990)
F. Shahrokhi
University of Tennessee Space Institute
G. Hazelrigg
National Science Foundation
R. Bayuzick
Vanderbilt University
ISBN 0-930403-76-2

128. Space Commercialization: Satellite Technology (1990)
F. Shahrokhi
University of Tennessee Space Institute
N. Jasentuliyana
United Nations
N. Tarabzouni
King Abulaziz City for Science and Technology
ISBN 0-930403-77-0

129. Mechanics and Control of Large Flexible Structures (1990)
John L. Junkins
Texas A&M University
ISBN 0-930403-73-8

130. Low-Gravity Fluid Dynamics and Transport Phenomena (1990)
Jean N. Koster
Robert L. Sani
University of Colorado at Boulder
ISBN 0-930403-74-6

131. Dynamics of Deflagrations and Reactive Systems: Flames (1991)
A. L. Kuhl
Lawrence Livermore National Laboratory
J.-C. Leyer
Université de Poitiers
A. A. Borisov
USSR Academy of Sciences
W. A. Sirignano
University of California
ISBN 0-930403-95-9

132. Dynamics of Deflagrations and Reactive Systems: Heterogeneous Combustion (1991)
A. L. Kuhl
Lawrence Livermore National Laboratory
J.-C. Leyer
Université de Poitiers
A. A. Borisov
USSR Academy of Sciences
W. A. Sirignano
University of California
ISBN 0-930403-96-7

133. Dynamics of Detonations and Explosions: Detonations (1991)
A. L. Kuhl
Lawrence Livermore National Laboratory
J.-C. Leyer
Université de Poitiers
A. A. Borisov
USSR Academy of Sciences
W. A. Sirignano
University of California
ISBN 0-930403-97-5

134. Dynamics of Detonations and Explosions: Explosion Phenomena (1991)
A. L. Kuhl
Lawrence Livermore National Laboratory
J.-C. Leyer
Université de Poitiers
A. A. Borisov
USSR Academy of Sciences
W. A. Sirignano
University of California
ISBN 0-930403-98-3

135. Numerical Approaches to Combustion Modeling (1991)
Elaine S. Oran
Jay P. Boris
Naval Research Laboratory
ISBN 1-56347-004-7

136. Aerospace Software Engineering (1991)
Christine Anderson
U.S. Air Force Wright Laboratory
Merlin Dorfman
Lockheed Missiles & Space Company, Inc.
ISBN 1-56346-005-5

137. High-Speed Flight Propulsion Systems (1991)
S. N. B. Murthy
Purdue University
E. T. Curran
Wright Laboratory
ISBN 1-56347-011-X

(Other Volumes are planned.)